Lecture Notes in Bioengineering

Lecture Notes in Bioengineering (LNBE) publishes the latest developments in bioengineering. It covers a wide range of topics, including (but not limited to):

- Bio-inspired Technology & Biomimetics
- Biosensors
- Bionanomaterials
- Biomedical Instrumentation
- Biological Signal Processing
- Medical Robotics and Assistive Technology
- Computational Medicine, Computational Pharmacology and Computational Biology
- Personalized Medicine
- Data Analysis in Bioengineering
- Neuroengineering
- Bioengineering Ethics

Original research reported in proceedings and edited books are at the core of LNBE. Monographs presenting cutting-edge findings, new perspectives on classical fields or reviewing the state-of-the art in a certain subfield of bioengineering may exceptionally be considered for publication. Alternatively, they may be redirected to more specific book series. The series' target audience includes advanced level students, researchers, and industry professionals working at the forefront of their fields.

Indexed by SCOPUS, INSPEC, zbMATH, SCImago.

Zulkifli Mohamed · Mohd Zamani Ngali ·
Suhizaz Sudin · Mohamad Fauzi Ibrahim ·
Alexander Casson
Editors

Proceedings of the 2nd International Conference on Innovation and Technology in Sports, ICITS 2023, 27–28 November, Malaysia

Innovation and Technology in Sports

Springer

Editors
Zulkifli Mohamed
School of Mechanical Engineering
Universiti Teknologi MARA
Selangor, Malaysia

Suhizaz Sudin
Universiti Malaysia Perlis
Perlis, Malaysia

Alexander Casson
University of Manchester
Manchester, UK

Mohd Zamani Ngali
Tun Hussein Onn University of Malaysia
Johor, Malaysia

Mohamad Fauzi Ibrahim
National Sports Institute Malaysia
Kuala Lumpur, Malaysia

ISSN 2195-271X ISSN 2195-2728 (electronic)
Lecture Notes in Bioengineering
ISBN 978-981-97-3743-7 ISBN 978-981-97-3741-3 (eBook)
https://doi.org/10.1007/978-981-97-3741-3

This Springer imprint is published by the registered company Springer Nature Singapore Pte Ltd.
The registered company address is: 152 Beach Road, #21-01/04 Gateway East, Singapore 189721, Singapore

Preface

Welcome to the compilation of papers presented at the 2nd *RevealDNA* International Conference on Innovation and Technology in Sports (ICITS 2023). This collection brings together a diverse array of cutting-edge research and insights presented by scholars, researchers, and practitioners at the forefront of the evolving intersection between sports, innovation, and technology.

In the dynamic realm of sports, continuous advancement is not only a tradition but a necessity. ICITS 2023 served as a nexus for experts and enthusiasts from around the globe to explore, discuss, and share their groundbreaking work in various facets of sports science and technology. The breadth of topics covered in this compilation reflects the rich tapestry of ideas that were exchanged during the conference.

The conference encompassed an array of pivotal themes, including sports instrumentation, sports materials, sports biomechanics, sports equipment design, simulation in sports, data analytics in sports, and performance analysis in sports. These themes collectively underscore the pivotal role that innovation and technology play in reshaping the landscape of sports, transcending traditional boundaries, and unlocking new frontiers of performance, safety, and overall athlete well-being.

As we delve into the pages of this compilation, readers will encounter a treasure trove of research papers that encapsulate the latest advancements in sports technology. From the intricacies of biomechanics shaping athletic prowess to sophisticated instrumentation enhancing sports performance, each paper contributes to the growing body of knowledge that propels the sports industry forward.

Moreover, the interdisciplinary nature of the conference is evident in the fusion of engineering, sports science, and data analytics. The synergistic collaboration of these domains is essential for driving innovation, creating novel solutions, and addressing the multifaceted challenges faced by athletes, coaches, and sports organizations.

This compilation is not merely a reflection of the conference proceedings but a testament to the dedication, passion, and collective efforts of the global community committed to advancing the frontiers of sports innovation. It is our hope that these papers will serve as a source of inspiration and a valuable reference for researchers, practitioners, and enthusiasts alike, fostering continued collaboration and pushing the boundaries of what is achievable in the world of sports through innovation and technology.

We extend our heartfelt gratitude to all the authors, presenters, reviewers, and organizers whose contributions have made this compilation possible. May this work contribute to the ongoing dialogue that propels the ever-evolving field of sports innovation to new heights.

Selangor, Malaysia — Zulkifli Mohamed
Johor, Malaysia — Mohd Zamani Ngali
Perlis, Malaysia — Suhizaz Sudin
Kuala Lumpur, Malaysia — Mohamad Fauzi Ibrahim
Manchester, UK — Alexander Casson

Contents

Assessing the Effectiveness of AFA Online Booking and Payment Gateway Services for Sports Facilities: A Customer Feedback Survey

Rohan Ganason and Rosalind Joyce Sinnappan

Abstract This study aimed to assess the effectiveness of AFA's online booking and payment gateway services for sports facilities from the customer's perspective. The research objectives included evaluating customer satisfaction, ease of use, reliability, areas of improvement, and the impact on loyalty and repeat bookings. A survey questionnaire was administered to 52 users of AFA's online booking and payment gateway services for sports facilities. Descriptive statistics, such as graphs, charts, and percentages, were employed to analyse the data collected from the survey. The findings indicate that a majority of customers (62%) were satisfied or very satisfied with AFA's online booking and payment gateway services. The services were perceived as easy to use by over 53% of customers. However, customers have varying levels of confidence in the AFA app's reliability, with some experiencing more issues than others. The net promoter score (NPS) of -15% revealed that enhancing services is necessary to drive customer loyalty and advocacy. While the study acknowledges limitations such as sample size, self-reported data, and the absence of causal relationships, it provides valuable insights for improving AFA's online booking and payment gateway services. Future research should explore the impact of external factors, such as competing online booking and payment gateway services, on customer satisfaction and loyalty.

Keywords AFA app · Online booking · Payment gateway services · Customer satisfaction

---

R. Ganason (✉) · R. J. Sinnappan
Lincoln University College (LUC), Kuala Lumpur, Malaysia
e-mail: rohan@isn.gov.my

R. Ganason
ISN Niaga, National Sports Institute of Malaysia, Kuala Lumpur, Malaysia

Z. Mohamed et al. (eds.), *Proceedings of the 2nd International Conference on Innovation and Technology in Sports, ICITS 2023, 27–28 November, Malaysia*, Lecture Notes in Bioengineering, https://doi.org/10.1007/978-981-97-3741-3_1

1 Introduction

The digitalisation of sports facility booking has been a key aspect of the trend towards private sector involvement in the sports industry. Digital platforms have allowed private operators to streamline the booking process, manage inventory, and market their facilities to a wider audience. Additionally, digitalisation has made it easier for consumers to access and book sports facilities, increasing convenience and accessibility.

As Mataruna-Dos-Santos et al. (2020) note in their study of digitalisation in sports facilities management, "the adoption of digital platforms in sports facilities management is seen as a way of improving efficiency, reducing costs, increasing revenue, and enhancing customer experience". They go on to highlight the benefits of digitalisation for both operators and consumers, including increased transparency, improved communication, and greater flexibility in scheduling and pricing.

Furthermore, digitalisation has also played a role in the commodification of sports experiences, as it allows operators to offer personalised and exclusive experiences to consumers. As Phillips (2015) notes in his book on the commercialisation of sport, "digitalisation has enabled operators to create more customised experiences for fans, offering VIP packages and access to behind-the-scenes content".

The AFA's (AFA Sports 2023) online booking and payment gateway services for sports facilities were introduced in response to the growing demand for a more convenient and efficient booking system. The system allows customers to book and pay for sports facilities online, rather than having to make a reservation over the phone or in person. While the system has been generally well-received, there has been a lack of validation on its effectiveness, particularly regarding ease of use and reliability.

As more and more customers choose to use this system, it becomes essential to assess its effectiveness from the perspective of users. This study aims to provide a comprehensive evaluation of the AFA's online booking and payment gateway services for sports facilities, with a focus on customer satisfaction, ease of use, reliability, areas of improvement, and impact on customer loyalty and repeat bookings.

2 Methods

The research question for this study is "How effective are AFA's online booking and payment gateway services for sports facilities from the perspective of consumers?" The study aims to investigate this question by setting five objectives.

Firstly, the study is measuring the level of customer satisfaction with AFA's online booking and payment gateway services for sports facilities. Secondly, it is assessing the ease of use of these services for consumers. Thirdly, the study is evaluating the reliability of AFA's online booking and payment gateway services for sports facilities.

Fourthly, the study is collecting customer feedback to identify areas for improvement in AFA's online booking and payment gateway services for sports facilities. Finally, the study is examining the impact of AFA's online booking and payment gateway services on repeat booking and customer loyalty.

To achieve these objectives, four tools were used in measuring customer experience (CX), namely Customer Satisfaction (CSAT), Customer Effort Score (CES), SERVQUAL, and Net Promoter Score (NPS). These tools will be used to establish conclusions regarding the objectives and infer customer satisfaction (CS), engagement (CE), and loyalty (CL) to understand the overall customer experience.

The research methodology for this study involves survey design, data collection, descriptive analysis, and interpretation of results to draw conclusions and make recommendations for improving AFA's online booking and payment gateway services for sports facilities from the perspective of consumers.

3 Results

See (Figs. 1, 2, 3, 4, 5, 6, 7, 8, 9 and 10).

Demographic Profile		**N**	**%**
Gender	Male	42	80.8
	Female	10	19.2
Age	< 25 year old	18	34.6
	25 - 40 year old	29	55.8
	41 - 50 year old	5	9.6
Education level	Higher Schooling	12	23.1
	Professional Certificate / Diploma	17	32.7
	Graduate (Bachelor Degree)	17	32.7
	Post Graduate (Master/Phd)	6	11.5
Income level	B40 (<RM5,000)	40	76.9
	M40 (RM 5,001 - 9,999)	10	19.3
	T20 (RM 10,000 >)	2	3.8
Employment status	Student	16	30.8
	Employed	28	53.8
	Self-employed	8	15.4
Geographical location	Urban	46	88.5
	Sub-urban	10	7.9
	Rural	2	3.6

Fig. 1 Demographic Profile

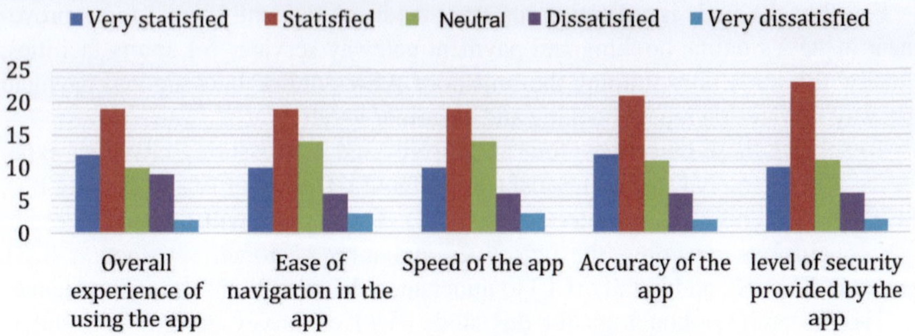

Fig. 2 Customer Satisfaction (CSAT) for overall experience using the app

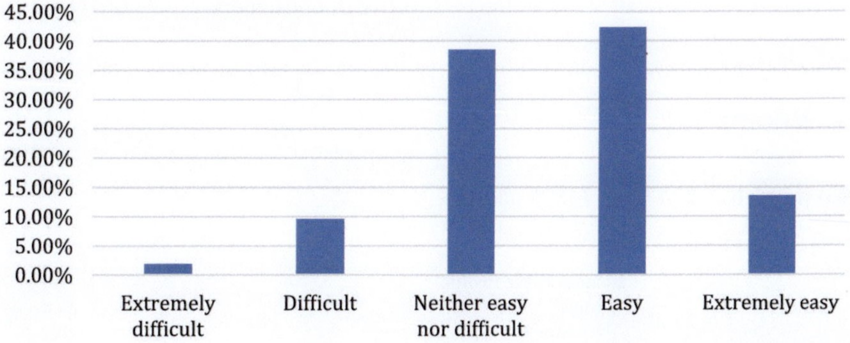

Fig. 3 Ease of using the app

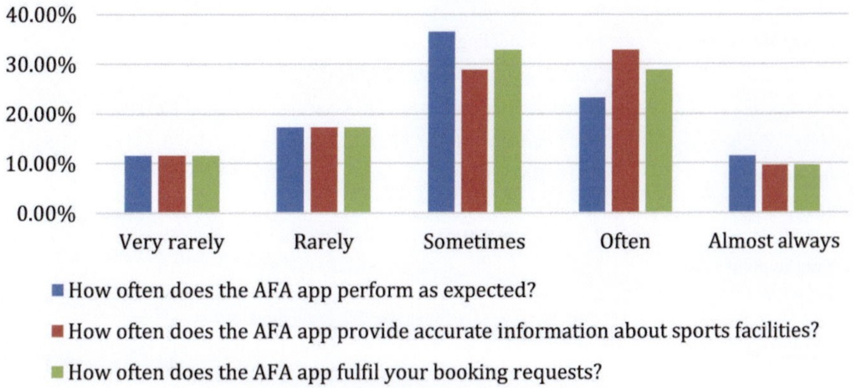

Fig. 4 Result of the SERVQUAL questions on the reliability dimension

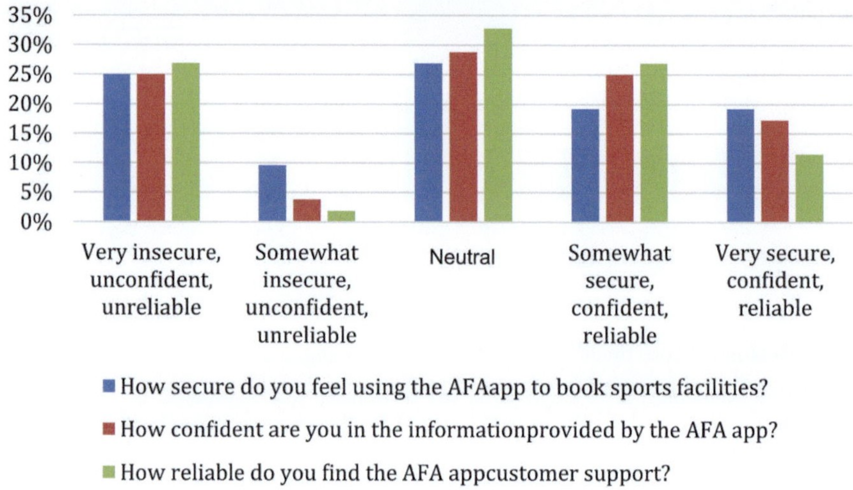

Fig. 5 Result of the SERVQUAL questions on the assurance dimension

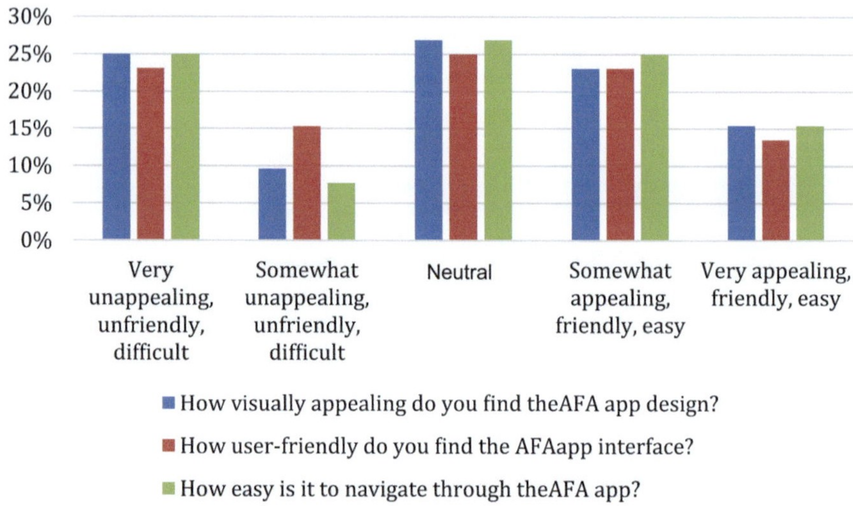

Fig. 6 Result of the SERVQUAL questions on the tangible dimension

4 Discussion

Based on the results of the study, we can draw the following inference related to the research question: "how effective are AFA's online booking and payment gateway services for sports facilities from the perspective of customers?" and the research objectives:

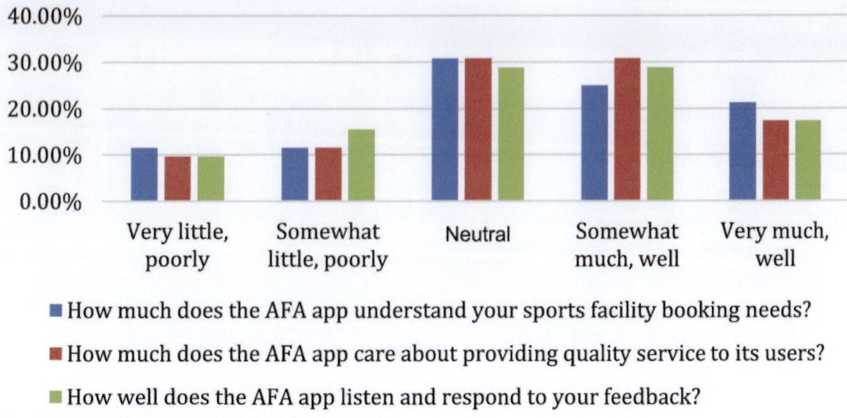

Fig. 7 Result of the SERVQUAL questions on the empathy dimension

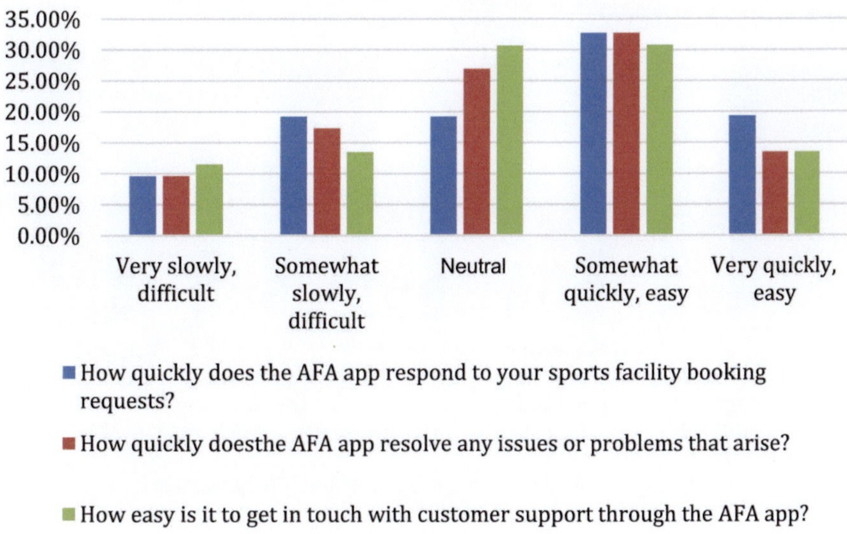

Fig. 8 Result of the SERVQUAL questions on the responsiveness dimension

4.1 *Level of Customer Satisfaction*

According to a study by Ahmad and Khawaji (2020), user satisfaction is a critical factor in the success of mobile apps. The authors note that user satisfaction is strongly related to the usability and functionality of the app, as well as the quality of customer service provided. Similarly, a study by Chu and Kim (2020) found that users' perception of the app's usability and functionality significantly affected their satisfaction level.

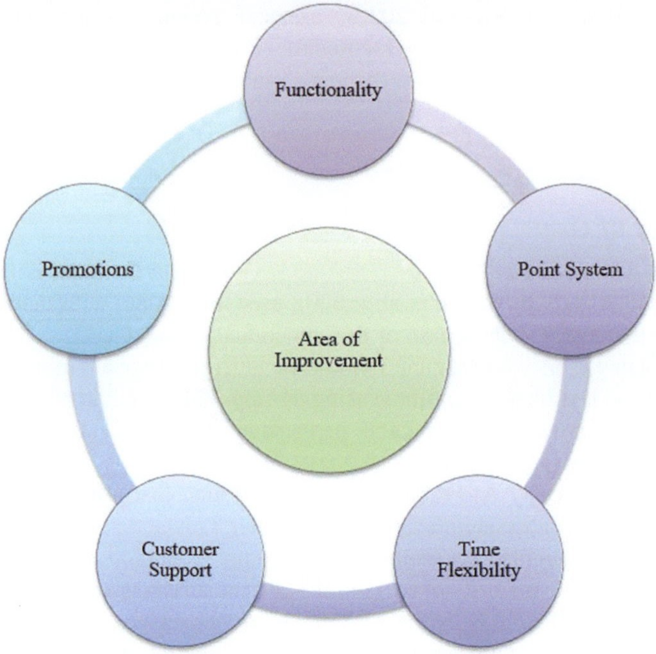

Fig. 9 Key area of improvement based on feedback given on CSAT for AFA app online booking and payment gateway services

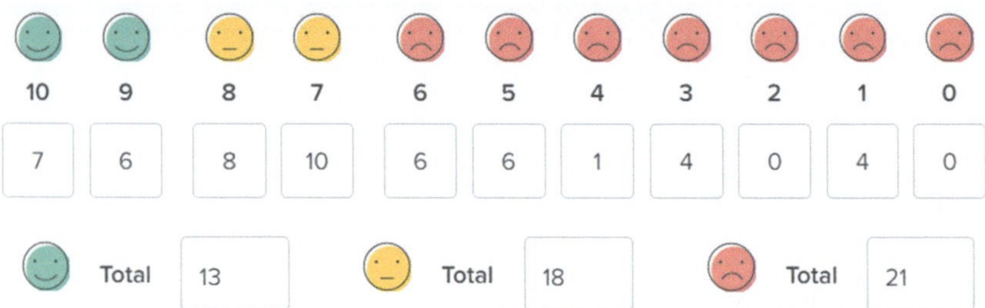

Fig. 10 The number of respondents for a scale of 0–10 in the NPS question asked in the survey and promoters to detractors

A majority of customers (62%) were satisfied or very satisfied with AFA's online booking and payment gateway services for sports facilities. This suggests that the services are effective in meeting the needs of customers and contributing to their overall satisfaction.

Overall, the findings of the survey are consistent with existing scholarly trends and findings on user satisfaction and customer service in mobile app development.

As noted by Ahmad and Khawaji (2020), "user satisfaction is a critical factor for the success of mobile apps". Thus, it is essential for developers to consider user feedback and implement changes to improve the app's usability, functionality, and customer service.

4.2 *Ease of Use*

According to a study by Sathar et al. (2020), user satisfaction with mobile apps is influenced by factors such as ease of use, functionality, and customer service. The study found that users who encounter problems with app usage are more likely to be dissatisfied and less likely to continue using the app. This is in line with the findings of the current survey, where a notable percentage of users encountered problems during the booking process.

Similarly, a study by Al-Fraihat et al. (2020) found that users' perceptions of customer service quality significantly impact their satisfaction with mobile apps. The study also highlighted the importance of prompt resolution of user issues. This is consistent with the majority of users in the current survey who reported that their issues were resolved quickly by the AFA app customer service team.

Over 53% of customers found AFA's online booking and payment gateway services easy or very easy to use. This suggests that the services are user-friendly and accessible, which is an important factor in determining their effectiveness. The findings of the current survey are consistent with existing scholarly trends regarding user satisfaction with mobile apps. They highlight the importance of factors such as ease of use, functionality, customer service quality, and prompt issue resolution.

4.3 *Reliability*

Based on the results provided in the SERVQUAL dimensions, it appears that the AFA app could benefit from improvements in various areas to enhance customer satisfaction and trust. These findings are consistent with existing scholarly research on the importance of service quality in enhancing customer loyalty and retention.

In their study on customer satisfaction with mobile apps, Kim and Lee (2020) found that perceived service quality was a significant predictor of customer loyalty towards mobile apps. Similarly, in a study on sports services, Ko and Pastore (2005) found that service quality was positively related to customer satisfaction and loyalty.

The results from the reliability dimension suggest that the AFA app's performance is perceived as inconsistent by customers. This finding is consistent with previous research that highlights the importance of reliability in enhancing customer satisfaction and trust (Parasuraman et al. 1988). Therefore, the AFA app could benefit from continuous monitoring and improvements to ensure consistent and reliable service.

In summary, the results from the SERVQUAL dimension graphs provide valuable insights into the areas where the AFA app could improve to enhance customer satisfaction and trust. These findings are consistent with existing scholarly research on the importance of service quality in enhancing customer loyalty and retention. Therefore, the AFA app could benefit from implementing strategies to improve its service quality, such as monitoring and continuous improvement, understanding customer needs, and enhancing responsiveness.

4.4 *Area of Improvement*

Improving user experience and addressing customer feedback is a crucial aspect of app development and maintenance. According to a study by Fundera, 42% of users will delete an app if they encounter a problem or issue, while 23% will abandon an app after only one use if the user experience is not satisfactory (Fundera 2022).

In terms of improving app functionality, previous studies have shown that slow app performance and technical issues can negatively impact user experience (Jung et al. 2016; Wang et al. 2017). Therefore, it is important for app developers to prioritise optimisation of app performance to improve user satisfaction.

Regarding the AFA points system, studies have shown that gamification and reward systems can increase user engagement and retention (Hamari et al. 2014). Therefore, addressing issues with the point system and improving its functionality could incentivise users to continue using the AFA app.

Customer service is also a crucial element in improving user satisfaction and retention. As per a study conducted by American Express, 33% of customers would consider switching to a different company after a single instance of poor customer service, while 60% would switch after multiple instances (American Express 2021). Therefore, providing excellent customer service is critical for retaining customers and maintaining a positive reputation. Companies should focus on improving customer support and making it more accessible to users to address any issues promptly and effectively.

In terms of app design and user interface, previous research has highlighted the importance of user-centred design and intuitive interfaces to improve user satisfaction (Kujala et al. 2011). Therefore, addressing user interface issues and making the app more user-friendly could help increase user satisfaction and retention.

In conclusion, addressing the key areas of improvement identified by customer feedback for the AFA app, such as payment gateway issues, points system, user interface, time availability and flexibility, customer service, app performance, and refunds and cancellations, could help increase user satisfaction and retention. App developers should prioritise optimisation of app performance, improving customer support, and adopting user-centred design principles to improve the user experience.

4.5 Impact on Loyalty and Repeat Booking

NPS score AFA (−15%). Research has shown that a high NPS score is associated with increased customer loyalty and profitability for a company (Morgan and Rego 2016). Therefore, it is crucial for AFA to focus on improving its NPS score by addressing customer concerns and improving the overall user experience.

In summary, the NPS score for AFA indicates that there is room for improvement in the app's functionality, point system, time flexibility, customer support, and promotions. By addressing these issues and improving the overall user experience, AFA can increase customer satisfaction and loyalty, ultimately leading to increased profitability for the company.

5 Conclusion

Based on the study, the following conclusion and recommendations can be made for AFA's online booking and payment gateway services for sports facilities:

5.1 Continue to Prioritise Customer Satisfaction

While the majority of customers are satisfied with the services, there is still room for improvement. AFA should prioritise ongoing evaluation and improvement to ensure that the services continue to meet the evolving needs and expectations of customers.

5.2 Address Areas of Improvement

The study revealed that 25% of customers identified areas where improvement could be made. AFA should prioritise addressing these areas to improve the overall customer experience and increase satisfaction.

5.3 Maintain Focus on User-Friendliness and Reliability

The study found that customers highly valued the user-friendliness and reliability of AFA's online booking and payment gateway services. AFA should continue to prioritise these factors in the development and maintenance of the services.

5.4 Use NPS Scores as a Measure of Customer Loyalty

The NPS score of -15% indicates that there is room for improvement in terms of customer loyalty. AFA should use NPS scores as a measure of customer loyalty and work to improve it over time.

5.5 Communicate Improvements to Customers

As AFA makes improvements to the online booking and payment gateway services, it is important to communicate these changes to customers. This will help to build trust and confidence in the services and contribute to higher levels of satisfaction and loyalty.

Acknowledgements I would like to express my heartfelt gratitude to Athletes for Athletes Sdn Bhd (AFA) for their kind support in allowing me to survey their mobile app users and providing valuable data for my study. I am truly thankful for their cooperation and willingness to assist in this research. Additionally, I would like to extend my sincere appreciation to all the participants who took the time to complete the survey and share their experiences. Their valuable contribution has been instrumental in making this study possible and has provided valuable insights for further analysis. Without their active involvement and dedication, this research would not have been successful.

References

Ahmad S, Khawaji M (2020) Analyzing user satisfaction factors for mobile apps using a fuzzy ANP-based approach. IEEE Access 8:247–256

Al-Fraihat D, Joy M, Al-Shehri S (2020) Investigating the impact of customer service quality on user satisfaction with mobile applications: a quantitative study. J Theor Appl Electron Commer Res 15(4):72–87

American Express (2021) The 2021 global customer service barometer. https://www.americanexpress.com/content/dam/amex/us/credit-cards/features-benefits/2021-Global-Customer-Service-Barometer.pdf

Court Booking App|AFA Sports (2023) AFA court booking. https://www.athleteforathletes.com

Chu SC, Kim Y (2020) Understanding users' satisfaction and continuance intention towards mobile apps: a UX approach. Inf Syst Front 22(2):347–362

Fundera (2022) How to make an app: a beginner's guide. https://www.fundera.com/blog/how-to-make-an-app

Hamari J, Koivisto J, Sarsa H (2014) Does gamification work?-a literature review of empirical studies on gamification. In: Proceedings of the 47th Hawaii international conference on system sciences

Jung Y, Lee S, Park H (2016) The effects of mobile app quality on user satisfaction and behavioral intention: an empirical study on hybrid apps. Telematics Inform 33(2):342–355

Kim J, Lee H (2020) The effect of online customer feedback on service quality, satisfaction, and loyalty in the airline industry. J Travel Res 59(1):144–159

Ko YJ, Pastore DL (2005) Current issues and future directions in sport consumer research. Sport Mark Q 14(1):5–11

Kujala S, Walsh T, Lahtinen T, Boano C, Obrist M (2011) User experience evaluation methods for ambient intelligence environments. Ambient Intell Smart Environ 7:263–280
Mataruna-Dos-Santos L, Almeida-Junior JA, Nascimento JA (2020) Digital platforms in sports facilities management: a literature review. Int J Inf Manag 53:166–177
Morgan NA, Rego LL (2016) The value of different customer satisfaction and loyalty metrics in predicting business performance. Mark Sci 35(3):426–439
Parasuraman A, Zeithaml VA, Berry LL (1988) SERVQUAL: a multiple-item scale for measuring consumer perceptions of service quality. J Retail 64(1):12–40
Phillips D (2015) The commercialization of sport. Routledge
Sathar A, Gupta S, Srivastava S (2020) Evaluating the factors influencing user satisfaction with mobile apps: an empirical study. J Theor Appl Electron Commer Res 15(8):183–200
Wang D, Zhang X, Guo X (2017) Factors affecting customer satisfaction with online booking: evidence from Chinese airline websites. J Air Transp Manag 59:13–20

Difference Aiming Time Toward Shooting Accuracy

Nur Suryani Mohamed Taibi, Hosni Hasan, and Mujahid Ramli

Abstract This study was conducted to investigate the effect of fast and slow shooting on shooting a 10 m air rifle among skilled shooters. Two national-level shooters was involved in this study, and both shooters used their own rifles for the shooting test. They were given a total of 90 shooting trials each for the fast and slow shooting, and the sessions were conducted on two separate days. The key variables analyzed were time, score, and aiming trace speed (mm/s), using a specially designed optoelectronic device and the SCATT system. Data analysis for this study was conducted using Paired Sample T-Tests with the significant level set at $p < 0.05$. Results revealed that there were significant differences between fast and slow shooting on the aiming time and score (both showed $p < 0.05$) where fast shooting techniques show better values in both variables. However, there was no significant different between fast and slow shooting on the aiming trace speed ($p > 0.05$). Our findings underscore the paramount importance of the holding and aiming ability in shooting performance, with implications for training methodologies in competitive shooting. These findings highlight the importance of efficient aiming time in achieving high shooting accuracy. The study's results contribute to a deeper understanding of the factors influencing shooting performance, with implications for shooters. The use of fast-fire shooting is suggested as an effective approach to improve shooting accuracy while conserving energy. Further research with a larger participant group can provide valuable insights into individual aiming time preferences and their impact on shooting accuracy.

Keywords Fast-fire shooting · Slow-fire shooting · 10 m air rifle · SCATT software

N. S. M. Taibi (✉) · H. Hasan · M. Ramli
Faculty of Sports Science and Recreation, Universiti Teknologi MARA (UiTM) Shah Alam, Selangor Darul Ehsan, 40450 Shah Alam, Malaysia
e-mail: 2022270266@isiswa.uitm.edu.my

H. Hasan
e-mail: hosnihasan@uitm.edu.my

Z. Mohamed et al. (eds.), *Proceedings of the 2nd International Conference on Innovation and Technology in Sports, ICITS 2023, 27–28 November, Malaysia*, Lecture Notes in Bioengineering, https://doi.org/10.1007/978-981-97-3741-3_2

1 Introduction

Many people are well-introduced to shooting activities, including police officers, soldiers, security guards, shooters, and, not to mention, hunters. Shooting is a precision sport requiring a high level of motor control to hit a target. Shooting is seldom related to sports played on the mental game; it demands high accuracy and precision, which any respective age and gender can perform.

The military and police officers commonly use these firearm types, including a revolver, pistol, shotgun, rifle or carbine, assault rifle, submachine gun, and machine gun. Shooting sports have a competitive and recreational environment, and several guns use a small bore or small arms according to the shooting distances, target, and time limit. Especially in the event of the 0.177 air rifle, many factors are described as performance influences. Thus, physical factors, such as stability, strength, endurance, and fitness (Mon-López et al. 2019), and psychological effects, such as anxiety, stress, doubt, and lack of confidence or self-esteem, could be related to shooting performance.

The 10 m Air Rifle is an ISSF event in which competitors use a rifle with a maximum weight of 5.5 kg, or 12.13 libbers, and shoot across a distance of 10 m, or 10.94 yards, while standing. The target's diameter is 45.5 mm, with its center situated 1.40 m above the ground. The tenth ring is 0.5 mm in diameter, while the fourth is 30.5 mm. In order to improve the stability of the shooting postures, the usage of specialist clothing equipment is permitted. Qualification round: Each athlete has one hour and fifteen minutes to fire sixty bullets during this round. The qualification score is expressed in decimal points, and the highest possible score per shot is 10.9 due to an extra set of 10 rings inside the tenth circle, which raises the score by 0.1 points as the shot gets closer to the target's center. The qualifying round has a maximum score of 654.0 points.

Aiming times are different for every shooter. One may squeeze the trigger after 2–3 s when they see a perfect aiming picture, while one will squeeze the trigger after a second or third picture which may differ for 15–20 s of aiming. As recommended by International Shooting Sports Federation (ISSF), it only takes 4–6 s to release a shot (Maksimovic 2020). Aiming too long will produce a distortion of the picture, eye muscle fatigue, a mirage, and even muscle fatigue. The focus is on mastering the aiming traveling plot toward the perfect time to trigger a perfect image picture in a plan toward high shooting accuracy.

There is minimal research examining and differentiating the various phases of motor control during aiming, despite the hypothesis that the final stage of aiming, which occurs just before pulling the trigger, influences shot accuracy (Schnipke et al. 2019). Sustained attention is the capacity to remain attentive and engaged throughout time, especially in situations of repetition and monotony (Unsworth and Robison 2020). It is the capacity to stay focused and awake for an extended period, concerning various components of human movement and diverse motor activities. A study by Wulf and Su (2007) describes that the internal focus induces conscious control and causes people to restrict their motor system by interfering with automatic control processes. In contrast, an external focus of attention promotes a more automatic mode for motor control through unconscious, quick, and reflexive control processes. The multielement array must be presented for a short period ($\leq$200 ms) to enable attention deployment while preventing helpful eye movements to ensure parallel processing (Valdois et al. 2019).

Many shooters, especially amateur and novice level, did not understand the right aiming time and tried to compensate rather than make it naturally by the body to reach the zero position. Therefore, the purpose of this study was conducted to investigate the effect of fast-fire and slow-fire shooting techniques on 10 m air rifle shooting performance.

2 Methodology

2.1 Participants

The participants for this study were two national shooters (mean age of 29.5 years old, SD 14.85 years old). The participants were classified as skilled shooters as they have involved in competitive and training experiences for more than four years (Hasan et al. 2016). They were not under medication and had no history of neurological, cardiovascular, or other significant disorders.

2.2 Equipment

The Air Rifle Precharged Pneumatic Single Shot with rear sight: 4 × 4 Vario-Sight is adjustable between 10.4–60 mm (measured from the barrel's center). The vario sight is also adjustable to five different heights and equipped with a removable air reservoir and integral manometer (air pressure gauge). The gun is individual-allocated used for personal training and competition. SCATT MX Version 20.05.31 is an analytical software and an advanced optical sensor mounted to the rifle. The shooter can view immediate and in-depth visual feedback on the aiming process through the laptop installed with the SCATT software. The shooter used a custom jacket and pants specifically with different manufacturers from Marksman House Korea and Capapie from India. The scoring system for this testing is Electronic Scoring Target (EST) Meyton Elektronik GmbH 2012 Version Roman Geile. The control PC assesses the measured data from the measuring frame. The target with the hits, the shot values and overall results, the amount of time left in the competition, the competition title, the position and fire point number, and the shooter's name are all shown on the screen.

2.3 Testing Protocol

The shooting task consisted of 30 shots on each trial, where the trial should be completed three times, each time per day. For each shooting task, the participants shot for 90 trials and was conducted in two separate days. The participants were also given trials for the shooting (as baseline), and they can shoot within their own time a day before the testing began. However, the analysis for this study was only conducted for fast-fire and slow-fire shooting techniques. The participant aimed to use a 0.177 caliber competition air rifle. We recorded aiming performance data with the SCATT software training system, which uses a barrel-mounted optical device (Fig. 1) to be recorded in two-dimensional space (PC and EST) as a function of time throughout each trial. Similar software was used by Ball et al. (2003) in investigating the body sway, aim point fluctuation, and performance in rifle shooters among six elite shooters performed under competition conditions. In this study, the participants need to complete the shooting trials in a 30-min period which is the time allowed under International Shooting Sports Federation rules.

The green line trajectories entered the fourth score line, marking the beginning of the timing for both the slow- and fast-fire protocols. The athletes were warned for fast-fire protocol within the final five seconds. Regarding the slow-fire protocol, the athletes could only fire once the timer proclaimed "go," which ended after fifteen seconds. The exact position of each score was graded at two decimal places when

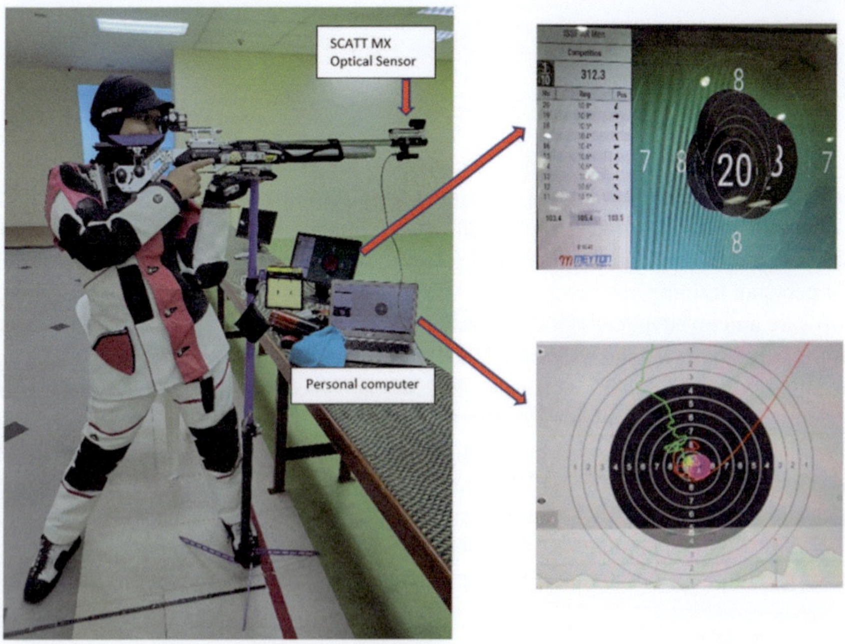

Fig. 1 Located device

the scores were reviewed in decimal. The athletes' cumulative score will rise in proportion to their accuracy in hitting the target's center. For this study, the shooter must strike the inner ten, or 10.4 and above, till the perfect maximum score of 10.9 (bullseye) as a mark is counted (an additional 0.1 points as it approaches the center of the target).

A few variables can be extracted from the software, including average aim point, shot coordinate, aiming zone, shot index, shot score, aiming time, the ratio of holding, and aiming trace feed result. However, our aim for this study was only to evaluate the time (s), score (percentage for score 10.4 and above), and aiming trace speed (mm/s).

2.4 *Shooting Protocol*

Ten minutes were allotted for warm-up and preparation after ATL: athletes to the line. The preparation and sighting period began; another fifteen minutes would follow, with endless shots. Then, the participants were given thirty minutes to execute thirty shots. All testing strategies followed this protocol.

2.5 *Data Analysis*

The data used in this study was using Paired T-Test to investigate the effect of fast-fire and slow-fire on shooting performances. All data was analyzed using IBM SPSS Statistics 20 and were presented as percentage score, means, and standard deviations. The alpha values for the significant level in this study were set at $p < 0.05$.

3 Result

3.1 *Score*

The mean values of percentage score for fast-fire and slow-fire shooting were 46% with SD of 0.32% and 30.50% with SD of 0.34%. Results revealed that there was a significant different ($p < 0.05$) in the percentage of score above 10.4 between fast-fire and slow-fire shooting techniques (Fig. 2).

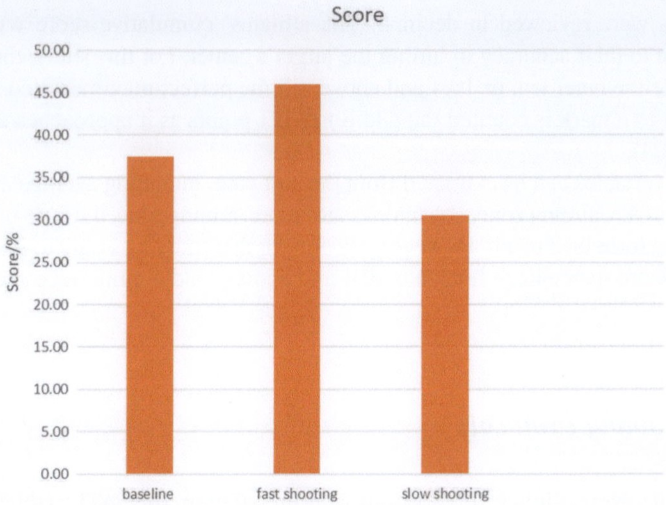

Fig. 2 Scores for fast-fire and slow-fire shooting

3.2 Aiming Time

The mean values of aiming time for fast-fire and slow-fire shooting were 6.936 s with SD of 1.84 s and 19.59 s with SD of 2.54 s. Results revealed that there was a significant different ($p < 0.05$) in the aiming time values between fast-fire and slow-fire shooting techniques (Fig. 3).

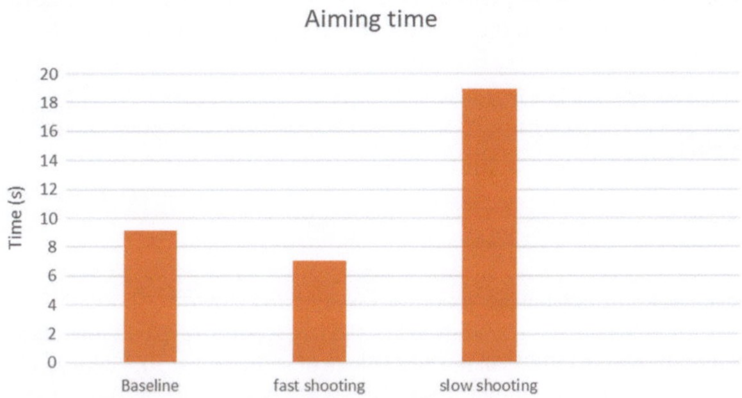

Fig. 3 Aiming time for fast-fire and slow-fire shooting

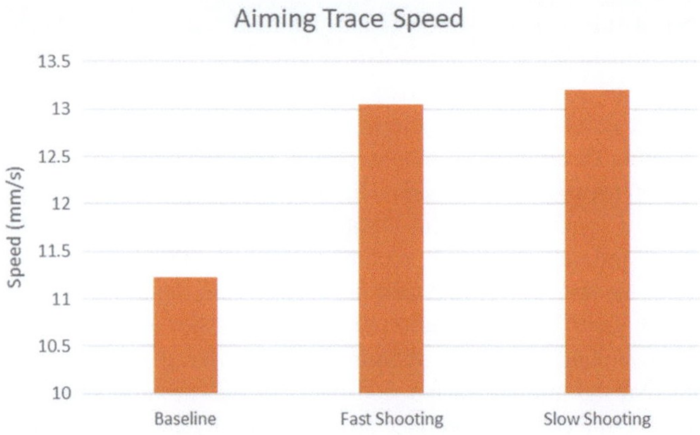

Fig. 4 Aiming trace speed for fast-fire and slow-fire shooting

3.3 Aiming Trace Speed

The mean values of aiming trace speed for fast-fire and slow-fire shooting were 13.1 mm/s with SD of 2.49 mm/s and 13.21 s with SD of 2.28 mm/s. Results revealed that there was not significant different ($p > 0.05$) in the aiming trace speed values between fast-fire and slow-fire shooting techniques (Fig. 4).

4 Discussions

This study aims to investigate fast-fire and slow-fire shooting techniques on a 10 m air rifle shooting performance. The fast-fire shooting was introduced to participants to shoot within the range of 10 s after the green line touched ring number 5 at the target. The same goes for slow-fire shooting, where the participant must shoot only after 15 s. From the data, we can see that the stability toward the aiming picture differs from each protocol, which may affect the percentage of holding the gun in the area of ring 10 mm/s, showing the steadiness of the rifle during the final aiming procedure. It should be lower than 15 mm/s if preferred to achieve high results (Maksimovic 2020). In addition, further analysis on both participants show a median of 13.20 mm/s and 13.05 mm/s lower than suggested. It shows that both participants were at their most stable performance, which may increase the stability in the future, as from the data, both can even shoot at nine mm/s in certain shots. While learning capacities do not deteriorate with age, older persons can still make significant performance improvements despite the tendency for their motor function to degrade. The study shows that fast-fire protocol can be adaptive not only to young athletes but also to older athletes. Most studies (Mason et al. 1990; Zanevskyy et al. 2012) have concentrated on the significance of hold stability for shooting performance.

A number of studies have demonstrated that hold stability was the most crucial technical element in air rifles (Ihalainen et al. 2015, 2016, 2018) and air pistol shooting (Hawkins and Bertrand 2015). However, according to Olsson and Laaksonen (2021), it is interesting to note that, contrary to findings from past studies on other shooting disciplines, aiming accuracy was regarded as the most crucial factor. The aforementioned knowledge limits studying performance in enhancing the perception of somatosensory information. The alertness of the shooter toward the aiming area and the best-aiming picture to execute a shoot are critical.

The current finding shows that fast-fire shooting is the most accurate time in achieving a high score. This observation suggests that response reactions are synchronized with movement if fast-fire shooting were adaptive. Results for this study also showed that the aiming trace speed between fast-fire and slow-fire shooting was similar as both techniques produce similar postural balance stability (Mononen et al. 2007) in aiding the shooting performances. It will be an interesting study in the future to further investigate the effects of fast-fire and slow-fire shooting on anterior/posterior and medial/lateral postural stability of shooters during shooting performance.

In this study, we proposed and only selected the score of $\leq$10.5 as it is required to perform and be in the top ten in finals for international competition with a $\leq$105 for every accumulated ten shots per series. Statistical data and graphics will allow the shooter and the coach to get a complete analysis of the performance of the shooting technique. A few efforts focus on the effects of shooting fast and shooting slow on emergent movement patterns of athletes during performance sports-related tasks. Data collection should be prioritized in shooting research since competitive shooting performance should be more pertinent to shooters. The restriction on dry fire during the qualifying round was one of the study's limitations. Some shooters in the current study had to make adjustments during the testing environment since they often dry fire during competition.

5 Conclusions

Using the fast-fire protocol seemed to improve the aiming time toward shooting accuracy and shooting stability itself. Evidence of two shooters enhancing their performance by scoring above 10.5 is proven. Applying these methods improved triggering stability, body stability, and confidence in a perfect aiming picture within 2–3 s. Our study revealed that shooting in a fast-fire protocol will not drain much energy, and energy efficiency is in good shape. Based on our findings, fast-fire shooting is ideal to be used by a shooter in performing their level best in scoring a high standard of a world-class shooter. An interesting question for future research is to gather more participants to complete this task to understand their own pace of perfect aiming pictures toward shooting accuracy.

References

Ball K, Best R, Wrigley T (2003) Body sway, aim point fluctuation and performance in rifle shooters: inter-and intra-individual analysis. J Sports Sci 21(7):559–566

Hasan H, Davids K, Chow JY, Kerr G (2016) Compression and texture in socks enhance football kicking performance. Hum Mov Sci 48:102–111

Hawkins R, Bertrand P (2015) Relationship between twelve mechanic measures and score for national-level pistol shooters. Int J Perform Anal Sport 15(1):332–342

Ihalainen S, Kuitunen S, Mononen K, Linnamo V (2015) Determinants of elite-level air rifle shooting performance. Scand J Med Sci Sports 26(3):266–274. https://doi.org/10.1111/sms.12440

Ihalainen S, Linnamo V, Mononen K, Kuitunen S (2016) Relation of elite rifle shooters' technique-test measures to competition performance. Int J Sports Physiol Perform 11(5):671–677. https://doi.org/10.1123/ijspp.2015-0211

Ihalainen S, Mononen K, Linnamo V, Kuitunen S (2018) Which technical factors explain competition performance in air rifle shooting? Int J Sports Sci Coach 13(1):78–85. https://doi.org/10.1177/1747954117707481

Maksimovic G (2020) ISSF national coach course rifle shooting positions. by Goran Maksimovic. Library. Accessed 29 Oct 2022. https://1library.net/document/q77jw1oq-national-coach-course-rifle-shooting-shooting-positions-maksimovic.html

Mason BR, Cowan LF, Gonczol T (1990) Factors affecting accuracy in pistol shooting. Aust Inst Sports 6(4):2–6

Mon-López D, Moreira da Silva F, Calero Morales S, López-Torres O, Lorenzo Calvo J (2019) What do olympic shooters think about physical training factors and their performance? Int J Environ Res Public Health 16(23):4629. https://doi.org/10.3390/ijerph16234629

Mononen K, Konttinen N, Viitasalo J, Era P (2007) Relationships between postural balance, rifle stability and shooting accuracy among novice rifle shooters. Scand J Med Sci Sports 17:180–5. https://doi.org/10.1111/j.1600-0838.2006.00549.xPMID: 17394480.

Olsson E, Laaksonen MS (2021) Key technical components for air pistol shooting performance. Int J Perform Anal Sport 21(3):348–360

Schnipke JR, LaFiandra ME, Freisinger GM, Hasselquist L, Brown SAT (2019) Determining the critical aiming window duration based on Shooter Skill Level. Proc Hum Factors Ergon Soc Annu Meet 63(1):1269–1273. https://doi.org/10.1177/1071181319631074

Unsworth N, Robison MK (2020) Working memory capacity and sustained attention: a cognitive-energetic perspective. J Exp Psychol Learn Mem Cogn 46(1):77–103. https://doi.org/10.1037/xlm0000712

Valdois S, Roulin JL, Line Bosse M (2019) Visual attention modulates reading acquisition. Vision Res 165:152–161. https://doi.org/10.1016/j.visres.2019.10.011

Wulf G, Su J (2007) An external focus of attention enhances golf shot accuracy in beginners and experts. Res Q Exerc Sport 78(4):384–389

Zanevskyy I, Korostylova Y, Mykhaylov V (2012) Aiming point trajectory as an assessment parameter of shooting performance. Hum Mov 13(3):211–217. https://doi.org/10.2478/v10038-012-0024-3

A Comparative Assessment of Static Balance Among Malaysian Blind Football Athletes

Sohaila Ishak, Hosni Hasan, and Ahmad Zaki Mohd Rum

Abstract Athletes in high-performance sports must maintain appropriate postural control despite numerous destabilization factors. A blind athlete may negatively impact static and dynamic balance, which is related to movement dysfunction and inability to balance. The study objective is to assess static balance among national blind football athletes; nevertheless, they prefer leg choosing on types of surfaces associated with several falls. Method: There are ten male blind national football athletes aged 18–29 with about 6–10 years of training experience. The material used is a balance beam, weight scale, stopwatch, and stature meter. Participants performed a single-leg standing test (flamingo position) with a preferred and non-preferred legs on the ground and the beam. Result: Surface type significantly affects the number of falls ($p = 0.022$), with significantly more falls occurring on the beam compared to the ground. However, the study found no statistically significant main effect of leg choice ($p = 1.000$) or interaction between surface type and leg preference ($p = 1.000$). In addition, there was no significant difference found in the number of falls between the preferred and non-preferred legs on the beam ($p = 0.911$) or on the ground ($p = 0.800$). Conclusions: The study concluded that there was a significant impact of surface type on the frequency of falls among national blind football athletes, notably more falls occurring on the beam compared to the ground. However, the research found no significant effect based on leg preference or an interaction between leg preference and surface type to falls. Moreover, no significant differences were observed in the number of falls between the preferred and non-preferred legs, neither on the beam nor the ground. These findings emphasize the critical influence of surface type while indicating that leg preference might not be a significant factor in the frequency of falls among these athletes. These findings also affect training and interventions to improve balance and performance in blind football.

S. Ishak (✉) · H. Hasan · A. Z. M. Rum
Faculty of Sports Science and Recreation, Universiti Teknologi MARA (UiTM) Shah Alam, 40450 Shah Alam, Selangor Darul Ehsan, Malaysia
e-mail: sohaila3729@uitm.edu.my

H. Hasan
e-mail: hosnihasan@uitm.edu.my

Z. Mohamed et al. (eds.), *Proceedings of the 2nd International Conference on Innovation and Technology in Sports, ICITS 2023, 27–28 November, Malaysia*, Lecture Notes in Bioengineering, https://doi.org/10.1007/978-981-97-3741-3_3

Keyword Static balance · Preference leg · Surfaces

1 Introduction

The nervous system is essential for human well-being, integrating sensory input, thinking, balance, and coordination (Bednarczuk et al. 2017). However, individuals with visual impairment face challenges adapting to their environment and adjusting their body capabilities. In elite sports, neural activation is crucial for real action, while environmental stability and predictability are essential for motor skills. Blind football athletes encounter difficulties in processing inputs and developing exceptional sports skills. Nevertheless, according to (Alotaibi et al. 2016), individuals with a small visual field can correct imbalances effectively due to spatial orientation benefits.

The balance system compromises two main components: the somatosensory input and the vestibular system. The somatosensory input is crucial in regulating vision direction and concentrating while the head and body are in motion. On the other hand, the vestibular system is responsible for providing information regarding head position and external pressures. In blind football, good static balance is essential for performing skills like dribbling, passing, and shooting (Bednarczuk et al. 2017). Previous studies have demonstrated that athletes with visual impairments exhibit impaired static balance compared to sighted individuals (Bednarczuk et al. 2017; Kurz et al. 2021). Moreover, balance training has shown the potential to improve sport-specific performance for blind athletes (Kurz et al. 2021).

Thus, numerous factors can influence static balance in sighted and non-sighted conditions. Age, agility, coordination, stability, body weight, height, and preferred leg selection contribute to a body control system (Panjan & Sarabon 2012; Ponchillia et al. 2002). The impact of leg preference on static balance in blind football athletes remains poorly understood. Additionally, different surface types used for balance tasks can affect stability, with uneven surfaces demanding increased proprioception and vestibular system engagement (Panjan & Sarabon 2012). However, the specific impact of surface types on static balance in blind football athletes requires further investigation.

This study examined the influence of leg preference and surface type on the number of falls among national blind football athletes, with all classified as B1 category (Explanatory Guide to Paralympic Classification Paralympic Summer Sports 2016; Rami & Prabhakar 2018). The hypothesis posits that there will be no significant difference in the number of falls between the preferred and non-preferred legs on either the beam or the ground. However, a significant main effect of the surface is expected, with more falls occurring on the beam compared to the ground. Thus, there is no significant effect based on leg preference or an interaction between leg preference and surface type to falls. Moreover, this study fills existing knowledge gaps. The results of this study will provide a valuable contribution toward advancing specific interventions and training methods to improve sports performance among blind football athletes.

2 Methodology

2.1 *Participants*

In this study, the participants involved were male athletes of national blind football, aged between 18 and 29 years old, in B1 (totally blind) and B2 (partially blind) categories, and experienced with about 6–10 years of training. The exclusion criteria included athletes with lower limbs or head injuries, sighted athletes (goalkeepers), and female athletes.

2.2 *Material and Method*

Before conducting the tests, the researchers recorded each participant's anthropometric measurements, such as height and weight. The researcher tested on two distinct surfaces: the ground and the beam. The researcher evaluated the participant's preferred and non-preferred legs on both surfaces. Before the test, the researcher also demonstrated to the participants to avoid all errors by giving clear verbal instructions, tactile guidance, and concise language. Researchers described the test's purpose and emphasized maintaining stability during the trial. To prevent the participant from sustaining injury throughout the test, researchers allow all participants to practice the test beforehand to become familiar with the procedure and environment surrounding it. Researchers gently guided participants' hands and legs to demonstrate the desired flamingo position. Moreover, the researcher would stand closely beside the participants to monitor any possible excessive body sway or imbalance during the test. Throughout the test, researchers provided continuous verbal reinforcement, reminding the participants to maintain body posture by keeping their legs steady and engaging their core muscles for stability.

The researchers applied a fall prevention plan to prevent injury during the test, such as clear surroundings from potential hazards or obstacles that may harm participants. Before conducting the study, researchers would have obtained ethical clearance from the ethics committee to ensure the study adheres to ethical guidelines and prioritizes participant safety. Moreover, researchers would strictly select based on the exclusion criteria mentioned in this study. Clear instructions and medical preparation are complied with throughout the test.

For the flamingo balance test, participants had to stand without shoes on one leg. The participant raised their hand on the same side as the standing leg to a 90-degree angle while flexing the other leg at the knee. They supported this leg with the same hand, mimicking the stance of a flamingo. The experiment was conducted for 60 s, during which the frequency of falls was quantified. The researcher would disqualify the test if they detected more than 15 falls in the first 30 s and applied the same procedure to the other leg. It was done first on the ground, followed by on the beam.

A metal beam measuring 50 cm long, 5 cm in height, and 3 cm wide was used for this assessment (Panjan & Sarabon, 2012).

The duration of the test was measured using a stopwatch. Timing commenced when the participant assumed the flamingo position and paused when off balance. Timing resumed once the participant returned to the tested position, and this process continued until 60 s had elapsed (Panjan & Sarabon, 2012). Additionally, the researchers closely monitored and recorded any excessive body sway exhibited by the participants during the test.

2.3 Data Analysis

The data obtained was analyzed using Statistical Package for the Social Sciences (SPSS) version 28. The researcher used Fisher's Exact test to assess the association between leg preference and the number of falls on the beam and the ground among national blind football athletes. The test was applied because the sample size was small or when the assumptions of other statistical tests, such as the Chi-Squared test, were not met.

Additionally, the test was employed when the expected cell count of a contingency table was low. Fisher's Exact test was considered a more precise and trustworthy method for analyzing contingency table data. The non-parametric test relied on the likelihood that chance caused the observed data to occur. Therefore, the researcher used it to evaluate if there was a meaningful relationship between these two categorical variables. The two categorical variables compared in this study were leg preference and the number of falls on the beam and the ground.

3 Result

3.1 Number of Falls Between the Preferred and Non-preferred

Fisher's Exact test (a two-way non-parametric test) was used to compare the number of falls between the preferred and non-preferred legs on the beam and the ground. The results of the test revealed a significant main effect of surface type, $p = 0.022$ ($p < 0.05$), still no significant main impact of leg preference, $p = 1.000$ ($p > 0.05$), or interaction between surface type and leg preference, $p = 1.000$ ($p > 0.05$). This finding indicates that there was no statistically significant interaction between leg preference and surface type to the number of falls among blind national football athletes.

3.2 *Interaction Between Surface Type and Leg Preference in the Number of Falls*

The researcher used Fisher's Exact test (the two-way non-parametric test) to compare the number of falls on the beam and the ground between the preferred and non-preferred legs. The results of the test revealed no significant difference in the number of falls between the preferred and non-preferred legs on the beam, $p = 0.911$ ($p > 0.05$), or on the ground, $p = 0.800$ ($p > 0.05$). This suggests that the number of falls on the beam and the ground did not differ significantly between preferred and non-preferred legs among national blind football athletes. The comprehensive data were assembled and presented in Tables 1, 2 and 3 to furnish a detailed overview of the variables under scrutiny. Table 1 offers insights into the mean and standard deviation of age, height, and weight among the national blind football athletes, providing foundational demographic information.

Meanwhile, Table 2 delineates the mean and standard deviation for the number of falls according to leg preference and surface types, elucidating the specific number of falls observed on the ground and the beam, segmented by preferred and non-preferred legs. Table 3 summarizes Fisher's Exact test outcomes for all significant variables, facilitating a concise overview of the statistical significance across various comparisons. The indicator abbreviations employed in the table—Preferred Leg on the Ground (PLOG), Non-Preferred Leg on the Ground (NPLOG), Preferred Leg on the Beam (PLOB), Non-Preferred Leg on the Beam (NPLOB)—facilitate a clear

Table 1 The mean and standard deviation of age, height, and weight of national blind football athletes

Variables	Mean	Standard deviation
Height (cm)	170.780	8.9663
Weight (kg)	80.50	12.5100

Table 2 The mean number of falls of the leg preference and surface types

Surface type	Leg preference	Mean fall	S.D. fall
On the ground	Preferred leg	1.80	3.910
On the ground	Non-preferred	0.4	0.966
Beam	Preferred leg	10.9	4.977
Beam	Non-preferred	11.4	5.082

Table 3 Fisher's exact test for all variables

Comparison	Fisher's exact test (p-value)
PLOG-NPLOG	0.022
PLOB-NPLOB	1.000
PLOG-PLOB	0.911
NPLOG-NPLOB	0.800

Indicator PLOG (Preferred Leg on the Ground), NPLOG (Non-Preferred Leg on the Ground, PLOB (Preferred Leg on the Beam), NPLOB (Non-Preferred Leg on the Beam)

understanding of the comparisons made and the associated p-values, enabling a comprehensive interpretation of the statistical significance to leg preference and surface type concerning the incidence of falls among national blind football athletes.

4 Discussions

The study's use of a small sample size consisting of 10 national blind football athletes may have constrained the statistical power of the analysis. Despite this limitation, Fisher's Exact test revealed a significant relationship ($p = 0.022$) between leg preference and the number of falls on the ground. It also suggests a difference in the number of falls between the preferred and non-preferred legs on the ground. This finding is consistent with the hypothesis, which stated that surface type would have a significant main effect on the number of falls, with a significantly higher number of falls on the beam than on the ground. This finding is in line with the study by (Alotaibi et al. 2016), which reported that individuals with visual impairments tend to have a preferred leg that they use more frequently and with more stability when standing on the ground.

Furthermore, the result shows meaningful distinctions in the frequency of falls between the preferred and non-preferred legs among blind football athletes, even within a limited sample size. However, it's vital to acknowledge that a larger sample size could have yielded more robust outcomes, enhancing the generalizability of findings to a broader population of blind football athletes. Additionally, employing higher power statistical tests like the Chi-Square test could have further explored the association between leg preference and falls on both the beam and ground surfaces.

Considering the confined sample size and the data's specific structure, Fisher's Exact test was deemed appropriate for this study despite the limitations imposed by the sample size. As a non-parametric test, it didn't assume a normal data distribution, rendering it best suited for evaluating categorical data within small sample sizes. The statistical analysis provided valuable insights into the relationship between leg choice and falls among national blind football athletes. In addition, Fisher's Exact test indicated no significant relationship between leg preference and the frequency of falls on the balance beam ($p = 1.000$) (Fig. 1). The distinct characteristics of the beam and ground surfaces significantly influenced balance and stability, which might have required varying balancing strategies (Jazi et al. 2012; Panjan & Sarabon 2012; Ponchillia et al. 2002; Urbaniak-Olejnik et al. 2022). These findings supported the initial hypothesis, indicating no significant difference in falls between preferred and non-preferred legs on the beam. This finding was also consistent with prior studies (Daneshmandi et al. 2021; Gamonales et al. 2018) suggesting that leg preference had no substantial impact on performance in 5-a-side football for individuals with visual impairments.

Conversely, a significant correlation emerged between leg preference and the number of falls on the ground ($p = 0.022$), confirming the substantial differences in

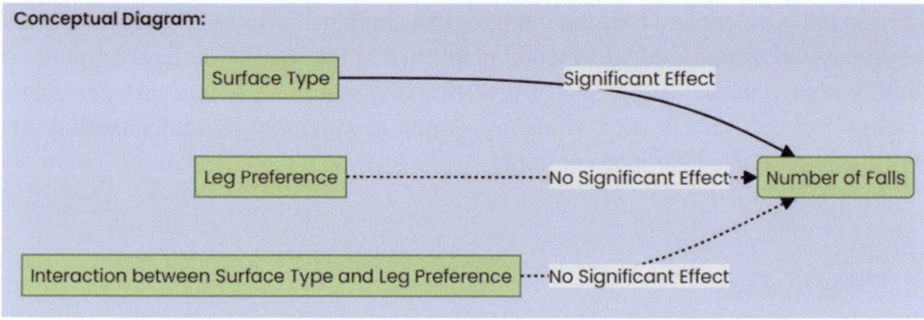

Fig. 1 The conceptual diagram of the relationship between the number of falls toward surface types, the leg preference, and the interaction between surface type and leg preference types

falls between preferred and non-preferred legs on this surface. This observation highlights the influence of surface type on fall occurrences, with the beam exhibiting a higher incidence of falls than the ground. These findings align with previous research (Alotaibi et al. 2016; Daneshmandi et al. 2021), where individuals with visual impairments leaned toward a preferred leg, demonstrating increased frequency and stability when standing on the ground.

The subsequent elucidation aligns with the association between leg preference (preferred and non-preferred) and surface types (on the ground and the beam). The results of Fisher's Exact test revealed that there was no significant difference in the number of falls between the preferred leg on the ground and the preferred leg on the beam, $p = 0.911$ ($p > 0.05$). This suggests that the number of falls on the beam and the ground did not differ significantly between the preferred leg among national blind football athletes. Similarly, the results of Fisher's Exact test revealed that there was no significant difference in the number of falls between the non-preferred leg on the ground and the non-preferred leg on the beam, $p = 0.800$ ($p > 0.05$). Overall, these results supported the hypothesis that there is no significant difference in the number of falls between the preferred and non-preferred legs on the beam and on the ground among national blind football athletes. It suggests that leg preference may not strongly predict the number of falls on different surface types in this population. It is crucial to consider the impact of surface type on balance and stability in this context.

Moreover, the beam size is narrow and unstable, which may require a different balance strategy than the ground (Panjan & Sarabon 2012). In contrast, the ground is a broader and more stable surface, which may allow for a different balance strategy (Ponchillia et al. 2002). It shows that there is a possibility that surface type may have contributed to the differences in the number of falls between the preferred and non-preferred legs on the beam and on the ground.

In addition to the statistical outcomes, this research study emphasizes the critical need for tailored interventions and training programs. Such initiatives should enhance balance strategies on different surfaces to reduce falls among blind football athletes. Moreover, considering the observed significant influence of surface types on fall occurrences, training modules that simulate the unique characteristics of varying surfaces could be instrumental in improving the stability and performance of blind football athletes. Additionally, investing in research exploring surface-specific training regimens could be a valuable avenue to enhance the understanding and development of effective interventions.

5 Conclusions

In summary, the results of this study demonstrated a significant correlation between leg preference and the number of falls on the ground among national blind football athletes. However, the finding did not find a significant association between leg preference and the number of falls on the beam. These findings have important implications for the training and performance of blind football athletes and the development of effective balance interventions for individuals with visual impairments. Coaches and trainers should consider integrating specific balance training exercises that target the challenges of different surface types, particularly the beam surface. Beyond surface type, other variables such as the degree of visual impairment, specific balance training, and sport-specific skills may have influenced the relationship between leg preference and the number of falls on different surface types. Further investigation is warranted to comprehend these factors and their variations across diverse surface conditions. While this study contributed to the existing knowledge base concerning the balance and stability of national blind football athletes, it also revealed potential avenues for future research. Further research is needed to understand better the factors that may contribute to this relationship and how they may vary across different surface types, as well as the role of other variables such as the level of visual impairment, balance training, and sport-specific skills. This study adds to the growing knowledge of balance and stability in blind football athletes and can potentially inform future studies.

References

Alotaibi AZ, Alghadir A, Iqbal ZA, Anwer S (2016) Effect of absence of vision on posture. J Phys Ther Sci 28(4):1374–1377. https://doi.org/10.1589/jpts.28.1374. (Society of Physical Therapy Science (Rigaku Ryoho Kagakugakkai)

Bednarczuk G, Wiszomirska I, Marszałek J, Rutkowska I, Skowroński W (2017) Static balance of visually impaired athletes in open and closed skill sports. Polish J Sport Tourism 24(1):10–14. https://doi.org/10.1515/pjst-2017-0002

Daneshmandi H, Norasteh AA, Zarei H (2021) Balance in the blind: a systematic review. Phys Treat Specif Phys Ther J 11(1):1–12. https://doi.org/10.32598/ptj.11.1.430.2

Explanatory guide to Paralympic classification Paralympic summer sports (2016). www.paralympic.org

Gamonales JM, Muñoz-Jiménez J, León-Guzmán K, Ibáñez SJ (2018) 5-A-side football for individuals with visual impairments: a review of the literature. Eur J Adapt Phys Act 11(1). https://doi.org/10.5507/EUJ.2018.004. (European Federation of Adapted Physical Activity)

Jazi SD, Purrajabi F, Movahedi A, Jalali S (2012) Effect of selected balance exercises on the dynamic balance of children with visual impairments. J Visual Impair Blind 106(8):466–474. https://doi.org/10.1177/0145482x1210600803

Kurz A, Lauber B, Franke S, Leukel C (2021) Balance training reduces postural sway and improves sport-specific performance in visually impaired cross-country skiers. J Strength Cond Res 35(1):247–252. https://doi.org/10.1519/JSC.0000000000002597

Panjan A, Sarabon N (2012) Review of methods for the evaluation of human body balance. Sport Sci Rev 19(5–6). https://doi.org/10.2478/v10237-011-0036-5

Ponchillia PE, Strause B, Ponchillia SV (2002) Athletes with visual impairments: attributes and sports participation. J Visual Impair Blind 96(4):267–272. https://doi.org/10.1177/0145482x0209600408

Rami PV, Prabhakar MM (2018) Comparison of static balance in male football and basketball players by using flamingo balance test. Int J Physiother 5(5). https://doi.org/10.15621/ijphy/2018/v5i5/177432

Urbaniak-Olejnik M, Loba W, Stieler O, Komar D, Majewska A, Marcinkowska-Gapińska A, Hojan-Jezierska D (2022) Body balance analysis in the visually impaired individuals aged 18–24 years. Int J Environ Res Public Health 19(21). https://doi.org/10.3390/ijerph192114383

Riding Skills Comparison Between Skilled and Less-Skilled Mountain Bike Cyclists

Amirul Aiman Mohd Noor, Haziq Herul Anuar, and Hosni Hasan

Abstract One of the major factors contributing to success in mountain bike racing is the demonstration of exceptional skills while navigating the track. This research was carried out to measure and compare the riding skills of mountain bikers, especially on a real trail simulation that incorporated various technical elements within the track. The comparison of riding skills in real simulation trails was conducted between the skilled (men's junior) and less-skilled (men's youth) riders among the local mountain bike association. Ten ($n = 10$) male mountain bike athletes were separated into two groups: skilled (under 21 years old) ($n = 5$) and less-skilled (under 16 years old) ($n = 5$) groups for field data collection at Putrajaya Challenge Park (PCP). Action cameras (GoPro) were mounted on the athletes' chests and used to analyze trail riding skills, such as shifting gears, climbing, descending, cornering, and braking, while bike computers (Garmin) were used for data collection on average and maximum speed. In one loop of the PCP track, all participants rode the blue, green, and red trails for three kilometers. Skilled-cyclist categories significantly scored higher ($p < 0.05$) than those less skilled during climbing, descending, cornering, and braking in mountain bike performance skills. These findings can be useful for mountain bike coaches to implement better training experiences for their athletes in improving mountain bike riding skills.

Keywords Motor skill · Cycling · Mountain bike · Riding skills

A. A. M. Noor (✉) · H. H. Anuar · H. Hasan
Faculty of Sports Science and Recreation, Universiti Teknologi MARA (UiTM) Shah Alam, 40450 Shah Alam, Selangor Darul Ehsan, Malaysia
e-mail: aiman991029@gmail.com

H. Hasan
e-mail: hosnihasan@uitm.edu.my

Z. Mohamed et al. (eds.), *Proceedings of the 2nd International Conference on Innovation and Technology in Sports, ICITS 2023, 27–28 November, Malaysia*, Lecture Notes in Bioengineering, https://doi.org/10.1007/978-981-97-3741-3_4

1 Introduction

Cycling is a popular physical activity in Malaysia, encompassing various categories such as leisure cycling, competitive cycling, cycling for health, and cycling for fitness. The overall goals of cycling include promoting cycling within the community through sustainable transportation, creating liveable communities, and encouraging physical activity for recreation, fitness, and health purposes (Götschi et al. 2015). Cycling could be broadly classified into two main categories: leisure cycling and performance cycling. Leisure cycling refers to activities that people engage in during their leisure time for relaxation and do not require long-distance rides. These leisure rides are often done in loops, starting and ending at one's home, and could connect multiple charging points located in parks, pedestrian zones, or along the coast (Khalaj et al. 2020). Performance cycling, also known as competitive cycling, involves individuals or groups competing against each other with the aim of winning the competition by exerting maximum energy and performance. The performance of cyclists is influenced by physiological variables that affect mechanical power output, as well as mechanical and environmental factors that influence power demand (Jeukendrup & Martin 2001).

Mountain biking is a sport that offers both competitive and leisurely challenges. There are various categories of mountain bike competitions, including Olympic cross-country (XCO), hill climb/uphill, downhill, four-crossed and free, dual slalom, mountain crossed, point-to-point racing, stage races, observed trails, ultra-endurance events, and short track races (Impellizzeri et al. 2005; Impellizzeri & Marcora 2007). Cross-country Olympic (XCO) racing, for instance, involves off-road tracks with a specified number of circuit loops (usually 5–7 laps for elite riders) and encompasses different terrains, including uphill climbs and downhill sections.

While mountain biking involves certain risks, many individuals enjoy these activities for the thrilling experiences they offer (Lopes 2017). Some people participate in mountain biking to enhance their skills, while others engage in it to exchange techniques with fellow riders. Regardless of the motivations, participating in mountain biking as a recreational activity offers several advantages. Although there has been vast research on the physiological aspects of mountain biking (Lee et al. 2002; Impellizzeri & Marcora 2007), research on the mountain biking skills itself were still limited, particularly regarding motor skills and riding techniques. Therefore, this research was conducted to assess the mountain biking skills among skilled and less-skilled cyclists. This research had two main objectives: first, to assess the perceived difficulty of the trail by different age groups, and, second, to analyze the factors that may hinder the performance of riders in trail simulations, focusing on the five technical sectors of mountain biking, namely shifting gears, climbing, descending, cornering, and braking. This research would serve as a reference for future analysis of motor skills among young athletes in Malaysia, particularly in the context of mountain biking. It would also provide valuable insights to educate athletes, coaches, and organizations on how to improve their riding skills and determine suitable training methods to enhance their ability to ride a mountain bike flawlessly, especially on trails and off-road tracks. Ultimately, this research aims to enhance awareness of the importance of technical skills in mountain biking and improve the overall understanding of motor skills in mountain biking.

2 Methodology

2.1 Participants

Ten (n = 10) male mountain bike athletes from the local mountain bike association took part in this study and were divided into men's junior (n = 5) and youth (n = 5) groups based on the age category in mountain bike competition in Malaysia. Men's juniors could be described as athletes aged 17–21 years old or below, and men's youth could be described as athletes aged 13–16 years old. The selection of the cyclists was in agreement and discussions with their coaches and also approval for participants from their parents/guardians. Men's junior participants have been involved in mountain bike training and competitions for more than four years and can be categorized as skilled participants (Nabil et al. 2022).

2.2 Instrument

This investigation was carried out at the Putrajaya Challenge Park (PCP), a popular mountain bike park located in Malaysia. It offers several trails that cyclists can choose from based on their skill levels and preferences. These trails are rated with varying degrees of difficulty. Three levels of the trail are categorized as blue (easy), green (medium), and red (advanced) and were used during the real-time simulation.

All participants used their own mountain bikes during the testing session. Testing instruments involve one action camera (GoPro), a bike computer (Garmin), and a digital stopwatch throughout the trail. Action cameras were used to analyze riding skills on the trail, including shifting gears, climbing, descending, cornering, and braking, while a bike computer was mounted on the bike stem for data collection on average and maximum speed. The scoring for shifting, climbing, descending, braking control, and cornering includes five components of riding skill: shifting, climbing, descending, controlled braking, and cornering. Additionally, this test consists of a four-point Likert scale. The researchers provided the scores to the riders based on their performance using the video recordings taken throughout the trail ride. A rider that achieves a high score is considered proficient in mountain bike riding skills.

2.3 Data Collection

Before experimenting, all participants were given the PAR-Q sheet, and warmed up for 10 min before proceeding to the real trail simulation on the PCP trail. All participants rode the blue, green, and red trails for three kilometers in one loop of the PCP trail. The loop consists of three sections (Fig. 1): the blue section (climbing and shifting), green section (cornering and braking), and red section (downhill) to evaluate riders' skills. The participants took turns in the testing session and were released one at a time from the starting point, and the researcher recorded the best time of the whole loop while the action camera (GoPro) recorded from the rider's chest along the way to the trails.

There were five challenging aspects of riding skill on the course throughout the three-kilometer ride on the trail, each with unique obstacles and riding skills to overcome these technical challenges. Each technical aspect contains three sections the researcher would analyze by utilizing the camera footage on the athlete's chest. At the end of the experiment, the researcher gathered the data and provided a score to the riders based on their performance in prior video recordings taken throughout the trail ride. Once the data were collected by filling out the score of riding skill in a real trail simulation, the researchers compared the scores between the men's junior and the men's youth cyclists.

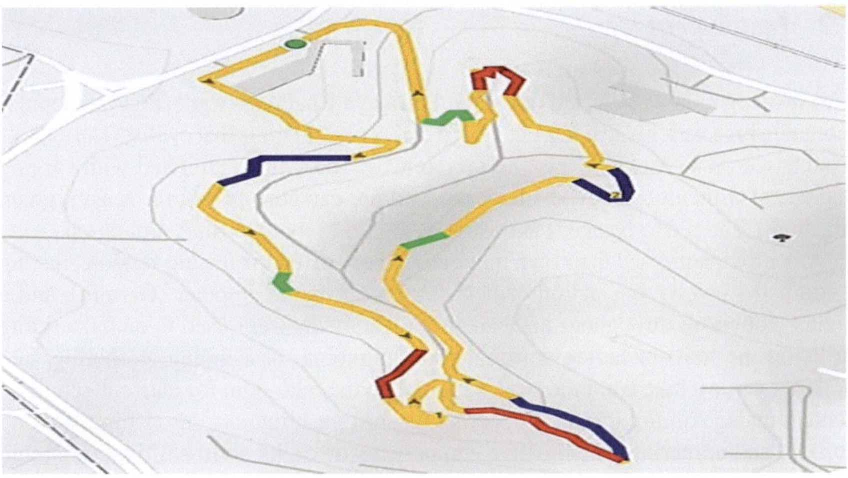

Fig. 1 Three sections on real simulation trail to evaluate riding skills

2.4 Data Analysis

For the statistical analyses, the data were analyzed using IBM SPSS Statistics (SPP Inc., Chicago, IL, USA). Normality tests were conducted by the researchers to determine the normality of the results. The data showed a non-normal distribution with a p-value less than 0.05, leading to the use of nonparametric tests due to the limited number of participants in this research. To observe the differences in riding skills in real simulation trails between men's youth and men's juniors from Putrajaya Cycling Association, the statistical analysis of this study was analyzed using the Mann–Whitney test. This statistic was described by using the value of U, the sample sizes, and the significance level. The significance level for this test was set at $p < 0.05$.

3 Results

3.1 Mountain Bike Riding Skills

Table 1 shows the results for mountain biking skills between men's junior and men's youth cyclists. There was no significant difference ($z = -1.940$, $p > 0.05$) between men's juniors and men's youth on shifting scores (Table 2). Results also revealed that there were significant differences between men's juniors and men's youth on climbing ($z = -2.460$, $p < 0.05$), descending ($z = -2.545$, $p < 0.05$), braking ($z = -2.495$, $p < 0.05$), and cornering ($z = -2.300$, $p < 0.05$) scores.

Table 1 Mean score for mountain bike riding skills among men's juniors and men's youth

Rank				
Skills	Group	Mean rank	Sum of rank	Median
Shifting	Less-skilled	3.70	18.50	2.00
	Skilled	7.70	36.50	4.00
Climbing	Less-skilled	3.30	16.50	2.00
	Skilled	7.70	38.50	3.00
Descending	Less-skilled	3.20	16.00	2.00
	Skilled	7.80	39.00	4.00
Braking	Less-skilled	3.20	16.00	2.00
	Skilled	7.80	39.00	4.00
Cornering	Less-skilled	3.40	17.00	2.00
	Skilled	7.60	38.00	4.00

Table 2 Mean score and statistical analysis for mountain bike riding skills

	Shifting	Climbing	Descending	Braking	Cornering
Mann–Whitney U	3.500	1.500	1.000	1.000	2.000
Wilcoxon W	18.500	16.500	16.000	16.000	17.000
Z	−1.940	−2.460	−2.545	−2.495	−2.300
Sig. (2-Tailed)	0.052	0.014	0.011	0.013	0.021

*p-value is significant when $p < 0.05$

4 Discussions

The purpose of this study was to compare mountain bike riding skills between men's junior and men's youth cyclists. Results revealed that there was no significant difference in shifting scores between skilled and men's less-skilled participants. Both groups frequently shifted gears to navigate terrain smoothly. Participants tended to shift into easier gears early before the climbing section to maintain a steady cadence. Cadence, measured in revolutions per minute (RPM), is crucial for determining when to shift into a different gear, with a recommended range of 80 to 95 RPM (Volynski 2022). Additionally, participants expressed the desire to shift to a higher gear during the downhill section for increased speed during pedaling. A significant difference was observed in climbing scores between those skilled and less skilled, with skilled participants scoring higher. Men's juniors (skilled) demonstrated a preference for riding out of the saddle during the climbing section, leveraging body weight for maximum power. This approach, utilizing upper body anaerobic power, allows for more power generation when standing and pedaling, particularly in high-resistance situations. Conversely, participants preferred a seated position during technical climbing sections, providing more control and keeping the center of gravity closer to the back to prevent the chance of the back wheel slipping.

In descending scores, a significant difference was noted between the groups, with men's juniors (skilled) scoring higher. Skilled riders exhibited greater confidence during descents, leading to better control and speed. Effective descending skills involve keeping eyes on the exit for a smooth, steady line (Shilton et al. 2022). However, men's youth (less-skilled) riders tended to steer toward whatever they were looking at, potentially leading to decreased speed. For braking scores, skilled participants had a higher score than those less skilled, showcasing a better understanding of ratio braking and when to apply each brake. Nevertheless, less-skilled riders tended to use more front brakes than rear brakes. Front brakes, having more braking power, are primarily responsible for stopping, while rear brakes are used to regulate speed (Miller et al. 2016, 2017).

In cornering scores, a significant difference was observed between the groups, with men's juniors (skilled) scoring higher. Men's junior (skilled) riders tended to lean more during cornering to maintain momentum and speed. Additionally, their focus on the outside leg enhanced control over traction. A stable riding position, lower center of gravity, and added weight to the front of the bike contribute to improved cornering skills (Brett 2022). Conversely, men's youth (less-skilled) riders tended to misjudge corners, corner too fast, and rely on brakes. Additionally, they did not take the optimum line or apex point that could benefit them during cornering. The ideal line revolves around the corner of the curve where the apex is located, marking the transition from entering to exiting while maintaining a calm posture for better braking control can help to make quick turns in a tight corner.

5 Conclusions

In summary, good riding skills among cyclists play a crucial role in achieving superior performance and success in competitions, particularly in technically challenging races such as mountain biking. This notable distinction between men's juniors (skilled) and men's youth (less-skilled) in this study arises from variations in the years of experience in cycling, with men's juniors (skilled participants) participating more frequently in competitions than in the youth category. Additionally, the type of bicycle used had a substantial impact, with the men's junior category utilizing semi-high-end equipment such as carbon components, while the youth category continued to use aluminum and medium-level equipment. Furthermore, the non-significant difference in shifting scores can be attributed to the fundamental nature of shifting skills in cycling, where cyclists routinely shift gears up or down, especially in competitive cycling. This research provides valuable insights for sports coaches to plan future athletes' sports training programs, particularly focusing on enhancing their riding skills on real simulation tracks. Future studies may involve more technical terrains, and the mountain biking skills among the skilled cyclists can be documented as references to younger cyclists.

References

Brett M (2022). 11 tips for better cornering—get round bends faster. road.cc. (2022, July 18). https://road.cc/content/feature/11-tips-better-cornering-217155

Götschi T, Garrard J, Giles-Corti B (2015) Cycling as a part of daily life: a review of health perspectives. Transp Rev 36(1):45–71

Impellizzeri FM, Rampinini E, Sassi A, Mognoni P, Marcora S (2005) Physiological correlates to off-road cycling performance. J Sports Sci 23(1):41–47

Impellizzeri FM, Marcora SM (2007) The physiology of mountain biking. Sports Med 37:59–71

Jeukendrup AE, Martin J (2001) Improving cycling performance: how should we spend our time and money. Sports Med 31:559–569

Khalaj F, Pojani D, Sipe N, Corcoran J (2020) Why are cities removing their freeways? A systematic review of the literature. Transp Rev 40(5):557–580

Lee H, Martin DT, Anson JM, Grundy D, Hahn AG (2002) Physiological characteristics of successful mountain bikers and professional road cyclists. J Sports Sci 20(12):1001–1008

Lopes B (2017) Mastering mountain bike skills. Human Kinetics

Miller MC, Fink PW, Macdermid PW, Perry BG, Stannard SR (2017) Validity of a device designed to measure braking power in bicycle disc brakes. Sports Biomech 17(3):303–313. https://doi.org/10.1080/14763141.2017.1338744

Miller MC, Macdermid PW, Fink PW, Stannard SR (2016) Performance and physiological effects of different descending strategies for cross-country mountain biking. Eur J Sport Sci 17(3):279–285. https://doi.org/10.1080/174613912016.1237550

Nabil MA, Hasan H, Ismail MSI, Azhan MAN (2022) Effects of wheel depth on individual time trial performance among skilled cyclists. In: International conference on movement, health and exercise. Springer Nature Singapore, Singapore, pp 41–48

Shilton AC, Hurford M, Stieda A. (2022). How to ride descents in all kinds of conditions like a pro. Bicycling. https://www.bicycling.com/skills-tips/a34920007/how-to-ride-descents/

Volynski V (2022) Bike gear shifting. How to change gears for dummies. Vuly Play Australia. https://www.vulyplay.com/en-AU/blog/bike-gear-shifting-how-to-change-gears-for-dummies

Does Altering the Longitudinal Bending Stiffness of the Sprint Spike Affect Sprinting Performance?

Muhammad Hasif Hilmi M. Hazman, Motohide Arayama, Hiroyuki Nunome, and Shariman Ismadi Ismail

Abstract This study aimed to clarify whether altering the longitudinal bending stiffness of sprint spike could substantially affect sprinting performance among 100-m and 200-m sprinters. A total of five (n = 5) male sprint athletes volunteered in this study (shoe size UK 8.5; age = 20.4 ± 1.5 years old; height = 170.8 ± 4.7 cm; weight = 66.2 ± 3.9 kg). A Decathlon-made (model Kalenji AT Start-Ref. 8,504,517) sprint spike (UK size 8.5) has been chosen as a standard spike used in this study. The spike longitudinal bending stiffness was altered into three different stiffness (S1: lower; S2: original; S3: higher) by changing the spike insoles property (lowering the stiffness using PU foam insoles and increasing the stiffness using a 3-mm vinyl board). To test the sprint performance, all participants were requested to perform a 30-m fly sprint test using three different spike conditions in randomized order. It was found that on average, participants performed faster using S3 spike (3.42 ± 0.13), followed by S2 and S1 respectively (S2: 3.48 ± 0.19; S1: 3.53 ± 0.27). However, an analysis of variance indicated that statistically there were no significant differences in the sprint performance among the three-spike condition [F (2,44) = 1.06 (p = 0.36)]. In addition, individual results have shown only 3 out of 5 participants improved their sprint performance using the S3 spike. Differences in improvement achieved among the sprinters when using different spike stiffness in this study could suggest that there may be an optimum bending stiffness property that suits each person individually.

Keywords 100-m · 200-m · Footwear · Mechanical · Forefoot bending stiffness

---

M. H. H. M. Hazman · S. I. Ismail (✉)
Faculty of Sports Science & Recreation, Universiti Teknologi MARA, Shah Alam, Malaysia
e-mail: shariman_ismadi@uitm.edu.my

M. Arayama
Dream GP Inc., Osaka, Japan

H. Nunome
Faculty of Sports and Health Science, Fukuoka University, Fukuoka, Japan

Z. Mohamed et al. (eds.), *Proceedings of the 2nd International Conference on Innovation and Technology in Sports, ICITS 2023, 27–28 November, Malaysia*, Lecture Notes in Bioengineering, https://doi.org/10.1007/978-981-97-3741-3_5

1 Introduction

The longitudinal bending stiffness of sprint spikes has been a subject of interest in recent years due to its function to assist in reducing energy loss at the metatarsophalangeal joint (MTPJ) (Smith et al. 2014, 2016; Stefanyshyn and Nigg 2000). By increasing the bending stiffness, the sole will limit the amount of dorsiflexion at the metatarsophalangeal joint that occurs during the stance phase of sprinting, while shifting the amount of application of force anteriorly (Willwacher et al. 2014). The amount of force reduction at the MTPJ has also been observed less in subjects wearing sprint spikes compared to barefoot sprinting (Smith et al. 2014). These findings from recent studies have demonstrated the mechanism of increasing the longitudinal bending stiffness of sprint spike that would help to improve sprinters' performance. However, previous authors (Stefanyshyn and Fusco 2004) have shown there have been significant improvements in sprinting performance with sprint spikes with increased bending stiffness, while several authors (Smith et al. 2016; Nagahara et al. 2017) have observed little to no improvements in sprinting performance while using sprint spikes with increased bending stiffness. Therefore, the results of these previous studies have been inconclusive and still warrant further investigation. Furthermore, so far limited investigation has been conducted on the effects of reducing the longitudinal bending stiffness of sprint spikes on sprinters' performance. We aimed to clarify whether altering the longitudinal bending stiffness of the sprint spike could substantially affect sprinting performance. It was hypothesized that longitudinal bending stiffness for the tested sprint spike in this study would have a substantial impact on the 30-m fly start sprint test resultant performance.

2 Methodology

2.1 Participants

A total of five (n = 5) male sprint athletes that participated in 100-m and 200-m events volunteered in this study (shoe size UK 8.5; age = 20.4 ± 1.5 years old; height = 170.8 ± 4.7 cm; weight = 66.2 ± 3.9 kg). At the time of data collection, all the participants that volunteered in this study had been actively involved in inter-university level competition for the past 12 months and trained regularly for 3 to 4 times per week. They also had no lower limb injury history in the past 6 months. The study was approved by the Institution Research Ethics Committee, and all participants have read and signed an informed consent document prior to the start of the study.

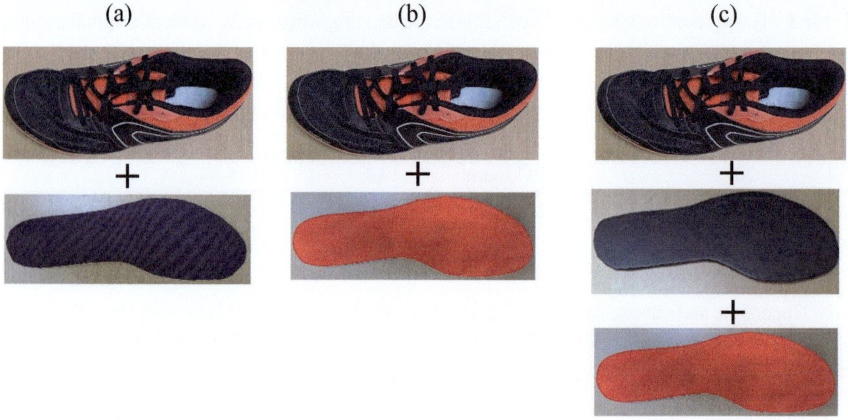

Fig. 1 The sprint spike and insoles—**a** S1: PU foam; S2, **b** EVA; **c** S3: vinyl board + EVA)

2.2 *Sprint Spike and Longitudinal Bending Stiffness Properties*

In order to prepare three sprint spikes (S1, S2 and S3) with different longitudinal bending stiffness properties, the insoles property of the spike has been altered. For S1, the original insoles supplied by the manufacturer (EVA foam insole) are replaced with a much softer and less-stiff insole made from PU foam to reduce the original longitudinal bending stiffness of the spike. For S2, the original condition of the spike has been maintained. For S3, the bending stiffness has been increased by placing a customized 3 mm vinyl board under the original insoles inside the spike. The vinyl board shape is customized to perfectly fit in the spike by fabricating it to the actual shape of the original insoles (Fig. 1).

A Decathlon-made (model Kalenji AT Start-Ref. 8,504,517) sprint spike (UK size 8.5) has been chosen as a standard spike used in this study (Fig. 1). The specification and material properties of the spike are shown in Table 1.

2.3 *Longitudinal Bending Stiffness Measurement*

The longitudinal bending stiffness for each spike (S1, S2 and S3) was measured using a similar method described in the study of Ismail et al. (2021), Worobets and Wannop (2015). Each spike was applied with forefoot torque (rotational force) up to 5 Nm in five stages until all spikes reached a maximum bending angle of more than 40 degrees. Data collected from these measurements were then plotted on a torque–angle graph. A regression line was fit to these data points. The slope of the regression line produced from the data plots is taken as the longitudinal bending

Table 1 The properties of sprint spike (S1: lowered bending stiffness; S2: control/original bending stiffness; S3: increased bending stiffness) used in the study

Spike	Mass per shoe with insoles (gram)	Insoles material	Insoles thickness (mm)	Spike longitudinal bending stiffness (Nm/deg.)
S1	202	PU foam	4	0.18
S2	204	EVA foam	3	0.34
S3	227	EVA foam + vinyl board	6	0.37

Note Upper sole and outsole material—Upper of 30.0% Polyurethane, 70.0% Polyester; Outer sole of 30.0% Synthetic Rubber, 40.0% Ethylene Vinyl Acetate, 30.0% Thermoplastic Polyurethane; Spike: 5 pieces of 6-mm spike

stiffness values. Each spike's bending stiffness was measured 3 times, and the mean values were computed as representing values.

2.4 30-M Fly Start Sprint Test

The test for sprinting performance is performed at an outdoor sports facility, where all participants were requested to run on a Tartan track used for athletics events (Fig. 2). After performing their own warmup procedure, the participants are requested to perform nine maximum 50-m sprints from a standing start. This consists of three trials in each sprint spike condition (S1, S2 and S3). The participants are also instructed to sprint at their maximal effort. The rest period given for each trial would be ten minutes to avoid fatigue (Nagahara et al. 2017). In the 30-m fly start sprint test, the times are recorded right after the participants have passed the 20-m mark, where the 20–50-m split times would be taken by single beam timing gates (Brower Timing Systems, South Fort Street, Draper, Utah, USA). During the test, the sprint spike insole condition is randomized in order.

2.5 Statistical Analyses

Statistical analyses were performed using IBM SPSS Statistics 26 (SPP Inc., Chicago, IL, USA). The mean time recorded of the 30-m fly start for each spike condition was compared using a one-way ANOVA repeated measure to identify differences between the sprinting performance under three different spike properties. Statistical significance was set at $p < 0.05$.

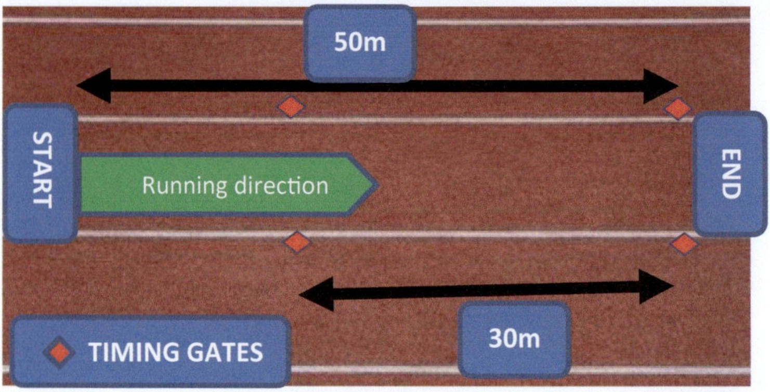

Fig. 2 Test set up for 30-m fly start sprint

3 Results

The results obtained from the 30-m fly start sprint test are presented in Fig. 3. On average, an improvement of time recorded when the longitudinal bending stiffness is increased was observed as shown in Fig. 3 [S1: 3.53 ± 0.27; S2: 3.48 ± 0.19; S3: 3.42 ± 0.13]. However, it was found that there were no significant differences in sprint performance among the three-spike conditions [$F(2,44) = 1.06$ ($p = 0.36$)].

Meanwhile, the individual sprint performance for each sprinter when sprinting under different spike conditions is shown in Fig. 4. It was found that three sprinters performed the best time in the S3 spike (participants 2, 3 and 4), while Participant 1 performed the best in the S1 spike and Participant 5 performed the best time in the S2 spike.

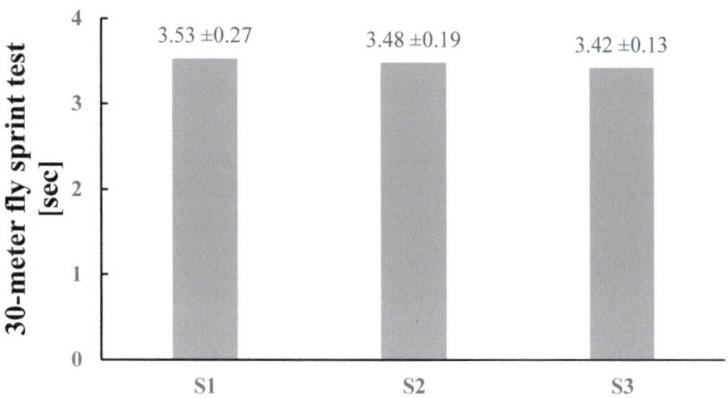

Fig. 3 The results from 30-m fly start sprint test

	S1	S2	S3
Participant 1	3.16 (±0.10)	3.24 (±0.18)	3.28 (±0.05)
Participant 2	3.34 (±0.05)	3.40 (±0.12)	3.28 (±0.06)
Participant 3	3.69 (±0.11)	3.55 (±0.09)	3.50 (±0.10)
Participant 4	3.65 (±0.04)	3.74 (±0.04)	3.53 (±0.03)
Participant 5	3.84 (±0.12)	3.49 (±0.05)	3.52 (±0.02)

Fig. 4 Individual performance results for the 30-m fly start sprint test for each sprinter (recorded time ± sd)

4 Discussion

This study aimed to verify longitudinal bending stiffness efficacy on resultant performance in a 30-m fly start sprint test. We hypothesized that different longitudinal bending stiffness properties would have a substantial impact on the outcomes of the sprint performance tests. It was demonstrated that statistically, higher longitudinal bending stiffness is considered not to have a substantial impact on the sprint performance tests conducted in this study. In addition, it was also observed that lowering the longitudinal bending stiffness would also not statistically influence the sprint performance. Thus, these findings most likely rejected our hypothesis that longitudinal bending stiffness for the tested sprint spike in this study would have a substantial impact on the 30-m fly start sprint test resultant performance.

This study found that using a sprint spike with an increased longitudinal bending stiffness did not statistically provide an improvement in the participant's sprinting performance. This finding was in line with the observation reported by Nagahara et al. (2017). However, it is also worth mentioning that in the same study by Nagahara et al. (2017), sprinters did actually improve their 30-m sprint performance when using a higher bending stiffness spike, lowering the recorded time by 0.01 s on average. Though this improved performance may not be statistically significant as observed by that particular study, t, the marginal gain of an improved record among the sprinters must also be acknowledged. In addition, in this present study, it can be seen by the individual improvements in sprinting performance with the S3 spike when compared to the control (S2) and lowered stiffness spike (S1), in particular among participants 2, 3 and 4. Therefore, partially, findings from this study are also supporting what previous studies found with regard to the improvement in sprint performance using a higher longitudinal bending stiffness spike (Stefanyshyn and Fusco 2004). Contradicting outcomes observed in this study in Participants 1 and 5 are rather interesting. This sparks another question: why do certain sprinters benefit from increasing the bending stiffness of the spike and some do not? At this time, we only have collected data from five participants. Collecting more data from other sprinters would help to increase our understanding of these findings.

Our study has found that increasing the longitudinal bending stiffness of sprint spikes could potentially help to improve sprint performance among some of the sprinters. Differences in the effects of bending stiffness among the sprinters could be related to their differences in foot characteristics. In a recent study, it was reported that the foot arch height is an important parameter that is associated with soccer player jumping performance, specifically for horizontal jump (Ismail et al. 2023). This could be true as well among sprinters, but so far, no such study has been conducted. This contradicting finding on the effects of the longitudinal bending stiffness of sprint spike on sprinting performance may be explained by further investigating the anatomical aspect of the participant's foot, namely the foot characteristics such as the foot width, length, arch height index and arch stiffness. Thus, the benefits gained by the sprinters upon the increased stiffness around the MTPJ, that potentially could help to reduce joint force around the MTPJ and reduce the contact time of the spike and running track during sprinting, may not be a universal truth for all sprinters. Other factors such as foot characteristics may account for it.

5 Conclusion

In this study, it was demonstrated that longitudinal bending stiffness may not have a systematic influence on the resultant performance of the 30-m fly start sprint test for all the sprinters. Although half of the sprinters that participated in this study (3 over 5) did improve their sprint performance when using higher stiffness spike, the same outcome was not observed for others. Differences of improvement achieved among the sprinters when using different spike stiffness in this study could possibly suggest that there may be an optimum bending stiffness property that suits each person individually. It was suggested that the foot characteristic differences among the sprinters may account for this contradicting outcome.

Acknowledgements The authors would like to thank Universiti Teknologi MARA for partially supporting this study.

References

Ismail SI, Nunome H, Tamura Y (2021) The influence of forefoot bending stiffness of futsal shoes on multiple V-cut run performance. Front Psychol 12:625079

Ismail SI, Nor Azian FS, Adnan R, Kamil ŞZ, Nun Arimanshah AA, Sulaiman N, Arayaman M, Raja Azidin RMF, Hamidon JK, Mustafa MS, Mohd Nor MA (2023) Association of arch height of the foot and jump capacity in youth soccer athletes. In: Proceedings of the 8th international conference on movement, health and exercise (MOHE 2022). Springer Nature Singapore, Malaysia. Singapore. (2023, July)

Nagahara R, Kanehisa H, Fukunaga T (2017) Influence of shoe sole bending stiffness on sprinting performance. J Sports Med Phys Fitness 58(12):1735–1740

Smith G, Lake M, Lees A (2014) Metatarsophalangeal joint function during sprinting: a comparison of barefoot and sprint spike shod foot conditions. J Appl Biomech 30(2):206–212

Smith G, Lake M, Sterzing T, Milani T (2016) The influence of sprint spike bending stiffness on sprinting performance and metatarsophalangeal joint function. Footwear Sci 8(2):109–118

Stefanyshyn D, Fusco C (2004) Athletics: increased shoe bending stiffness increases sprint performance. Sports Biomech 3(1):55–66

Stefanyshyn DJ, Nigg BM (2000) Influence of midsole bending stiffness on joint energy and jump height performance. Med Sci Sports Exerc 32(2):471

Willwacher S, König M, Braunstein B, Goldmann JP, Brüggemann GP (2014) The gearing function of running shoe longitudinal bending stiffness. Gait Posture 40(3):386–390

Worobets J, Wannop JW (2015) Influence of basketball shoe mass, outsole traction, and forefoot bending stiffness on three athletic movements. Sports Biomech 14(3):351–360

Insignificant Influence of Sport Participation on Inhibitory Control Among Student Athletes in the Stop Signal Task

Mohd Azim Nural Azhan, Aizreena Azaman, Syed Haziq Syed Mohd Khairizam, Mohd Syafiq Miswan, Hosni Hasan, and Nurul Farha Zainuddin

Abstract Cognition was initially believed to be entirely governed by the central nervous system, with long-term potentiation and neurogenesis contributing to memory development and storage. In addition, cognitive function, particularly in sports, has gained popularity among researchers, with the current emphasis on comparing cognitive function and sports performance between different types of athletes. Therefore, this study aims to compare the cognitive functions of the inhibition stop signal task between elite and non-elite student athletes. A total of 48 student athletes from the Kedah State Sports Council and Universiti Teknologi MARA participated in this study. A preliminary survey and a computerized stop signal task from E-prime 3.0 were used in this study. Descriptive and statistical inferential methods were used in conjunction with Microsoft Excel 2020 and the Statistical Package for the Social Sciences (SPSS 26.0 version). An equal variance t-test found insignificant statistical differences for mean go trial response time (GTRT) and mean stop signal response time (SSRT) between elite and non-elite participants. This result suggests that the criteria used to differentiate elite athletes from non-elite athletes, such as the number of hours or years of training completed, may need to be more precise to truly distinguish between elite and non-elite student athletes. For future studies, the stop

M. A. N. Azhan · A. Azaman
Department of Biomedical Engineering and Health Sciences, Faculty of Electrical Engineering, Universiti Teknologi Malaysia, 81310 Skudai, Malaysia

S. H. S. M. Khairizam · M. S. Miswan · N. F. Zainuddin (✉)
Faculty of Sports Science and Recreation, Universiti Teknologi MARA, Cawangan Perlis, Kampus Arau, 02600 Arau, Perlis, Malaysia
e-mail: nurulfarha@uitm.edu.my

H. Hasan
Faculty of Sports Science and Recreation, Universiti Teknologi MARA, 40450 Selangor, Malaysia

M. A. N. Azhan
Faculty of Computer Media and Technology Management, University College TATI, Ke- Maman, 24000 Terengganu, Malaysia

Z. Mohamed et al. (eds.), *Proceedings of the 2nd International Conference on Innovation and Technology in Sports, ICITS 2023, 27–28 November, Malaysia*, Lecture Notes in Bioengineering, https://doi.org/10.1007/978-981-97-3741-3_6

signal task used in this study may require several modifications to ensure that a more reliable SSRT is achieved.

Keywords Cognitive inhibition · Stop signal response time · Cognitive ability · Athletes

1 Introduction

Outstanding athletes have been more proficient than inferior athletes in making judgments and extrapolating essential information from the environment to forecast future results. Consequently, improved brain functioning in athletes may increase athletic performance by making fast and correct decisions during the game. The ability to manage many tasks and adjust to shifting environments are all necessary for strong cognitive function in sports (Hernández-Mendo et al. 2019). While non-elites spent more time taking part in soccer-related activities, Ahmad et al. (2020) found that they were less likely to take part in organized activities. It is important to examine the activities in which successful elite athletes participated before achieving their peak performance. Competitive athletes who train intensively and for an extended period tend to have high physical and motor fitness levels. It is reasonable to assume that these athletes demonstrate cognitive adaptations (Chang et al. 2017). High cognitive function will enable players to perform at their very best physically and mentally. High-level competitive sports require a different kind of cognitive capacity than low-level sports because they need the athlete to retain quick reflexes and quick analytical thinking. When athletes are not able to focus, their sport-specific motor talents may be impaired, resulting in losses and unsuccessful execution of strategy or skill. The demand for cognitive abilities in sports is increasing because of the changing competitiveness in sports, which requires excellent physiological skills and outstanding motor control, perception, and cognitive functioning (Scharfen and Memmert 2019a).

Athletes participating in externally paced open-skill sports, such as basketball or tennis, are required to promptly react and adjust to a dynamic and constantly evolving environment. In contrast, closed-skill sports are typically characterized by self-paced activities, predetermined movement patterns, and take place in venues with predictable weather conditions (Koch and Krenn 2021). Research has demonstrated that athletes participating in open-skill sports have superior problem-solving ability (Jacobson and Matthaeus 2014) and inhibition tasks (Ballester et al. 2019) compared to those involved in closed-skill sports. The influence of adapting to the distinct environments of open- and closed-skill sports on cognitive function remains uncertain. It is unclear whether athletes enhance their cognitive abilities through sports participation or if they already possess a crucial cognitive role prior to excelling in their respective sports. Therefore, conducting extended research or looking back at past assessments of athletes' physical activity, sports involvement, training duration, and cognitive function assessment could provide novel perspectives on this issue. In recent years, there has been an increase in interest in the role of cognition and neuroscience in understanding, predicting, and potentially enhancing top sports performance (Walsh 2014). Previous research has shown that elite athletes exhibit faster performance and make less errors on cognitive tasks compared to non-elite athletes (Voss et al. 2010). These findings have been supported by research indicating that an individual's initial cognitive capacity may serve as an indicator of their future athletic performance (Vestberg et al. 2017). Therefore, this study aims to compare the cognitive function of elite and non-elite athletes among university students.

1.1 Participant Selection

A total of 48 participants aged 18–26 years old were recruited for this study (M = 20.7, SD = 1.5 years old). The participants were divided into two groups: elite athletes and non-elite athletes. The participants were selected from the Kedah State Sports Council and Universiti Teknologi MARA Perlis athletes. In the case of the current research, the inclusion criteria for elite athletes were as follows: (1) participate actively in a structured training program for more than 3 years; (2) represent the Kedah State Sports Council. Meanwhile, the inclusion criteria for non-elite athletes were as follows: (1) participate actively in structured training programs for less than 3 years; (2) represent Universiti Teknologi MARA Perlis. The exclusion criteria for both groups were as follows: (1) Have neurological problems, (2) are currently on psychoactive medication, (3) are visually impaired, and (4) have color blindness. Other than that, it was compulsory for all participants to be free from alcohol or any drug consumption 24 h before the start of the test.

1.2 Procedure

All participants were required to complete a preliminary survey to determine their neurological health, dominant hand, medication, and vision status. The participants with a neurological problem, currently on psychoactive medication, who were visually impaired and had color blindness were excluded from the study. All participants were provided with consent forms and were informed of their rights and obligations before participating in the experiment. They were required to complete the Physical Activity Readiness Questionnaire (PAR-Q) (American College of Sports Medicine [ACSM], 2014). Other demographic data such as gender, age, height, weight, years of formal training experience, hours of training, and amount of exercise per week were also collected. In this study, we divided elite and non-elite athletes on the basis of the frequency of training per week and years of the structured training program. The elite athletes train more than four times per week, whereas the non-elite athletes train less than four times per week. Elite athletes participate in the structures training

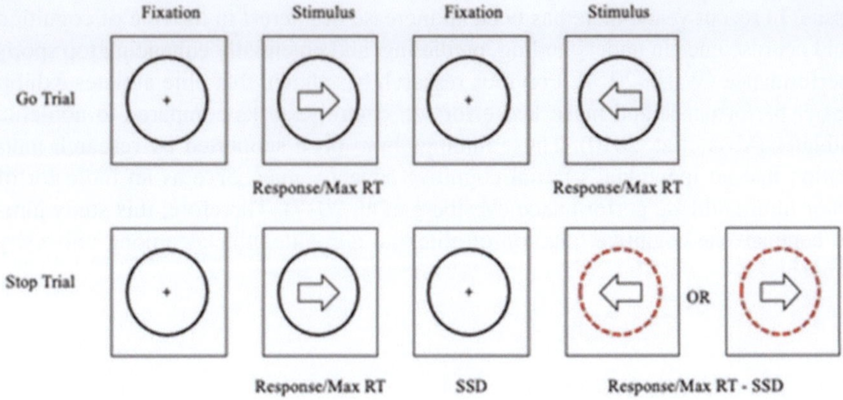

Fig. 1 Sequence of events in a stop signal task

program for more than three years, whereas non-elite athletes participate in the structures training program for less than three years. Elite athletes represent Kedah state, whereas non-elite athletes represent the university. The study was conducted on the basis of the guideline set by the board of ethics committee Universiti Teknologi MARA (REC/02/2022 (ST/MR/10)). All participants were informed that they would be performing cognitive tests to study the cognitive function of elite and non-elite athletes. Their participation would take approximately 8–10 min. The computerized test was divided into two phases: a practice phase of 10 trials and an experimental phase of 1 block of 150 trials. The participants were given a 1-min rest between blocks and counteract the possible fatigue impact. The stop-signal paradigm explored motor inhibition in a laboratory setting using a computerized test. Visual cues were used in the task (Fig. 1).

A visual stimulus consisting of a circle with a small cross and an arrow (either pointing right or left) will be displayed at the center of the circle. The paradigm comprised 10 trial practise blocks and a 150-trial experiment block. The trials are arranged in a random sequence, with an equal number of go and stop trials in each block, maintaining a 1:1 ratio. During the go trial, a fixation stimulus consisting of a circle with a small cross was displayed in the center of the screen for a duration of 250 ms. In contrast, the stop trial utilized a stop signal delay (SSD) determined by the staircase model. The stop signal delay is initiated at 200 ms and will progressively vary by 50 ms, dependent upon the participant's reaction. If response inhibition is successful, an additional 50 ms will be added to the prior stop signal delay value. In contrast, if stop inhibition fails, 50 ms will be subtracted from the previous stop signal delay value. Upon the presentation of the fixation stimulus on the screen, a task stimulus will promptly appear, either as a go or stop task, in accordance with a randomized sequence. The task stimulus was presented for a duration of 500 ms, or until the individual provided a response. During the go task, participants were instructed to press the "q" key for the left arrow and the "p" key for the right arrow when they encountered a circle with an arrow. During the stop task, the circle will change its color to red, indicating to the participants that they should inhibit their actions and refrain from responding.

1.3 Statistical Analysis

The data collected from this investigation were analyzed using IBM SPSS Version 26, with a significance level set at $p < 0.05$. This study employs two forms of data analysis: descriptive and inferential statistical analysis. Descriptive statistics provide an overview of demographic data, including age, weight, height, and gender. The data is reported as the mean and standard deviation (M ± SD). The interferential statistic was employed to analyze the value obtained from the stop signal task by an independent t-test, which compared the two groups. The computerized test was developed for this study using E-Prime 3.0 software. The test was adapted from the stop signal task template included in the Psychology Software Tool library. It was then enhanced by including a staircase algorithm to provide delays in the stop signals. To streamline the process of compiling data, the collected data were merged using E-Merge 3. Subsequently, the data underwent filtration and analysis using E-DataAid 3, followed by the utilization of SPSS Version 26 to compute the differences between the groups. SSRT, a measure of inhibition latency, was used to assess reactive inhibition. The SSRT can be computed via the integration technique (Verbruggen and Logan 2009). In the integration approach, the nth reaction time (RT) was adjusted by subtracting the mean sum of squared differences (SSD) and multiplying it by the overall probability of responding to a signal (p[respond/signal]).

2 Results

Table 1 summarizes the demographic data of the 48 participants (N = 48). Eight of the 48 participants (N = 8) were female. The 48 participants were divided into two groups, elite and non-elite, with 24 participants for each group.

This study conducted an independent t-test to compare two groups with the aim of investigating the differences between the cognitive function of the elite and non-elite groups. However, no significant difference was found for stop signal reaction time (SSRT) between the groups, $t(46) = 0.046$, $p = 0.964$, $a = 0.05$, despite elite (M = 75.9, SD = 21.4) attaining better SSRT than non-elite (M = 76.2, SD = 17.6). Besides that, there were also no differences in the Go Trial Reaction Time (GTRT) between group $t(46) = 0.496$, $p = 0.622$, $a = 0.05$; interestingly, elite (M = 345.26, SD = 22.77) performed worse than the non-elite (M = 342.20, SD = 19.92). There were also no differences in the Go Trial response accuracy $t(46) = 0.535$, $p = 0.596$, $a = 0.05$ for non-elite (M = 0.97, SD = 0.05) and elite (M = 0.96, SD = 0.04) (Figs. 2, 3 and 4).

Table 1 Demographic data

Variables	Elite ($M \pm SD$)	Non-elite ($M \pm SD$)
Age	20.4 ± 1.8	21.0 ± 1.1
Starting age of involvement in sport	13.9 ± 2.8	11.3 ± 2.1
Duration of training hours	5.4 ± 1.9	2.0 ± 0
Duration of involvement in specific sports	6.7 ± 2.3	10.7 ± 2.6
Frequency of training	5.1 ± 1.3	4.5 ± 1.1
Stop Signal Reaction Time (SSRT)	75.9 ± 21.4	76.2 ± 17.6
Go Trial Reaction Time (GTRT)	345.3 ± 22.8	342.2 ± 19.9

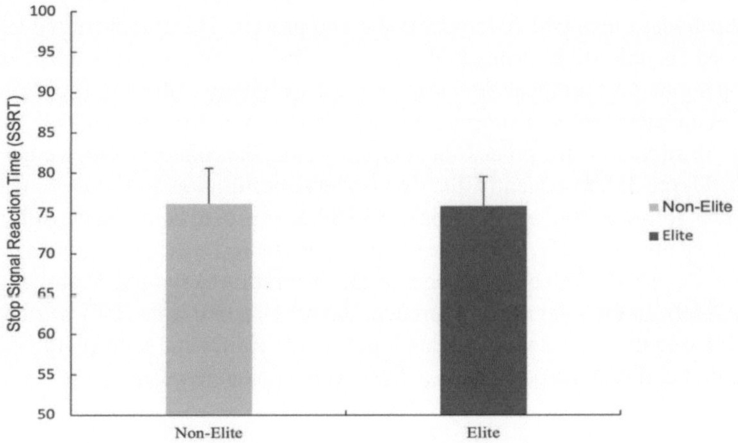

Fig. 2 Stop signal reaction time (SSRT)

3 Discussion

During this study, the cognitive function of elite athletes and non-elite athletes was compared. It was found that both groups had a similar cognitive function and inhibitory control, as measured by their ability to withhold a typically executed response. Faster and more accurate task performance was also observed in both groups. Several studies that used cross-sectional designs have demonstrated that people who exercise regularly or show high fitness levels have superior cognitive abilities compared to those who exercise infrequently or have low fitness levels (Chaddock et al. 2011; Netz et al. 2011). From the findings, the mean score of SSRT and GTRT for this study shows that elite athletes were faster than non-elite athletes. However, this study failed to find any statistically significant differences in SSRT and GTRT between the non-elite and elite groups. Based on these results, it has been shown that there were no differences in terms of cognitive inhibition between elite and

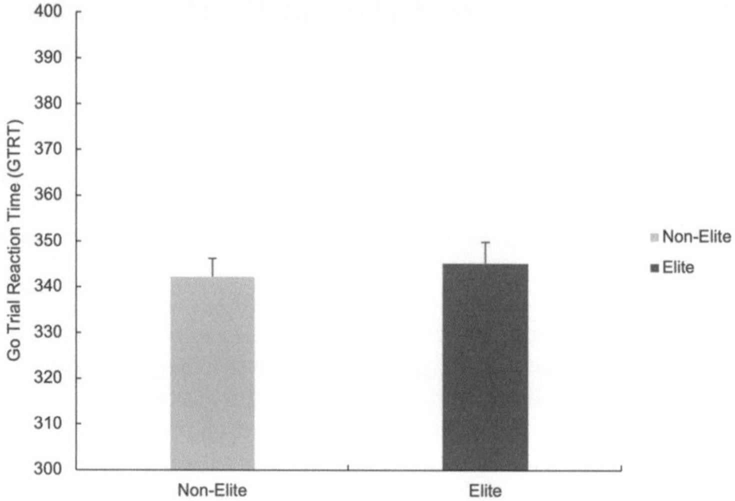

Fig. 3 Go trial reaction time (GTRT)

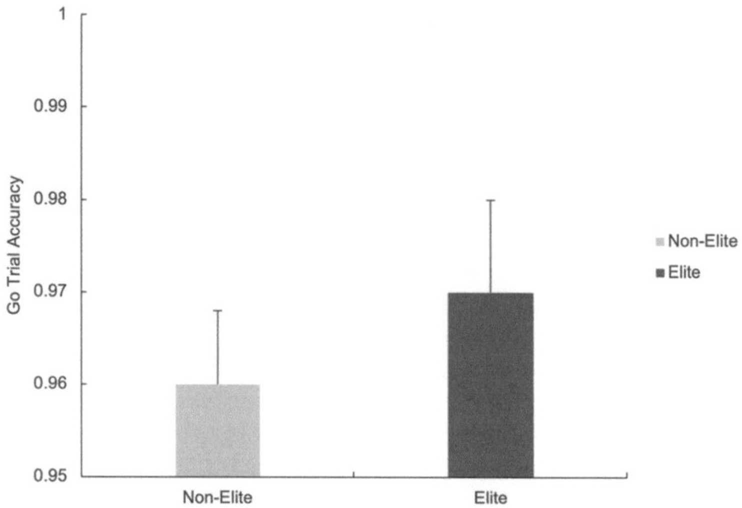

Fig. 4 Go trial response accuracy

non-elite groups. Elite athletes perform slightly better but with no statistical differences between the two groups. These conclusions were supported by a previous study that was not able to find any positive relationship between expertise in sports and general cognitive abilities (Chang et al. 2017). However, another researcher demonstrated that expertise in sports was associated with greater inhibitory control, and long-term participation in sports can influence inhibitory control, thereby improving performance in sports (Hagyard et al. 2021a). In the present study, elite athletes were recruited from the Kedah State Sports Council, and the non-athlete was recruited from UiTM Perlis. Although elite athletes compete in a higher level of the tournament, in our record, non-elite athletes have more years of experience competing in their particular sport but in a lower level competition. It is possible that the experience absorbed in lower level competition would also result in cognitive adaptation similar to what occurs in elite groups. Therefore, our results demonstrate considerable debate about whether the level of expertise accounts for specific aspects of general cognition.

There were several limitations in this study. Firstly, the group was separated based on their training experience. There may need to be more criteria to differentiate between the non-elite and elite groups. Data from the descriptive study indicate that elite athletes undergo more structured training than non-elite athletes. In this study, the elite and non-elite athletes were categorized based on their years of structured training. The problem with this study method of recruiting research participants is that the number of hours practiced and the number of years practiced can vary substantially between the participants. This approach is in contrast to classifying elite athletes according to their current level of competition or division participation, eliminating the chance of incorrectly classifying athletes. Although athletes with high-performance status can be distinguished by the number of hours or years they have spent training, more is needed to provide specificity to distinguish elites from non-elites (Scharfen and Memmert 2019b). Thus, the definition of elite and non-elite must be more precise in order for the result to be more reliable. Besides that, other studies have demonstrated that sports knowledge can strengthen inhibitory control and improve sports performance (Hagyard et al. 2021b). It is important to note that the majority of non-elite athletes were recruited from universities, whereas the elite athletes were recruited from the SUKMA Kedah State Sports Council, which may have influenced the results of this study. Furthermore, there was no recruitment selection based on the sport played by participants. The two groups of athletes are composed of a mix of open- and closed-skill sports. A previous study found that open-skill sports had higher inhibition control than closed-skill sports (Zhu et al. 2020). To better distinguish between non-elites and elites, future studies should define the recruitment process thoroughly. A previous study that showed the difference in inhibitory control between badminton athletes and non-athletes has found out that badminton athletes displayed superior inhibitory control, inhibiting their responses and reengaging alternative movements quicker than non-athletes (Chen et al. 2019). Participating in open-skill sports has been linked to improved conflict monitoring and better performance on tasks requiring inhibitory control (Ludyga et al. 2022). Thus, open-skill sports showed superior control of cognitive inhibition than close-skill sports.

This study employed the stop signal task, which was created using E-prime 3.0. The task design was derived from the E-prime stop signal task template found in the PST collection. We implemented an adaptive tracking technique, specifically the staircase approach, to dynamically modify the stop signal delay based on the participant's reaction. Moreover, in the current investigation, the ratio of the go task to the stop task was equal, with a 1:1 ratio. According to the consensus guide from a prior study, the stop signal trials should only be provided in a minority of trials (Verbruggen et al. 2019). When participants deliberately delay their actions until a stop signal is sent, the characteristics of both the stop signal procedure and the task itself will be altered. This will make it more challenging to compare conditions or groups. For instance, the variations in SSRT groups might be attributed to either differential deceleration or strategic adaptations. Delaying the occurrence of the stop signal will further diminish the dependability of SSRT estimates (Verbruggen et al. 2013). The recommended guideline for the stop signal task is to include a stop task in 25% of the total trials. Additionally, there is a substantially larger probability of receiving the stop response when the number of total stop tasks exceeds 25% of the whole trial. The limited number of blocks and trials in this study was insufficient to illustrate any significant improvements, as it only consisted of 75 go trials and 75 stop trials. Limited data exists to support the notion that reducing the number of trials can yield dependable estimates of SSRT (Verbruggen et al. 2019). A minimum of 200 trials, including 25% stop trials, should be presented for the stop signal task. Furthermore, the stop signal task involved the athlete responding to right and left arrows by either pressing or inhibiting their response, but this particular sort of movement is not closely related to the execution of sports skills. Therefore, the assessment should be modified in order to replicate sport-specific tasks more accurately.

4 Conclusion

In conclusion, the comparison of cognitive function between elite athletes and non-elite athletes revealed comparable cognitive functions in both groups. The study reveals no significant differences between the two groups in terms of cognitive inhibition. The notion that sports expertise does not uniformly translate to superior general cognitive abilities is supported by a body of literature (Chang et al. 2017; Qiu et al. 2019; Lappi 2015). The intricate interaction between different types of cognitive function and athletic expertise highlights the necessity for a comprehensive strategy in future research endeavors. It is paramount to further explore how specific aspects of sports expertise may correspond to particular cognitive capabilities. This will enhance our understanding of the complex link between sports expertise and cognitive function, a topic that is currently only partially explored by existing research.

Acknowledgements This study was supported by Universiti Teknologi MARA with the grant name MyRA and grant number 600-RMC/GPM LPHD 5/3 (071/2021).

References

Ahmad MF, Low JFL, Nadzalan AM, Aziz AA (2020) Developmental practice activities of youth soccer players. Eur J Mol Clin Med 7(2):5874–5882.

Ballester R, Huertas F, Pablos-Abella C, Llorens F, Pesce C (2019) Chronic participation in externally paced, but not self-paced sports is associated with the modulation of domain-general cognition. Eur J Sport Sci 19:1110–1119. https://doi.org/10.1080/17461391.2019.1580318

Chaddock L, Hillman CH, Buck SM, Cohen NJ (2011) Aerobic fitness and executive control of relational memory in preadolescent children. Med Sci Sports Exerc 43:344–349. https://doi.org/10.1249/MSS.0b013e3181e9af48

Chang EC-H, Chu C-H, Karageorghis CI, Wang C-C, Tsai JH-C, Wang Y-S, Chang Y-K (2017) Relationship between mode of sport training and general cognitive performance. J Sport Health Sci 6:89–95. https://doi.org/10.1016/j.jshs.2015.07.007

Chen J, Li Y, Zhang G, Jin X, Lu Y, Zhou C (2019) Enhanced inhibitory control during re-engagement processing in badminton athletes: An event-related potential study. J Sport Health Sci 8:585–594. https://doi.org/10.1016/j.jshs.2019.05.005

Hagyard J, Brimmell J, Edwards EJ, Vaughan RS (2021a) Inhibitory control across athletic expertise and its relationship with sport performance. J Sport Exerc Psychol 43:14–27. https://doi.org/10.1123/jsep.2020-0043

Hagyard J, Brimmell J, Edwards EJ, Vaughan RS, Hagyard J, Brimmell J, Edwards EJ, Vaughan RS (2021b) Inhibitory control across athletic expertise and its relationship with sport performance. J Sport Exerc Psychol 43:14–27

Hernández-Mendo A, Reigal RE, López-Walle JM, Serpa S, Samdal O, Morales-Sánchez V, Juárez-Ruiz de Mier R, Tristán-Rodríguez JL, Rosado AF, Falco C (2019) Physical activity, sports practice, and cognitive functioning: the current research status. Front Psychol 10. https://doi.org/10.3389/fpsyg.2019.02658

Jacobson J, Matthaeus L (2014) Athletics and executive functioning: How athletic participation and sport type correlate with cognitive performance. Psychol Sport Exerc 15:521–527. https://doi.org/10.1016/j.psychsport.2014.05.005

Koch P, Krenn B (2021) Executive functions in elite athletes—comparing open-skill and closed-skill sports and considering the role of athletes' past involvement in both sport categories. Psychol Sport Exerc 55:101925. https://doi.org/10.1016/j.psychsport.2021.101925

Lappi O (2015) The racer's brain—how domain expertise is reflected in the neural substrates of driving. Front Hum Neurosci 9. https://doi.org/10.3389/fnhum.2015.00635

Ludyga S, Mücke M, Andrä C, Gerber M, Pühse U (2022) Neurophysiological correlates of interference control and response inhibition processes in children and adolescents engaging in open- and closed-skill sports. J Sport Health Sci 11:224–233. https://doi.org/10.1016/j.jshs.2021.01.001

Netz Y, Dwolatzky T, Zinker Y, Argov E, Agmon R (2011) Aerobic fitness and multidomain cognitive function in advanced age. Int Psychogeriatr 23:114–124. https://doi.org/10.1017/S1041610210000797

Qiu F, Pi Y, Liu K, Zhu H, Li X, Zhang J, Wu Y (2019) Neural efficiency in basketball players is associated with bidirectional reductions in cortical activation and deactivation during multiple-object tracking task performance. Biol Psychol 144. https://doi.org/10.1016/j.biopsycho.2019.03.008

Scharfen H-E, Memmert D (2019a) The relationship between cognitive functions and sport-specific motor skills in elite youth soccer players. Front Psychol 10. https://doi.org/10.3389/fpsyg.2019.00817

Scharfen H, Memmert D (2019b) Measurement of cognitive functions in experts and elite athletes: a meta-analytic review. Appl Cogn Psychol 33:843–860. https://doi.org/10.1002/acp.3526

Verbruggen F, Logan GD (2009) Models of response inhibition in the stop-signal and stop-change paradigms. Neurosci Biobehav Rev 33:647–661. https://doi.org/10.1016/j.neubiorev.2008.08.014

Verbruggen F, Chambers CD, Logan GD (2013) Fictitious inhibitory differences. Psychol Sci 24:352–362. https://doi.org/10.1177/0956797612457390

Verbruggen F, Aron AR, Band GP, Beste C, Bissett PG, Brockett AT, Brown JW, Chamberlain SR, Chambers CD, Colonius H, Colzato LS, Corneil BD, Coxon JP, Dupuis A, Eagle DM, Garavan H, Greenhouse I, Heathcote A, Huster RJ, Jahfari S, Kenemans JL, Leunissen I, Li C-SR, Logan GD, Matzke D, Morein-Zamir S, Murthy A, Paré M, Poldrack RA, Ridderinkhof KR, Robbins TW, Roesch M, Rubia K, Schachar RJ, Schall JD, Stock A-K, Swann NC, Thakkar KN, van der Molen MW, Vermeylen L, Vink M, Wessel JR, Whelan R, Zandbelt BB, Boehler CN (2019) A consensus guide to capturing the ability to inhibit actions and impulsive behaviors in the stop-signal task. Elife 8. https://doi.org/10.7554/eLife.46323

Vestberg T, Reinebo G, Maurex L, Ingvar M, Petrovic P (2017) Core executive functions are associated with success in young elite soccer players. PLoS ONE 12:e0170845. https://doi.org/10.1371/journal.pone.0170845

Voss MW, Kramer AF, Basak C, Prakash RS, Roberts B (2010) Are expert athletes "expert" in the cognitive laboratory? A meta-analytic review of cognition and sport expertise. Appl Cogn Psychol 24. https://doi.org/10.1002/acp.1588

Walsh V (2014) Is sport the brain's biggest challenge? Curr Biol 24:R859–R860. https://doi.org/10.1016/j.cub.2014.08.003

Zhu H, Chen A, Guo W, Zhu F, Wang B (2020) Which type of exercise is more beneficial for cognitive function? A meta-analysis of the effects of open-skill exercise versus closed-skill exercise among children, adults, and elderly populations. Appl Sci 10:2737. https://doi.org/10.3390/app10082737

Thermophysiological Evaluation Properties of Sport Hijab Knitted Fabrics for Thermoregulatory Comfort

Aznin Baharudin, Atiyyah Musa, Nurul Hidayah Abdul Aziz, Raja Mohammed Firhad Raja Azidin, and Mohd Rozi Ahmad

Abstract Sport hijab is a garment worn by female Muslim athletes to cover their aurah that includes the hair and chest area during physical activities. The comfort of the athletes is a priority so that the performance of the athletes is not affected by any discomforts from their sport attire. This calls for the investigation of sport hijab materials. In this study, thermophysical evaluation for thermoregulatory comfort was measured using SDL Atlas Sweating guarded hot plate (SGH). Ten knitted fabric samples are used. The test gave thermal resistance (Rct) and vapour resistance (Ret) value where calculation for water vapour permeability index (IMT) was made. The study also measured the relationship between IMT with fabric density, fabric thickness and fabric weight. Fabric I gave the best thermal comfort and breathability values with lighter weight (170 gm^{-2}), higher IMT value (0.11093) and suitable density (64 $\times$ 50). Fabric E was found to give the least IMT value among all 10 samples as it has the close IMT value to 0 (0.01117).

Keywords Sport hijab · Thermal comfort · Thermal resistance · Water vapour permeability · Breathability

A. Baharudin (✉) · A. Musa
School of Industrial Technology, Faculty Applied Sciences, Universiti Teknologi MARA Kuala Pilah Campus, Negeri Sembilan 72000 Kuala Pilah, Malaysia
e-mail: aznin905@uitm.edu.my

N. H. Abdul Aziz
Science and Technology Research Institution for Defense (STRIDE), 4300 Kajang, Selangor, Malaysia

R. M. F. Raja Azidin
Faculty of Science and Recreation, Universiti Teknologi MARA Shah Alam, 40450 Shah Alam, Selangor, Malaysia

A. Baharudin · M. R. Ahmad
School of Industrial Technology, Faculty of Applied Sciences, Universiti Teknologi MARA Shah Alam, 40450 Shah Alam, Selangor, Malaysia

Z. Mohamed et al. (eds.), *Proceedings of the 2nd International Conference on Innovation and Technology in Sports, ICITS 2023, 27–28 November, Malaysia*, Lecture Notes in Bioengineering, https://doi.org/10.1007/978-981-97-3741-3_7

1 Introduction

Thermophysiological comforts relate to thermal conductivity, thermal heat transfer, water vapour permeability and air permeability. This heat that is created by the metabolism of the athletes when they move must be transferred out from skin to the environment to ensure that the sport attire, especially sport hijab does not feel wet, sticky and in the end gives uncomfortable feeling (Onofrei et al. 2011; Classen 2018). Masood et al. (2020) elaborated that the thermophysiological comfort relates with the thermoregulations from human body heat transfer to the outer fabric, which reduces the risk of hypothermia or heat stroke and maintaining ideal human body temperature (37 °C) in extreme weather conditions.

Machado-Moreira et al. (2008) studied that during exercise, the heat dispersion around the head differs. The area without hair gives the highest sweat percentage compared to the area with hair. This means wearing head gear or a sport hijab will add another layer to the head, hence influenced the effectiveness of the thermophysiological comfort of the athletes. Awais et al. (2020) then confirm that when wearing a sport hijab, it will become a barrier or an interface which influences the thermal transmission between the human body and the environment.

Though donning a sport hijab may create a barrier to heat conductivity and heat transfer, Muslim female athletes see great significance in doing so. They can choose what to wear during the competition, and it complies with Islamic law that requires women to cover their aurah. For Muslim women, aurah entails covering everything but the face and hand palms from head to toe (Othman et al. 2016).

It is crucial to understand the effect of wearing a sport hijab during sport activity, especially in a country where the weather is hot and humid. In this study researchers will see the relationship between thermal insulation and breathability to the fabric properties—fabric density, fabric weight and fabric thickness. The breathability of the fabric are measures as thermals resistance (Rct), vapour resistance (Ret) and water vapour permeability index (IMT). In conclusion, our study aims to shed new light on sport hijab thermal comforts and take a first step towards dispelling common misconceptions about sport hijab.

2 Methods

All the tests were conducted at University Teknologi MARA, Shah Alam and Science and Research Institute for Defence (STRIDE). The process started with sample preparation and fabric physical characterizations. In fabric characterized, the sport hijab fabric densities are determined through the course and wale densities according to MS ISO 7211/2—2003 standard method. The weight per unit area and the thickness of the fabric are measured according to ASTM D 3779–1996 and ASTM D 3776 96/2002, respectively. Finally, the thermal resistance and thermal conductivity are measured using SDL Atlas Sweating Guarded Hotplate (SGH) where the machine specifications are according to GB/T 11,048 and ISO 11092.

2.1 Materials

Sport hijab materials used for the study have variations in material compositions and knitted structures significantly with respect to their thermal properties, being the object of evaluation. The major composition of the fabrics is nylon, cotton, Tencel and minimum percentages of spandex. Observing the tabulated fabric weight of the fabric samples are between 150 gm^{-2} and 225 gm^{-2}. These weight falls in a medium weight fabric (150 gm^{-2}–350 gm^{-2}) which typically this range of fabric weight are used for t-shirt, apparel, and sport hijab. The sample description can be observed in Table 1.

Table 1 Sample description and basic parameters of knitted sport hijab fabric samples

	Sample	Composition	Fabric weight (g^{-2})	Fabric thickness (mm)	Fabric density (Wale × Course)	
					Wale	Course
1	Fabric A	75% Nylon, 25% Spandex	225	0.524	72	78
2	Fabric B	80% Nylon, 20% Spandex	180	0.612	61	90
3	Fabric C	81% Nylon, 19% Spandex	180	0.502	60	64
4	Fabric D	83% Nylon, 17% Spandex	160	0.316	67	32
5	Fabric E	85% Nylon, 15% Spandex	150	0.342	55	94
6	Fabric F	87% Nylon, 13% Spandex	200	0.534	60	90
7	Fabric G	51% Nylon, 12% Spandex, 37% Polyester	185	0.378	60	95
8	Fabric H	40% Nylon, 10% Spandex, 35% Rayon, 15% Tencel	160	0.362	60	60
9	Fabric I	35% Nylon, 10% Spandex, 40% Cotton, 15% Tencel	170	0.374	64	50
10	Fabric J	5% Spandex, 65% Cotton, 30% Polyester	200	0.484	70	42

The samples fabric selections are from a knitted fabric structures in references to Kanakaraj and Ramachandran (2015) where he elaborated that knitted fabric able to gives an ergonomic comfort or fits to wear especially near to skin. Under a low loading condition, the material can have high extensibility which allows it to fit snuggly and without discomfort on any form on which it is pulled. Hence, the knitted fabric does not change in shape, able to stretch and move according to the athlete's movement.

2.2 *Thermophysiological Testing*

The SGH was calibrated to 35 °C for the test plate and the test environment at 25 °C, where relative humidity (RH) is 65%. The thermal resistance reading test should range between 0.001 ~ 2.0 m^2K/W and vapour resistance test should range between 1.0 ~ 1000 m^2Pa/W. The reading of the test is given in thermal resistance test (Rct), water vapour resistance test (Ret) and water vapour permeability index (IMT). Water vapour permeability index (IMT) was then calculated using the results from Rct and Ret. The formula for the calculations is as follows.

The thermal conductivity is calculated with the Eq. (1) (Salopek Čubrić et al. 2022).

$$\text{Rct}\left(\text{m}^2\text{KW}\quad 1\right) = ((\text{Ts} - \text{Ta}))/\text{Q} \times \text{A} \tag{1}$$

where,

Ts = Temperature at the skin surface (in K)

Ta = Temperature of the environment (in K)

Q = the heat flux through the fabric (in W)

A = the area of the fabric (in m^2).

Water vapour resistance is calculated according to ISO 11092 by Eq. (2) (Marolleau et al. 2017).

$$\text{Ret}\left(\text{m}^2\text{Pa/W}\right) = (\text{Tm} - \text{Ta}).\,\text{A})/\,(\text{H}.\Delta\,\text{He}) - \text{Ret0} \tag{2}$$

Tm = the water vapour partial pressure (Pa) at the surface.

Ta = air temperature in the test enclosure

A = the area of the fabric

H = the heating power supplied to the measuring unit (W)

ΔHe = the correction term for heating power (W).

Rct and Ret information can be used to calculate IMT (3) (Gibson et al. 1994). IMT values are dimensionless hence there is no unit in the definition.

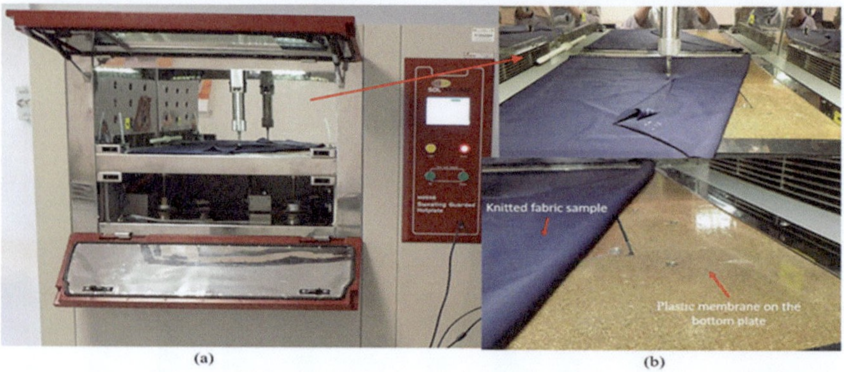

Fig. 1 **a** SDL sweating guarded hotplate testing machine, **b** the layout of the plastic membrane and knitted fabric samples in the testing chamber

$$\mathrm{IMT} = \mathrm{S}\,(\mathrm{Rct}/\mathrm{Ret}) \tag{3}$$

S = Lewis constant (60 pa/K).

Figure 1 shows the SDL Sweat guarded hotplate testing machine. In preparing the fabric sample on the apparatus, it is important to ensure that the plastic membrane is laid flat without any bubble or uneven surfaces (Fig. 1b). Extra care should be taken as the membranes are fragile and easily torn. Once the membrane is torn, during the test, the water pump under the membrane will pass through it and wet the fabric sample. True results will not be collected.

3 Results and Discussion

The comfort of the sport hijab made for the research is analysed using the statistical software platform (SPSS). This enables the determination of contributions on textile properties for the thermoregulatory characteristic of the selected sport hijab fabric samples.

Figure 2 displays the bar chart graph for Rct results from SGH tester. Thermal resistance (Rct) determines if the fabric holds the heat. The higher the number, the better thermal resistance of the fabric. From the results, it can be observed that Fabric I has the highest Rct value (0.0163), while Fabric E has the lowest Rct value (0.0018).

Figure 3 shows the results for water vapour permeability in Ret values. This value determines the fabric's ability to allow vapour (perspiration) to pass through it. The lower the value obtained, the more breathable the fabric is. From Fig. 3, Fabric G (8.3698) has the highest fabric perspiration and Fabric D (10.5416) has the lowest fabric perspiration ability. Nevertheless, the values of Ret are between 8 and

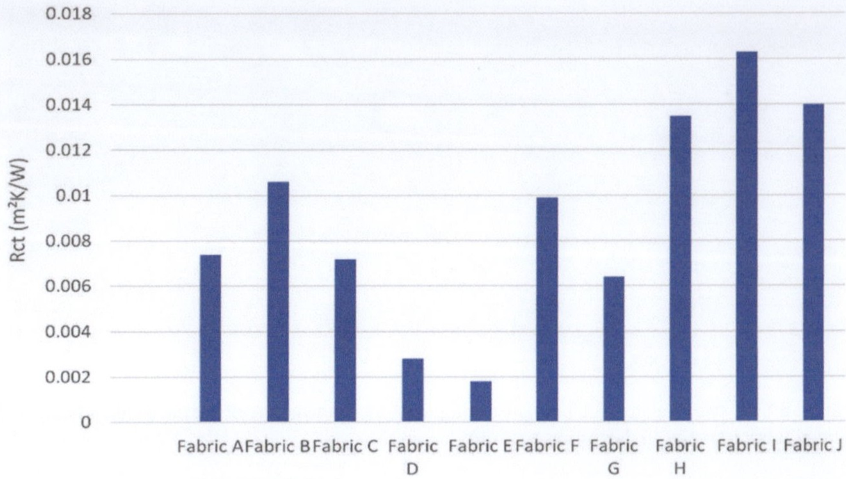

Fig. 2 The thermal resistance of fabric samples

10. Overall the value is not giving a distinctive or difference between each fabric sample. It can be said that all the fabric has a good perspiration value.

The results of water vapour permeability index for all the sport hijab knitted fabrics are shown in Fig. 4. Water vapour permeability index (IMT) expresses the equilibrium (ratio) between the insulation (Rct) and the breathability (Ret), with value from 0 (impermeable fabric) to 1 (permeable fabric) (Marolleau et al. 2017). This means the closer the IMT value to 1, the better comfort the material is to be worn. Fabric I (0.11093) show the closest IMT value to 1 compared to other fabric samples.

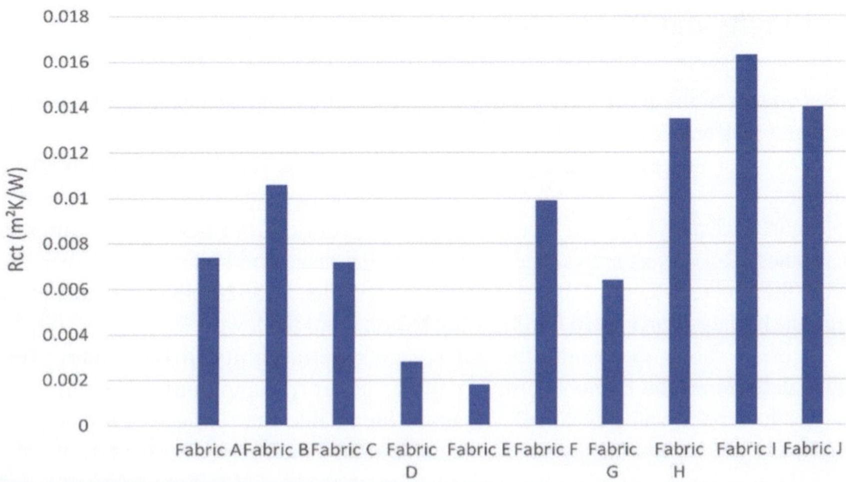

Fig. 3 Water vapour permeability for fabric samples

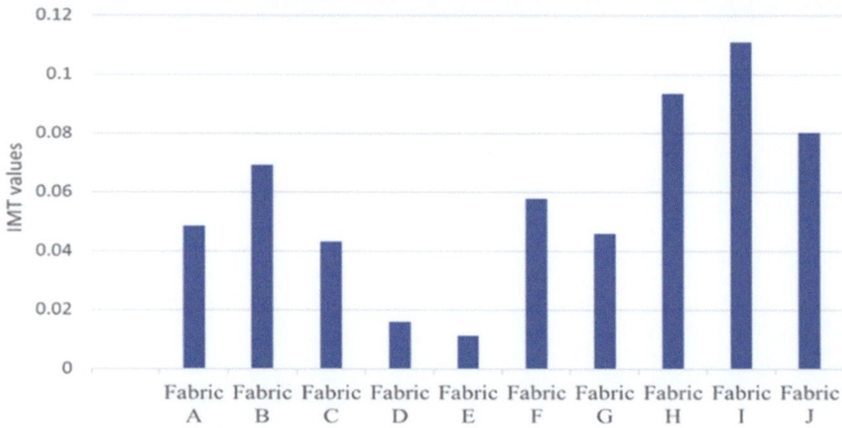

Fig. 4 Water vapour permeability index (IMT) for sport knitted fabric samples

Fabric E has the lowest IMT index (0.01117) hence predicted to be uncomfortable among all the fabric samples. The second-best material among the fabric samples is Fabric H (0.09354).

3.1 IMT Thermal Comfort Influenced by the Fabric Characteristics.

The thickness of the fabric has a strong correlation which is portrayed in Fig. 5. The dispersion of the points shows that most of the fabric sample increases in thickness as the weight increases. Lesser weight gives thinner thickness that enables the fabric to have water regulation with moisture-wicking and quick drying processes. Thinner and lesser weight may be due to the usage of finer yarn. Finer yarn has better wicking properties than a coarser yarn (Kumar et al. 2017). The description mentioned are proof through the observation of Fabric E, as the fabric weight is lower (150 gm^{-2}) and thickness (0.342 mm) the IMT value is 0.01117 indicate that movement of water and moisture management (water vapour and water resistance) are very small and if the fabric are used during a physical activity, it should give dry feeling to it wearer.

It can be observed from Fig. 6, there is a medium correlation between both IMT values toward fabric thickness and fabric weight. The value of the correlation is slightly positive along the trendline showing that the IMT can be influenced by the fabric weight and fabric thickness. However because the variation of weight is very similar, the distinctive gap of IMT value between the fabric samples is not easy to see. Using the ANOVA analysis, the hypothesis of probability and R-value are very low, some are not applicable. The value of correlation between IMT, fabric thickness and fabric weight is tabulated in Table 2.

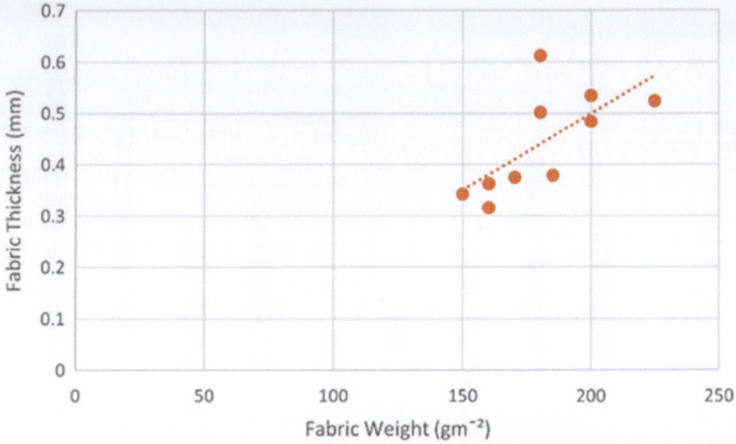

Fig. 5 Relationship between fabric weight (gm^{-2}) and fabric thickness (mm)

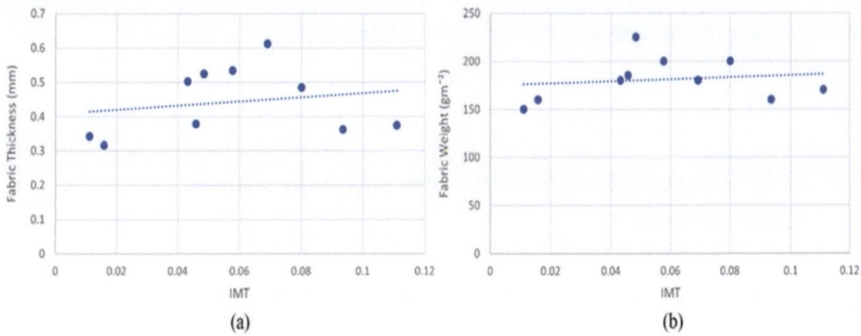

Fig. 6 **a** Fabric thickness (mm) to IMT, **b** fabric weight (gm^{-2}) to IMT

Table 2 Correlation of the fabric thickness and fabric weight to IMT values

	IMT	Fabric thickness (mm)	Fabric weight (g^{-2})
IMT	1		
Fabric thickness	0.189617	1	
Fabric weight	0.147396	0.675703	1

The thermoregulatory comfort of the fabric initially is influenced by the fabric material, surface change, fibre and fibre finishing, fibre type and thickness of the knitted fabric (Rathinamoorthy 2017). Aside from the thickness and weight of the fabric, the material of the fabric also may strongly influence the ability of the fabric to manage moisture. In Fabric J (65% cotton, 5% Spandex, 30% polyester), the composition of cotton is the highest. This resulted in the 0.0801 IMT value regardless the thickness is 0.484 mm with 200 gm^{-2} weight. Compared to Sample H (40% Nylon, 10% Spandex, 35% rayon, 15% Tencel) with 0.0935 IMT value, but with lesser weight and thickness respectively (160 gm^{-2}, 0.362 mm). This shows that having cotton in the material composition for Sample J gives better thermoregulatory management. Cotton has the tendency to absorb and retain moisture (Kejkar and Dhore 2019), and cotton fibres' moisture regains readily between 5 and 10% (Soomro 2014). This explains why Fabric J as fabric with higher moisture absorption able to provide a microclimate between body and the atmosphere and acts as a courier for heat and vapour between the skin and the environment (Kejkar and Dhore 2019).

4 Conclusion

In conclusion, it can be said that in county that is humid and hot, the best material should not hold onto heat in its structure or product, hence Fabric E (0.0018) with the lowest Rct values is suggested to be the best sport hijab material comparing to Fabric I (0.0163). As sport hijab are worn during high-intensity activities and games, it should be light, agile and comfortable. Overall Fabric I (0.11093) shows the best IMT value with 170 gm^{-2} weight which is considered in the middle range of 150 gm^{-2}–200 gm^{-2}, and thinner thickness (0.374 mm) compared to all the fabric samples. Fabric I also contains 40% cotton material. Fabric I shows the most comfortable knitted fabric material suitable for high intensity sport hijab as it is in compliance with the characteristic discussed.

Acknowledgements The authors would like to thank the Ministry of Higher Education for the funding under the Fundamental Research Grant Scheme (FRGS/1/2018/STG07/UITM/02/1). The authors would also like to acknowledge the assistance given by the Research Management Centre (RMC), Universiti Teknologi MARA and Science and Technology Research Institution for Defence (STRIDE), Kompleks Induk Stride as where all the testing takes place.

References

Awais M, Krzywinski S, Wölfling B, Classen E (2020) Thermal simulation of close-fitting sportswear

Classen E (2018) Comfort testing of textiles. In: Advanced characterization and testing of textiles. https://doi.org/10.1016/b978-0-08-100453-1.00004-0

Gibson P, Auerbach M, Giblo J, Teal W, Endrusick T (1994) Interlaboratory evaluation of a new sweating guarded hot plate test method (ISO 11092). J Therm Envelope Build Sci 18(2):182–200. https://doi.org/10.1177/109719639401800207

Kanakaraj P, Ramachandran R (2015) Active knit fabrics—functional needs of sportswear application. J Textile Appar Technol Manag 9(2)

Kejkar V, Dhore R (2019) Review article active sportswear fabrics. Trends Textile Eng Fash Technol 5(2):603–608. https://doi.org/10.31031/TTEFT.2019.05.000606.Copyright

Machado-Moreira CA, Wilmink F, Meijer A, Mekjavic IB, Taylor NAS (2008) Local differences in sweat secretion from the head during rest and exercise in the heat. Eur J Appl Physiol 104(2):257–264. https://doi.org/10.1007/s00421-007-0645-y

Marolleau A, Salaun F, Dupont D, Gidik H, Ducept S (2017) Influence of textile properties on thermal comfort. IOP Conf Ser: Mater Sci Eng 254(18). https://doi.org/10.1088/1757-899X/254/18/182007

Masood R, Jamshaid H, Khubaib MA (2020) Development of knitted vest fabrics for human body thermoregulation. J Therm Anal Calorim 139(1):159–167. https://doi.org/10.1007/s10973-019-08430-2

Onofrei E, Rocha AM, Catarino A (2011) The influence of knitted fabrics' structure on the thermal and moisture management properties. J Eng Fibers Fabr 6(4):10–22. https://doi.org/10.1177/155892501100600403

Othman R, Muhamad SN, Mohd Yusof F, Wan Harun MA (2016) Fesyen Wanita Menurut Perspektif Islam dan Kesannya Terhadap Kesejahteraan Fizikal dan Spiritual. Sains HumIka 8(3). https://doi.org/10.11113/sh.v8n3.953

Rathinamoorthy R (2017) Moisture management characteristics of knitted casein fabric. Indian J Fibre Text Res 42(4):488–494

Salopek Čubrić I, Potočić Matković VM, Pavlović Ž, Pavko Čuden A (2022) Material and structural functionalization of knitted fabrics for sportswear. Materials 15(9):3306. https://doi.org/10.3390/ma15093306

Experimental Investigation of Dynamic Properties of Sepak Takraw Balls

Nik Mohd Haikal Mohamed Hassan, Nasrul Hadi Johari, Mohd Hasnun Arif Hassan, Mohd Nadzeri Omar, and Zulkifli Ahmad

Abstract The purposeful heading of a sepak takraw ball as the game's primary movement, as intensively practiced by sepak takraw players, can result in head injuries such as concussions. The head injury prediction can be practically measured, yet the mechanical properties of the sepak takraw ball are not well studied. The primary aim of this study is to investigate the stiffness and damping coefficient of two types of sepak takraw balls, GE511 and MT908. Each ball has been subjected to drop tests from three distinct heights, precisely 2.5, 7.0, and 11.5 m. It was found that MT908 has a greater stiffness and damping coefficient than GE511 with proposed valvalues stiffness and damping coefficient for GE511 arbeing6673.23 N/m and 18.492, and for MT908, 30,672.2 N/m and 20.249 Ns/m, respectively. These discoveries are particularly relevant in determining head injuries as well as the creation of head protectors.

Keywords Sepak takraw ball · Drop ball test · Ball stiffness

N. M. H. Mohamed Hassan (✉) · N. H. Johari · M. H. A. Hassan · M. N. Omar · Z. Ahmad
Faculty of Mechanical and Automotive Engineering Technology, Universiti Malaysia Pahang, 26600 Pahang, Malaysia
e-mail: haikalhassan5@gmail.com

N. H. Johari
e-mail: nhadi@umpsa.edu.my

M. H. A. Hassan
e-mail: mhasnun@umpsa.edu.my

M. N. Omar
e-mail: nadzeri@umpsa.edu.my

N. H. Johari
Centre for Advanced Industrial Technology, Universiti Malaysia Pahang, Pahang, Malaysia

Z. Mohamed et al. (eds.), *Proceedings of the 2nd International Conference on Innovation and Technology in Sports, ICITS 2023, 27–28 November, Malaysia*, Lecture Notes in Bioengineering, https://doi.org/10.1007/978-981-97-3741-3_8

1 Introduction

In Southeast Asia, sepak takraw is a stimulating point-based team sport played outdoors or indoors. A player must make a strong, fast, and accurate serve ball in order to make a good attacking play (Hanafi 1945). To obtain a good return ball from the powerful serve ball, sepak takraw players train intensively on ball control and heading with approximately a sufficient ball head from a serve ball training session, which can be as many as 100 balls per session (Hasanuddin 2013).

Previous studies showed that the purposeful heading of a soccer ball is associated with a concussion (Peek et al. 2021). Considering this study, a purposeful ball head is only necessary for football and sepak takraw. Additionally, a fastball and a strong serve complement one another. A study by Hasanuddin et al. (2015) found that the maximum velocity of a sepak takraw ball before heading impact is 13.581 m/s. Furthermore, a sepak takraw ball is harder than a soccer ball. Consequently, receiving a sepak takraw serve ball while constantly using the head may lead to MTBI (Hasanuddin et al. 2015). A suggestion of the need for protective equipment arises from South Korean elite athletes' sports injury incidence (Lhee et al. 2018).

There are several takraw balls used in the sepak takraw community. The most popular in takraw society are Marathon Brand and Gajah Emas Brand. The Marathon is made in Thailand, while the Gajah Emas is from Malaysia. Takraw ball is also categorized into men, women, and school. Each category comes with two specifics; tournament and training. For instance, for the Gajah Emas Brand, there are three types of balls for men; GE511 for tournament specifics, GE411 for training specific, and GE 311 for school.

A soccer ball's stiffness and damping coefficient that results in brain linear accelerations has been studied by Taha et al. (2015) based on linear accelerations of the head's center of gravity by its mathematical model. The level of head injury can be determined using Head Injury Criteria (HIC). The higher the stiffness and damping coefficient of the ball, the higher the peak force experienced by players and riskier to have concussion (Nevins and Smith 2013). Therefore, this paper plans and tests the corresponding dynamic stiffness of the sepak takraw ball.

2 Methodology

The GE511 (Men) and MT908 (Men) takraw balls were selected for the study. The GE511 ball was manufactured using specialized waterproof synthetic polymeric fibers, while the MT908 ball was constructed with rubber-coated synthetic polymeric fibers. As depicted in Fig. 1a and b, these balls are made by combining four distinct fiber shapes and sizes and then woven into a circular shape.

The International Sepak Takraw Federation (ISTAF)-approved, tournament-specific GE511, and MT908 sepak takraw balls were used for the drop tests. GE511 was constructed from special waterproof synthetic polymeric fibers with 172.8 g of

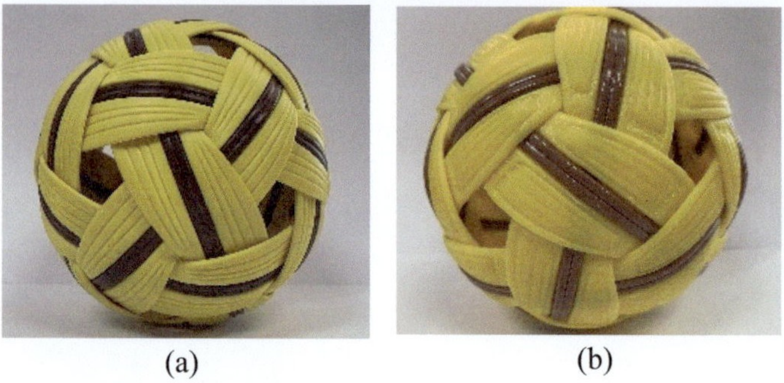

Fig. 1 Unique woven sepak takraw ball, **a** GE511, **b** MT908

weight, whereas MT908 was constructed from rubber-coated synthetic polymeric fibers with 179.6 g of weight, as shown in Fig. 1a and b.

Drop ball tests were performed vertically for three different heights, h (2.5, 7.0, and 11.5). The ball is assumed to have elastic deformation; therefore, the tests were performed with three repetitions using the same ball for each height, h (Ahmad et al. 2014). The three different heights resulted in different inbound velocities. A string was attached to the balls through one of its stitches (Taha et al. 2015). The balls were dropped with the same face at the initial position (Ahmad et al. 2009). A horizontal concrete floor was used during the experiments in an area with no wind, calibrated using an Anemometer before every test started.

A high-speed video camera (nac MEMRECAM HX-7 s) at a speed of 2000 frames per second was used to record the dynamic behavior before, during, and after impact. A plywood with checkerboard paper on it with a height of 58 mm was used as the background (see Fig. 2). To ensure a consistent measure of the drop ball location and the camera, the checkerboard position was set close to the drop-ball point, i.e. 0.05 m. These picture frames were measured using a Tracker Video Analysis and Modelling Tool (Ahmad et al. 2019) and the ball velocities before and after the impacts were manually calculated after a definite quantity of frames were determined. The high-speed camera was set to 2000 frames per second. Thus, the time for 1 frame is 0.0005 s.

Given the ball traveled through the bottom checkerboard, the ΔT was calculated using frame number from the ball that started touches the ground to the final touch of the ball leaving the ground. The total frame is then multiplied by 0.0005s. The inbound and rebound velocities were calculated by dividing the checkerboard height (0.058 m) with the total time frames (number of frames X 0.0005 s).

The coefficient of restitution, *e* value was determined by the ratio of the ball velocity before (inbound velocity) and after (rebound velocity) the impact (Cross 1999; Aryaei et al. 2010; Has 2020):

$$e = \frac{v_o}{v_i}$$

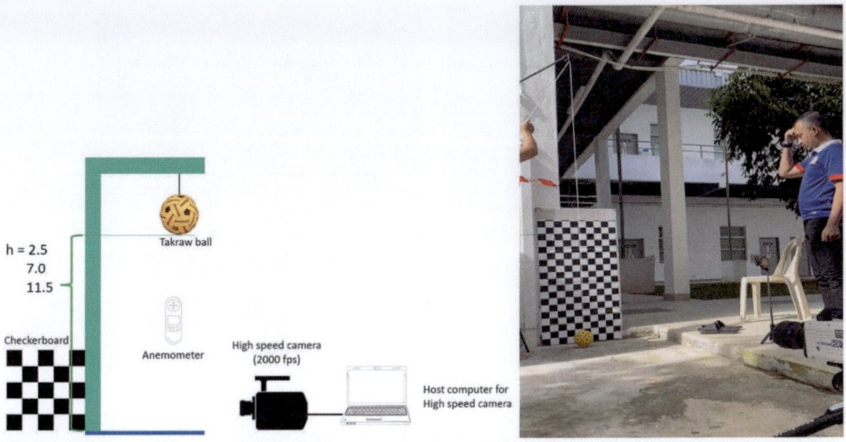

Fig. 2 Drop ball experiment setup

The average e was calculated (Haron and Ismail 2012). The contact time of impact, ΔT between a sphere and a rigid surface was given by Zhang and Vu-Quoc (2002). Image frames from the high-speed camera determined the e and ΔT values as calculated by Ahmad et al. (2009).

From the coefficient of restitution, e and the contact time, ΔT obtained the stiffness, k and damping coefficient, c of the sepak takraw ball which were calculated using the following equations (Nagurka and Huang 2004):

$$k = \frac{m}{(\Delta T)^2}\left\{\pi^2 + (\ln e)^2\right\} \tag{1}$$

$$c = -\frac{2m}{\Delta T}(\ln e)^2 \tag{2}$$

3 Result and Analysis

Understanding the dynamic properties of a ball is critical in determining head injury level. The drop ball test explains some properties of the ball, such as its dynamic stiffness and damping coefficient.

The sample calculation of inbound velocity, v_i Rebound Velocity, v_o and contact time of impact, ΔT are shown in Table 1, and the results delivered from the testing values are summarized in Table 2.

Inbound velocity and coefficient of restitution are discussed in this section, aided by Tables 2 and 3, while the mean stiffness and mean damping coefficient between drop ball test heights are discussed, aided by Fig. 3a and b. The average values of the

Table 1 Sample calculation of velocity, v and contact time, ΔT

Ball	Height	Trial	Frame inbound number		Inbound velocity, v_i [m/s]	Frame inbound number		Rebound velocity, v_o [m/s]	Frame ball contact		Contact time, ΔT [s]
			Start	End		Start	End		Initial	End	
GE 511	2.5	Trial 1	145	162	6.82	202	227	4.64	172	189	0.0085
		Trial 2	107	124	6.82	166	190.5	4.73	135	152	0.0085
		Trial 3	97	115	6.44	158	182.5	4.73	126	142	0.008

ball stiffness and damping coefficient are listed in Table 2 for GE511 and MT908 in three heights.

It is clear from Table 2 that the value of e decreases with the increasing inbound velocity of the ball for both balls, which is in agreement with (Ahmad et al. 2009; Aryaei et al. 2010). The value of e also decreases with the increasing drop test height as described by Haron and Ismail (2012). This phenomenon is possibly due to the energy loss during impact and similar with (Arakawa et al. 2006). Interestingly, even though MT908 has a bigger diameter than GE501, the value of e is also higher than GE501, in contrast with (Aryaei et al. 2010; Has 2020). The material of synthetic polymeric fibers perhaps is the critical factor undermining the diameter influence in the producing coefficient of the restitution.

From the conservation of energy for object falls from rest, the potential energy is converted to kinetic energy; therefore, the velocity of the drop object can be determined using:

$$v = \sqrt{2gh} \tag{3}$$

Thus, the theoretical velocity and its difference in percentage are shown in Table 3.

The maximum velocity from the test meets the value suggested by Hasanuddin et al. (2015). All experimental inbound velocities are lower than the theoretical value with the difference in percentage from 1 to 6%. Thus, the values are tolerable to describe the actual value of the inbound velocity. The losses of velocity may be due to aerodynamic losses as depicted by Mubin et al. (2015).

The graphs for Stiffness Height and Damping Coefficient Height are plotted in Fig. 3a and b for GE511 and MT908. It is assumed that a linear pattern occurred as shown in both graphs, with the values of stiffness and damping coefficient increasing with the drop test height. The values are in agreement with (Taha et al. 2015). This shows that a higher velocity will increase the ball stiffness and damping coefficient, which leads to a higher possibility of head injury to occur.

The MT908 ball has a greater value of stiffness and damping coefficient than the GE511 ball. For stiffness, MT908 is higher at 3617.96 N/m, 4561.27 N/m, and

Table 2 Dynamic properties of GE511 and MT908

Type of Ball	Height [m]	Trial	Inbound velocity, v_i [m/s]	Rebound velocity, v_o [m/s]	Mean coefficient of restitution, e	Contact time, ΔT [s]	Mean stiffness, k [N/m]	Mean damping coefficient, c [Ns/m]
GE 511	2.5	Trial 1	6.82	4.64	0.702	0.00850	24,857.83	14.577
		Trial 2	6.82	4.73		0.00850		
		Trial 3	6.44	4.73		0.00800		
	7.0	Trial 1	11.60	7.25	0.633	0.00850	26,697.96	19.551
		Trial 2	11.60	7.03		0.00800		
		Trial 3	11.60	7.73		0.00775		
	11.5	Trial 1	14.50	9.67	0.617	0.00800	28,463.89	21.349
		Trial 2	14.50	8.59		0.00800		
		Trial 3	14.50	8.59		0.00750		
MT 908	2.5	Trial 1	6.63	4.83	0.717	0.00800	28,475.79	15.011
		Trial 2	6.82	4.83		0.00800		
		Trial 3	6.63	4.73		0.00775		
	7.0	Trial 1	11.05	7.25	0.655	0.00775	31,259.23	20.031
		Trial 2	11.60	8.00		0.00750		
		Trial 3	11.05	6.82		0.00750		
	11.5	Trial 1	14.50	8.29	0.583	0.00750	32,281.58	25.705
		Trial 2	14.50	8.29		0.00750		
		Trial 3	13.65	8.29		0.00750		

Table 3 The difference in velocity value

Height [m]	Type of ball	Mean inbound experimental velocity, v_i [m/s]	Theoretical velocity, v_i [m/s]	Difference percentage
2.5	GE511	6.70	7.00	4.37
	MT908	6.73		3.96
7.0	GE511	11.23	11.72	1.02
	MT908	11.42		4.16
11.5	GE511	14.50	15.02	3.47
	MT908	14.22		5.36

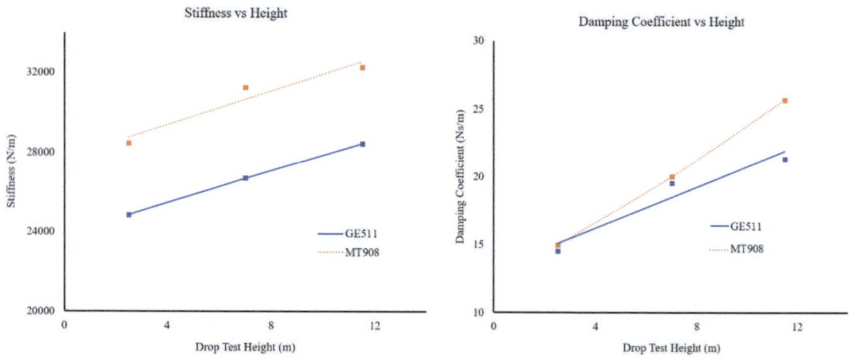

(a) (b)

Fig. 3 The plot of **a** stiffness versus test height (**a**) damping coefficient versus test height

3817.69 N/m for drop heights of 2.5 m, 7.0 m, and 11.5 m, respectively. The differences are huge but almost constant in all drop heights. In the meantime, the damping coefficient difference values are 0.434 Ns/m, 0.479 Ns/m, and 4.356 Ns/m for drop heights of 2.5 m, 7.0 m, and 11.5 m, respectively. A huge difference is shown at the drop ball height of 11.5 m. This could lead to a bigger difference if the test is continued to a higher height. The higher value of stiffness and damping coefficient for MT908 means a higher possibility of head injury to be experienced.

To calculate the linear acceleration experienced by the brain using the mathematical model by Taha et al. (2015), the overall height mean stiffness and damping coefficient are used. The injury level can be determined using Head Injury Criteria (HIC). Therefore, the proposed sepak takraw ball stiffness and damping coefficient are shown in Table 4.

Table 4 Proposed sepak takraw ball stiffness and damping coefficient

Ball type	Stiffness, k [N/m]	Damping coefficient, c [Ns/m]
GE511	26,673.23	18.492
MT918	30,672.2	20.249

4 Conclusion

This paper has presented an overview of two main sepak takraw balls that are mainly used in official tournaments, which are made from woven synthetic polymeric fibers. The ball stiffness is essential in determining the head injury experienced by a sepak takraw player.

The inbound velocities using a drop test meet the suggested velocity by Hasanuddin et al. (2015) and are in tolerance with a 1%–6% difference from the theoretical values. A higher value of stiffness and damping coefficient for the MT908 ball shows a higher possibility of head injury occurrence than the GE511 ball. This paper also proposed that the values of stiffness and damping coefficient for GE511 are 26,673.23 N/m and 18.492, and for MT908, 30,672.2 N/m and 20.249 Ns/m, respectively. This leads to the future development of modern sepak takraw balls as players' head protection.

Acknowledgements The authors would like to thank the Universiti Malaysia Pahang for providing financial support under Product Development Grant (PDU203226), Post Graduate Research Grant Scheme (PGRS210344) and to Ministry of Higher Education under Hadiah Latihan Persekutuan (HLP 2021-2023).

References

Ahmad N, Taha Z, Ujihashi S, Tanaka K (2009) An experimental study of the impact of a sepak takraw ball on a flat surface. Impact Technol Sport III, no. III, pp 447–452

Ahmad N, Hassanudin I, Yik WC (2014) A proposed standard method for measuring the coefficient of restitution of a takraw ball

Ahmad M, Ismail KA, Mat F, Nawi MAHM, Hanid MHM (2019) Velocity measurement of a sports ball during the drop test by a high-speed camera. Int J Mech Eng Robot Res 8(2):273–278. https://doi.org/10.18178/ijmerr.8.2.273-278

Arakawa K et al (2006) Dynamic contact behavior of a golf ball during an oblique impact. Exp Mech 46(6):691–697. https://doi.org/10.1007/s11340-006-9257-4

Aryaei A, Hashemnia K, Jafarpur K (2010) Experimental and numerical study of ball size effect on restitution coefficient in low velocity impacts. Int J Impact Eng 37(10):1037–1044. https://doi.org/10.1016/j.ijimpeng.2010.04.005

Cross R (2013) The bounce of a ball 222(1999). https://doi.org/10.1119/1.19229

Hanafi H (2016) Perbandingan Latihan Bola Digantung Dan Bola Dilambungkan Terhadap Hasil Belajar Sepak Mula Dalam Permainan Sepak Takraw Pada Siswa Putra Kelas X-Is 1 Sma 17 Agustus 1945 Surabaya. Wahana 67(2):71–76. https://doi.org/10.36456/wahana.v67i2.504

Haron A, Ismail KA (2012) Coefficient of restitution of sports balls: a normal drop test. IOP Conf Ser Mater Sci Eng 36(1): 0–8. https://doi.org/10.1088/1757-899X/36/1/012038.

Hasanuddin I et al (2015) Investigation of the head impact power of a sepak takraw ball on sepak takraw players 4(2):45–56

Hashemnia K (2020) Experimental study of the effect of temperature on the coefficient of restitution of steel balls impact to some industrial metal sheets at elevated temperatures. Powder Technol 368:170–177. https://doi.org/10.1016/j.powtec.2020.04.053

Hasanuddin I (2013) Modeling and analysis of impact of sepak takraw ballplayer's player's head. Thesis: University Malaya

Lhee SH et al (2021) Sports injury and illness incidence among South Korean elite athletes in the 2018 Asian games: a single-physician prospective study of 782 athletes. BMJ Open Sport Exerc Med 7(1). https://doi.org/10.1136/bmjsem-2019-000689

Mubin ASA, Ahmad N, Mamat A, Taha Z, Hasanuddin I (2015) CFD study of drag and lift of sepak takraw ball at different face orientations. Adv Mech Eng 7(1). https://doi.org/10.1155/2014/369137

Nagurka M, Huang S (2004) A mass-spring-damper model of a bouncing ball. Proc Am Control Conf 1(2):499–504. https://doi.org/10.23919/acc.2004.1383652

Nevins D, Smith L (2013) Influence of ball properties on simulated ball-to-head impacts. Procedia Eng 60:4–9. https://doi.org/10.1016/j.proeng.2013.07.002

Peek K et al (2021) The effect of ball characteristics on head acceleration during purposeful heading in male and female youth football players. Sci Med Footb 5(3):195–203. https://doi.org/10.1080/24733938.2021.1897657

Taha Z, Arif Hassan MH, Hasanuddin I, Hasnun M, Hassan A (2015) Analytical modelling of soccer reading. Sadhana—Acad Proc Eng Sci 40(5):1567–1578. https://doi.org/10.1007/s12046-015-0383-5

Zhang X, Vu-Quoc L (2002) Modeling the dependence of the coefficient of restitution on the impact velocity in elasto-plastic collisions. Int J Impact Eng 27(3):317–341. https://doi.org/10.1016/S0734-743X(01)00052-5

Analysis of Muscle Activation of Badminton Player's Forward Serving Technique Using sEMG

Syahmie Idham Mohd Sharulnizam, Riyan Shahaby Shuib, Mohammad Azzeim Mat Jusoh, Sukarnur Che Abdullah, and Nursalbiah Nasir

Abstract Serving is the fundamental skill needed by every badminton player. There are two common types of serving in badminton; the backhand serve and the forehand serve. Both serves are legal and widely used by professionals or novices during badminton tournaments. This project aims to study the muscle activation of badminton players using the forehand serving technique. The main focus is to analyze the muscle activation of the flexor carpi radialis. The experiment was conducted using the sEMG to collect the muscle activation data of the specified muscle. Five respondents consisting of badminton players between the age of 20 and 25 from various sports experiences have been chosen. Based on the result obtained, the average RMS value during the front forehand serve is 0.130 V, while the flick serve is around 0.163 V. As the flick serve requires higher muscle activation compared to the normal serve, proper strategy is needed to prevent unnecessary muscle fatigue, especially during long gameplay.

Keywords Badminton · Forehand serving · Flexor muscle · sEMG · RMS

1 Introduction

Serving is a fundamental skill needed by every badminton player, either professional or amateur. Each badminton rally must start from the serving. Every player must execute a proper serving to avoid fouls and losing points. A solid serving provides an advantage if played correctly, as the opponent also requires additional mental and physical energy to return it.

Badminton players rely on their muscles during match or training session, especially the dominant hand. As the service action needs to be performed repeatedly, it

S. I. M. Sharulnizam · R. S. Shuib · M. A. M. Jusoh (✉) · S. C. Abdullah · N. Nasir
School of Mechanical Engineering, College of Engineering, Universiti Teknologi MARA (UiTM), 40450 Shah Alam, Selangor, Malaysia
e-mail: m_azzeim@uitm.edu.my

Z. Mohamed et al. (eds.), *Proceedings of the 2nd International Conference on Innovation and Technology in Sports, ICITS 2023, 27–28 November, Malaysia*, Lecture Notes in Bioengineering, https://doi.org/10.1007/978-981-97-3741-3_9

will cause muscle weariness and fatigue in the long run. The effect, such as injuries, results from a poor and continuous serving technique, which usually happens late in the game (Sheng et al. 2020).

This study will focus on forearm muscle activation, as it is the primary muscle involved in the event of serving. In general, the superficial anterior forearm muscles are a collection of five muscles found in the forearm's anterior (flexor) segment, consisting of the pronator teres, flexor carpi radialis, flexor carpi ulnaris, palmaris longus, and flexor digitorum superficialis (Fuller et al. 2013). This study will focus on the flexor carpi radialis muscle. The radial hand flexor is one of the four muscles on the surface of the anterior compartment of the forearm. This muscle comes from the medial epicondyle of the humerus as part of the common flexor tendon. It runs directly to the lateral side of the flexor digitorum superficialis, starting from the anterior part of the base of the second metacarpal bone, with minor extensions to both the third metacarpal bone and the trapezoidal nodule (Falou et al. 2005).

The sEMG (surface electromyography) is a commonly used device for studying muscle activity and can provide non-invasive measurements. Several parameters shall be used to study the changes in the amplitude and frequency range of the EMG signal. These include the medium frequencies for investigating changes in the frequency spectrum of the EMG signals and RMS (Root Mean Square) values for determining amplitude changes (Felici and Vecchio 2020).

In this study, five respondents consisting of badminton players between the age of 20 and 25 from various sports experiences have been chosen. The respondents' background has been checked, with no history of critical injuries. The sEMG device used in this project is safe as it is a certified and non-invasive type (Konrad 2005). It can detect, monitor, and interpret electrical impulses from muscle groups on the skin's surface. One concern is that it is susceptible to noises, which can affect the accuracy of the readings. A strict procedure is required to prevent unwanted noise. Also, any aggressive movement could create a problem, such as detachment of the electrode. It is crucial to follow the specified procedure to gain a reliable output.

2 Methodology

In this experiment, the wireless Myon 320 sEMG is used to analyze the muscle activation during the badminton serving (Refer Fig. 1). The sEMG device consists of a transmitter, the bass system unit (BSU), and the analog-to-digital converter. The procedure needs to be repeated to get the most accurate result. The device will detect the electrical activity produced by the muscle during any exercise and project it into sets of data and graphs (Chowdhury et al. 2013; Norali et al. 2009). This study will focus on the flexor carpi radialis muscle, as shown in Fig. 2. The sEMG sensor shall be placed on the specified muscle, following guidelines established by the European project organization SENIAM (Surface Electromyography for the Non-Invasive Assessment of Muscles) (Hermens et al. 1999). The data of the muscle activation shall be extracted using the proEMG software. As the sEMG sensor is a

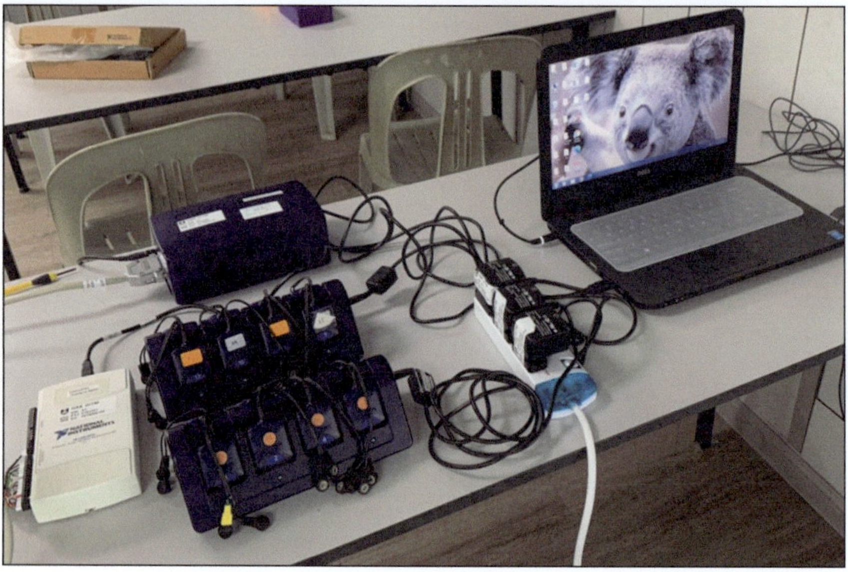

Fig. 1 The sEMG setup

sensitive device, preventing any noises affecting the accuracy of the data obtained is essential.

The experiment must be conducted in a controlled and spacious environment, such as in a private badminton court or domestic lab. This experiment was conducted in the Ergonomics Lab of the School of Mechanical Engineering, UiTM Shah Alam. The temperature and humidity were monitored within the ambient range to maintain the stability of the result. The flow of the experimental procedure can be viewed in Fig. 3.

The application of EMG variables includes utilizing the raw EMG signal to determine the time of a muscle's activation (on and off), calculate the degree of stimulation, and calculate the force that a muscle will generate. The median or mean frequency is used to track the rate at which a muscle is activated or becomes fatigued. The term RMS refers to the root mean square of the value of the voltage signal. Since the voltage value cannot be directly used to calculate an AC signal, the RMS value is the equivalent DC voltage. This is because it gives the amount of power that an electric device can draw from a given source of DC (Fukuda et al. 2010).

The EMG signal's amplitude has a Gaussian distribution with a stochastic (random) range of 0–10 mV (peak to peak). For the majority of muscles, the power density distribution of the sEMG signal falls between 0 and 400 Hz. The root mean square (RMS) and mean absolute (MA) values are two metrics that are frequently employed to quantify amplitude. In the RMS equation, n represents the number of measurements and x_i represents each value, as shown in Eq. 1.

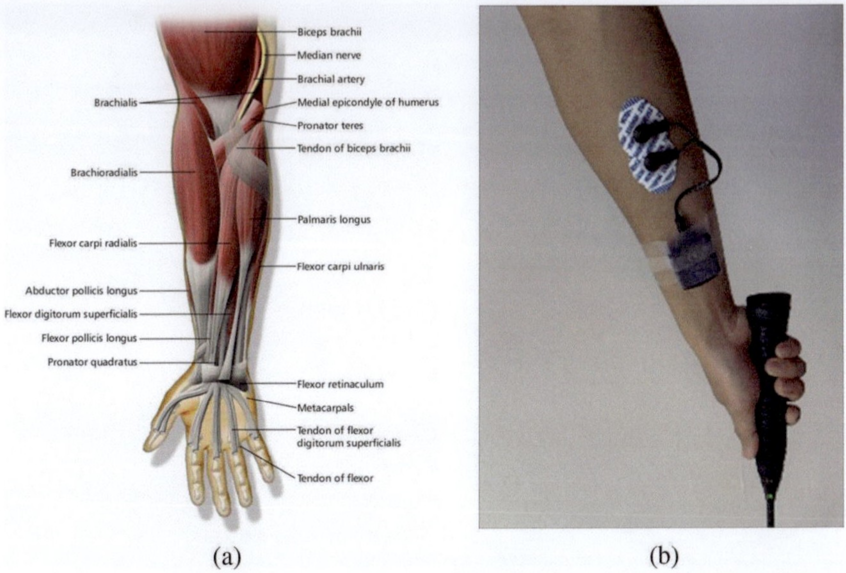

Fig. 2 **a** The flexor muscle anatomy ("RCSI—Drawing Flexor muscles and tendons of forearm—English labels") versus. **b** The electrode setup on the test subject

$$RMS = \sqrt{\frac{1}{n}\sum x_i^2} \quad (1)$$

Muscle activation refers to a muscle's capacity to turn on and function. The mind-muscle link is an important factor. For it to be more active, there has to be a more robust mental and physical link. It indicates the ability of the brain to activate a specific muscle by engaging it at the appropriate time. This is crucial to perform an activity effectively. Indirectly, the muscles can also become overactive, causing other muscles to become less active.

The accuracy of an EMG reading is highly dependent on preparing the skin surface and the placement of the electrodes. A good electrode connection and low surface resistance are the primary methods of skin prepping. Before the electrodes can be placed, skin prepping is often required. Hair removal and skin cleaning are the most basic procedures to prepare the skin for applying the electrode.

Another necessary procedure is the muscle palpation procedure. The most crucial component of palpation is determining the condition of the muscle (Kvåle et al. 2003; Fu et al. 2016). The flexor carpi radialis is the target muscle in this experiment. Firstly, muscular stimulation was combined with palpation. The muscles were felt while the respondent was instructed to make motions. The muscular belly is the region of tissue where the muscles bulk up most when performing the serving action. Two reference locations have been established. The first is the midway of the flexor carpi radialis muscle, which is identified as the origin by a center line. The secondary fixed reference point is needed because the skin extends when the body is in motion.

Fig. 3 Flowchart of the experiment

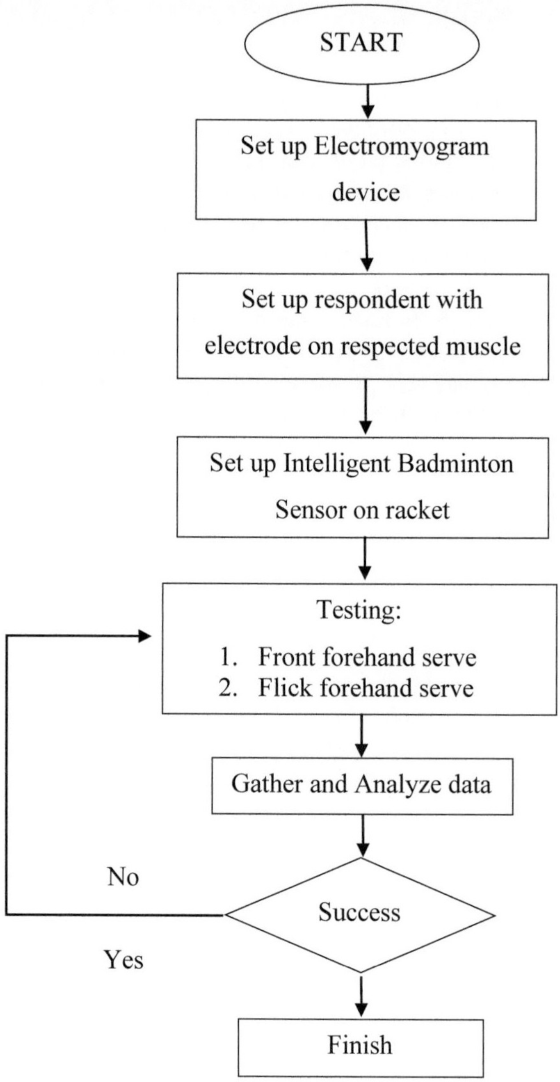

The elbow's olecranon point and its distance from the origin assist in defining this point. Nearer to the elbow, the skin and supporting muscles are less likely to move upon rotation. Once the center location of the flexor carpi radialis muscle has been determined, it was marked with a marker initiating the exact place to place the electrode.

3 Result

This study aimed to examine the flexor muscular activation during the forehand serving. To ensure the reliability of the sEMG data, all test subjects were required to act for a total number of three times in under 6 s. Figure 4 shows the raw signal amplification data extracted from the sEMG device. To study the amount of muscle activation, the data needs to be converted to the RMS value, as shown in Fig. 5. The RMS value for both flick and front serve that was obtained from all five respondents shall be evaluated.

Figure 6 shows the summary of the average RMS value during the forehand front serve and the forehand flick serve. From this figure, the average RMS value of Respondent 1, Respondent 2, Respondent 3, Respondent 4, and Respondent 5 during the forehand front serve, which are around 0.131V, 0.161V, 0.114V, 0.116V, and 0.131V, respectively. From the observation, Respondent 2 has a slightly higher value than the other respondents. Some factors might cause the reading to escalate.

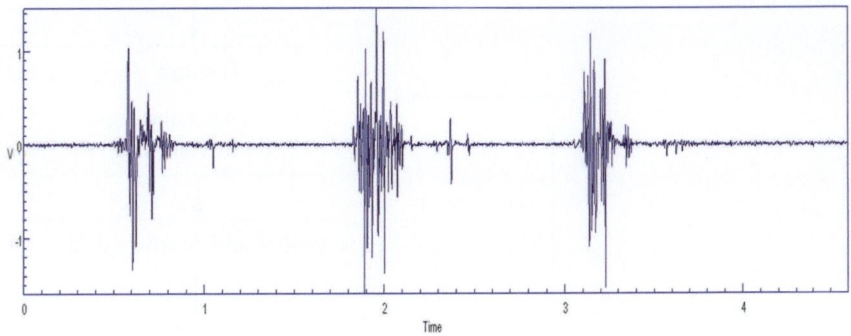

Fig. 4 Sample of the raw signal amplification (from first respondent)

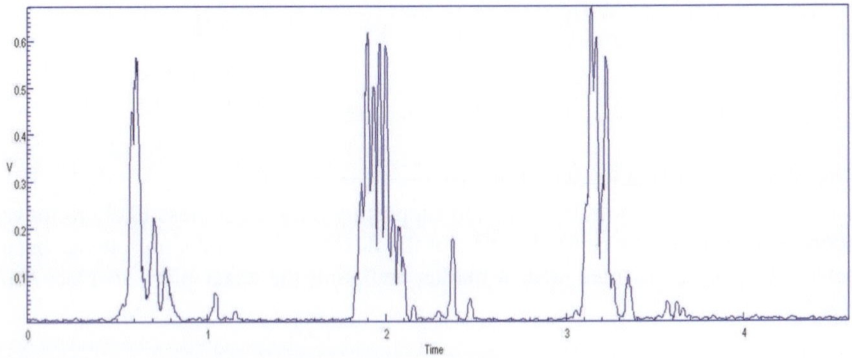

Fig. 5 Sample of the RMS graph (from first respondent)

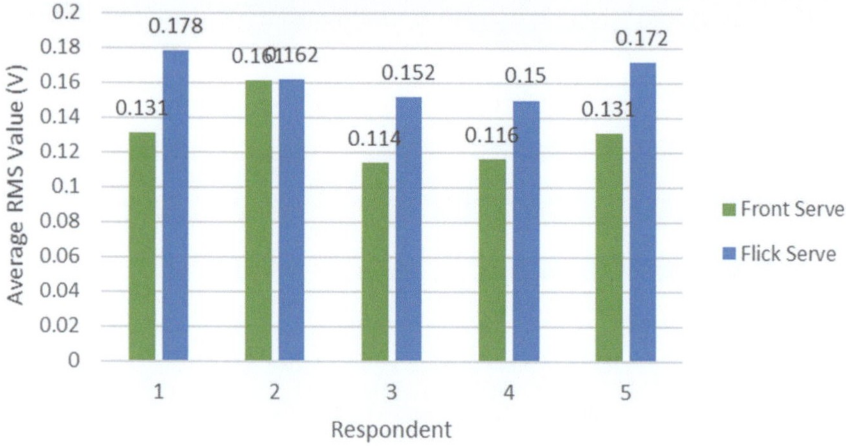

Fig. 6 The average RMS value between the front and flick forehand serve

It can be because of unwanted motions during the test, noises, and uneven swing power by the respondent.

Whereas the average RMS value during the forehand flick serve of Respondent 1, Respondent 2, Respondent 3, Respondent 4, and Respondent 5 are around 0.178V, 0.162V, 0.152V, 0.150V, and 0.172V, respectively. The side-by-side comparison clearly shows that the average RMS value of each respondent during forehand flick serve is higher than forehand front serves. By considering all those factors, the result is considered valid and acceptable.

As shown in Fig. 7, each respondent's average RMS values of flick forehand serve are slightly higher than front forehand serve. The average RMS value for the front forehand serve is 0.130V, while the average RMS value for the flick forehand serve is 0.163V. It can be said that the ratio of the average RMS value of flick serve to front serve is almost 6:4 ratio. The comparison suggested that the flexor carpi radialis muscle activity during flick serve is higher than the front serve. Theoretically, more power is needed to drive the shuttlecock from one court to the back of another court. This causes the muscle to work harder, thus generating more prominent electrical impulses during forehand flick serve. All five respondents, except Respondent 2, have the same muscle activity. Respondent 2 has an almost equal value of RMS for both front and flick serve, respectively, resulting in similar muscle activity during both actions.

In conclusion, the forehand flick serve requires higher muscle activation of the flexor carpi radialis muscle than the forehand front serve. Even with a slight difference, this could lead to a higher consequence in the long run. Each badminton player should strategize on choosing the most opportune moment to use either the flick or the front forehand serve during the game to avoid unnecessary muscle energy, which may cause fatigue and injuries if overworked.

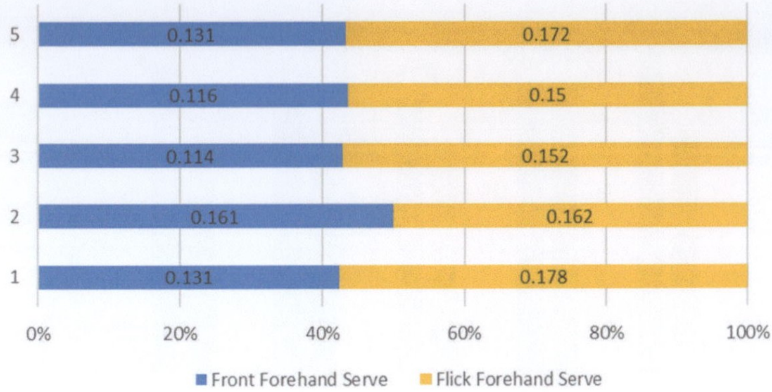

Fig. 7 Distribution between the front forehand versus the flick forehand serve

4 Discussion and Conclusion

The capability of the sEMG should be explored, especially in sports science. Since sEMG is not fully adapted to every sport, there are chances to study this. Therefore, there is a need to assess sEMG's readiness as a stand-in endpoint for muscle growth and strengthening. The initial foundation is there, but the solid result is still insufficient. In the future, more ideas and studies are needed to improve this research. For example, to explore the capability of sEMG to predict strength enhancements. Also, the applications of new technologies and advanced signal-processing techniques to improve research value.

An example on how the performance could be improved is in noise reduction. Any unnecessary movement will affect the consistency and accuracy of the reading. To overcome the problem, a special jig or positioner can be designed to control the motion of hands during serving. The design of the sEMG device can also be improved to be more portable and compact to ease any sports activity.

In conclusion, the study on muscle activation of the flexor carpi radialis muscle during forehand badminton serve was successfully performed. With the help of the sEMG, muscle activation can be identified. It can be concluded that the RMS value (which also reflects the muscle activation) during the flick serve is higher than the front serve. For long rallies, the front forehand serve is more beneficial in terms of stamina and muscle fatigue management. Hopefully, the outcome of this study will help badminton players to improve their game quality in the future.

References

Chowdhury RH, Reaz MB, Ali MABM, Bakar AA, Chellappan K, Chang TG (2013) Surface electromyography signal processing and classification techniques. Sensors 13(9):12431–12466. https://doi.org/10.3390/s130912431

El Falou W, Duchêne J, Hewson D, Khalil M, Grabisch M, Lino F (2005) A segmentation approach to long duration surface EMG recordings. J Electromyogr Kinesiol 15(1):111–119
Felici F, Del Vecchio A (2020) Surface electromyography: what limits its use in exercise and sport physiology? Front Neurol 11:578504. https://doi.org/10.3389/fneur.2020.578504
Fu Z, Hashim AYB, Jamaludin Z, Mohamad I S. (2016) Using palpation for identification and discrimination of optimal points for EMG harvesting on the lower forearm muscles. https://doi.org/10.18178/ijsps.4.5.437-441
Fukuda TY, Echeimberg JO, Pompeu JE, Lucareli PRG, Garbelotti S, Gimenes RO, Apolinário A (2010) Root mean square value of the electromyographic signal in the isometric torque of the quadriceps, hamstrings and brachial biceps muscles in female subjects. J Appl Res 10(1):32–39
Fuller JR, Fung J, Côté JN (2013) Posture-movement responses to stance perturbations and upper limb fatigue during a repetitive pointing task. Hum Mov Sci 32(4):618–632
Hermens HJ, Freriks B, Merletti R, Stegeman D, Blok J, Rau G, Hägg G (1999) European recommendations for surface electromyography. Roessingh Res Dev 8(2):13–54
Konrad P (2005) The ABC of EMG. A Pract Introd Kinesiol Electromyogr 1(2005):30–35
Kvåle A, Ljunggren AE, Johnsen TB (2003) Palpation of muscle and skin is this a reliable and valid procedure in assessment of patients with long-lasting musculoskeletal pain? Adv Physiother 5(3):122–136. https://doi.org/10.1080/14038190310016526
Norali AN, Som M, Kangar-Arau J (2009) Surface electromyography signal processing and application: a review. In: Proceedings of the international conference on man-machine systems (ICoMMS) (No. 11–13)
"RCSI—Drawing Flexor muscles and tendons of forearm—English labels" at Anatomy-TOOL.org by Royal College of Surgeons of Ireland, license: creative commons attribution-NonCommercial-ShareAlike
Sheng WY, Ginanjar A, Wei GT (2020) The effects of teaching badminton practice on improving badminton capabilities of sports department students. Eur J Mol Clin Med 7(1):3853–3866

Impact of Countdown Audio Cue in the Modified Bleep Test on Sedentary Male Youth

Md Safwan Samsir, Abdul Mu'iz Nor Azmi, Zuliana Bacho, Hasnol Noordin, and Jorrye Jakiwa

Abstract This study investigated the impact of a countdown audio cue in the Modified Bleep Test Group on aerobic capacity, measured by estimated VO2Max scores. Forty male youth participants (N = 40) aged 18–20 were randomly divided into the "Modified Bleep Test Group" (MBTG) with the audio cue and the "Normal Bleep Test Group" (NBTG) without the modification. The "Modified Bleep Test Group" achieved significantly higher estimated VO^{2Max} scores (mean ± SD: 42 ± 3 ml/kg/min) compared to the "Normal Bleep Test Group" (mean ± SD: 38 ± 4 ml/kg/min). The audio cue likely improved temporal awareness and pacing, leading to better aerobic capacity. Auditory feedback in fitness testing protocols could optimize exercise experiences and enhance performance outcomes for sedentary male youth. Further research should validate and expand upon these findings in diverse populations, exploring how audio cues influence VO2Max scores during the bleep test. This study highlights the potential of audio cue feedback to improve fitness assessments and tailor exercise prescriptions to individual needs and capacities.

Keywords Bleep test · Cardiorespiratory · Sedentary

1 Introduction

1.1 Field Test for VO2Max Estimation

Cardiorespiratory Endurance (CRE) is a fitness aspect that needs to be looked into when choosing or developing athlete in most sporting activity (Senanayake et al. 2023). Thus, during preparatory phase in periodization of training, muscular and

M. S. Samsir (✉) · A. M. N. Azmi · Z. Bacho · H. Noordin
Universiti Malaysia Sabah, Kota Kinabalu, Malaysia
e-mail: safwan.samsir@ums.edu.my

J. Jakiwa
Universiti Pertahanan Nasional Malaysia, Kuala Lumpur, Malaysia

Z. Mohamed et al. (eds.), *Proceedings of the 2nd International Conference on Innovation and Technology in Sports, ICITS 2023, 27–28 November, Malaysia*, Lecture Notes in Bioengineering, https://doi.org/10.1007/978-981-97-3741-3_10

cardiovascular endurance will be trained initially. Commonly, there are two types of cardiorespiratory endurance, which are aerobic and anaerobic (Aditia et al. 2018). Both muscular endurance and cardiovascular endurance are usually emphasized to establish a solid base of fitness. Through the enhancement of both endurance components, athletes have the opportunity to augment their holistic stamina, elevate their capacity for sustained physical activity, and establish a foundation for more targeted training in subsequent stages (Bompa 2017).

To ensure the planned training has its desired effect, a test should be implemented to see if it is adequate or needs some modification. Maximal oxygen consumption, often known as VO2 max or maximal aerobic capacity, represents the capacity to transport and utilize oxygen during the highest level of muscular exertion and is a gold standard as an indicator for cardiorespiratory endurance. VO2Max is strongly linked to the endurance of the cardiovascular system. The actual value of VO2Max relies on the efficiency of oxygen transportation through the blood vessels and the cell's capacity to receive and utilize this oxygen to generate the energy required by the body (Hasea Purba et al. 2021).

VO2Max could be assessed using a lab test or various field test batteries that could predict the VO2 max, depending on factors such as equipment availability, testing economy, and number of participants. Multi-stage 20-m shuttle run test or also called bleep test is one of the popular field tests used to find the estimated value of VO2Max (Waluya et al. 2018). In addition to its simplicity and minimal equipment requirements, the bleep test's benefit lies in its ability to accommodate a large number of participants simultaneously. This has made it a popular choice in at least 50 countries due to the low cost of equipment, flexibility in testing, convenience for the participants, and practicality in evaluating large groups of children and youth (Dimarucot and Macapagal 2021; Hasea Purba et al. 2021).

The Bleep test also can serve as High-Intensity Interval Training (HIIT) to enhance VO2Max. HIIT involves engaging in brief (10–60 s) or extended (1–5 min) bursts of high-intensity exercise, close to the velocity at which maximal oxygen uptake (VO2Max) is achieved. The participants engaged in a 10-min High-Intensity Interval Training (HIIT) session, which involved 15-s shuttle running at either 100% or 110% of their v20mSRT (velocity at 20 m shuttle run test). After each running interval, there was a 15-s period of passive recovery. To ensure that the running intensity was correct, the total distance covered during the 15-s run was calculated based on each participant's v20mSRT. These intense intervals are alternated with active or passive recovery periods of similar or shorter length. The study found that 110% of their v20mSRT was more effective to be used (Bok et al. 2023).

2 Literature Review

2.1 Validity and Reliability of Bleep Test

In recent years, several studies have been conducted to improve the testing battery, such as developing software that could improve the validity, reliability and assist in calculating the estimated VO2Max using computerized technology (Waluya et al. 2018). The result also varies from one research to another. Some of previous research found that Beep Test overestimate the actual VO2Max, when compared to Cardiopulmonary exercise testing (CPET) which is the "gold standard" test for measuring real-time maximal oxygen uptake (Magee et al. 2021; Senanayake et al. 2023). However, some other research also found that it is a reliable maximal test for cardiovascular endurance (Dimarucot and Macapagal 2021).

The bleep test has also been adapted to several testing environments and participants' needs. For example, bleep test for the visually impaired and physically disabled. It also has been modified to measure cardiorespiratory endurance among swimmers. However, some studies show that bleep test may not accurately predict the swimmers' VO2Max because of the nature of the swimming event (Aditia et al. 2018) Studies on various populations, especially among non-athletes also being debated by researchers (George-Florin Băiţan 2020). The bleep test is reliable and valid for measuring cardiorespiratory endurance among children, youth, and freshmen (Dimarucot and Macapagal 2021; Lang et al. 2017; Tomkinson et al. n.d.).

Extensive research has also been done in calculating the VO2Max estimation, especially in Asia, as the formula favors the Westerner. Thus, some researchers from Asia were trying to develop a new formula based on Asian characteristics (Mardius et al. 2020). Factors such as weight, age, gender, genetics, body mass index, type of exercise, highest heart rate, hemoglobin, lactate, temperature and humidity, total shuttle, and pulmonary function are keys to be considered in formulating VO2Max estimation value (Hasea Purba et al. 2021).

Even though the bleep test seems simple, the subject tends to run faster from one point to another at the initial level. Beginners may need help to predict when they should reach the other point. So, they will not run faster from the beginning of the test, where the speed should increase from level to level. This issue could affect their VO2Max final estimation, which makes it not a reliable test. Thus, the study focuses on improving the testing experience, especially for those new to the bleep test. By providing auditory feedback and supplementary details about the body's state, one can enhance motor learning and the effectiveness of physical exercise (Pugliese and Takala 2015). This study hypothesized that audio cues would improve estimated VO2Max scores in the bleep test among sedentary youth.

3 Methodology

3.1 Modification of Bleep Test

In this study, the well-established procedure for conducting the bleep test was adhered to, with a minor modification incorporated into the instrumentation, namely, the introduction of audio cue feedback. A limitation identified in the existing version of the bleep test was the absence of cues signaling when participants should reach the 20-m point during each level. A countdown was implemented between the "beep" sounds for each level to address this limitation and enhance participants' temporal awareness.

The countdown duration was determined based on the pre-defined time allocation for completing each level in the bleep test. Upon the commencement of a new level with the "beep" sound, the countdown initiated, providing participants with a time reference for reaching the 20-m point. This feature offered participants a more structured approach to running speed and pacing strategies.

The introduction of the countdown was anticipated to yield several benefits. The implementation of more accurate timing cues aimed to enable participants to better control their running pace, potentially resulting in overall performance enhancements. Moreover, the enhanced feedback provided by the countdown was anticipated to create a more engaging and motivating experience for the participants, which could positively impact their enjoyment and satisfaction with the bleep test.

3.2 Sample

We conducted a study and tested 40 youth participants aged between 18 and 20. The inclusion criteria required that the participants had a sedentary lifestyle, meaning they engaged in minimal physical activity or exercise. Additionally, they were required to be free from any known health-related problems and not currently involved in regular exercise or training programs. Moreover, we excluded individuals who fell outside the specified age range or led an active lifestyle with regular moderate to vigorous physical activity. Participants with pre-existing health conditions or recent injuries that could be compromised by physical activity were also excluded. Another important criterion is that the participant has never participated in a bleep test before or six months prior to the test.

Furthermore, we did not include individuals with respiratory issues such as asthma to ensure their ability to perform the test safely. Lastly, for safety reasons, pregnant individuals were excluded from the study. Before conducting the test, we conducted a thorough screening process and obtained informed consent from all participants to ensure their suitability and compliance with the established inclusion and exclusion criteria.

After the screening process, the 40 recruited participants were randomly divided into the "Modified Bleep Test Group" and the "Normal Bleep Test Group." Each group comprised an equal number of participants to ensure a balanced distribution of characteristics within the study. Both groups then participated in the bleep test session, which took place in a suitable indoor or outdoor location with a flat and non-slippery surface for running.

3.3 Procedure

The "Modified Bleep Test Group" experienced the bleep test with the introduced countdown between the "beep" sounds for each level, as described in the earlier section of the research procedure. This audio cue feedback aimed to give participants more precise timing cues during the test, potentially improving their temporal awareness and pacing strategies.

Conversely, the "Normal Bleep Test Group" underwent the traditional bleep test without the countdown modification. As per the standard protocol, participants in this group relied solely on the auditory "beep" signals to regulate their running speed and transitions between levels. Participants' running performances were recorded during the bleep test, including the level achieved and the total distance covered. The estimated VO2Max of each subject was calculated using the online calculator available on topendsports.com and recorded as the final score for analysis.

Following the bleep test sessions, the data collected from each group were analyzed separately using appropriate statistical methods. This analysis evaluated participants' performance and potential differences between the "Modified Bleep Test Group" and the "Normal Bleep Test Group."

To compare the performance of the "Modified Bleep Test Group" and the "Normal Bleep Test Group," several statistical analysis methods were employed. The Independent Samples t-test was used to compare the means of continuous variables, such as total distance covered, heart rate, and perceived exertion, between the two groups. This test was appropriate since the data followed a normal distribution, and the variances were equal.

4 Results

The study results showed significant differences in the estimated VO2Max scores between the "Modified Bleep Test Group" and the "Normal Bleep Test Group." Participants in the "Modified Bleep Test Group," who experienced the bleep test with the introduced countdown audio cue, demonstrated higher estimated VO2Max scores (mean $\pm$ SD: 42 $\pm$ 3 ml/kg/min) compared to the "Normal Bleep Test Group" (mean $\pm$ SD: 38 $\pm$ 4 ml/kg/min). This difference was statistically significant ($p < 0.05$) based on the Independent Samples t-test.

Group	Estimated VO2Max (ml/kg/min)	Mean ± SD	p-value
Modified bleep test group	42	42 ± 3	< 0.05
Normal bleep test group	38	38 ± 4	

5 Discussion and Conclusion

The results of the study provide valuable insights into the impact of introducing a countdown audio cue in the "Modified Bleep Test Group" on aerobic capacity, as measured by estimated VO2Max scores. The significantly higher VO2Max scores observed in the "Modified Bleep Test Group" suggest that the audio cue feedback played a crucial role in improving participants' performance during the bleep test. The enhanced temporal awareness and pacing strategies provided by the audio cue likely allowed participants to better regulate their running speed and transitions between levels, leading to more efficient energy utilization and improved aerobic capacity (Nylander et al. 2014; Pugliese and Takala 2015).

On the other hand, participants in the "Normal Bleep Test Group," who relied solely on the auditory "beep" signals without the countdown modification, achieved lower estimated VO2Max scores. The absence of the audio cue feedback may have resulted in less precise timing cues and pacing, potentially leading to suboptimal performance during the test.

The observed difference in VO2Max scores between the two groups underscores the significance of incorporating auditory feedback in fitness testing protocols. The countdown audio cue appears to be a valuable tool in helping participants optimize their exercise experiences and achieve better outcomes in aerobic capacity assessments. This finding has practical implications for fitness testing and training programs, especially in sedentary youth populations.

In conclusion, including the countdown audio cue in the bleep test positively impacted participants' estimated VO2Max scores, suggesting an improvement in aerobic capacity. These results highlight the potential of audio cue feedback as a means of enhancing fitness testing protocols and optimizing performance outcomes (Pacholek 2023). Nevertheless, further research is warranted to confirm and extend these findings, particularly with more extensive and diverse populations. Exploring the specific mechanisms by which the audio cue feedback influences VO2Max scores during the bleep test will provide deeper insights into its potential benefits for individuals across various fitness levels and age groups. Ultimately, the integration of audio cue feedback in fitness assessments may contribute to more accurate and effective exercise prescriptions tailored to individual needs and capacities.

References

Bok D, Gulin J, Gregov C (2023) Accuracy of the 20-m shuttle run test for individualizing exercise intensity of high-intensity interval training. Kinesiology 55(1):3–12. https://doi.org/10.26582/k.55.1.1

Dimarucot HC, Macapagal LS (2021) The validity and reliability of three field tests for assessing college freshmen students' cardiovascular endurance. Int J Hum Mov Sport Sci 9(2):363–374. https://doi.org/10.13189/saj.2021.090226

George-Florin Băiţan L (2020) Using the beep test as an alternative to measuring the military's aerobic capacity. https://doi.org/10.12753/2284-9378-20-55

Hasea Purba R, Tulaar ABM, Yunus F, Abdullah M, Tamin TZ, Tangkudung J, Purba RH (2021) Model prediction maximum oxygen intake (Vo2 max) using the bleep test in male junior athletes correspondence. In: Systematic reviews in pharmacy, vol 12, Issue 1

Lang JJ, Belanger K, Poitras V, Janssen I, Tomkinson G, Tremblay MS, Langa JJ, Tomkinson GR (2017) Systematic review of the relationship between 20m shuttle run systematic review of the relationship between 20m shuttle run performance and health indicators among children and youth performance and health indicators among children and youth recommended citation recommended citation systematic review of the relationship between 20 m shuttle run performance and health indicators among children and youth. https://commons.und.edu/ehb-fac

Magee MK, White JB, Merrigan JJ, Jones MT (2021) Does the multistage 20-m shuttle run test accurately predict Vo2max in Ncaa division I women collegiate field hockey athletes? Sports 9(6). https://doi.org/10.3390/sports9060075

Mardius A, Rezika Siswara M, Astuti Y (2020) Sosialisasi tes kemampuan daya tahan cardiovascular (Vo2max) dengan menggunakan audio bleep test Pada Perguruan Perisai Diri. J Widya Laksana 9(2).

Nylander S, Jacobsson M, Tholander J (2014) Runright—real-time visual and audio feedback on running. In: Conference on human factors in computing systems—proceedings, pp 583–586. https://doi.org/10.1145/2559206.2574806

Pugliese R, Takala T (2015) Sonic trampoline: how audio feedback impacts the user's experience of jumping. IEEE Multimed 22(1):74–79. https://doi.org/10.1109/mmul.2015.13

Senanayake SP, Gunerathne BH, Weerasinghe RD, Wijesundara WW, Dabare HP, Senanayake SP (2023) A validation study to evaluate the accuracy of the beep test in measuring cardiopulmonary endurance using maximum oxygen uptake (Vo2max) as the gold standard. https://www.researchgate.net/publication/370934944

Terapan Ilmu Keolahragaan J, Aditia L, Tantra Paramitha S, Studi Ilmu Keolahragaan P, Pendidikan Kesehatan Dan Rekreasi D, Pendidikan Olahraga Dan Kesehatan F (2018) Perbandingan hasil tes Vo2max atlet renang menggunakan williams swimming beep test dan bleep test 3(1). http://ejournal.upi.edu/index.php/jtikor/

The influence of verbal encouragement on heart rate, maximum oxygen uptake, and distance covered in young male adults during beep test. J Men's Health (2023). https://doi.org/10.22514/jomh.2023.001

Tomkinson GR, Lang JJ, Blanchard J, Léger LA, Tremblay MS (n.d.). The 20-M shuttle run: assessment and interpretation of data in relation to youth aerobic fitness and health. Health Sci Commons. https://commons.und.edu/ehb-fac

Waluya MZ, Fauzan A, Turjamil R (2018) Development of bleep test team software

Self-competition Virtual Reality Cycling for Enhanced Motivation: Interactive Feedforward Adjustments

Abdul Syafiq Bahrin, Mohd Shahrizal Sunar, and Hadafi Fitri Mohd Latip

Abstract Virtual Reality (VR) cycling exergames serve as an enticing platform for physical activity. A prominent feature in recent studies is the use of the "interactive feedforward" psychological training technique in self-competition VR cycling exergames to elevate players' intrinsic motivation. Traditional implementations let players compete against past performances, visualized as a "past ghost" avatar based on two sprints in the initial session. In subsequent sessions, players compete against this "past ghost" including an enhanced "future ghost" which moves faster by 5% extra speed, ensuring the player witnesses achievements they haven't yet attained. However, this feedforward adaptation is session-based, missing out on intra-session dynamics. Given that each session contains two sprints, this study advocates for a more nuanced adjustment approach: enabling feedforward modifications within sessions, tailored to each sprint. To actualize this concept, a VR exergame prototype was developed. A two-session pilot study, encompassing twenty-nine healthy participants, utilized a semi-structured interview approach to glean qualitative reflections on player experiences. Concurrently, the System Usability Scale (SUS) quantified the prototype's usability. Preliminary results suggest a positive reception, with participants relishing the engaging blend of exercise and gaming, validating the merits of this innovative approach.

A. S. Bahrin · M. S. Sunar (✉)
Department of Emergent Computing, Faculty of Computing, Universiti Teknologi Malaysia, 81310 Johor Bahru, Johor, Malaysia
e-mail: shahrizal@utm.my

Media and Game Innovation Centre of Excellence, Institute of Human Centered Engineering, Universiti Teknologi Malaysia, 81310 Johor Bahru, Johor, Malaysia

H. F. M. Latip
School of Education, Faculty of Social Science and Humanity, Universiti Teknologi Malaysia, 81310 Johor Bahru, Johor, Malaysia

Sport Innovation Technology Centre, Institute of Human Centered Engineering, Universiti Teknologi Malaysia, 81310 Johor Bahru, Johor, Malaysia

A. S. Bahrin
School of Creative Industry Management and Performing Arts, College of Arts and Sciences, Universiti Utara Malaysia, 06010, Bukit Kayu Hitam, Kedah, Malaysia

Z. Mohamed et al. (eds.), *Proceedings of the 2nd International Conference on Innovation and Technology in Sports, ICITS 2023, 27–28 November, Malaysia*, Lecture Notes in Bioengineering, https://doi.org/10.1007/978-981-97-3741-3_11

Keywords Interactive feedforward · Self-competition · Virtual reality · Cycling · Exergame

1 Introduction

Sedentary lifestyles present numerous health challenges, with many individuals avoiding exercise due to constraints like lack of time and diminished interest. To combat this, the advent of Virtual Reality (VR) exergames has proven to be a powerful tool, making exercise more engaging and fun, thus effectively reducing the risks associated with obesity (Yoo et al. 2018; Rincker and Misner 2017). At the intersection of "exercise" and "game," exergames offer users an immersive fitness experience.

The term VR exergame is best understood as exercise-oriented games designed to not only increase but also maintain users' motivation for indoor exercise. These games leverage the immersive nature of a virtual world, using gamification factors to enhance the health benefits (Bahrin et al. 2022). Crucially, these gamification elements are keys to the success of VR exergame interventions. They serve to elevate the user's motivation during physical activity (Nor et al. 2020), while also providing a framework to guide and improve their performance (Nor et al. 2020).

VR cycling exergames, in particular, have gained significant attention, propelled by events like Zwift's Kiss Super League (Zwift launches the first esports Pro Cycling League | Zwift 2022). However, as with many competitive platforms, not all users thrive in this environment. For some, external competition can be more demotivating than encouraging. To address this, the present research emphasizes the concept of self-competition in VR exergames, introducing a "ghost" competitor, a virtual representation of a player's past performance.

Key contributions of this paper include:

- Development of an enhanced self-competition VR cycling exergame.
- Presentation of empirical evidence that supports improved usability and enriched player experiences.

2 Related Work

Past research on VR cycling exergames integrated self-competition using the interactive feedforward or "ghost" approach. Here, virtual trainers modulate speed based on the player's Heart Rate (HR), offering competitive and cooperative modes (Shaw et al. 2016a). Players also receive feedback based on their proximity to these virtual trainers, enhancing motivation. Another study introduced two types of "ghosts": a replay of a player's past performance and an optimal gameplay and exercise trainer (Shaw et al. 2016b). The self-competition concept was further expanded upon by Whaley (2017) and adopted/adapted by Koulouris et al. (2020), Farrow et al. (2019), Barathi et al. (2018). These studies, with variations, implemented a High-Intensity

Interval Training (HIIT) based on the Wingate protocol, which has been proven to be efficient for VR exergames. Further innovation was introduced by Michael and Lutteroth (2020), where multiple "past ghosts" were used, culminating in a multi-ghost competition system. This longitudinal study spanning a month added a new "past ghost" per session, including the interactive feedforward: an enhanced version of the player's best part performance by an extra 5% speed called "future ghost." The study reported improved intrinsic motivation, flow, and exercise self-efficacy.

3 VR Exergame Design and Gameplay

This section discusses the design and gameplay mechanisms of the multi-ghost competition VR exergame, innovating to amplify the feedforward technique for players.

3.1 Initial Gameplay Setup

Players start by registering their details and pairing devices. In the first sprint, they cycle without competition. Upon completion, performance data are logged. Two avatars, or "ghosts," are generated (Refer to Fig. 1):

- The "past ghost" mirrors the player's initial performance.
- The "future ghost" performs at a rate of 5% faster than the player's baseline.

Fig. 1 Second sprint moment with 'past ghost' and 'future ghost'

Fig. 2 On the left: instructor guiding a player in the game menu. On right: sprinting phase with ghost competitors

3.2 Subsequent Gameplay

In the next session, gameplay is adjusted based on performance:

- New Record: If players beat their previous best in Sprint 1, the future ghost updates to be 5% faster than the new record. Two past ghosts (from the initial session and Sprint 1) are also visible.
- No New Record: If no records are broken, the future ghost remains at its previous speed.

4 Pilot Testing

From 34 participants recruited via word-of-mouth and WhatsApp, 5 were excluded due to motion sickness and technical errors. The final group had 29 participants (21 males, 7 females; age range 21–23, mean age 22.07).

Using a cross-sectional study, the game was tested in a day per participant (Refer to Fig. 2). Emphasizing usability and player experience, the study drew from HIIT protocols (Whaley 2017; Barathi et al. 2018). To mitigate familiarization and fatigue, participants had a familiarization phase followed by the main session, urging maximum effort in sprints.

5 Result

The primary objective of this study, outlined in the introduction, was to progress the development of VR cycling exergames, emphasizing usability and player experiences. The findings are summarized from observations, semi-structured interviews, and the System Usability Scale (SUS).

5.1 Observation Results

- **Engagement:** Players exhibited high engagement, notably during sprints.
- **Technical Issues:** Initial sessions experienced Bluetooth connectivity problems, with interference from other devices suspected. Another bug created resistance issues in the smart trainer, making cycling overly strenuous. These were subsequently addressed.
- **Auditory Cues Needed:** Given that some players, especially VR first-timers, became so immersed they missed phase changes, there's a need for differentiating music or sounds for sprint and slow phases.

5.2 Semi-structured Interview Results

- **Initial Reactions:** Excitement dominated, with a mix of anticipation and nervousness. Some felt the game was challenging beyond their expectations.
- **Balance Concerns:** First-time VR users initially felt unbalanced on the stationary bike but adapted quickly.
- **Ghost Interaction:** The "ghost" feature, representing past performance, was a significant motivator. While some were caught off-guard by their ghost's speed, others tactically moderated their initial sprints. Despite varied strategies, the ghost was universally seen as a beneficial motivator and challenge.
- **Overall Feedback:** Players expressed enthusiasm for the exergame and showed interest in revisiting post-improvements.

5.3 System Usability Scale (SUS)

The SUS scores were recalibrated and mapped onto a 0–100 scale, with higher scores indicating enhanced usability. Scores ranged between 57.5 (players P8 and P10) and 87.5 (player P11), with an average score of 71.47 (SD = 7.98). Players P9 and P32 also showcased scores above 80, signifying high system satisfaction. Regarding individual items, 65.5% of participants strongly agreed with the fourth SUS item, possibly reflecting the need for technical support (refer to 4.2.3.1). Additionally, 65.5% felt they needed extensive understanding to begin cycling, though this primarily related to the initial sign-up process. Subsequent logins were straightforward, requiring only a Bluetooth connection to initiate the 5-min game. In summary, the feedback from participants indicates positive usability perceptions for the VR cycling exergame system.

6 Conclusion and Future Work

The exploration into VR cycling exergames revealed promising potential for enhancing user engagement. Observational and interview data highlighted the motivating impact of the "ghost" feature and the initial excitement surrounding gameplay. While there were initial technical challenges and a learning curve for first-time VR users, the overall usability score from the SUS suggests a positive reception. Continued refinement and user feedback will be crucial for the future development of such exergames.

Acknowledgements This research received financial support from the Ministry of Education Malaysia and Universiti Utara Malaysia. The primary author conducted this study during their study leave as part of the 2020 Academic Training Scheme for Bumiputera (SLAB) scholarship program.

We extend our gratitude to Universiti Teknologi Malaysia, Skudai, Johor, Malaysia, for their generous research grant provided through the Malaysia Research University Network Grant Scheme MRUN/R.J130000.7809.4L870. We also express our appreciation to the XRlab team members at the Media and Game Innovation Centre of Excellence, Institute of Human-Centered Engineering, Universiti Teknologi Malaysia, for their valuable assistance and support.

References

Bahrin AS, Sunar MS, Mohd Latip HF, Zamri MN (2022) Chronology of self-competition in virtual reality exergames: a historical review. Entertain Comput 43:100513. https://doi.org/10.1016/j.entcom.2022.100513.

Barathi SC et al (2018) Interactive feedforward for improving performance and maintaining intrinsic motivation in VR exergaming. In: Conference on human factors in computing systems, pp 1–14. https://doi.org/10.1145/3173574.3173982

Farrow M, Lutteroth C, Rouse PC, Bilzon JLJ (2019) Virtual-reality exergaming improves performance during high-intensity interval training. Eur J Sport Sci 19(6):719–727. https://doi.org/10.1080/17461391.2018.1542459

Koulouris J, Jeffery Z, Best J, O'Neill E, Lutteroth C (2020) Me vs. Super(wo)man: effects of customization and identification in a VR exergame. In: Conference on human factors in computing systems, pp 1–17. https://doi.org/10.1145/3313831.3376661.

Michael A, Lutteroth C (2020) Race yourselves: a longitudinal exploration of self-competition between past, present, and future performances in a VR exergame. In: Conference on human factors in computing systems, pp 1–17. https://doi.org/10.1145/3313831.3376256

Nor N, Sunar M, Kapi A (2020) A review of gamification in virtual reality (VR) sport. EAI Endorsed Trans Creat Technol 6(21):163212. https://doi.org/10.4108/eai.13-7-2018.163212

Nor NN, Sunar MS, Kapi AY (2020) User experience of gamified virtual reality (VR) in sport: a review. In: Lecture notes of the institute for computer sciences, social informatics and telecommunications engineering LNICST, vol 323, pp 440–449. https://doi.org/10.1007/978-3-030-51005-3_36

Rincker M, Misner S (2017) The jig experiment: development and evaluation of a cultural dance active video game for promoting youth fitness. Comput Sch 34(4):223–235. https://doi.org/10.1080/07380569.2017.1387468

Shaw LA, Buckley J, Corballis PM, Lutteroth C, Wuensche BC (2016b) Competition and cooperation with virtual players in an exergame. PeerJ Comput. Sci. 2016(10):1–20. https://doi.org/10.7717/peerj-cs.92

Shaw LA, Tourrel R, Wunsche BC, Lutteroth C, Marks S, Buckley J (2016a) Design of a virtual trainer for exergaming. In: Proceedings of the Australasian computer science week multiconference, New York, NY, USA: ACM, pp 1–9. https://doi.org/10.1145/2843043.2843384

Whaley A (2017) 'Evercycle'—a virtual reality exercise game to accelerate exercise performance improvement using the 'feedforward effect.' University of Bath, Bath, U. K

Yoo S, Carter M, Kay J (2018) VrMove: design framework for balancing enjoyment, movement and exertion in VR games. In: CHI Play 2018—Proceedings of 2018 annual symposium on computer-human interaction in play companion extended abstracts, pp 295–307. https://doi.org/10.1145/3270316.3272054.

Zwift launches the first esports Pro Cycling League | Zwift. https://www.zwift.com/news/10822-kiss-super-league. Accessed 26 Sept 2022

Blockchain Technology: Reshaping the Sports Industry

Suhizaz Sudin, Mohammad Shahril Salim, and R. Badlishah Ahmad

Abstract Blockchain technology is poised to revolutionize the sports industry through its decentralized, transparent, and immutable ledger systems. This research paper explores how blockchain can address the challenges related to data integrity, transparency, and trust within the sports ecosystem, mitigating fraud and manipulation concerns. It also highlights the impact of technology on sports governance and fan engagement by streamlining contract negotiations, transfers, and royalty payments through smart contracts, as well as enabling secure ticketing, merchandise authentication, and peer-to-peer engagement. Moreover, blockchain has the potential to combat doping and enhance sports integrity through tamper-proof recording and verification systems. Additionally, it offers transparency in sports licensing, intellectual property management, and content distribution, empowering stakeholders with frictionless transactions and revenue opportunities. Embracing blockchain technology has the potential to drive innovation, streamline operations, and create a more connected and secure sports industry, albeit with challenges such as scalability and regulatory considerations that require attention.

Keywords Sports · Technology · Blockchain · Cryptocurrency

S. Sudin (✉) · M. S. Salim · R. B. Ahmad
Faculty of Electronic Engineering and Technology, University Malaysia Perlis, 02600 Arau, Perlis, Malaysia
e-mail: suhizaz@unimap.edu.my

S. Sudin · M. S. Salim
Sports Engineering Research Centre (SERC), University Malaysia Perlis, 02600 Arau, Perlis, Malaysia

R. B. Ahmad
Centre of Excellence for Advanced Computing (AdvComp), University Malaysia Perlis, 02600 Arau, Perlis, Malaysia

Z. Mohamed et al. (eds.), *Proceedings of the 2nd International Conference on Innovation and Technology in Sports, ICITS 2023, 27–28 November, Malaysia*, Lecture Notes in Bioengineering, https://doi.org/10.1007/978-981-97-3741-3_12

1 Introduction

The sports industry encompasses various sectors such as professional sports leagues, teams, athletes, governing bodies, sponsors, and fans. While sports play a significant role in entertainment, competition, and cultural events, the industry faces several challenges. First, transparency issues arise in areas such as ticket sales, athlete transfers, and sponsorship deals. Lack of visibility and trust can lead to fraudulent activities, counterfeiting, and unfair practices, eroding confidence in the industry (Feulner et al. 2022).

Besides, sports organizations often struggle with complex governance structures, slow decision-making processes, and disputes over contracts and regulations (Hulme et al. 2019; Petronel and Florentina 2013). This can result in inefficiencies, delays, and conflicts that hinder the smooth operation of the industry. Traditional fan engagement methods face limitations in terms of ticketing processes, merchandise authentication, and interaction with sports organizations (Huettermann et al. 2022). Fans seek enhanced experiences, direct engagement, and assurance of authenticity, which the current systems may not adequately provide.

On the other hand, preserving the integrity of sports competitions and combating issues like doping, match-fixing, and cheating is a continuous challenge (Clancy et al. 2022; Vidal 2014; Carpenter 2012). Ensuring fair play, protecting athletes' rights, and maintaining the credibility of the sports industry are paramount concerns.

Blockchain technology is a decentralized and distributed ledger system that enables secure and transparent transactions between parties without the need for intermediaries (Holland 2023). It operates on the principles of decentralization, immutability, and consensus, making it highly secure and resistant to tampering. Blockchain achieves this by recording transactions in blocks that are linked together in a chronological and cryptographic manner, forming an unalterable chain of data.

The integration of blockchain technology into the sports industry holds immense potential to revolutionize various aspects, including transparency, security, and efficiency. Blockchain technology offers solutions to the challenges faced by the sport industry, which firstly, efficient governance. With blockchain, sports organizations can enhance governance by creating transparent and immutable records of decisions and streamlining contract management through smart contracts (Glebova and Mihaľová 2023).

Additionally, blockchain-based ticketing systems provide secure and transparent transactions, ensuring authenticity and preventing ticket fraud (Tackmann 2017). The technology also enables direct fan engagement and loyalty programs, facilitating peer-to-peer interactions and seamless transactions for merchandise and experiences. Moreover, blockchain's tamper-proof nature can safeguard the integrity of sports competitions by recording and verifying anti-doping measures, ensuring fair play, and maintaining the credibility of the industry (Chanda Jindal et al. 2021; Pinto et al. 2020).

Furthermore, blockchain simplifies licensing processes, royalty tracking, and content distribution, empowering sports organizations to protect their intellectual property and efficiently manage licensing agreements (Conrad 2022; Hugendubel 2021). Overall, by leveraging blockchain technology, the sports industry can enhance transparency, fan engagement, sports integrity, and streamline various operations, ultimately shaping a more efficient and secure sports ecosystem.

2 History of Blockchain Technology

In 1991, research scientists Stuart Haber and W. Scott Stornetta introduced a groundbreaking concept known as blockchain technology (Devi and Sharma 2022). Their aim was to provide a practical solution for time-stamping digital documents, ensuring their integrity and preventing unauthorized modifications. To achieve this, they devised a system based on the idea of a cryptographically secured chain of blocks, specifically designed for storing these time-stamped documents securely.

A significant development took place in 1992 when they incorporated Merkle Trees into the blockchain design (Bayer et al. 1993). This addition enhanced the efficiency of the blockchain by enabling the consolidation of multiple documents into a single block. Merkle Trees, serving as the foundation for the secure chain of blocks, allowed for the storage of a series of interconnected data records. Each data record was linked to the preceding one, establishing a continuous chain. The most recent record in this chain encapsulated the complete history of the entire chain.

In 2004, Hal Finney, a computer scientist and cryptographic activist, made a significant contribution to the history of cryptocurrencies by introducing a system called Reusable Proof Of Work (RPoW) (Finney 2004). This system served as an early prototype for digital cash and played a crucial role in the development of cryptocurrencies. RPoW operated by exchanging non-exchangeable Hashcash-based proof of work tokens for RSA-signed tokens that could be transferred from person to person. To address the issue of double-spending, RPoW implemented a mechanism where token ownership was registered on a trusted server. This server allowed users worldwide to verify the correctness and integrity of ownership in real time.

Going forward, in 2008, Satoshi Nakamoto introduced the idea of distributed blockchains, marking a significant progression in the field (Casino et al. 2019). Nakamoto enhanced the design by enabling the addition of blocks to the chain without the necessity for authentication by trusted parties. These revised structures stored a secure record of data exchanges and functioned on a peer-to-peer network for timestamping and validating every transaction. The blockchain could operate autonomously, eliminating the requirement for a central authority. These developments proved immensely advantageous, solidifying blockchains as the cornerstone of cryptocurrencies. Presently, blockchains function as the public ledger for all cryptocurrency transactions.

The evolution of blockchains has been steady and promising. Initially, the terms "block" and "chain" were used separately in Nakamoto's original paper, but they eventually became popularized as a single word, "Blockchain," by 2016. In recent times, the file size of cryptocurrency blockchains, which contain records of all network transactions, has grown from the ground to over 226Gbytes in the past 10 years from January 2009 to June 2019 (Xu and Huang 2020), highlighting the continued expansion and adoption of blockchain technology.

3 Blockchain Applications in Sport

Numerous blockchain applications have been proposed by researchers. Initially, blockchain technology can be employed to establish a decentralized and transparent system for handling sports logo licensing, anti-counterfeiting, and tracking counterfeit products (Lv et al. 2022). The distributed ledger of blockchain facilitates improved tracking of products, ensuring their authenticity and mitigating the circulation of counterfeit goods. Blockchain technology operates on a distributed ledger system, wherein information is stored across multiple nodes in a network. Each node preserves a copy of the complete blockchain, and any alterations or additions to the data require approval through a consensus mechanism agreed upon by the network participants.

The key to long-term development is to improve financing ability and solve financing difficulties in order to improve the performance of professional sports club (Wang 2022). Conventional financing methods often involve intermediaries like banks or financial institutions, introducing delays, costs, and constraints. By harnessing blockchain technology, sports clubs can secure funding directly from investors, bypassing the need for intermediaries. This direct peer-to-peer interaction enables quicker transactions and has the potential to eliminate hidden costs associated with traditional financing.

In the sports industry, fan tokens and NFTs (Non-Fungible Tokens) have emerged as innovative tools to enhance fan engagement strategies (Baker et al. 2022). These tokens offer fans access and opportunities to participate in club decision-making, such as selecting uniform designs or player honors. Furthermore, NFTs can represent transferable membership or ownership stakes in sports teams or clubs. Integrating fan tokens and NFTs into fan engagement strategies allows clubs to establish a stronger connection with their supporters. Fans holding these tokens gain exclusive access to content, experiences, and rewards, fostering a sense of involvement and empowerment.

The utilization of tickets as NFTs (Non-Fungible Tokens) presents a solution to various concerns associated with traditional ticketing systems, by leveraging the capabilities of blockchain technology (Heilmann et al. 2023). By adopting NFT-based tickets, multiple challenges can be addressed. Firstly, the issue of verifying ticket authenticity is resolved, as each ticket is uniquely minted as an NFT on the blockchain, ensuring its integrity and preventing fraudulent duplication. Furthermore, the transferability and resale of tickets are simplified, eliminating barriers for ticket holders unable to attend an event, while maintaining a secure and transparent transaction process through blockchain-based ownership transfers. Additionally, the immutable nature of blockchain records ensures a transparent and accountable ticketing ecosystem, reducing fraudulent activities and promoting trust among stakeholders.

4 Limitations of Blockchain Technology

Blockchain technology, although promising, comes with its set of limitations. Several important constraints should be taken into account when employing blockchain. Scalability poses a primary challenge, given that the decentralized nature of blockchain networks can lead to slower transaction processing and reduced efficiency as the number of participants and transactions grows (Pinto et al. 2022). Additionally, concerns arise about the energy consumption related to certain consensus algorithms, particularly Proof of Work (PoW), raising environmental impact worries regarding blockchain networks (Lopez-Barreiro et al. 2022). Storage and bandwidth requirements present additional obstacles, as the expanding size of the blockchain demands substantial storage capacity and the transmission of large amounts of data.

Governance and compliance introduce complex issues, with the absence of clear governance frameworks hindering decision-making, protocol upgrades, and dispute resolution. Furthermore, ensuring privacy and confidentiality within a transparent blockchain environment remains challenging. Achieving interoperability between different blockchain networks and traditional systems proves difficult due to the lack of standardized protocols.

Despite these limitations, ongoing research and development efforts are focused on addressing these challenges to unlock the full potential of blockchain technology and mitigate its drawbacks.

5 Conclusion

This research paper delves into the utilization of blockchain technology to address various challenges within the sports ecosystem, particularly in relation to data integrity, transparency, and trust. The paper discusses several existing blockchain applications proposed by other researchers in the field. One such application is the use of blockchain for managing sports logo licensing, anti-counterfeiting, and counterfeit product tracking. By leveraging the decentralized and transparent nature of blockchain, this application enables better tracking of products to ensure their authenticity and prevent the circulation of counterfeit items.

Another area where blockchain can make a significant impact is in improving the financing ability and addressing financing difficulties faced by professional sports clubs. Traditional financing methods often involve intermediaries, leading to delays, costs, and constraints. However, by leveraging blockchain technology, sports clubs can directly access funding from investors without the need for intermediaries. This peer-to-peer interaction enables quicker transactions and potentially eliminates hidden costs associated with traditional financing.

Fan tokens and NFTs have emerged as innovative tools for enhancing fan engagement strategies in the sports industry. These tokens provide fans with exclusive access and opportunities to participate in club decision-making processes, such as selecting uniform designs or player honors. Additionally, NFTs can represent transferable membership or ownership stakes in sports teams or clubs. By incorporating fan tokens and NFTs into their fan engagement strategies, clubs can establish a stronger connection with their supporters, fostering a sense of involvement and empowerment.

Furthermore, the utilization of tickets as NFTs addresses various concerns associated with traditional ticketing systems. By minting tickets as NFTs on the blockchain, the authenticity of each ticket can be easily verified, preventing fraudulent duplication. NFT-based tickets also simplify the transferability and resale processes, enabling secure and transparent ownership transfers. Additionally, the immutable nature of blockchain records ensures transparency and accountability in the ticketing ecosystem, reducing fraudulent activities and building trust among stakeholders.

While blockchain technology offers promising solutions, it is crucial to consider its limitations. Scalability remains a primary challenge, as the decentralized nature of blockchain networks can result in slower transaction processing as the network grows. Energy consumption associated with certain consensus algorithms, storage and bandwidth requirements, governance and compliance issues, privacy and confidentiality concerns, and interoperability challenges are among the limitations that need to be addressed. However, ongoing research and development efforts are dedicated to overcoming these limitations and fully unlocking the potential of blockchain technology in the sports industry.

References

Baker B, Pizzo A, Su Y (2022) Non-fungible tokens. Sport Innov J 3:1–15. https://doi.org/10.18060/25636

Bayer D, Haber S, Stornetta WS (1993) Improving the efficiency and reliability of digital time-stamping. In: Sequences II, New York, NY, Springer New York, 1993, pp 329–334. https://doi.org/10.1007/978-1-4613-9323-8_24

Carpenter K (2012) Match-fixing—the biggest threat to sport in the 21st century? Int Sport Law Rev 2:13–24

Casino F, Dasaklis TK, Patsakis C (2019) A systematic literature review of blockchain-based applications: current status, classification and open issues. Telemat Inf 36:55–81. https://doi.org/10.1016/j.tele.2018.11.006

Chanda Jindal SO, Chanda S, Star S (2021) Contouring e-doping: a menace to sportsmanship in Esports. Turk Online J Qual Inq (TOJQI) 12(8):966–981. https://www.researchgate.net/publication/354325314

Clancy S, Owusu-Sekyere F, Shelley J, Veltmaat A, De Maria A, Petróczi A (2022) The role of personal commitment to integrity in clean sport and anti-doping. Perform Enhanc Health 10(4):100232. https://doi.org/10.1016/j.peh.2022.100232

Conrad M (2022) Non-fungible tokens, sports, and intellectual property law issues: a case study applying copyright, trademark, and right of publicity law to a non-traditional ownership vehicle. J Leg Asp Sport 32(1):132–152. https://doi.org/10.18060/26091

Devi V, Sharma A (2022) A sophisticated analysis of the blockchain technology and its applications domain. In: 2022 international conference on advances in computing, communication and applied informatics (ACCAI), IEEE, pp 1–4. https://doi.org/10.1109/ACCAI53970.2022.9752624

Feulner S, Sedlmeir J, Schlatt V, Urbach N (2022) Exploring the use of self-sovereign identity for event ticketing systems. Electron Mark 32(3):1759–1777. https://doi.org/10.1007/s12525-022-00573-9

Finney H (2004) Reusable proofs of work (RPOW). Agosto 15:2004

Glebova E, Mihaľová P (2023) New currencies and new values in professional sports: blockchain, NFT, and fintech through the stakeholder approach. J Phys Educ Sport 23(5):1244–1252. https://doi.org/10.7752/jpes.2023.05153

Heilmann D et al (2023) Mitigation of scaling effects in NTF-based ticketing systems, pp 110–119. https://doi.org/10.1007/978-3-031-21229-1_11

Holland B (2023) Handbook of research on advancements of contactless technology and service innovation in library and information science. In advances in library and information science. IGI Global. https://books.google.com.my/books?id=zbjFEAAAQBAJ

Huettermann M, Uhrich S, Koenigstorfer J (2022) Components and outcomes of fan engagement in team sports: the perspective of managers and fans. J Glob Sport Manag 7(4):447–478. https://doi.org/10.1080/24704067.2019.1576143

Hugendubel J (2021) Blockchain technology and intellectual property—a basic introduction. Int J Cryptocurrency Res 1(1):5. https://doi.org/10.51483/ijccr.1.1.2021.5-25

Hulme A, McLean S, Read GJM, Dallat C, Bedford A, Salmon PM (2019) Sports organizations as complex systems: using cognitive work analysis to identify the factors influencing performance in an elite netball organization. Front Sports Act Living 1(November):1–12. https://doi.org/10.3389/fspor.2019.00056

Lopez-Barreiro J, Alvarez-Sabucedo L, Garcia-Soidan JL, Santos-Gago JM (2022) Use of blockchain technology in the domain of physical exercise, physical activity, sport, and active ageing: a systematic review. Int J Environ Res Public Health 19(13):8129. https://doi.org/10.3390/ijerph19138129

Lv C, Wang Y, Jin C (2022) The possibility of sports industry business model innovation based on blockchain technology: evaluation of the innovation efficiency of listed sports companies. PLoS ONE 17(1):e0262035. https://doi.org/10.1371/journal.pone.0262035

Petronel M, Florentina M (2013) Sports organization management: between constraints and objectives. Procedia Soc Behav Sci 81:95–99. https://doi.org/10.1016/j.sbspro.2013.06.394

Pinto F, Rahulamathavan Y, Skinner J (2022) Blockchain for doping control applications in sports: a conceptual approach. Future Internet 14(7):210. https://doi.org/10.3390/fi14070210

Pinto F, Rahulamathavan Y, Kondoz A, Skinner J (2019) Blockchain technology for trustworthy doping control data management, June 2020

Tackmann 2017Tackmann B (2017) Secure event tickets on a blockchain. https://doi.org/10.1007/978-3-319-67816-0_26

Vidal L (2014) Protecting the integrity of sport competition - the last bet for modern sport. ICSS.https://doi.org/10.13140/RG.2.1.2546.0325

Wang S (2022) Improved blockchain technology for performance optimization model design of sports clubs. J Electr Comput Eng 2022:1–8. https://doi.org/10.1155/2022/4436471

Xu Y, Huang Y (2020) Segment blockchain: a size reduced storage mechanism for blockchain. IEEE Access 8:17434–17441. https://doi.org/10.1109/ACCESS.2020.2966464

ANN-PSO Modelling and Optimization to Reduce Surface Roughness in Additive Manufactured Sports Equipment

Mohd Zakimi Zakaria, Mohd Sazli Saad, Mohamad Ezral Baharudin, Azuwir Mohd Nor, Nashrul Fazli Mohd Nasir, and Geak Eng Chua

Abstract Fused Deposition Modelling (FDM) is an additive manufacturing (AM) technology that creates a wide variety of structures and complex 3D model data geometries. Due to the layered manufacturing process, poor surface roughness is the primary issue with FDM, limiting the direct applicability of FDM-printed parts to a significant number of applications such as sports equipment. There were few new meta-heuristic approaches for modelling and optimizing FDM surface roughness. As a result, a newly hybrid method ANN-PSO is developed to effectively assist practitioners in the prediction and optimization of FDM process parameters for improving fabricated part surface roughness. The experimental data from the previous study were used to train, validate, and test the artificial neural network (ANN) model, which finely maps the relationship between input process control factors and output response. The best network has a single hidden layer of 15 neurons. Thus, the 4-15-1 NN model was optimized by using the Particle Swarm Optimization (PSO) algorithm to find the best combination of parameter settings to improve the surface roughness of fabricated parts. The best values for layer height, print speed, print temperature, outer shell speed, and surface roughness were obtained by PSO, which were 0.2674 mm, 19.66 mm/s, 190.51 °C, 22.29 mm/s, and 2.0113 μm, respectively. As a result, the integrated ANN-PSO is a perfect combination for improving the surface finish of FDM printed parts especially for sports equipment.

Keywords Fused deposition modelling (FDM) · Surface roughness · Particle swarm optimization (PSO) · Artificial neural network (ANN) · Sports equipment

M. Z. Zakaria (✉) · N. F. Mohd Nasir
Centre of Excellence (COE), Sports Engineering Research Centre (SERC), Universiti Malaysia Perlis, Main Campus Pauh Putra, 02600 Arau, Perlis, Malaysia
e-mail: zakimizakaria@unimap.edu.my

M. S. Saad · M. E. Baharudin · A. Mohd Nor · G. E. Chua
Department of Manufacturing, Faculty of Mechanical Engineering & Technology, Universiti Malaysia Perlis, Main Campus Pauh Putra, 02600 Arau, Perlis, Malaysia

Z. Mohamed et al. (eds.), *Proceedings of the 2nd International Conference on Innovation and Technology in Sports, ICITS 2023, 27–28 November, Malaysia*, Lecture Notes in Bioengineering, https://doi.org/10.1007/978-981-97-3741-3_13

1 Introduction

Fused Deposition Modelling (FDM) is a subset of additive manufacturing and a type of 3D printing technology (AM). AM refers to the process of constructing a physical thing by adding material layer by layer. FDM is one of the technologies that fall within the additive manufacturing category. Other well-known AM processes include Stereolithography (SLA), Selective Laser Sintering (SLS), and Directed Energy Deposition (DED). Each of these technologies adds material layer by layer to build a three-dimensional object. FDM is a widely utilized, cost-effective, and user-friendly technique for prototyping, manufacturing small parts, sports equipment, making models and figurines. Even though FDM is a productive and effective AM technology, its large-scale application has not received much consideration due to the compatibility of existing materials with AM technologies. This constraint is typically circumvented in two ways: improvement of material qualities and their compatibility with the process, and proper modification of FDM input parameters to improve product quality and achieve great mechanical properties (Saad et al. 2019). Because the mechanical qualities of 3D-printed items can be adjusted rapidly and affordably by adjusting the FDM process parameters, the second method is frequently utilized.

Artificial Neural Networks (ANN) have been researched for the prediction of dimensional accuracy since the output performance of the FDM is difficult to forecast by mathematical equations with conflicting factors and process phenomena affecting one another (Noriega et al. 2013). The performance of an ANN is influenced by its structural factors and network training. In general, there is not a single ANN model that can be used in every situation. The relevant parameter values for each scenario must therefore be established experimentally (Alafaghani et al. 2021). The design and development of mathematical models based on a physical understanding of the FDM process is crucial in optimizing the FDM system and process parameter selection to improve the performance characteristic of the 3D model (Vahabli and Rahmati 2016). One performance characteristic to be optimized is surface roughness. The best output value can be obtained by optimizing the predefined input target. It can be applied in a variety of areas, including accounting, manufacturing, and others (Chen et al. 2018). Additionally, there are two categories into which optimization techniques are typically divided: (i) classical algorithms and (ii) meta-heuristics algorithms. Table 1 displays an example of an optimization algorithm. But because it is more adaptable and controlled randomly for combinations, the current (meta-heuristic) approach is better than the traditional (classical) algorithm (Sedghi et al. 2016).

Table 1 Optimization method example (Sedghi et al. 2016)

Classical algorithm	Meta-heuristic algorithm
Transportation algorithm	Genetic algorithm
Branch and bound	Ant colony optimization
Mixed Integer programming	Particle swarm optimization

Due to the complexity of the parameters, it can be challenging to achieve the ideal input parameter in many situations. Thus, the modelling and optimization process has been used to circumvent this issue. Numerous studies have investigated how some conventional methods can contribute to the optimization processes. The precision and robustness of the optimization process are being reduced in local optimal solutions due to the limitations of this methodology (Singh et al. 2018). As a result, the meta-heuristic approach is considered in this study in order to produce global, precise, and reliable optimal solutions. Since the underlying nonlinear complex relationship between decision variables and output responses was not clearly defined in the Regression model, Artificial Neural Network (ANN) method was used in this study to predict the ideal surface roughness value (Zain et al. 2012).

In the past few years, there have been a lot of studies that focused on the FDM optimization process. By optimizing the different input parameters of FDM, researchers have tried to improve the quality of the surface finish, the accuracy of the dimensions, and the mechanical behaviour of the parts. From the study by Jaisingh Sheoran and Kumar (Jaisingh Sheoran and Kumar 2020), it was found that the layer thickness was the most variable parameter that affected how well FDM parts fit together and how rough their surfaces were. If you choose a lower value for layer thickness, the surface finish will be better. Also, a lower value for the extrusion temperature, outer shells, and print speed can help improve the performance of the output. Also, Yadav and Mingareev (2023) drew the conclusion from their experiments that only a few variable parameters affect the roughness of the printed parts' surfaces. So, the optimization process is needed to find the right input parameters that will make FDM work better in terms of accuracy and repeatability. In this study, the layer thickness, air gap, raster width, contour width, and raster orientation were chosen to see how they affected the surface quality.

Saad et al. (2019) investigated the use of the FDM process in determining the optimal machining factors for a better surface roughness of an FDM-printed part using the RSM, PSO, and SOS approaches. The machine settings considered included layer height, print speed, print temperature, and outer shell speed, with surface roughness as the desired outcome. Yang et al. (2019) conducted a study on the tensile strength, surface roughness, and manufacturing time of FDM fabricated components, analysing five related factors: nozzle diameter, liquefier temperature, velocity of extrusion, velocity of filling, and layer thickness. The experimentation number has been reduced using Central Composite Design (CCD). Furthermore, the Non-dominated Sorting Genetic Algorithm II (NSGA-II) was used to optimize the related factors that maximized the tensile strength while minimizing surface roughness and manufacturing time of the printed components. Response Surface Methodology (RSM) was also used to investigate the relationship between the control factor and the output factor.

Previous research has not extensively investigated the utilization of a hybrid Artificial Neural Network and Particle Swarm Optimization (ANN-PSO) for enhancing surface roughness in the Fused Deposition Modelling (FDM) process, as indicated by a review of the pertinent literature. Therefore, this novel addition to the field of combination can serve to address the gap left by the previous research. This study aims to utilize an artificial neural network (ANN) for the purpose of modelling the Fused Deposition Modelling (FDM) machine. The objective is to predict the optimal surface roughness value. Additionally, the Particle Swarm Optimization (PSO) method will be employed to optimize the input parameters of the FDM machine. The input parameters encompass the velocity of the outer shell, the temperature at which printing occurs, the thickness of each layer, and the velocity at which printing is conducted. In conclusion, the objective of this study is to ascertain the most favourable FDM machine parameters that exert a substantial impact on the surface roughness of the printed component in the domain of sports equipment.

2 Methodology

This study starts with searching the related previous study to collect the data. Then, the relationship between input variables and output variables was established using Artificial Neural Network (ANN). Surface roughness value then being optimized using Particle Swarm Optimization (PSO). This process was continued with the experimental validation by comparing the simulation value and actual value. Finally, the overall process and the result were discussed and concluded. The general flowchart for the study is shown in Fig. 1.

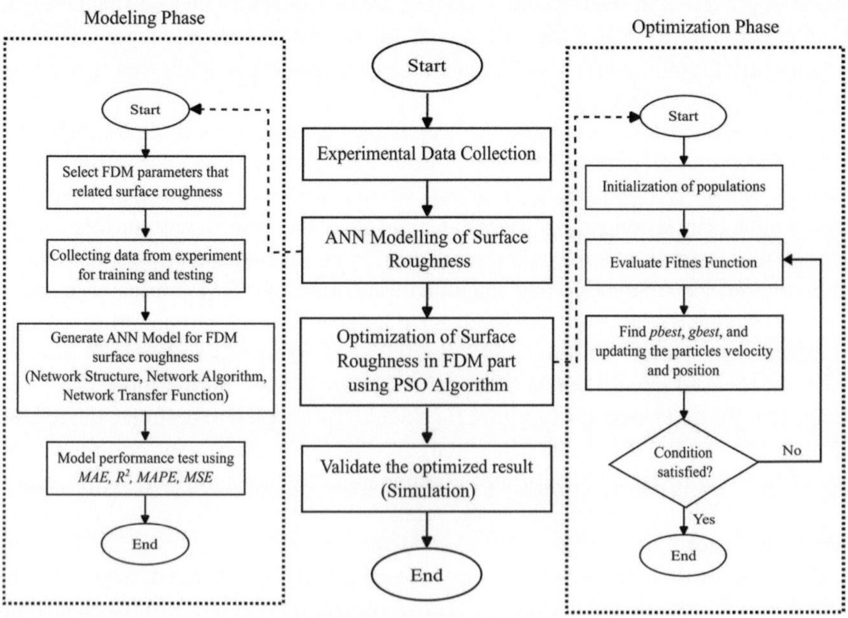

Fig. 1 Overall research methodology for this study

2.1 Data Collection

The data for this study were collected experimentally from previous work (Saad et al. 2019), which was then modelled and optimized the surface roughness using a conventional method called the response surface method (RSM). First, the experimental works were established, followed by the experiment design in order to prepare a set of reliable data of surface roughness values.

Experiment Setup. In the experimental work, 3D printer machine (Tarantula Prusa 13) was used to print the number of specimens. The technical specification of Tarantula Prusa 13 is stated in Table 2. Poly-lactic acid (PLA) material is used for printing due to corresponding to its material properties. As shown in Table 3, tensile strength, tensile modulus, elongation of break, flexural strength, and flexural modulus are involved in a range of properties of PLA material.

Furthermore, the sample in the experiment has been drawn by using CATIA software. Then, the drawing was exported as STL file and imported in the 3D printing software to print each of the samples. The dimension of the specimen is 3.2 mm × 12.7 mm × 125 mm. Figure 2 shows the CAD design and dimensions of the specimen.

Design of Experiment. The experiment was held by using Response Surface Methodology (RSM) based on Central Composite design (CCD). Table 4 represents the optimization limitation for the parameter setting of FDM machine from the

Table 2 Technical specification for 2016 Newest TEVO Tarantula Prusa 13

Technical specifications	Tarantula Prusa I3	Units
Layer resolution	50	microns
Printing speed	150	mm/s
Build size	200 × 200 × 200	mm^3
Material type	PLA, PETG, Wood, PVA, ABS and Flexible filaments	subject
Nozzle diameter	0.4	mm
Material diameter	1.75	mm
Position accuracy	Z = 0.004, XY = 0.012	mm
Ambient temperature	Above > 25	°C
Extruder temperature	Recommended = 210, Maximum = 260	°C

Table 3 Typical range of mechanical properties for PLA materials

Mechanical properties	PLA material
Tensile strength (MPa)	15.5–72.2
Tensile modulus (GPa)	2.020–3.550
Elongation at break (%)	0.5–9.2
Flexural strength (MPa)	52–115.1
Flexural modulus (GPa)	2.392–4.930

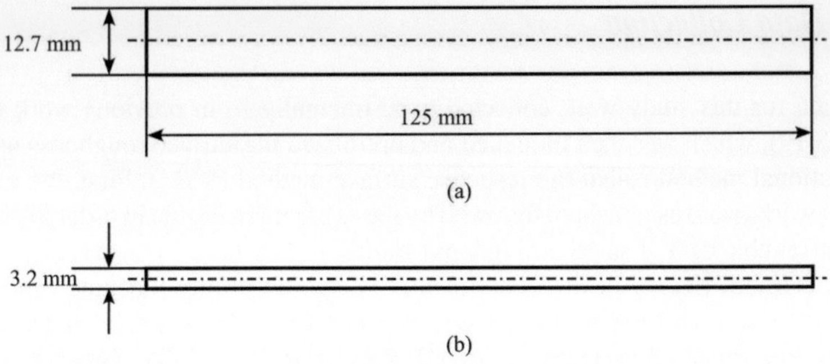

Fig. 2 Computer-aided design (CAD) design specimen: (**a**) Top view; and (**b**) Side view

previous study. Layer thickness (A), print speed (B), print temperature (C), and outer shell speed (D) are parameter settings of FDM to be considered. Next, based on design of experiment (DOE) process, five range of parameter settings in 3D printing machine has been obtained using Design Expert software as listed in Table 5. However, level 2 until level 4 is the FDM parameter setting that in the range of lower and upper boundaries. After all samples had been printed, the surface roughness value was measured using SV-C4500 Mitutoyo Formtracer at each level of experiments and recorded as shown in Table 6.

From Table 6, 46 samples were printed, and the surface roughness value of each specimen is being measured. Specimen number 40 is resulting the lowest surface roughness value which is 2.299 μm where the layer thickness = 0.10 mm, print speed = 48.00 mm/s, print speed = 195.00 °C, and outer shell speed = 34.50 mm/s meanwhile specimen with highest surface roughness is specimen number 19 which is 3.911 μm.

Table 4 3D printer parameters and their levels

Symbol	Parameter	Unit	Lower boundary	Higher boundary
A	Layer height	mm	0.18	0.3
B	Print speed	mm/s	36	60
C	Print temperature	°C	185	205
D	Outer shell speed	mm/s	29	40

Table 5 Five levels of FDM parameter setting using RSM

Symbol	Parameter	Unit	Level 1	Level 2	Level 3	Level 4	Level 5
A	Layer height	mm	0.10	0.18	0.24	0.30	0.38
B	Print speed	mm/s	19.46	36.00	48.00	60.00	76.54
C	Print temperature	°C	171.22	185.00	195.00	205.00	218.78
D	Outer shell speed	mm/s	21.42	29.00	34.50	40.00	47.58

Table 6 Experimental data

Number of Sample	Parameter setting				Surface roughness, Ra (μm)
	A (mm)	B (mm/s)	C (°C)	D (mm/s)	
1	0.18	60.00	185.00	40.00	3.546
2	0.38	48.00	195.00	34.50	2.954
3	0.18	60.00	205.00	40.00	2.788
4	0.24	19.46	195.00	34.50	2.880
5	0.24	48.00	195.00	21.42	3.584
6	0.30	36.00	185.00	40.00	3.126
7	0.30	36.00	185.00	40.00	3.117
8	0.24	48.00	195.00	34.50	3.298
9	0.18	60.00	205.00	29.00	3.626
10	0.18	36.00	205.00	29.00	3.420
11	0.30	36.00	205.00	29.00	3.222
12	0.18	60.00	205.00	29.00	3.826
13	0.30	36.00	205.00	29.00	3.215
14	0.24	48.00	218.78	34.50	3.190
15	0.18	36.00	205.00	40.00	2.392
16	0.24	48.00	195.00	34.50	3.014
17	0.18	60.00	205.00	40.00	2.732
18	0.30	60.00	185.00	29.00	3.869
19	0.30	60.00	185.00	40.00	3.911
20	0.30	36.00	205.00	40.00	3.217
21	0.18	36.00	185.00	40.00	2.989
22	0.24	48.00	195.00	34.50	3.647
23	0.30	36.00	185.00	29.00	2.788
24	0.24	48.00	195.00	34.50	3.565
25	0.30	60.00	205.00	40.00	3.474
26	0.18	36.00	185.00	40.00	3.232
27	0.18	60.00	185.00	29.00	2.754
28	0.24	76.54	195.00	34.50	3.666
29	0.30	36.00	205.00	40.00	3.289
30	0.24	48.00	171.22	34.50	2.856
31	0.30	60.00	205.00	29.00	3.752
32	0.24	48.00	195.00	34.50	3.606
33	0.30	60.00	185.00	29.00	3.773
34	0.18	36.00	205.00	29.00	3.299
35	0.24	48.00	195.00	34.50	3.417
36	0.30	60.00	205.00	29.0	3.777

(continued)

Table 6 (continued)

Number of Sample	Parameter setting				Surface roughness, Ra (μm)
	A (mm)	B (mm/s)	C (°C)	D (mm/s)	
37	0.30	60.00	185.00	40.00	3.906
38	0.24	48.00	195.00	47.58	3.110
39	0.30	36.00	185.00	29.00	2.584
40	0.10	48.00	195.00	34.50	2.299
41	0.18	36.00	205.00	40.00	2.354
42	0.18	36.00	185.00	29.00	2.967
43	0.18	60.00	185.00	29.00	2.303
44	0.18	36.00	185.00	29.00	2.697
45	0.18	60.00	185.00	40.00	3.514
46	0.30	60.00	205.00	40.00	3.550

2.2 *ANN Modelling*

According to Fig. 1, the ANN modelling procedure in this project began with data collection from previous experimental data, the FDM input parameters and output response were collected and grouped into three groups, with 70% of data used for training, 15% for testing, and 15% for validating purposes. The FDM parameters were layer height, print speed, print temperature, and outer shell speed, with the surface roughness, Ra, as the output response. Thus, the ANN model was created using the MATLAB software. In this study, single and double hidden layers for neural networks were investigated. The number of neurons used in this project's running model was determined using the research paper by Bagheri et al. (2019). In this paper, the neuron numbers used to predict the compressive strength of boron-based geopolymers were 10, 15, and 20. The results of the ANN model prediction with the regression coefficient were greater than 0.99. Following the generation of the NN models, the results were recorded, validated, and evaluated. The NN model was validated and evaluated using line pattern data and a performance test based on the values MSE, RMSE, MAE, and MAPE.

Generate Network Structure. The parameter within a neural network comprises of weights, which serve to quantify the strength of connections between neurons in consecutive layers. The determination of weight values is achieved through an iterative training process in neural networks, with the objective of minimizing the loss function that measures the discrepancy between the actual outputs and the predicted values. The backpropagation method is predominantly employed in this process to update the weights. This method utilizes the mathematical chain rule to iteratively compute the gradients for each layer. Upon completion of the training process, the neural network will possess the capability to deduce the outputs by leveraging previously unobserved inputs (Đurović et al. 2022). Figure 3 presents an illustrative example showcasing a neural network structure, which includes the number of layers and nodes.

Performance Analysis. For neural network model performance analysis, the error will be determined using the following performance functions: mean absolute error (MAE), root mean square error (RMSE), coefficient of determination (R^2), and mean absolute percentage error (MAPE). Priority consideration in selecting the performance function is the reduction of the error rate in the predicted value. The MAE is a metric that measures how close model outputs or predictions are to actual values. The MAE is given as:

$$MAE = \frac{1}{N}\sum_{i=1}^{N}(\overline{f}_i - E_i) \tag{1}$$

The RMSE measures the deviation between the predicted value and the target value. The greater the accuracy between the predicted value and the target value, the smaller the RMSE value. The formulation is shown as:

$$RMSE = \sqrt{\frac{\sum_{i=1}^{N}\left(E_i - \overline{f}_i\right)^2}{N}} \tag{2}$$

The coefficient of determination, R^2 indicates the model accuracy, and it can be obtained by:

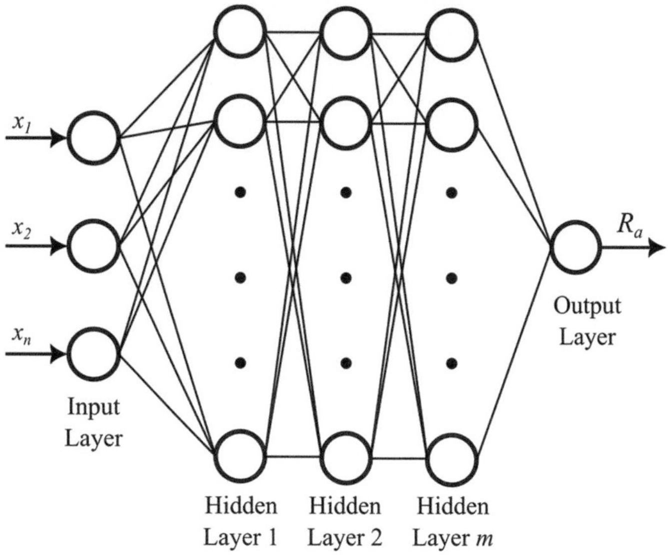

Fig. 3 Sample of a neural network structure

$$R^2 = 1 - \frac{\sum_{i=1}^{N}\left(E_i - \overline{f}_i\right)}{\sum_{i=1}^{N}\overline{f}_i} \tag{3}$$

The normalized error is obtained from MAPE. Hence, the derivation of nets from different data sets can be compared effectively, as presented as:

$$MAPE = \frac{1}{N}\frac{\sum_{i=1}^{N}\left|E_i - \overline{f}_i\right|}{\sum_{i=1}^{N}E_i} \tag{4}$$

3 PSO Algorithm

The two main parameters to consider in PSO are velocity update and position update. Each bird in the population is referred to as a particle. During the searching process, its velocity, position, and memory of the best location are all randomly initialized. Particle velocity is updated by three attributes: social, cognitive, and inertia. The social aspect, also known as the global best (*gbest*) position, expands the search agents to the best position, which is also known as the best position. The cognitive aspect, or personal best (*pbest*) position, accelerates particles to their best position, and the final aspect, inertia, is dependable in maintaining a balance between local (*pbest*) and global (*gbest*) search domain exploration (Raju et al. 2018). The updating process moves the particle swarm to a higher objective function value region, and each particle eventually obtains the highest objective value around the field. The PSO implementation method is continuously dependent on the number of iterations set to complete the minimum value of surface roughness, as shown in Fig. 1.

Implementation PSO. The integrated ANN-PSO has been developed to find the optimal FDM parameters setting in order to achieve the minimum surface roughness of printed part in fused deposition modelling (FDM). Figure 4 depicts the implementation of PSO optimization on surface roughness in FDM. This execution begins with the development of the feed forward back propagation (BP) neural network model using the ANN toolbox in MATLAB software and 46 experimental data. To determine the best ANN model, the *logsig* transfer function and *trainlm* training algorithm were used to train the network. In FDM, the best ANN model will be used as a fitness function to predict the smallest surface roughness.

Following that, the PSO algorithm was developed to determine the optimal value of design variables A, B, C, and D in terms of obtaining the best fitness value. The design variables A, B, C, and D are layer thickness, print temperature, print speed, and outer shell speed, in that order, whereas the optimal fitness value to evaluate is the minimum value of surface roughness, Ra.

In PSO algorithm, the optimal design variables are determined by comparing the overall best fitness value of populations (*pbest*) in each iteration. In other words, if

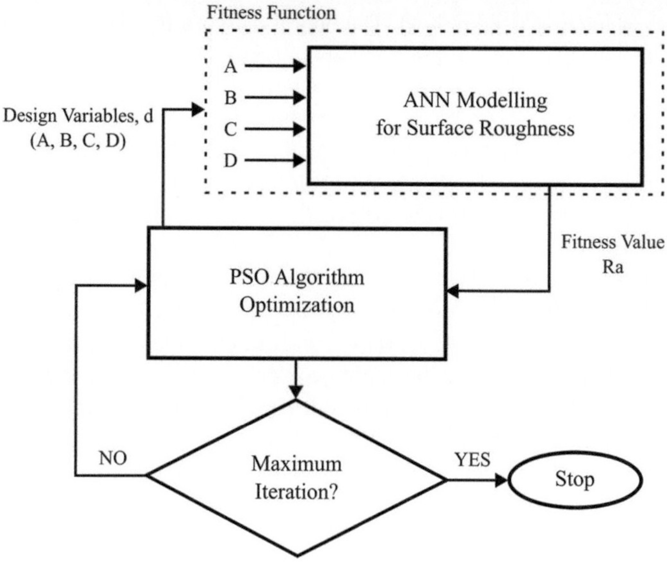

Fig. 4 PSO implementation on surface roughness in FDM

the new value of surface roughness is less than the current value, the new value will be chosen as the new global best (*gbest*) value. This process is repeated until the maximum number of iterations is reached.

4 Results and Discussion

The results of finding the best ANN model are shown using experimental data from a previous study (Saad et al. 2019). The experiment data are brought into the MATLAB workspace and normalized. The best ANN model is created by training, testing, and validating. Performance data such as RMSE, MAE, MAPE, and Regression Analysis are used to evaluate and validate the model to find the best ANN model. The data on the line pattern are also compared to what the ANN model predicted and what occurred. After determining which ANN model is best, PSO is used to advance it to the next stage of optimization. The best result is found at the end of optimization, which is the optimal value of FDM's input parameters for the smoothest surface.

The 46 data sets, which include input parameters and surface roughness, are listed in Table 6, and the values range between 0.1 and 0.9. Normalization of data made neural network model learning more flexible, and learning considers all input features to the same extent. The neural network is trained using the Levenberg–Marquardt backpropagation training function. The input layer contains four response parameters, whereas the output layer contains only one response, surface roughness. The network models were built using single and double hidden layers. First, the trial-and-error method is employed. The neural network models were generated at random in MATLAB with varying numbers of neurons in single and double hidden layers. Following that, the number of neurons was determined using a 5-variation ranging from 5 to 30 ($i = 5,10,15,20,25,30$ where $i =$ number of neurons).

There were six neural network model structures created for each under single and double hidden layers. The ANN model result that matches with the targeted value higher than 90% is filtered from the total of 12 neural network model structures, and the model is run three times more to get the optimal value. The 46 data sets were divided into three groups: 70% for training (from set 1 to set 32), 15% for validating (from set 33 to set 39), and 15% for testing (from set 40 to set 46). Tables 7 and 8 show the overall training, testing, and validating performance results of the single hidden layer neural network model and the double layer neural network model, respectively, along with the MSE value. Table 8 shows the overall training, validating, and testing performance error for a single-layer neural network. The smaller (approaching zero) the number of overall performance errors required to determine the best NN model. The 4-15-1 model is used in single hidden layer NN models because it has the lowest percentage of performance error among all single hidden layer NN models.

Specifically, the error is determined by applying the performance functions, such as RMSE, MAE, and MAPE. As mentioned before, the selection of the performance function is to reduce the error rate in the predicted value. So, to achieve the best NN model, the value should be at least as possible. Table 7 shows the results of the analysis in MATLAB, the value of overall MAE, MAPE, and RMSE for single hidden layer NN model and double hidden layer NN model, respectively. Besides,

Table 7 The final performance of selected ANN structure

ANN Structure	MAE	MAPE (%)	RMSE
4-5-1	7.6253	26.1996	0.1137
4-10-1	5.3576	21.0687	0.0932
4-15-1	4.0134	16.2959	0.0795
4-5-5-1	5.3028	17.2121	0.0748
4-10-10-1	4.8333	22.4300	0.1665
4-15-15-1	7.4295	25.1860	0.1136

Table 8 The final regression analysis of selected ANN structure

ANN Structure	Training	Testing	Validating	Overall
4-5-1	0.8931	0.6625	0.8692	0.8608
4-10-1	0.9489	0.9366	0.9927	0.9527
4-15-1	0.9500	0.9678	0.9890	0.9603
4-5-5-1	0.9623	0.8736	0.9878	0.9509
4-10-10-1	0.9623	0.9260	0.9930	0.9499
4-15-15-1	0.9622	0.8456	0.9933	0.9408

regression analysis was also conducted to increase the accuracy in determining the best NN model. Table 8 shows the results on the regression analysis in MATLAB, the correlation coefficient, R is presented. The higher the value of correlation coefficient, which means the higher the model accuracy. The determination of the best NN model is also based on the overall regression percentage analysis results shown in Table 8.

Figure 5 shows the regression plot of the 4-15-1 model's training, validation, testing, and overall data. Each regression plot's correlation coefficient, R, for training, validation, testing, and overall was 0.94999, 0.98903, 0.96776, and 0.96031, respectively. Instead, the results of neuron numbers 5, 10, 20, 25, and 30 in single hidden layer and double hidden layer, with single hidden layer and 15 number of neurons, demonstrated the best performance with the highest overall correlation coefficient, $R = 0.96031$ among the other models. Figure 6 depicts the plots of the predicted NN model data and the target experimental data, as well as the test data plot of 15 neuron numbers in a single hidden layer. The line pattern of the predicted NN model data and the target experimental data is nearly identical, with only a small variation in value, indicating a strong correlation between the target data from experiment and the predicted NN model.

5 PSO Optimization

The iteration numbers were kept constant at 100 because it was discovered that most runs converged between 40 and 80 iterations in the previous study. Because this study involves four input parameters, the parameter dimension (dim) has been set to four. Following that, C_1 is the acceleration factor for the cognitive parameter, pushing each particle to the personal best (*pbest*) position, and C_2 is the acceleration factor for the social parameter, pushing each particle to the global best (*gbest*) position. According to Chebrolu et al. (2023), the range of C_1 and C_2 values is between 0 and 2. C_1 and C_2 should have the same value because they affect the speed of particles to obtain the output response. As a result, C_1 and C_2 are both set to 1.5. Based on the analysis above, Table 9 tabulated the appropriate learning parameters that were set before running the simulation of the PSO algorithm.

The goal of PSO optimization implementation is to determine the optimal parameter setting in an FDM machine that results in the lowest value of surface roughness. As a result, the range of process parameters is important in this study because it provides upper and lower boundary values, as listed in Table 10. The result of PSO optimization in each process parameter and predicted surface roughness value is shown in Table 11. The process parameter results have been generated according to the range in Table 10 where the layer thickness, print speed, print temperature, and outer shell speed are 0.2674 mm, 19.6556 mm/s, 190.5087 °C, and 22.2931 mm/s, respectively. This optimal value of process parameter produced the minimum surface roughness value which is 2.0113 μm. Figure 7 shows the predicted surface roughness of the 4-15-1 NN structure.

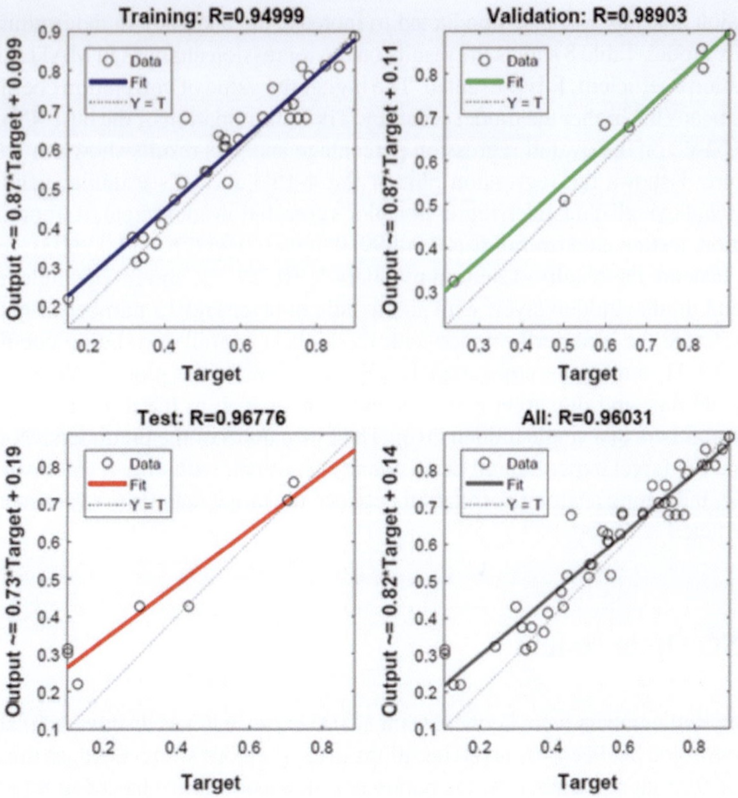

Fig. 5 Linear regression plot for NN model 4-15-1

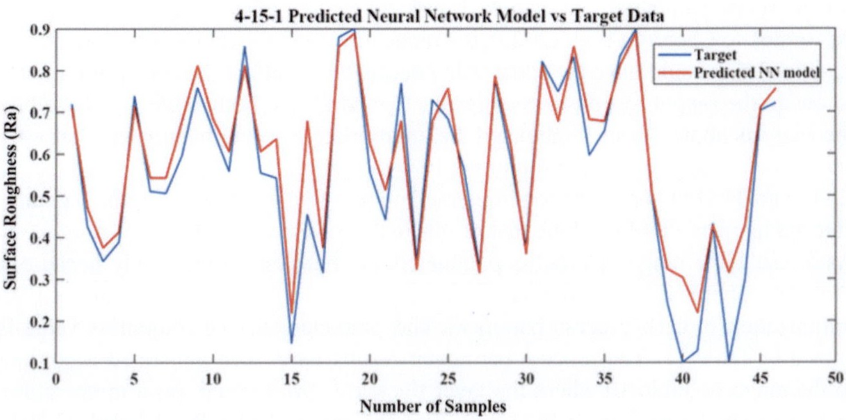

Fig. 6 Comparison plots of the predicted and the target experimental data

Table 9 The initial learning parameters of PSO

Subject of learning parameter	Parameter value
Number of particles (NP)	50
Number of iterations (particle step)	100
Dimension of parameter (dim)	4
Inertia weight (w)	1.5
Cognitive parameter ($\boldsymbol{C}_1$)	1.5
Social parameter ($\boldsymbol{C}_2$)	1.5

Table 10 The range of parameter setting

FDM parameters	Range value
Layer thickness (A)	0.10 mm $\leq$ A $\leq$ 0.38 mm
Print speed (B)	19.46 mm/s $\leq$ B $\leq$ 76.54 mm/s
Print temperature (C)	171.22 °C $\leq$ C $\leq$ 218.78 °C
Outer shell speed (D)	21.42 mm/s $\leq$ D $\leq$ 47.58 mm/s

Table 11 The results of PSO optimization

Process parameter	Optimal value
Layer thickness (mm)	0.2674
Print speed (mm/s)	19.66
Print temperature (°C)	190.51
Outer shell speed (mm/s)	22.29
Predicted surface roughness (μm)	2.0113

Figure 8 shows the performance of each parameter setting generated by PSO optimization. From 31 to 100 iterations, each parameter in this optimization continues to converge to an optimal solution. Figure 9 shows the PSO optimization convergence profile. This optimization yielded a minimum surface roughness value of 2.0113 μm, which remained converged at 31 until 100 iterations. Finally, a combination of layer thickness, print speed, print temperature, and outer shell speed results in a better surface finish, and the optimization process is successful when the PSO algorithm is used. The modelling and optimization process was investigated using the Artificial Neural Network (ANN) and Particle Swarm Optimization (PSO) algorithms. The combination of these two methods, dubbed integrated ANN-PSO, has a significant impact on the performance of FDM process parameters for minimizing surface roughness in 3D printed parts. The ANN modelling is carried out by selecting the best ANN model to be an objective function in the PSO algorithm, which aids in achieving the lowest possible surface roughness. Meanwhile, the PSO method is used to determine the optimization output, which is the optimal value of process parameters for minimizing surface roughness.

Fig. 7 The predicted surface roughness from 4-15-1 NN structure

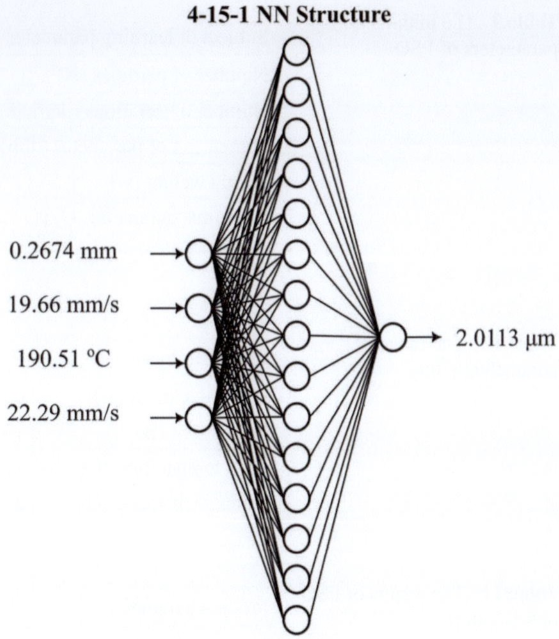

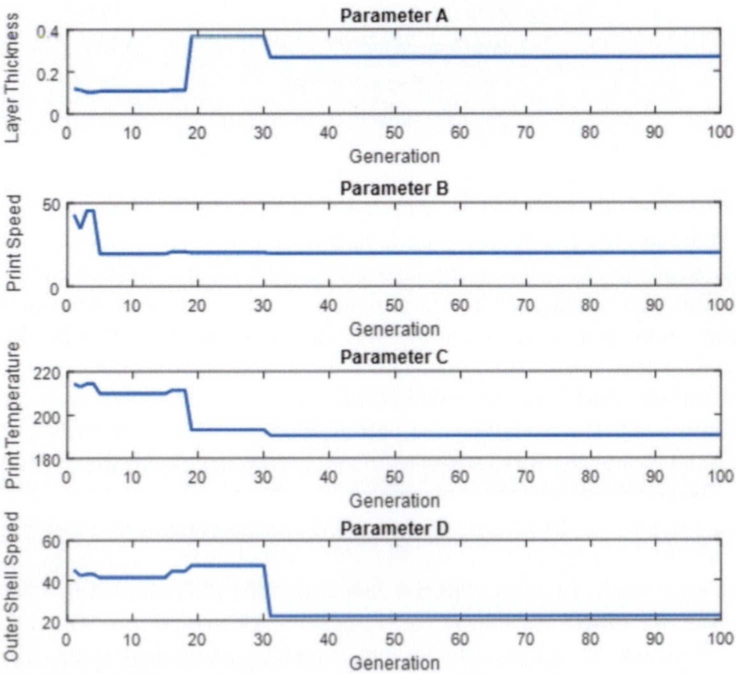

Fig. 8 FDM input parameters' convergence profile

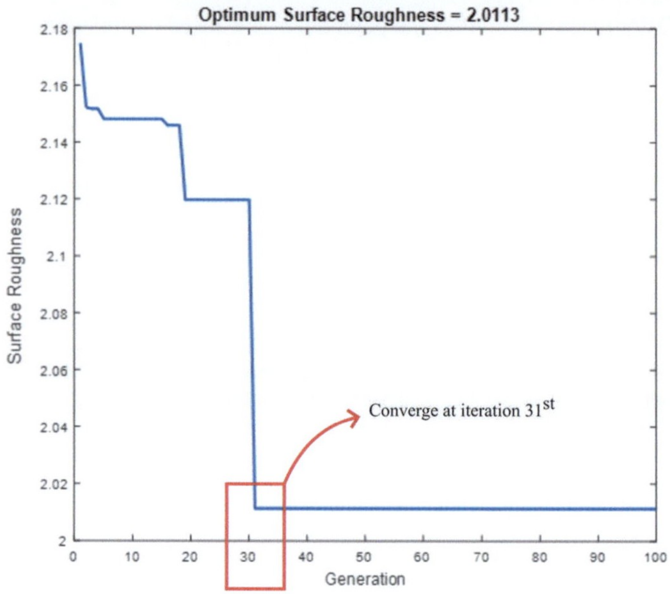

Fig. 9 Convergence profile for surface roughness

6 Conclusion

In conclusion, the use of Fused Deposition Modelling (FDM) in additive manufacturing (AM) has demonstrated enormous potential for creating intricate 3D geometries and structures. However, the issue of poor surface roughness caused by the layered manufacturing process has limited its direct application across a variety of domains, including sports equipment. To address this limitation, novel metaheuristic methodologies for modelling and optimizing FDM surface roughness have been investigated. A novel hybrid approach combining Artificial Neural Network (ANN) and Particle Swarm Optimization (PSO) was developed in this study. This method provides a comprehensive solution for practitioners looking to predict and improve FDM process parameters, resulting in better surface roughness in printed components. The ANN model was painstakingly trained, validated, and tested using experimental data. The PSO algorithm was used to fine-tune the ANN's optimal configuration, which included a single hidden layer with 15 neurons (4-15-1). The collaboration of ANN and PSO resulted in the identification of optimal parameter settings for FDM, which resulted in significant improvements in surface roughness. Layer height, print speed, print temperature, outer shell speed, and surface roughness were fine-tuned to 0.2674 mm, 19.66 mm/s, 190.51 °C, 22.29 mm/s, and 2.0113 μm, respectively. This ANN-PSO integration provides a powerful and effective strategy for improving the surface finish of FDM-printed parts, especially those catering to the intricate demands of sports equipment fabrication. The developed

ANN-PSO approach exemplifies the potential of advanced optimization methods in revolutionizing additive manufacturing by seamlessly blending cutting-edge computational techniques with real-world manufacturing challenges. As FDM continues to evolve and find applications in a variety of industries, this research paves the way for improved surface quality, broadening the scope of FDM technology's possibilities in producing high-performance components such as sports equipment and beyond.

References

Alafaghani A, Ablat MA, Abedi H, Qattawi A (2021) Modeling the influence of fused filament fabrication processing parameters on the mechanical properties of ABS parts. J Manuf Process 71:711–723

Chebrolu V, Koona R, Raju RSU (2023) Automated evaluation of surface roughness using machine vision based intelligent systems. J Sci Ind Res 82(1):11–25

Chen Y, Yuan Z, Chen B (2018) Process optimization with consideration of uncertainties—an overview. Chin J Chem Eng 26(8):1700–1706

Đurović S, Stanojković J, Lazarević D, Ćirković B, Lazarvić A, Džunić D, Šarkoćević Ž (2022) Modeling and prediction of surface roughness in the end milling process using multiple regression analysis and artificial neural network. Tribol Ind 44(3):540–549

Jaisingh Sheoran A, Kumar H (2020) Fused deposition modeling process parameters optimization and effect on mechanical properties and part quality: review and reflection on present research. Mater Today: Proc 21:1659–1672

Noriega A, Blanco D, Alvarez BJ, Garcia A (2013) Dimensional accuracy improvement of FDM square cross-section parts using artificial neural networks and an optimization algorithm. Int J Adv Manuf Technol 69(9–12):2301–2313

Raju M, Gupta MK, Bhanot N, Sharma VS (2018) A hybrid PSO–BFO evolutionary algorithm for optimization of fused deposition modelling process parameters. J Intell Manuf 30(7):2743–2758

Saad MS, Nor AM, Baharudin ME, Zakaria MZ, Aiman AF (2019) Optimization of surface roughness in FDM 3D printer using response surface methodology, particle swarm optimization, and symbiotic organism search algorithms. Int J Adv Manuf Technol 105(12):5121–5137

Sedghi M, Ahmadian A, Aliakbar-Golkar M (2016) Assessment of optimization algorithms capability in distribution network planning: review, comparison and modification techniques. Renew Sustain Energy Rev 66:415–434

Singh GR, Gupta MK, Mia M, Sharma VS (2018) Modeling and optimization of tool wear in MQL-assisted milling of Inconel 718 superalloy using evolutionary techniques. Int J Adv Manuf Technol 97(1–4):481–494

The use of machine learning in boron-based geopolymers 2019Bagheri A, Nazari A, Sanjayan J (2019) The use of machine learning in boron-based geopolymers: function approximation of compressive strength by ANN and GP. Meas: J Int Meas Confed 141:241–249

Vahabli E, Rahmati S (2016) Application of an RBF neural network for FDM parts' surface roughness prediction for enhancing surface quality. Int J Precis Eng Manuf 17(12):1589–1603

Yadav D, Mingareev I (2023) Utilizing ultrafast lasers for postprocessing to improve mechanical properties of 3D-printed parts. J Laser Appl 35

Yang L, Li S, Li Y, Yang M, Yuan Q (2019) Experimental investigations for optimizing the extrusion parameters on FDM PLA printed parts. J Mater Eng Perform 28(1):169–182

Zain AM, Haron H, Qasem SN, Sharif S (2012) Regression and ANN models for estimating minimum value of machining performance. Appl Math Model 36(4):1477–1492

Design and Development of an Automated Golf Scoring System

Ooi Teng Chow, Siti Azura Abuzar, Dayang Khadijah Hamzah, Aznor Hanah Abdul Halim, and Rohani Farook

Abstract Golf is a popular sport played by millions all around the world regardless of age and background. Due to the recent COVID-19 pandemic, the number of golf players has been rising steadily with both new and returning players. Therefore, this study will focus on the design and development of an automated scoring system for golf sports. The automated scoring system uses sensors for the detection of golf swing and golf ball hole entry. The data collected are then sent to a web application for score calculations before displaying the scores to the user. A sensor is attached to the golf club while two other sensors are placed in the golf hole to detect golf ball hole entry. All the sensors work together to measure the number of golf swings and indicate ball hole entry for different holes. A microcontroller is then used to send the collected data to a hosted web server using Wi-Fi, where the web application calculates and displays scores for the player. Golf players can focus on mastering their golf swing techniques while the system helps them record and calculate their scores.

Keywords Automated scoring system · Golf scoring sensors

1 Introduction

Today, golf is played with manual score calculation using a scorecard. There is a chance for the golfer to make errors while counting the number of strokes during play. Non-professional golf scorekeeping just requires accuracy while professional

O. T. Chow · S. A. Abuzar (✉) · D. K. Hamzah · A. H. A. Halim · R. Farook
Faculty of Electronic Engineering & Technology, Universiti Malaysia Perlis, Arau, Malaysia
e-mail: azuraabuzar@unimap.edu.my

O. T. Chow
e-mail: elvisooichow@gmail.com

S. A. Abuzar · D. K. Hamzah · A. H. A. Halim · R. Farook
Sports Engineering Research Centre, Universiti Malaysia Perlis, Arau, Malaysia

Z. Mohamed et al. (eds.), *Proceedings of the 2nd International Conference on Innovation and Technology in Sports, ICITS 2023, 27–28 November, Malaysia*, Lecture Notes in Bioengineering, https://doi.org/10.1007/978-981-97-3741-3_14

golf scorekeeping requires accuracy and integrity. Golf requires players to keep track of their own score and the opponent's score as they play (2023 Rules of Golf 2022). This fundamental skill can often divert a novice's attention and affect concentration during play. Thus, an automated scoring system that is user-friendly and safe to use can help new golf players learn scorekeeping techniques and focus on gameplay.

An automated scoring system for golf reduces the time taken to calculate the score for the golfers. The use of sensors for detecting input is also more reliable compared to golfers' memory. This system also is less likely to make errors like golfers would under the stress of the audience. With an automated scoring system, integrity can be enhanced as it is not easily manipulated through personal connections and biasing. By implementing this system in software, added security can be used to secure access for authorized users only to avoid tempering the scores without permission.

The first objective of the project is to build two prototypes, one for golf swing detection and another for golf ball hole entry detection. The next objective is to develop a web application for calculating and displaying user scores. The final objective is to verify the accuracy of the automated scoring system. A limitation of this project is that it only calculates golf scores for stroke play. Besides, the system can only identify up to four different golf holes. Lastly, the web application security depends on personal web browser configurations.

2 Methodology

2.1 Score Calculation

Score calculation in the automated scoring system requires three parameters for the web application: total strokes, total par, and handicap. Total strokes will be obtained from the web server. Players need to provide the total par that is usually given in the scorecard while the handicap relies on the average of the player's past golf games. For beginners to golf, the player can use the handicap provided in the scorecard of the golf course.

With this information fulfilled, the player's score can be calculated by using the Eq. 1:

$$\text{Effective Score} = (\text{Total Strokes} - \text{Total Par}) + \text{Handicap} \tag{1}$$

2.2 Project Design

Figure 1 shows the flowchart of the project which consists of sensor development and flow of the web server processing the data.

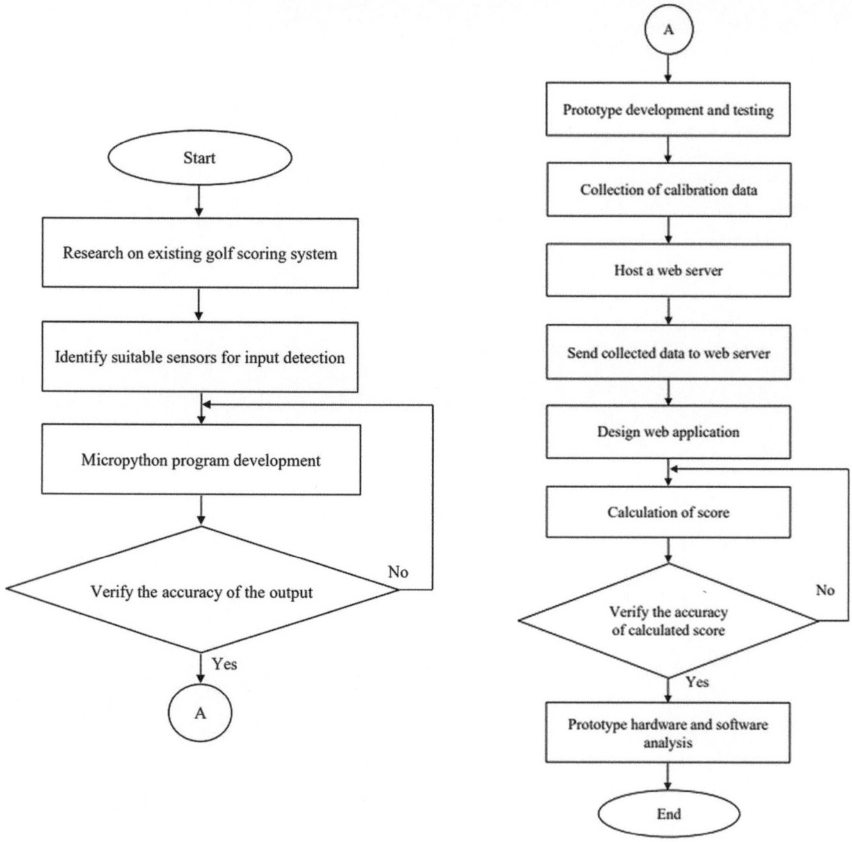

Fig. 1 Project flowchart

Figure 2 is the project block diagram where the Raspberry Pi Pico W has three input sensors, and the output of the processor is data being sent to the web server. Figure 3 shows the project flow diagram. Figure 3a shows the three types of sensors. First, is an inertial sensor, used for detecting golf swings, the middle one is a vibration sensor (Hsu et al. 2016; Burchfield and Venkatesan 2010), used for detecting golf ball hole entry, and the bottommost is a color sensor (Baird and Sparks 2023), used for differentiating the hole from each other. Figure 3b is the Raspberry Pi Pico W used as a microprocessor to execute programs to operate the three sensors (Peiro 2023). Besides, it also allows communication between golf swing detection and golf ball hole entry detection prototypes as they are interdependent. Figure 3c is the web application output that will calculate and display the final score of the player (Li and Xue 2014).

Figure 4a shows the components inside the Golf Swing Detection prototype. It contains an inertial sensor, a rechargeable battery, a Raspberry Pi Pico W, and a switch. The prototype is placed in a clear casing which protects its components from

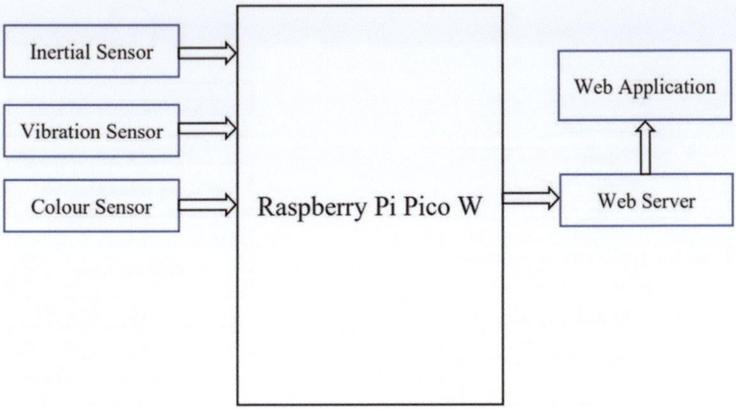

Fig. 2 Project block diagram

(a) (b) (c)

Fig. 3 Project flow diagram

external elements as depicted in Fig. 4b. The function of the switch is to switch modes between "On" and "Charging the battery". Figure 4c shows the attachment of the prototype to the golf club (King et al. 2008). The attachment is removable and non-slip. Besides, it also does not cause physical damage to the golf club.

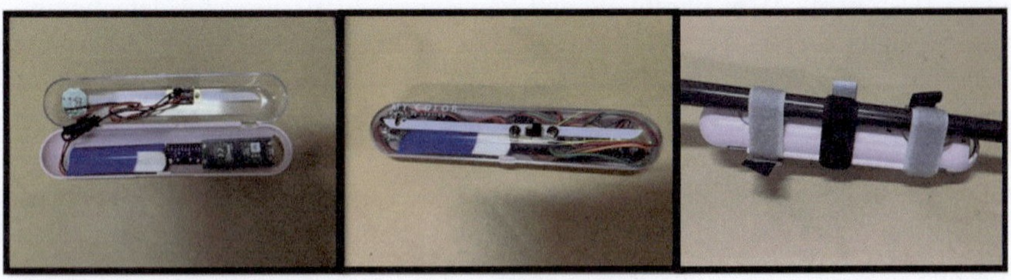

(a) (b) (c)

Fig. 4 Golf swing detection prototype

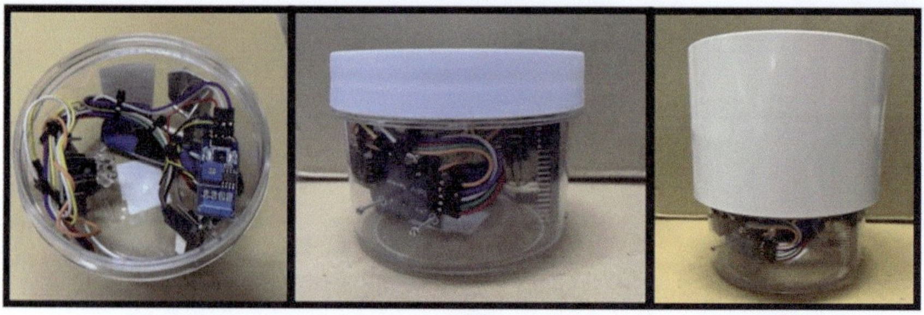

(a) (b) (c)

Fig. 5 Golf Ball Hole entry detection prototype

Figure 5a, b show the top and front view of the components inside the golf ball hole entry detection prototype. It contains a color sensor, a rechargeable battery, a Raspberry Pi Pico W, a switch, and a vibration sensor. Similarly, this prototype is also placed in a clear container to protect its components from external elements. The switch provides the same functionality to switch modes between "On" and "Charging the battery". Besides, Fig. 5c shows the prototype design according to the standard size as specified by golf rules. Furthermore, it would fit under the golf hole cup to allow easy field implementation.

3 Results and Discussion

This section covers the verification of the accuracy and functionality of the project. A golf player was invited to perform a verification test. To achieve this, the web server analyzes the input data detected by the sensors. From the analysis, the sensor inputs were accurate and the communication between the interdependent prototypes was verified.

To verify the accuracy of the scoring system, a simple setup was conducted as shown in Fig. 6 to test for the input detection during a shot putt. A shot putt was chosen for the test because golf players would perform a gentle hit to hole the ball. This gentle hit serves as a benchmark for the ability of the sensors to detect input accurately as compared to stronger golf swings.

When a golf swing was detected, the input data from the sensors were sent to the web server as displayed in Fig. 7a and the golf swing count was increased by one (MacHeso et al. 2021). When the ball was holed, the golf swing count was reset to zero as in Fig. 7b.

The data are then retrieved by the web application using an API. After the authorized log-in, the web application performs score calculations and displays the score to the user as shown in Fig. 8.

Fig. 6 Verification test setup

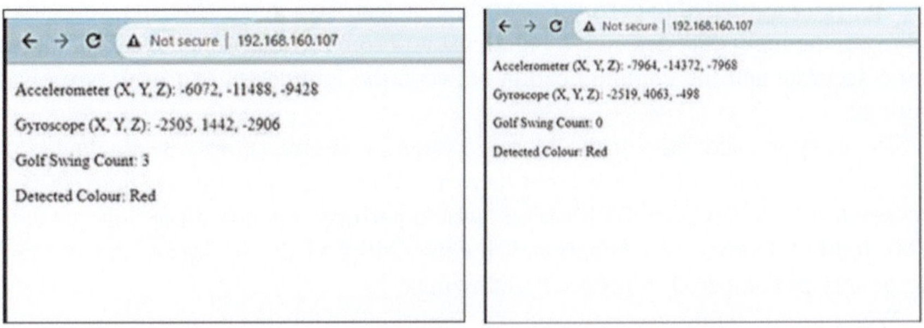

(a) (b)

Fig. 7 Display of input detected by sensors

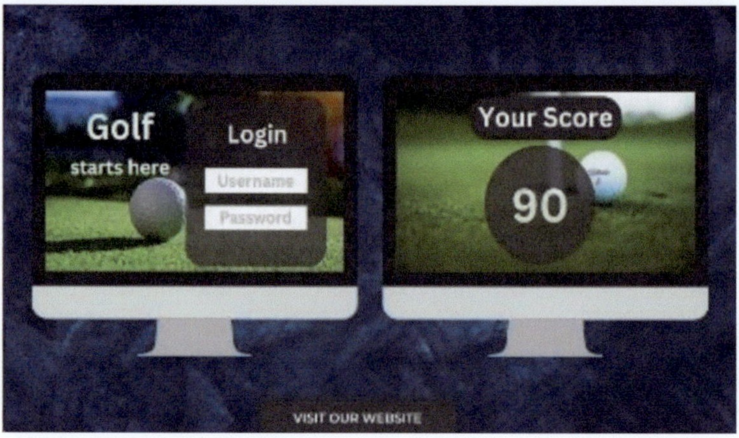

Fig. 8 Display of input detected by sensors

Figure 9 shows the project implementation on the golf field. The prototype of the golf hole is placed in the golf hole while the golf swing detection prototype is attached to the golf club. For the user interface implementation, it requires a device with internet access for accessing the web application. Besides, the device must be connected to the golf swing detection prototype via a LAN connection for data exchange between the sensors and the web application to occur. Display of the golf score can be accessed through the web application.

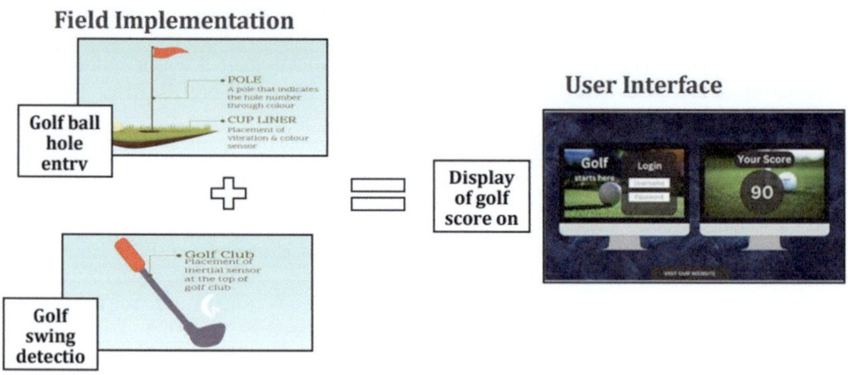

Fig. 9 Project implementation

4 Conclusion

In conclusion, the objectives of this project were achieved. Two prototypes were successfully developed, and the sensors can detect inputs. Through a web application, it can retrieve data from the sensors and calculate and then, display the player score. From the testing, the system has been verified to give accurate scores. Here are some suggestions for future work on this project. The prototype can be optimized for improved efficiency and design. The prototype can be made smaller with a user-friendly attachment. Moreover, this project can be integrated with other technologies. For example, this project can be integrated with a golf swing analysis device to provide more features. Finally, this project serves as a technological innovation in sports. In this case, this project bridges the study gap of automated scoring systems in golf sports.

References

Anvil | Anvil on the Pico W. https://anvil.works/pico#start-building-now. Accessed 29 Nov 2023

Baird SG, Sparks TD (2023) Building a 'Hello World' for self-driving labs: the closed-loop spectroscopy lab light-mixing demo. STAR Protoc 4(2):102329

Burchfield R, Venkatesan S (2010) A framework for golf training using low-cost inertial sensors. In: 2010 international conference on body sensor networks, BSN 2010, pp 267–272. https://doi.org/10.1109/BSN.2010.46

Hsu YL, Chen YT, Chou PH, Kou YC, Chen YC, Su HY (2016) Golf swing motion detection using an inertial-sensor-based portable instrument. In: 2016 IEEE international conference on consumer electronics-Taiwan, ICCE-TW 2016. Institute of Electrical and Electronics Engineers Inc. https://doi.org/10.1109/ICCE-TW.2016.7521016

King K, Yoon SW, Perkins NC, Najafi K (2008) Wireless MEMS inertial sensor system for golf swing dynamics. Sens Actuators Phys 141(2):619–630. https://doi.org/10.1016/J.SNA.2007.08.028

Li X, Xue Y (2014) A survey on server-side approaches to securing web applications. ACM Comput Surv (CSUR) 46(4). https://dl.acm.org/doi/10.1145/2541315

MacHeso P, Chisale S, Daka C, Dzupire N, Mlatho J, Mukanyirigira D (2021) Design of standalone asynchronous ESP32 web-server for temperature and humidity monitoring. In: 2021 7th international conference on advanced computing and communication systems, ICACCS 2021, pp 635–638

McAleer K. Build your own web server using a Raspberry Pi Pico W using Phew!. YouTube. https://www.youtube.com/watch?v=0sPPxIq4hg8&t=3317s. Accessed 29 Nov 2023

Peiro JAG (2023) An infinity of Pong: a Raspberry Pi Pico W handheld writes its own games. IEEE Spectr 60(2)

2023 Rules of Golf (2022) R&A Rules Limited and the United States Golf Association. https://assets-us-01.kc-usercontent.com/c42c7bf4-dca7-00ea-4f2e-373223f80f76/48712d47-76dc-4fd3-add1-53972c021580/2023%20Rules%20of%20Golf.pdf. Accessed 28 Nov 2023

A Study on the Perception of Insole Comfort During Running Using Kansei Engineering Method

Abdullah Akram Azman, Mohammad Azzeim Mat Jusoh, Hosni Hasan, and Anitawati Mohd Lokman

Abstract Proper selection of shoe insoles is critical, especially during athletic activities such as running, as it is directly in contact with the human foot. The purpose of this study is to identify suitable hardness properties based on users' perceptions of comfort during running, using the Kansei Engineering (KE) method. A semantic differential measurement was implemented to relate emotions and products. The KE Type 1 procedure was implemented in this study to translate human emotions into design attributes. In this study, seven pairs of the most relevant words were shortlisted in the form of semantically different Kansei words to relate to the subject's feelings. Five healthy men between the age of 20 and 25, with BMI between 18.5 and 25 kg/m^2, voluntarily participated in this study. The Crep Protect insoles were used as the main sample with three different sets of hardnesses: soft, medium, and hard. The Principal Component Analysis (PCA) and Factor Analysis (FA) were used for the statistical analysis. The result shows that the preferred hardness for an insole during running was the hard type. The *dense* and *flexible* keywords were shown to be the most appropriate words to describe insole comfort. Overall, the result supports that most participants prefer a higher-density insole to be used during running.

Keywords Kansei Engineering · Insole · Running · Principal component analysis · Factor analysis

A. A. Azman · M. A. M. Jusoh (✉)
School of Mechanical Engineering, College of Engineering, Universiti Teknologi MARA (UiTM), 40450 Shah Alam, Selangor, Malaysia
e-mail: m_azzeim@uitm.edu.my

H. Hasan
Faculty of Sports Science & Recreation (FSR), Universiti Teknologi MARA (UiTM), 40450 Shah Alam, Selangor, Malaysia

A. M. Lokman
School of Computing Sciences, College of Computing, Informatics and Media, Universiti Teknologi MARA (UiTM), 40450 Shah Alam, Selangor, Malaysia

Z. Mohamed et al. (eds.), *Proceedings of the 2nd International Conference on Innovation and Technology in Sports, ICITS 2023, 27–28 November, Malaysia*, Lecture Notes in Bioengineering, https://doi.org/10.1007/978-981-97-3741-3_15

1 Introduction

Running is a good example of a common activity that is performable by everyone with maximum health benefits. Humans can benefit greatly from running in many ways. It is also a good way to socialize and exercise. Selecting a comfortable insole design is essential to achieve full running potential and, at the same time provides good protection. The insoles have a wide range of uses, including providing insulation, odor control, and to relieve foot pain (Ribeiro and João 2022; Relph et al. 2022). Insoles would aid in reducing the additional stress and pressure placed on the feet. To get the most out of insoles, we need to know which insole design is best for us. In recent years, several studies have focused on the perception of the comfort of insoles during various activities, including marching, straight-line walking, turning, and standing for long periods of time (Lo et al. 2016, 2018; Shirvani et al. 2021; Anderson et al. 2020). As a key point of contact between the foot and shoes, insoles are strongly suggested for use during athletic activities and to treat a variety of foot problems. The purpose of this study is to find a suitable and appropriate insole hardness design based on users' perceptions of insole comfort during running by implementing the Kansei method during the process (Nagamachi 2016). A semantic differential measurement has been applied in Kansei engineering to address the relationships between emotions and products (Osgood et al. 1957). The relative correlations between the Kansei word variables and the product samples were determined using Principal Component Analysis (PCA) and Factor Analysis (FA).

1.1 Kansei Engineering

Kansei engineering, also known as emotional or affective engineering (Japanese: Kansei Kougaku), aims to develop or improve products and services by translating the psychological needs and desires of the customer into the field of product design. It was first proposed in the 1970s by Mitsuo Nagamachi. Kansei engineering parametrically connects a product's or service's attributes to the physical and psychological reactions a customer has to them. The psychological evaluation is also used as an impression approach to quantify Kansei, and as a result, products that reflect the intended emotion can be made. Kansei Engineering Type 1 is a popular approach within the Kansei Engineering framework. It involves gathering data through surveys or questionnaires to understand people's emotional reactions and perceptions of various product characteristics. Kansei engineering has proven to be an effective technology for minimizing the time and resources required for new product development and design (Nagamachi 2016; Ishihara and Ishihara 2011; Nagamachi and Lokman 2016).

1.2 Semantic Differential

Semantic difference is a measurement tool that has been frequently applied in user-centered design approaches to assess products' emotional value (Stevens 1957). The semantic differential technique was invented in the 1950s by psychologist Charles E. Osgood and has since been used in a variety of sectors, such as psychology, marketing, and user experience research. The semantic differential includes a set of bipolar adjectives or phrases that express various emotional or attitudinal aspects. The subject's perception of a product is quantified on a semantic difference scale to examine product semantics. These scales record the emotional aspects or semantic interpretations connected to the characteristics. Afterward, the gathered data is analyzed through statistical methods to determine the connection between the product features and users' emotional reactions. This data is valuable as guidelines in developing products that evoke the behavior and response of users (Osgood et al. 1957; Osgood and Suci 1955).

1.3 Principal Component Analysis (PCA)

In Kansei engineering, a large number of variables (emotional attributes) can be obtained from surveys or evaluations, which could lead to high-dimensional data. High-dimensional data can be challenging to interpret and visualize. PCA helps reduce the number of variables while preserving the most important information. PCA aims to transform the original correlated variables (emotional attributes) into a new set of uncorrelated variables called principal components. The first few principal components capture most of the data's variability. By analyzing these components, the key emotional dimensions that drive human perceptions and preferences can be identified. PCA enables the visualization of complex, high-dimensional data in a lower-dimensional space. This visualization allows us to see how different emotional attributes relate to each other and how respondents perceive products or experiences based on those attributes. Since PCA reduces the dimensionality of the data, it also acts as a data compression technique. It represents the data with fewer variables while maintaining a significant portion of the original information. This can be beneficial for simplifying models and computations without losing critical insights (Nagamachi 2016; Nagamachi et al. 2008; Mohd Lokman 2010; Yanai 1994).

2 Methodology

The Kansei engineering KE Type 1 procedure was applied to translate human emotions into design attributes, as shown in Fig. 1.

Fig. 1 Flowchart for experiment

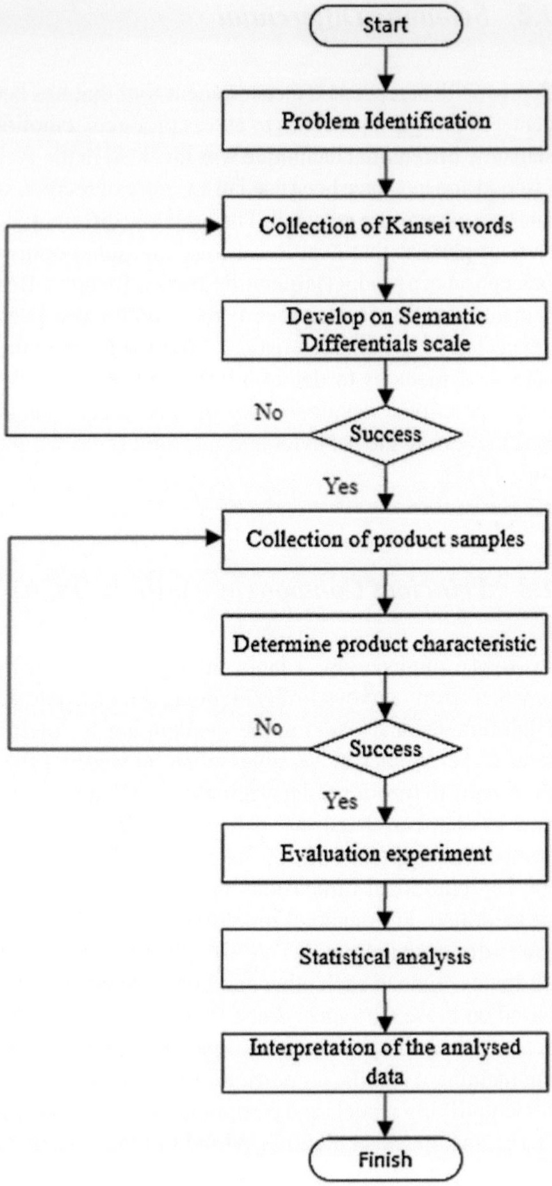

2.1 Kansei Words

The Kansei words were collected and chosen to represent emotions about the insole's hardness. Adjectives whose meaning and information related to the insole's hardness were used by people to express product characteristics were collected from various

Table 1 Selected words related to Kansei and physical state

Dense	Flexible	Light
Natural	Stable	Non-slippery
Prefer	Porous	Stiff
Heavy	Abnormal	Wobble
Slippery	Dismiss	

Table 2 7-point SD scale of selected keywords

Negative	−3	−2	−1	0	1	2	3	Positive
Porous								Dense
Stiff								Flexible
Heavy								Light
Abnormal								Natural
Wobble								Stable
Slippery								Non-slippery
Dismiss								Prefer

sources, including relevant literature, product advertisements, catalogs, magazines, product reviews, and user feelings (Ishihara et al. 2001a, 2008). This extensive collection aimed to create a comprehensive semantic description. A total of 43 adjective words were gathered. Following extensive brainstorming and discussions among the expert team, seven pairs of highly relevant words in the form of semantically different Kansei words were finally determined to improve participants' feelings during the Kansei word collection stage. Tables 1 and 2 depicted the collection of Kansei words and a 7-point SD Scale of Kansei words, respectively. In this study, the questionnaire incorporated the evaluation of Kansei words using the Semantic Differential (SD) scale, which utilized a 7-point scale as recommended by Nagamachi for ease of assessment (Nagamachi et al. 2008; Mohd Lokman 2010). Table 3 presents the scale used to identify the factors that influenced people's preferences regarding the hardness of insoles.

2.2 *Participants*

This study employed five healthy men (recreational runners) who voluntarily participated. They run 5–10 km per week, at an average pace of 6–8 min per km. The participant data and information were obtained before the experiment, as shown in Table 4. They were required to be free of foot pain at the time of testing and have no history of orthopedic or neurological conditions. Their body mass index (BMI)

Table 3 The scale of Preference using a 7-point SD scale

Score	Information
3	Sample is strongly related with the Kansei word in the right side of the scale
2	Sample is moderately related with the Kansei word in the right side of the scale
1	Sample is slightly related with the Kansei word in the right side of the scale
0	Sample is neutral, with equal association with the Kansei word on both the right and left side of the scale
−1	Sample is slightly related with the Kansei word in the left side of the scale
−2	Sample is moderately related with the Kansei word in the left side of the scale
−3	Sample is strongly related with the Kansei word in the left side of the scale

Table 4 Participant's information and data

Participant	Age	Height (cm)	Weight (kg)	BMI (kg/m^2)	Foot size (European)
1	21	173	71	23.7	42
2	21	168	64	22.7	42
3	24	182	66	19.9	42
4	24	172	73	24.7	41
5	25	169	70	24.5	42

ranged from 18.5 to 25 kg/m^2, and their ages ranged from 20 to 25. Their respective European sizes for their feet are between 41 and 42.

2.3 *Product Samples*

For this study, three types of insoles from Crep Protect were utilized as a sample, with varying levels of hardness: soft, medium, and hard, as indicated in Table 5. The hardness of the insoles was tested before the experiment using a shore durometer at three different points, and they were then customized to fit the participants' foot sizes. The hardness of the insoles was assessed using test procedures compliant with ASTM D2240 (American Society for Testing and Materials 2016). Throughout the tests, the same pair of N-brand Pegasus-38 running shoes (weight: 328 g, midsole: 10 mm heel drop), commonly used in running environments, was provided, and used, as shown in Fig. 2. To ensure the data's reliability, all participants were given identical pairs of running socks.

Table 5 Insoles sample types

No	Type	Hardness	Description
1	Crep Protect PORON Comfort 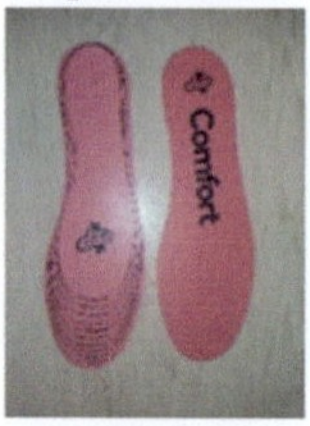 	34A (Soft)	• Providing reliable comfort and airflow to enhance daily performance • Allowing for better ventilation and temperature control through open-cell technology • Outstanding ability to absorb impacts, reducing or redirecting external forces from affecting the foot • Maintaining its original size, shape, and snug fit for a prolonged period
2	Crep Protect XRD Technology 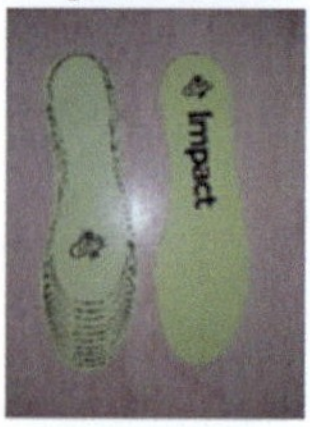 	52A (Medium)	• Protects against impacts by dispersing forces effectively • Uses breathable open-cell technology to improve ventilation and keep you dry • Incorporates a flexible and adaptable material that quickly diffuses force upon contact • Has the ability to absorb a substantial amount, up to 90%, of energy during high-speed impacts • Demonstrates a distinctive characteristic of solidifying upon strong impact, creating a comfortable and protective barrier
3	Crep Protect PORON Vive 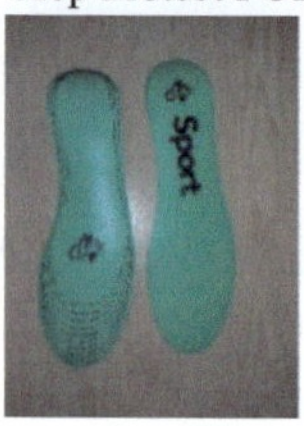 	75A (Hard)	• Provide both shock absorption and energy return in one technology • Use dynamic open-cell technology that reacts instantly to each step • Achieve a ball rebound rate of over 55%, enabling powerful push-offs and quick, responsive movements • Minimize or redirect forces away from the foot and create a dedicated "lift off" zone for maximum propulsion • Maintain consistent resilience and cushioning to support an energy-efficient stride and combat fatigue

2.4 Evaluation Experiment

A running test was conducted at the Taman Tasik Shah Alam running trail to investigate and evaluate the sample, as shown in Fig. 3. Participants ran for 2 min, covering approximately 200 m on the running trail using the original insoles, with a 1 min

Fig. 2 N-brand Pegasus-38 running shoe

rest interval while answering the provided questionnaire. The first and last running tests served as the baseline, using the participants' original insoles. Subsequently, participants continued running for 2 min on the trail for each provided insole (Crep Protect). They received a 1 min rest interval for every 2 min of running, during which they completed the provided form. To eliminate bias, the participants' vision was obstructed by a blindfold while putting on the running shoes, ensuring they remained unaware of the specific insole they were wearing. Throughout the test, participants wore the provided socks. Detailed explanations of the procedures, including how to complete the questionnaire and the meaning of Kansei words, were provided to all participants before the evaluation.

The data were obtained for all Kansei words collected from the insoles during the experiment, and the mean values were determined. The participants' original insoles for the first and last running tests were labeled as A and E, respectively. The soft, medium, and hard insoles were labeled as B, C, and D, respectively, as shown in Table 6. Principal Component Analysis (PCA) and Factor Analysis (FA) were used to determine the relative correlations between Kansei word variables and the product samples (Nagamachi 2016; Osgood and Suci 1955).

3 Result and Discussion

KE Type 1 was used to determine participant preferences for the hardness of the product samples. Data from the experiment were collected, and the overall mean value for all insoles is presented in Table 7, based on the positive Kansei words

Fig. 3 Taman Tasik Shah Alam running trail

Table 6 Insole labeling

Insole	Description
A	Participant's original insole for the first running test
B	Crep Protect PORON Comfort (Soft Insoles)
C	Crep Protect XRD Technology (Medim Insoles)
D	Crep Protect PORON Vive (Hard Insoles)
E	Participant's original insole for the last running test

(Nagamachi 2016; Nagamachi and Lokman 2016). Figure 4 illustrates the tabulated graph displaying the average values of the insole samples for all the semantics.

Table 7 Participants' mean values for all insole samples

Insole	Positive Kansei Word						
	Dense	Flexible	Light	Natural	Stable	Non-slippery	Prefer
A	0	0.2	−0.8	0.2	0.2	−0.2	0.4
B	0.4	1.4	1.4	1.6	1.4	1.6	0.6
C	1.6	1.8	1.2	1.8	1.2	2	1.4
D	2.2	2	1.8	1.6	1.6	1.4	1.6
E	0.2	−0.4	−0.2	0.2	0.4	0.4	0.4

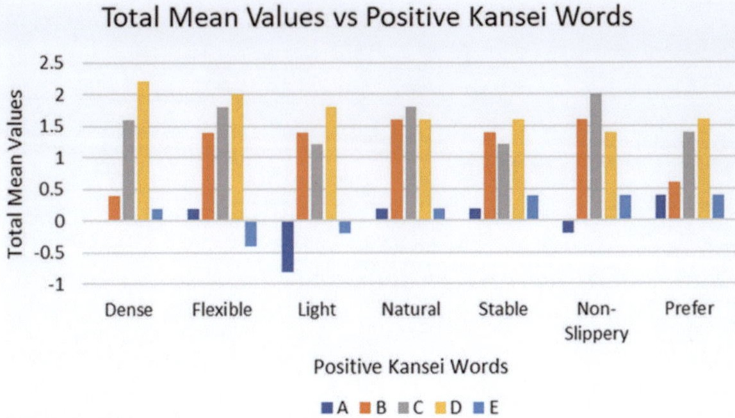

Fig. 4 Graph for participants' mean values

3.1 *Principal Component Analysis (PCA)*

The average values collected for each insole related to the Kansei words were statistically analyzed. This analysis involved the application of Principal Component Analysis (PCA), a mathematical technique used to identify data patterns and correlations (Nagamachi et al. 2008; Mohd Lokman 2010). PCA helped simplify the data by identifying a smaller set of variables known as principal components. In this study, PCA was used to define the underlying structure of emotions. To initiate the analysis, the participants' average scores were calculated using the XLSTAT software. Subsequently, as presented in Table 8, construct factors referred to as principal components (PC) were derived. These principal components represented the key dimensions that captured the variation and relationships within the dataset (American Society for Testing and Materials 2016).

Table 8 displays the eigenvalue scores, variances, and variability for each of the four construct factors: PC1 through PC4. Specifically, the eigenvalue scores for PC1 and PC2 were noted as 3.883 and 0.323, respectively, with corresponding levels of variability amounting to 88.501% and 7.353. The cumulative variance row indicated that PC1 and PC2 accounted for a total variation of 95.854%. This indicated that these two principal components played a significant role in the data analysis and had a substantial influence on the emotions being studied (Ishihara et al. 2001b).

Table 8 Principal component score

	PC1	PC2	PC3	PC4
Eigenvalue	3.883	0.323	0.097	0.085
Variability (%)	88.501	7.353	2.211	1.935
Cumulative %	88.501	95.854	98.065	100.000

Conversely, PC3 ~ 7 exhibited low eigenvalue scores and contributed to a relatively small percentage of variability. As a result, these principal components could be disregarded, and the focus could be directed toward the analysis of PC1 and PC2. By doing so, more than 90% of the total variance was adequately represented, making PC1 and PC2 the primary components of interest in the study.

(i) The Principal Component Loading

In this study, the analysis incorporated Principal Component Loading to investigate the distribution of emotions along the axis and aimed to define the concept of emotion and its influence on the samples. Figure 5 depicts the results of the Principal Component Analysis (PCA) for PC1 and PC2, where the emotions were distributed along two axes, namely the x-axis and the y-axis. On the x-axis, which only extended to the positive value side, the emotions associated with this axis included *dense, prefer, flexible, natural, non-slippery, light*, and *stable*. Subjectively, this axis was referred to as the attractive concept. The concept of emotion did not have a fixed naming convention, so words representing all the emotions observed in the data were encompassed within the concept name. Moving to the y-axis, which represented contrasting emotions, the upper side corresponded to the positive zone, comprising the emotions *dense, prefer*, and *flexible*. Conversely, the lower side represented the negative zone and included the emotions *light, stable, natural*, and *non-slippery*. Subjectively, the y-axis was referred to as adaptable, while the lower side, was associated with the concept of *balance*. By examining the distribution of emotions along these axes, the study gained insights into the relationship between emotions and the analyzed samples, helping to define and understand the emotional aspects within the dataset.

(ii) The Principal Component Score

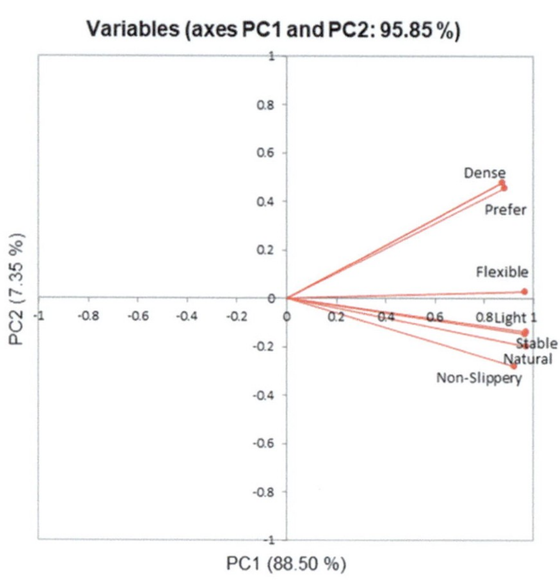

Fig. 5 Result of principal component loading

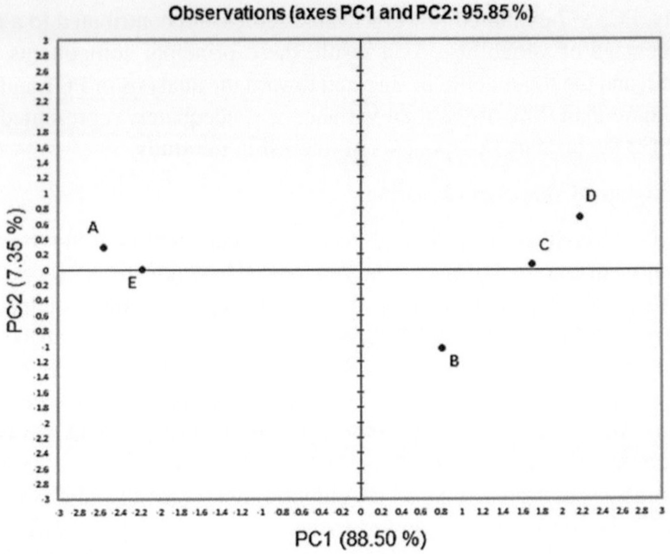

Fig. 6 Result of principal component score

Once the concepts of emotion had been established, the subsequent step involved performing Principal Component Scores. The purpose of PC Scores was to investigate the relationship between emotion and the samples as well as observe how the samples were distributed along the axis based on the concepts of *attractive, adaptable*, and *balance.* Figure 6 illustrates the distribution of insole samples on the axis, representing different conditions such as soft, medium, hard, and two of the participants' original insoles. This visual provided insight into how each sample aligned with the established concepts and allowed for a comprehensive understanding of their positioning concerning *attractive, adaptable*, and *balance.*

Samples A and E, located on the left side of the x-axis and the upper side of the y-axis, implied that participants perceived their respective insoles as unattractive and adaptable. Moving on to sample B, which represented a soft insole positioned on the right side of the x-axis and the lower side of the y-axis, it suggested that this soft insole was perceived as attractive but also well-balanced. In this context, *balance* indicated that the insole was *light, stable, natural*, and *non-slippery.* Notably, the lower hardness of the sample contributed to a feeling of lightness and naturalness compared to other insoles. As for sample C, a medium insole positioned on the right side of the x-axis and the upper side of the y-axis, it conveyed that this soft insole sample was both attractive and adaptable. The concept of adaptability in this study signified that the insole was *dense, prefer*, and *flexible.* Moving to sample D, which shared the same axis position as sample B (right side of the x-axis and upper side of the y-axis), it carried a higher value compared to sample C, indicating a more favorable perception.

(iii) The Principal Component Vector

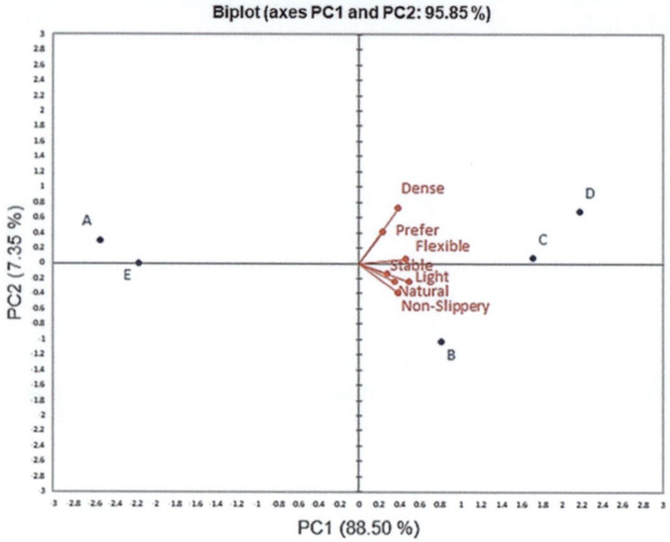

Fig. 7 Result of principal component vector

PC Vector is a valuable tool used to visualize the direction and intensity of emotions within the structure of emotion, aiding in the identification of Kansei areas. Figure 7 represents the outcomes of the PC vector analysis, presenting the distribution of emotions and samples. From this figure, it became apparent that samples A and E were situated farther away from the emotional cluster. This indicated that participants perceived their respective insoles as less comfortable when compared to the provided samples. On the other hand, the positive zone on the PC vector chart highlighted the Kansei area where Sample D (hard) received the highest preferences. Within this area, emotions such as *dense, prefer*, and *flexible* were closely associated. These findings held considerable significance as they assisted in determining the optimal level of hardness for insoles that aligned with participant preferences. The analysis and interpretation of the PC Vector results played a crucial role in identifying the emotional aspects related to insole comfort, allowing for informed decision-making in designing insoles that effectively catered to participant preferences.

3.2 Factor Analysis

Factor Analysis was employed to enhance and categorize the outcomes obtained from Principal Component Analysis. Varimax Rotation was applied within Factor Analysis to produce a precise outcome. The findings of Factor Analysis after Varimax Rotation are presented in Table 9.

Table 9 Factor analysis result after varimax rotation

	F1	F2	F3
Contribution (%)	85.867	5.385	0.775
Cumulative %	85.867	91.252	92.027

Table 9 shows that Factor 1 had the highest contribution level, while Factor 2 had a lower score compared to Factor 1. On the other hand, Factor 3 had the lowest score in terms of contribution level. Consequently, Factors 1 and 2 could be considered dominant factors in assessing emotional parameters, while Factor 3 could be disregarded due to its insignificant contribution level. When considering the cumulative percentage, Factors 1 and 2 together accounted for 91.252% of the total contribution. These factors were analyzed to determine the coefficient of emotion and generate various scores for each of the seven emotions outlined in Table 10.

The correlation scores provided in Table 10 offered valuable insights into the impact of emotions on the factors investigated in this study. To assess the significance of each emotion as a contributing factor, positive values (>0) were taken into consideration and highlighted in bold. In Factor 1, all emotions exhibited positive values (>0), namely *dense, prefer, flexible, natural, non-slippery, light*, and *stable.* These emotions were subjectively categorized under the concept of *attractiveness.* It is important to note that the naming of emotional concepts in this study did not adhere to a specific convention; instead, it encompassed words that represented the range of emotions observed in the analysis. Shifting the focus to Factor 2, three emotions demonstrated scores greater than 0: dense, like, and flexible. These emotions were subjectively grouped under the concept of *adaptability*. Similarly, in Factor 3, three emotions with scores higher than 0 were identified: *dense*, *prefer*, and *stable.* These emotions were subjectively grouped under the concept of *balance.*

Based on these findings, it can be inferred that Principal Component Analysis and Factor Analysis produce comparable results. The three emotion concepts derived, namely *attractiveness*, *adaptability*, and *balance*, remained consistent across both methods. The distinction arose in the third factor, which contributed to a lesser extent. As mentioned earlier, the purpose of Factor Analysis was to enhance the outcomes obtained from Principal Component Analysis by generating precise emotion coefficient scores for each emotion.

Table 10 Factor and emotion correlation score

	F1	F2	F3
Dense	**0.348**	**0.578**	**0.125**
Flexible	**0.393**	**0.070**	−0.444
Light	**0.394**	−0.237	**0.554**
Natural	**0.395**	−0.305	−0.480
Stable	**0.391**	−0.250	**0.459**
Non-slippery	**0.368**	−0.334	−0.197
Prefer	**0.354**	**0.581**	−0.013

4 Conclusion and Recommendation

In this study, a suitable and appropriate insole hardness design based on users' perceptions of insole comfort during running was investigated. The Kansei method was employed to explore the relationships between emotions and products. Kansei engineering Type 1 process steps were applied to translate human emotions into design attributes. Principle Component Analysis (PCA) and Factor Analysis (FA) were utilized as statistical methods to analyze the results. Based on the result, the hard type of insoles was the most preferred version, especially during running. The words *dense* and *flexible* were associated with preference and were determined to be the appropriate descriptors of insole comfort. It was observed that participants preferred using higher-density insoles to achieve greater flexibility during running. For future analysis, other types of multivariate statistical methods, for example, Regression Analysis or Cluster Analysis could be considered to interpret the data into an element design.

Overall, this study on insole design was solely focused on KE Type 1, which is based on human sensory perception. In future, other Kansei Engineering types such as the KE Type 2 (KE Computer Aided System) and KE Type 3 (KE Modeling) could also be used to determine users' perceptions of insole comfort. For example, by adapting the KE Type 2 method during the physiological examination, wearable equipment could be used to assess physiological responses such as heart rate, skin conductance, and muscle tension. By objectively assessing these reactions, the data produced shall be more precise, and any biases caused by subjective judgments can be eliminated. The KE Type 3 mathematical models could provide an integrated approach to insole design, by optimizing all factors to provide the most comfortable and functional insoles. By incorporating these suggestions, future studies will be able to contribute to the field of insole design, ultimately by improving user satisfaction and overall product quality.

References

American Society for Testing and Materials (2016) Standard test method for rubber property—Durometer hardness. ASTM. https://doi.org/10.1520/D2240-15

Anderson J, Williams AE, Nester C (2020) Development and evaluation of a dual density insole for people standing for long periods of time at work. J Foot Ankle Res 13:1–13. https://doi.org/10.1186/s13047-020-00402-2

Ishihara S, Ishihara K, Nagamachi M, Nishino T, Komatsu K (2001a) An analysis of non-linearity characteristics on Kansei data. Syst Soc Int Des Aspects Human-Comput Interact 2(2):320

Ishihara S, Ishihara K, Nagamachi M, Nishino T (2001b) Mathematical modeling of nonlinearity on form ratio Kansei evaluation data. In: Proceedings of the fourth international quality management and organizational development conference, Linköping, Sweden, pp 536–542

Ishihara S, Ishihara K, Nagamachi M, Sano M, Fujiwara Y, Naito M, Ozaki K (2008) Developments of home electric appliances with Kansei ergonomics: Sanyo cases: Kansei and kinematic considerations on washer-dryer and electric shaver. In: 11th QMOD conference, quality management and organizational development, attaining sustainability from organizational excellence to sustainable excellence, Helsingborg

Ishihara S, Ishihara K (2011) Psychological methods of Kansei engineering. In: Nagamachi M (ed) Kansei/affective engineering. CRC Press, Taylor and Francis Group, Boca Raton, FL, USA, pp 31–38

Lo WT, Wong DP, Yick KL, Ng SP, Yip J (2016) Effects of custom-made textile insoles on plantar pressure distribution and lower limb EMG activity during turning. J Foot Ankle Res 9(1):1–13. https://doi.org/10.1186/s13047-016-0154-5

Lo WT, Wong DP, Yick KL, Ng SP, Yip J (2018) The biomechanical effects and perceived comfort of textile-fabricated insoles during straight line walking. Prosthet Orthot Int 42(2):153–162. https://doi.org/10.1177/0309364617696084

Mohd Lokman A (2010) Design & emotion: the Kansei engineering methodology. Malays J Comput (MJOC) 1(1):1–14

Nagamachi M, Tachikawa M, Imanishi N, Ishizawa T, Yano S (2008) A successful statistical procedure on Kansei engineering products. In: Electronic conference proceedings

Nagamachi M (ed) (2016) Kansei/affective engineering. CRC press

Nagamachi M, Lokman AM (2016) Innovations of Kansei engineering. CRC Press

Osgood CE, Suci GJ (1955) Factor analysis of meaning. J Exp Psychol 50(5):325. https://doi.org/10.1037/h0043965

Osgood CE, Suci GJ, Tannenbaum PH (1957) The measurement of meaning, no 47. University of Illinois press

Relph N, Greaves H, Armstrong R, Prior TD, Spencer S, Griffiths IB, Dey P, Langley B (2022) Running shoes for preventing lower limb running injuries in adults. Cochrane Database Syst Rev (8). https://doi.org/10.1002/14651858.CD013368.pub2

Ribeiro AP, João SMA (2022) The effect of short and long-term therapeutic treatment with insoles and shoes on pain, function, and plantar load parameters of women with plantar fasciitis: a randomized controlled trial. Medicina 58(11):1546. https://doi.org/10.3390/medicina58111546

Shirvani H, Ghasemi MH, Shamsoddini A, Barati K (2021) Immediate effects of using insoles with various wedges on plantar pressure measurements and comfort level during marching. https://doi.org/10.21203/rs.3.rs-258589/v1

Stevens SS (1957) On the psychophysical law. Psychol Rev 64(3):153

Yanai H (1994) Multivariate data analysis method: theory and application. In: Asakura Bookstore, pp 77–99

Malaysia Super League Match Results Prediction with Football Rating System and Machine Learning Algorithms

Muhammad Nazim Razali and Aida Mustapha

Abstract This paper explores the application of football rating system and machine learning algorithms to predict match results in the Malaysia Super League (MSL). The dataset spans eight seasons of MSL, from 2015 to 2022, encompassing variables such as match round, match date, home and away team names, goals scored by each team, goal difference, and match outcomes (win, draw, lose). The dataset from seven seasons of MSL 2015 until MSL 2021 serves as a training set to predict match outcomes for the MSL 2022 season. The study implements a novel Football Rating System comprising two components: Elo rating and pi-rating. The Elo rating and pi-rating systems are utilized to evaluate team strengths and performance in terms of rating across the league. Machine learning algorithms, including Naïve Bayes (NB), Logistic Regression (LR), Support Vector Machine (SVM), and Random Forest (RF), are employed to predict match outcomes in terms of probability of win, draw, and lose. The performance of these algorithms is assessed using the Rank Probability Score (RPS) and Accuracy (ACC) metrics. A lower RPS indicates superior performance, while higher ACC values signify better predictive accuracy. The findings highlight that the combination of Elo rating with Logistic Regression (ELO + LR) achieves the most favorable outcome, attaining the lowest RPS score of 0.1979 in comparison to the other models. Additionally, Elo Rating with Random Forest (ELO + RF) exhibits the highest accuracy at 56.06%, albeit with the third highest RPS value. This underscores the intricate relationship between ACC and the RPS, suggesting the significance of a balanced evaluation of predictive models. In summary, this research highlights the possibilities of utilizing machine learning algorithms and an advanced football rating system for predicting match outcomes in the Malaysia Super League. The research contributes to the knowledge of predictive modeling in sports and offers valuable insights aimed at improving the accuracy of outcome predictions.

M. N. Razali
Faculty of Computer Science and Information Technology, Universiti Tun Hussein Onn Malaysia, Parit Raja, Malaysia

A. Mustapha (✉)
Faculty of Applied Sciences and Technology, Universiti Tun Hussein Onn Malaysia, Parit Raja, Malaysia
e-mail: aidam@uthm.edu.my

Z. Mohamed et al. (eds.), *Proceedings of the 2nd International Conference on Innovation and Technology in Sports, ICITS 2023, 27–28 November, Malaysia*, Lecture Notes in Bioengineering, https://doi.org/10.1007/978-981-97-3741-3_16

Keywords Malaysia Super League · Prediction · Pi-rating · Elo rating · Machine learning

1 Introduction

The use of machine learning techniques in sports analytics has gained significant attention in recent years, presenting new avenues for predicting match outcomes with heightened accuracy. In 2022, Bunker and Susnjak (2022) have reviewed studies and research related to sports results prediction using machine learning from 1996 to 2019 in a total of 31 papers with covered multiple sports and almost half the papers covered in association football (British English) or also known as soccer (American English). Predicting football matches' outcomes always becomes a hot topic among experts, pundits, and fans but it is very challenging due to the variation of potential variables and factors that can influence the matches' results.

Past literature related to football prediction commonly focuses on prediction in a specific tournament or league using various predictive modeling techniques with exception in the work of Hubacek et al. (2019), Constantinou (2019a, b), and Berrar et al. (2019) as they participate the Open International Soccer Prediction Challenge which contains over 216000 matches from 52 leagues and 35 countries as dataset (Dubitzky et al. 2019). Constantinou (2019a, b) categorized football prediction models into three clusters: statistical models, machine learning and probabilistic graphical models, and rating systems. Remarkably, the challenge's top performers achieved success by combining two specific predictive modeling approaches: machine learning and probabilistic graphical models along with rating systems. Figure 1 shows the classification of football prediction models.

This paper delves into the realm of the Malaysia Super League (MSL) by harnessing the power of both machine learning algorithms and a meticulously developed football rating system. As a result, this paper attempts to present football match outcomes prediction models based on the football rating system (Elo rating and pi-rating) and machine learning algorithms utilizing seven seasons (2015–2021) as

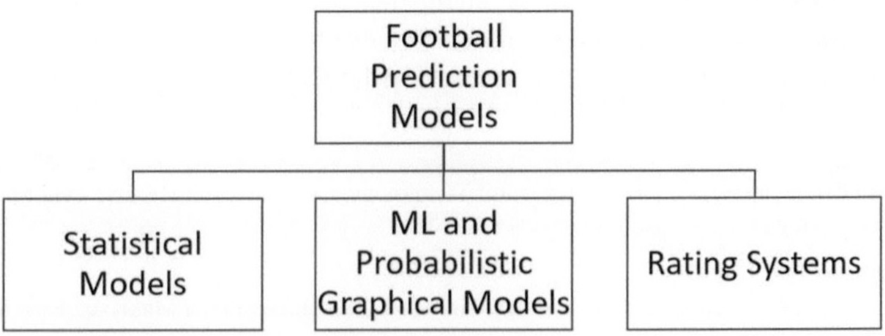

Fig. 1 A classification of football prediction models

a learning set, subsequently predicting the football match outcomes for the MSL season 2022 as the testing set. The remaining paper is organized as follows: Sect. 2 presents a literature review on football rating system; Sect. 3 describes the research methodology including the data collection and pre-processing, football rating system (Elo rating and pi-rating), Machine Learning (ML) algorithms and performance evaluation; Sect. 4 is about results and discussion; Finally, the conclusion in Sect. 5.

2 Literature Review

The application of rating systems in predicting football match outcomes has been a prominent approach in sports analytics. These systems assess team strengths by quantifying specific data, then comparing values between competing teams to estimate win and loss probabilities. Various rating systems have been proposed, including Elo rating, FIFA ranking, network-based ranking, Markovian rating, and Power Rank (Lasek et al. 2013; Lasek 2016). Notably, this section examines two notable rating systems from existing literature, namely the Elo rating, and pi-rating.

Originating in chess, the Elo rating model gauges the relative skill levels of players in one-on-one competitions. Next, the usage of Elo rating has since been extended to predict results for other sports including the association football by deriving the teams' strength based on their ranking. Runyan (2023) extended and refined the Elo rating in 1997 to suit the dynamics of association football. Interestingly, Leitner et al. (2010) and Hvattum and Arntzen (2010) applied Elo models differently based on tournament type (international vs. domestic leagues). Leitner et al. (2010) drew Elo ratings from various competitions, while Hvattum and Arntzen (2010) concentrated on top divisions in the English league system. In a subsequent study, Arntzen and Hvattum (2021) expanded their research by integrating player ratings into their football prediction model. This enhanced model demonstrated significantly improved performance compared to a prediction model solely relying on one type of rating. Meanwhile, Robberechts and Davis (2019) conducted a comparative analysis, using an ordered logit regression and a bivariate Poisson model, to predict outcomes for both tournament (FIFA World Cup) and domestic league football matches. They compared the effectiveness of result-based Elo ratings and goal-based Offense Defense Model ratings in the context of the Open International Soccer Prediction Database Challenge. As a result, the result-based Elo rating with ordered logit regression outperforms all prediction models in literature.

In 2013, Constantinou and Fenton introduced the innovative football rating system known as pi-rating. The proposal behind pi-rating incorporates three crucial football aspects often considered by experts: home advantage, the significance of recent performance over outdated data, and the relative importance of winning compared to goal difference. The performance of Pi-rating varies based on whether a team is playing at home or away, with a focus on prioritizing recent information, such as the latest home and away performances, over historical data. Constantinou and Fenton (2013) conducted a comparison study, evaluating Pi-rating against two versions of the Elo rating system proposed by Hvattum and Arntzen (2010). The assessment specifically aimed to predict the probabilities of home wins, draws, and away wins for English Premier League (EPL) seasons spanning from 2007–2008 to 2011–2012. The results indicated that the Pi-Rating approach outperformed both variations of the Elo rating proposed by Hvattum and Arntzen (2010) in terms of accuracy and profitability when compared against market odds. This suggests that pi-rating holds promise as a valuable tool for more accurate and profitable football match outcome predictions. In addition, the pi-rating has been used in prediction models proposed by Hubacek et al. (2019) and Constantinou (2019a, b) in Open International Soccer Prediction Database Challenge which they successfully obtained 1st and 2nd places in the challenge.

3 Experiment

The experiment section presents how the experiment was carried out, including the steps for data collection and pre-processing, implementation of the football rating system, applying Machine Learning (ML) algorithms using learning dataset, predicting, and measuring performance of prediction model using testing dataset. First, the MSL football data is collected and cleaned, and then the football rating system consisting of Elo rating and pi-rating is calculated. After that, ML algorithms are used to predict match outcomes using the updated dataset with football rating system. In essence, the modeling procedure for the proposed prediction model for football match outcomes comprises two subsystems: the implementation of the football rating system (Elo rating and pi-rating) and the application of machine learning algorithms such as Naïve Bayes (NB), Logistic Regression (LR), Support Vector Machine (SVM), and Random Forest (RF) on the football rating system for predicting the MSL season 2022. The entire process is depicted in Fig. 2, illustrating the collaborative functioning of the football rating system, inclusive of Elo rating and pi-rating, and machine learning algorithms to anticipate football match results in MSL 2022.

3.1 Data Collection and Pre-processing

For this paper, a comprehensive dataset spanning eight seasons of the Malaysia Super League (MSL) from 2015 to 2022 are collected and cleaned. This dataset contains basic information, including match dates, participating home and away teams, goals scored by each team, and the match results (win, draw, or lose). To facilitate the predictive modeling process, the data was divided into two distinct sets: seven seasons (2015–2021) for training and one season (2022) for testing the accuracy and rank probabilities score of predictions. A pre-processing phase was also undertaken, which involved tackling missing data, addressing outliers, and normalizing the dataset for

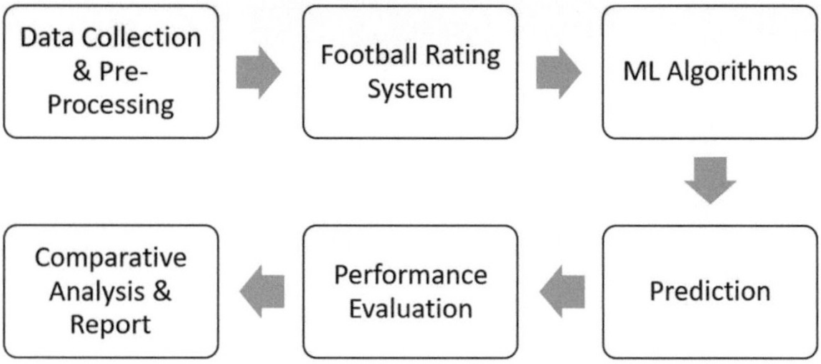

Fig. 2 An ML workflow for MSL football matches results prediction based on football rating system

consistent model performance. For example, MSL keeps changing each season due to various reasons such as team disbandment, reconstruction, withdrawals, regulatory and financial compliance issues, team privatization, and shortened league season due to the impact of COVID-19 pandemic.

The Malaysia Super League (MSL) dataset, sourced from online repositories (Stokkerman 2015), encompasses eight crucial attributes: the number of rounds in the league season (Round), date (Date), home team name (HomeTeam), away team name (AwayTeam), goals scored by the home team (FTHG), goals scored by the away team (FTAG), goal difference (GD), and match results (FTR) denoted as home win (H), draw (D), or away win (A). This dataset spans eight MSL seasons, ranging from 2015 to 2022. The primary aim of the experiment was to employ Machine Learning (ML) algorithms, specifically those based on football rating systems such as Elo rating and pi-rating, to predict match outcomes for the MSL 2022 season. The dataset was split into a learning set comprising 858 instances (covering MSL 2015 to MSL 2021) and a testing set with 132 instances representing the MSL 2022 season. An excerpt of the dataset used in the experiment is illustrated in Fig. 3.

3.2 Football Rating Systems

In the experiment, an innovative football rating system composed of two rating systems: the Elo rating and the pi-rating are introduced and implemented. These rating systems were leveraged to gauge the teams' prowess and competitiveness throughout the league. The football team's rating system varies depending on its performance in each match played.

- **Elo rating** is one prominent football rating system since it was introduced by Bob Runyan for football (Runyan 2023) which originated from chess games. Within the scope of this study, the Elo rating system, as outlined by Hvattum and Arntzen (2010), which is recognized as a result-based Elo rating, is employed.

Round	Date	HomeTeam	AwayTeam	FTHG	FTAG	GD	FTR
Round 1	31/1/2015	JDT	Sri Pahang FC	2	0	2	H
Round 1	7/2/2015	Young Lions	PDRM FA	5	3	2	H
Round 1	7/2/2015	Perak FC	Sime Darby FC	2	0	2	H
Round 1	7/2/2015	Terengganu FC	Selangor FC	2	0	2	H
Round 1	7/2/2015	Felda United FC	Sarawak FA	3	3	0	D
Round 1	7/2/2015	ATM FA	Kelantan FA	0	2	-2	A
Round 2	14/2/2015	Felda United FC	Sime Darby FC	3	0	3	H
Round 2	14/2/2015	Sri Pahang FC	Selangor FC	1	1	0	D
Round 2	14/2/2015	PDRM FA	JDT	1	0	1	H
Round 2	14/2/2015	Sarawak FA	Terengganu FC	3	1	2	H
Round 2	14/2/2015	Perak FC	ATM FA	2	0	2	H
Round 2	15/2/2015	Kelantan FA	Young Lions	2	0	2	H

Fig. 3 Excerpt of MSL dataset used in the experiment

This variant of the Elo rating system encapsulates multiple facets, including the number of goals scored, match outcomes, and the significant home advantage to accurately assess the capabilities of football teams.

- **Pi-rating**. Introduced in 2013 by Constantinou and Fenton, the Pi-rating is a football rating system that offers an innovative approach to assess the performance of football teams. This method analyzes the relative differences in goal scores between two competing teams within a match. The Pi-rating framework is based on three essential football insights: the impact of home advantage, the varying significance of recent versus outdated results in estimating current performance, and the differing importance assigned to match outcomes compared to goal differences.

Thus, the dataset needs to be updated according to the newly obtained football rating system (Elo rating and pi-rating) as the dataset will be used as a learning and testing set to modeling football prediction model using ML algorithms. Figure 4 shows the excerpt of updated dataset used for the experiment which consists of home team pi-rating (PRHT), away team pi-rating (PRAT), home team Elo rating (ELORHT), and away team Elo rating (ELORAT).

Round	Date	HomeTeam	AwayTeam	FTHG	FTAG	GD	FTR	PRHT	PRAT	ELORHT	ELORAT
Round 1	31/1/2015	JDT	Sri Pahang FC	2	0	2	H	0	0	1507.737	1492.263
Round 1	7/2/2015	Young Lions	PDRM FA	5	3	2	H	0	0	1507.737	1492.263
Round 1	7/2/2015	Perak FC	Sime Darby FC	2	0	2	H	0	0	1507.737	1492.263
Round 1	7/2/2015	Terengganu FC	Selangor FC	2	0	2	H	0	0	1507.737	1492.263
Round 1	7/2/2015	Felda United FC	Sarawak FA	3	3	0	D	0	0	1497.737	1502.263
Round 1	7/2/2015	ATM FA	Kelantan FA	0	2	-2	A	0	0	1487.737	1512.263
Round 2	14/2/2015	Felda United FC	Sime Darby FC	3	0	3	H	0	-0.11451	1505.326	1484.674
Round 2	14/2/2015	Sri Pahang FC	Selangor FC	1	1	0	D	-0.08016	-0.11451	1490	1494.525
Round 2	14/2/2015	PDRM FA	JDT	1	0	1	H	-0.08016	0.080156	1500.427	1499.573
Round 2	14/2/2015	Sarawak FA	Terengganu FC	3	1	2	H	0	0.080156	1510.15	1499.85
Round 2	14/2/2015	Perak FC	ATM FA	2	0	2	H	0.114509	-0.08016	1514.936	1480.539
Round 2	15/2/2015	Kelantan FA	Young Lions	2	0	2	H	0.080156	0.080156	1519.877	1500.123

Fig. 4 Excerpt of updated MSL dataset after implementation of football rating system

3.3 Machine Learning Algorithms

The experiment encompasses the utilization of four distinct Machine Learning (ML) algorithms, namely Naïve Bayes (NB), Logistic Regression (LR), Support Vector Machine (SVM), and Random Forest (RF), in conjunction with the Elo rating system to facilitate predictions. The application of these ML algorithms was executed through the default options provided by the Waikato Environment for Knowledge Analysis (WEKA) (Frank et al. 2016).

- **Naïve Bayes (NB)** entails a simplified implementation of Bayes' theorem, assuming independence and no relationship among the features.
- **Logistic Regression (LR)** inherently pertains to predictive analysis for binary classifications. Notably, the scope extends to multinomial logistic regression, which effectively addresses multiclass classification scenarios.
- **Support Vector Machine (SVM)** operates as a supervised machine learning algorithm, proficiently handling predictive analyzes for both classification and regression problems.
- **Random Forest (RF)** functions as a supervised machine learning technique, encompassing decision trees and adeptly accommodating predictive analyzes for both classification and regression tasks.

3.4 Performance Evaluation

To quantify the efficacy of predictive models, two key performance metrics were employed: The Rank Probability Score (RPS) and Accuracy (ACC) (Dubitzky et al. 2019). The RPS provided insights into the probabilistic predictions' quality, with lower values signifying superior performance. On the other hand, Accuracy measured the proportion of correctly predicted outcomes. These metrics collectively formed a robust framework for assessing the models' predictive prowess.:

- **Accuracy** can be defined as in Eq. (1), where the total number of correctly predicted results is divided by the total number of actual observed results. A higher value of score indicates better accuracy.

$$\text{Accuracy} = \frac{\text{The total prediction results}}{\text{The total observed results}} \tag{1}$$

- **Rank Probability Score (RPS)** can be defined as in Eq. (2), where r is the number of potential outcomes, p_j is the forecasted probability of outcome j and pe_j is the actual probability of outcome j. The smaller value of score indicates better forecast accuracy.

$$\text{RPS} = \frac{1}{r-1}\sum_{i=1}^{r-1}\left(p_j - e_j\right)^2 \tag{2}$$

4 Results and Discussion

The evaluation of machine learning algorithms combined with distinct football rating systems offers valuable insights into their predictive potential for Malaysia Super League (MSL) match outcomes. When paired with the pi-rating, Naïve Bayes achieved an accuracy of 51.52%, closely aligned with the Support Vector Machine's performance under the same rating system. These outcomes reflect moderate predictive accuracy, with corresponding Rank Probability Scores (RPS) of 0.2141 and 0.2475 respectively. Similarly, Logistic Regression, when integrated with the pi-rating system, displayed a slightly improved accuracy of 53.03% and a corresponding RPS of 0.2155. Notably, the Elo rating demonstrated its influence, yielding accuracies of 50% and 53.79% when combined with Naïve Bayes and Logistic Regression respectively. While Naïve Bayes maintained an RPS of 0.2103 under the Elo Rating, Logistic Regression's RPS dropped significantly to 0.1979, showcasing enhanced probabilistic prediction quality. In contrast, the Random Forest algorithm exhibited intriguing results. While its accuracy of 43.18% (pi-rating) and 56.06% (Elo rating) suggests variability, the corresponding RPS values of 0.2526 and 0.2285 hint at room for improvement in refining probabilistic predictions. Table 1 shows the comparative analysis results of the ML algorithms based on football rating system in terms of accuracy and RPS.

Among the algorithms tested, Naïve Bayes, Logistic Regression, Support Vector Machine, and Random Forest, the combination that consistently outshone the others was Elo Rating integrated with Logistic Regression (ELO + LR). This particular combination achieved the highest accuracy of 53.79% while maintaining an impressively low RPS of 0.1979, reflecting both predictive accuracy and the precision of its probabilistic estimations. This outcome underscores the significance of employing a sophisticated football rating system alongside machine learning algorithms to enhance the accuracy and reliability of match outcome predictions. It also emphasizes the pivotal role of informed model selection and performance evaluation metrics in optimizing predictive capabilities within sports analytics.

Table 1 The comparative analysis results of the ML algorithms based on football rating system in terms of accuracy and RPS

ML algorithms	Rating system	Accuracy	RPS
Naïve Bayes	PI	51.52	0.2141
	ELO	50	0.2103
Logistic regression	PI	53.03	0.2155
	ELO	53.79	0.1979
Support vector machine	PI	51.52	0.2475
	ELO	53.79	0.2227
Random forest	PI	43.18	0.2526
	ELO	56.06	0.2285

5 Conclusion

In conclusion, this paper has shed light on the potential of integrating machine learning algorithms and a sophisticated football rating system to predict match outcomes within the context of the Malaysia Super League (MSL). Through an extensive analysis, it became evident that the combination of the Elo Rating system with Logistic Regression, denoted as ELO + LR, emerged as a particularly promising approach. This combination achieved the highest accuracy of 53.79% among the various algorithm-rating system pairs tested. Notably, the corresponding Rank Probability Score (RPS) was impressively low at 0.1979, highlighting the reliability of its probabilistic predictions. The outcomes of the Naïve Bayes, Support Vector Machine, and Random Forest algorithms were also evaluated under different rating systems, revealing varying degrees of accuracy and probabilistic prediction quality. Overall, these findings underscore the potential of employing a strategic amalgamation of predictive methodologies for enhancing match outcome predictions in the MSL.

Looking ahead, there are several avenues for refining and extending the predictive modeling efforts applied in this study. First and foremost, exploring advanced football rating systems that capture intricate dynamics and team interactions could potentially yield more accurate predictions. Additionally, enriching the feature set by integrating player-specific statistics, contextual variables like weather conditions, and the influence of home-ground advantage could provide a more comprehensive basis for predictions. Ensemble methods, which combine the strengths of multiple algorithms, deserve further investigation as they might lead to enhanced accuracy and robustness. Incorporating temporal analysis to consider evolving team performance trends over the course of a season could provide deeper insights into match dynamics. Furthermore, the development of real-time adaptable models capable of responding to changing team conditions could ensure more up-to-date predictions. Lastly, the cross-application of predictive models across different football leagues could offer insights into the generalizability of the methods beyond the MSL. By delving into these avenues, future research can contribute to refining predictive accuracy and enriching decision-making processes for sports enthusiasts and stakeholders alike.

References

Arntzen H, Hvattum LM (2021) Predicting match outcomes in association football using team ratings and player ratings. Stat Model 21(5):449–470

Berrar D, Lopes P, Dubitzky W (2019) Incorporating domain knowledge in machine learning for soccer outcome prediction. Mach Learn 108(1):29–47

Bunker R, Susnjak T (2022) The application of machine learning techniques for predicting match results in team sport: a review. J Artif Intell Res 73:1285–1322

Constantinou AC, Fenton NE (2013) Determining the level of ability of football teams by dynamic ratings based on the relative discrepancies in scores between adversaries. J Quant Anal Sports 9(1):37–50

Constantinou AC (2019a) Dolores: a model that predicts football match outcomes from over the world. Mach Learn 108(1):49–75
Constantinou AC (2019b) Dolores: a model that predicts football match outcomes from over the world. Mach Learn 108(1):97–126
Dubitzky W, Lopes P, Davis J, Berrar D (2019) The open international soccer database for machine learning. Mach Learn 108(1):9–28
Frank E, Hall MA, Witten IA (2016) The WEKA Workbench. Online Appendix for "Data Mining: Practical Machine Learning Tools and Techniques". In: Morgan Kaufmann, 4th edn
Hubacek O, Sourek G, Zelezny F (2019) Learning to predict soccer results from relational data with gradient boosted trees. Mach Learn 108(1):29–47
Hvattum LM, Arntzen H (2010) Using ELO ratings for match result prediction in association football. Int J Forecast 26(3):460–470
Lasek J, Szlavik Z, Bhulai S (2013) The predictive power of ranking systems in association football. Int J Appl Pattern Recogn 1:27–46
Lasek J (2016) Euro 2016 predictions using team rating systems. CEUR Workshop Proceedings 1842:1–6
Leitner C, Zeileis A, Hornik K (2010) Forecasting sports tournaments by ratings of (prob)abilities: a comparison for the EURO 2008. Int J Forecast 26(3):471–481
Robberechts P, Davis J (2019) Forecasting the FIFA World Cup—Combining result- and goal-based team ability parameters. Mach Learn Data Mining Sports Anal 16–30
Runyan B. World football Elo ratings. Accessed 20 Aug 2023
Stokkermans K (2015) Rec.Sport. Soccer Statistics Foundation, Malaysia. http://www.rsssf.com/tablesm/malay2015.html#super. Accessed 20 Aug 2023

Development of Motion Tracking Device to Determine Swing Profile of a Golf Player

Atiqullah Huzaifah, Mohd Nadzeri Omar, Nasrul Hadi Johari, Mohd Hasnun Arif Hassan, Mahadzir Ishak, and Zuazizi Aziz

Abstract Understanding the biomechanics during a golf swing is crucial. One way to understand it is by looking at the swing profile. Most of the time, the swing profile is assessed qualitatively by golfers. However, the qualitative evaluation is very subjective, and golfers have no quantitative data which they might refer to for optimization. The quantitative assessment, meanwhile, is expensive, has limited access, and focuses on a single-point measurement. This research aims to create a motion tracking device that captures a golfer's swing profile. The device comprises an MPU6050, FireBeetle ESP32 microcontroller, and a 3.7v Li-Po battery. The commercial MMS sensor was the reference, and the MPU6050 was set to the nearest equivalent specifications. Based on the findings, the device matches the performance of the MMS, considering the bandwidth difference and the position of the MPU6050. Moreover, multiple devices can be used in which ESP-Now communication was embedded between the FireBeetle ESP32 boards to capture the swing profile of different body parts. The device offers golfers quantitative feedback on their golf swing, which can be analyzed for improvement.

Keywords Golf swing profile · ESP32 · ESP-Now · MPU6050 · Biomechanics of golf swing · Motion tracking device

1 Introduction

Golf is a sport that needs accuracy and technique. It is a complicated process in which various body elements move in straight lines and rotate while swinging a golf club (Bourgain et al. 2022). It is critical to understand biomechanics throughout the golf

A. Huzaifah (✉) · M. N. Omar · N. H. Johari · M. H. A. Hassan · M. Ishak
Faculty of Mechanical and Automotive Engineering Technology, Universiti Malaysia Pahang, 26600 Pekan, Pahang, Malaysia
e-mail: atiqullahhuzaifah09@gmail.com

Z. Aziz
ALJULF Golf Club Fitter, 25150 Kuantan, Pahang, Malaysia

Z. Mohamed et al. (eds.), *Proceedings of the 2nd International Conference on Innovation and Technology in Sports, ICITS 2023, 27–28 November, Malaysia*, Lecture Notes in Bioengineering, https://doi.org/10.1007/978-981-97-3741-3_17

swing. This includes examining golfers' movements and muscle activation patterns as they swing the club, as well as the internal and external forces at work (Brown et al. 2020; Balzerson et al. 2016). Understanding these aspects allows golfers and coaches to discover areas for improvement to assist golfers in optimizing their swing and shot-making abilities (Reveendran et al. 2023; Lamb and Pataky 2018).

Traditionally, a golfer's biomechanics are assessed qualitatively. It entails breaking down the swing into discrete phases (preparation, execution, and follow-through) and identifying crucial aspects within each phase. These fundamental elements are certain bodily motions critical to the swing's overall performance. Most of the time, video footage is used (Mun et al. 2015). The captured video will be edited in slow motion following the phases. Even though it can provide a clear visual throughout the golf swing, the qualitative method needs more consistency, and the visual feedback is restricted to 2D observations.

The qualitative method also includes observing the outcome of the golf swing. It relates the performance of the golf swing directly to the outcome. Even though it provides quick feedback, it is highly subjective, and golfers have no consistent data to which they may refer, which may result in incorrect analysis.

Technology can aid this process by providing quantitative data on the golfer's swing mechanics. Quantitative golf biomechanics analyzes the body angles and movements during the golf swing using numerical data. In general, there are two types of technological aids available in the market. The first type of technical aid focuses on the outcome of the golf swing. It captures the data during the impact and estimates the outcome. TrackMan, SkyTrak, Foresight, and FlightScope are among the well-known products that belong to this group. Although these devices can evaluate accurate outcomes, the data on body movement during the golf swing is missing.

The second technological aid device focuses on obtaining data during the golf swing. The most accurate devices use optical tracking, where several high-speed cameras are placed at different angles, and multiple markers are attached to the golfer's body and club (Corke et al. 2018). However, the availability of this device is limited, expensive, and uncomfortable, and it needs a particular space for the setup. On the other hand, wearable devices are getting more attention. Most of it includes Inertial Measurement Unit (IMU), Global Positioning System (GPS), or both (Jiao et al. 2018; Stančin and Tomažič 2013). It is either attached to the golfer's body or the club. Because of its portability, golfers can easily reach it, but it is still costly and is limited to a single-point measurement.

The goal of this research is to develop a low-cost, dependable device for measuring a golfer's motion during a golf swing. The study includes hardware development, system development, and performance evaluation.

2 Methodology

The methodology is divided into 3 sections: (1) Hardware development, (2) System development, and (3) Performance evaluation.

2.1 *Hardware Development*

The device has an IMU, a microcontroller, and a battery. For the IMU, the MPU6050 sensor is used (see Fig. 1). It is a 6-axis motion tracking unit that combines a 3-axis accelerometer and a 3-axis gyroscope. This sensor is reliable and accurate in sensing motion, making it ideal for capturing and analyzing movement data. In addition, the MPU6050 provides flexibility regarding the whole scale range of acceleration that may be monitored. It has four programmable settings for the acceleration scale range: ±2, ±4, ±8, and ±16 g, and the gyroscope scale range ±250, ±500, ±1000 and ±2000. This means the sensor can be calibrated to measure accelerations and rotational velocities within a specific range, depending on the application or requirements.

The FireBeetle ESP32 (DFR0478) board is the microcontroller used in the project due to its compatibility with the MPU6050 module and communication variety (see Fig. 2). The board integrates a Dual-Core ESP-WROOM-32 module that supports microcontroller units (MCU) and Wi-Fi & Bluetooth dual-mode communication. The

Fig. 1 MPU6050 module

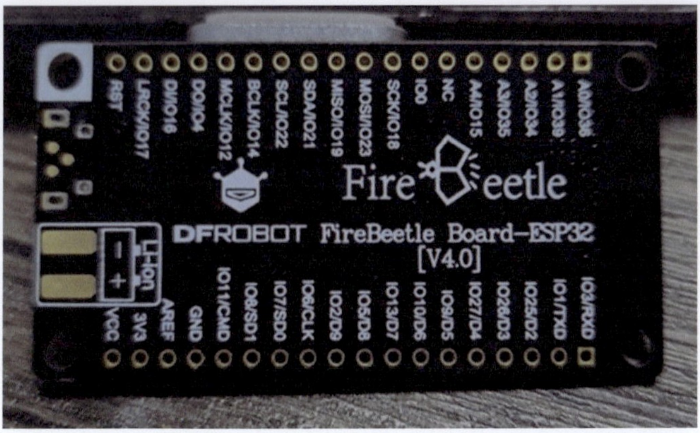

Fig. 2 FireBeetle ESP32 (DFR0478) board

board's Wi-Fi capability enhances the system's stability through wireless communication, enabling real-time data transmission and remote control. USB or 3.7v external lithium battery powers the compact board. Its ultra-low-power consumption makes it an ideal solution for IoT projects.

The MPU6050 communicates with the FireBeetle ESP32 board via the I2C bus. The schematic diagram of the communication is shown in Fig. 3. A 3.7v Li-Po battery is used for the power supply, which is directly connected to the battery input on the FireBeetle ESP32 board. All these components are organized and placed in a rectangular PLA box, as illustrated in Fig. 4.

2.2 *System Development*

The ESP-Now protocol was chosen to establish seamless communication between the components. ESP-Now is a lightweight communication protocol designed for low-power and low-latency applications. ESP-Now allows simple packet communication between ESP devices using the 2.4 GHz band, enabling direct wireless connection between 2 or more ESP devices without Wi-Fi. The project's system architecture comprises a receiver and three senders. The receiver serves as a central hub, collecting data from senders and providing a user interface for data acquisition. Figures 5 and 6 depict the flow-code of the receiver and sender boards.

To facilitate user interaction, the receiver incorporates a web server implemented using HTML and embedded within the receiver's sketch. The web server allows authorized users, identified by their IP address, to access the control interface, providing a secure platform for controlling data acquisition. As depicted in Fig. 7, the system's control interface provides Start and Stop buttons that enable users to

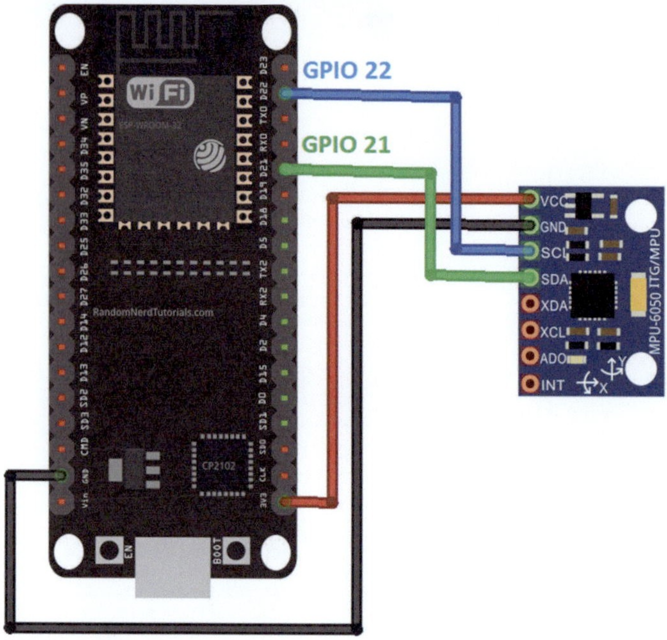

Fig. 3 Schematic diagram between the MPU6050 and FireBeetle ESP 32 board

Fig. 4 Device hardware

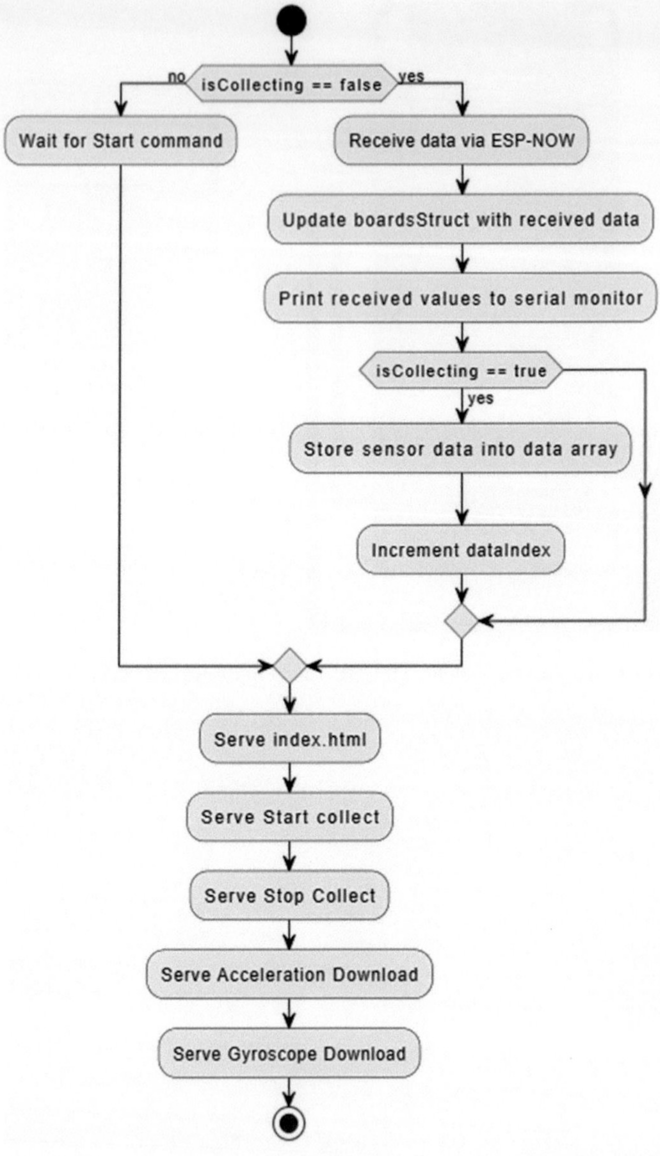

Fig. 5 Flow-code of the receiver board

initiate and terminate data collection from the senders. Additionally, the interface includes separate download buttons for acceleration and gyroscope data, allowing users to retrieve the collected data for further analysis. By developing a user-friendly web server interface, the system ensures convenient data management and enhances usability.

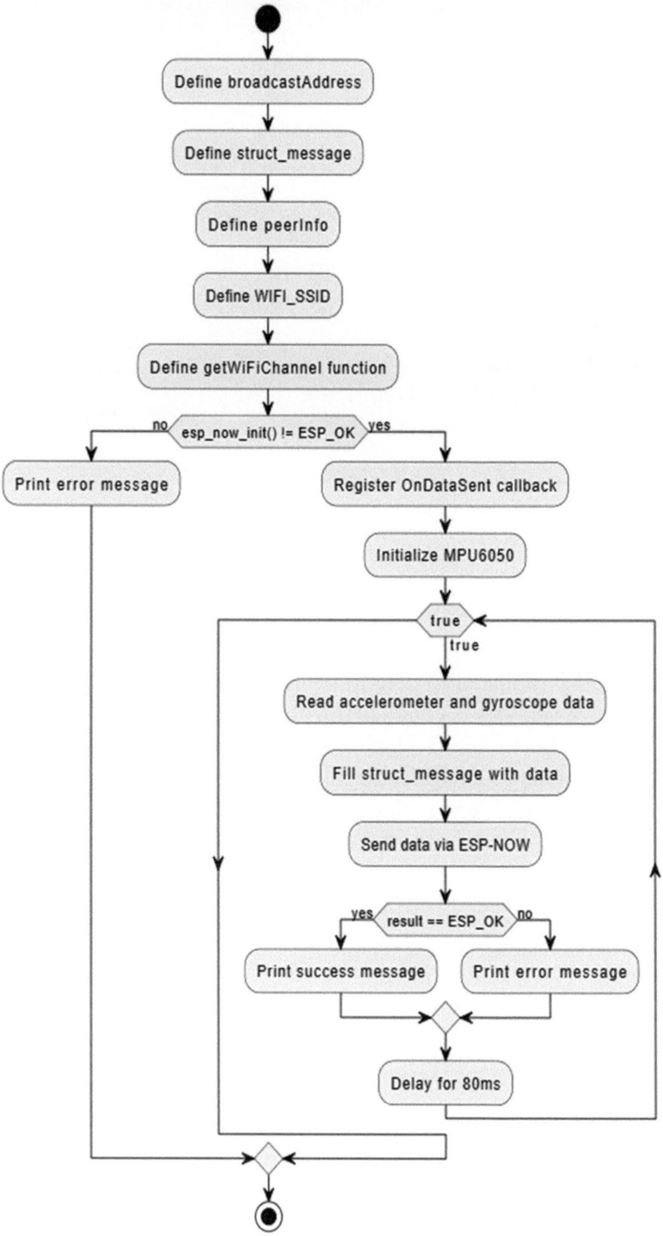

Fig. 6 Flow-code of the sender board

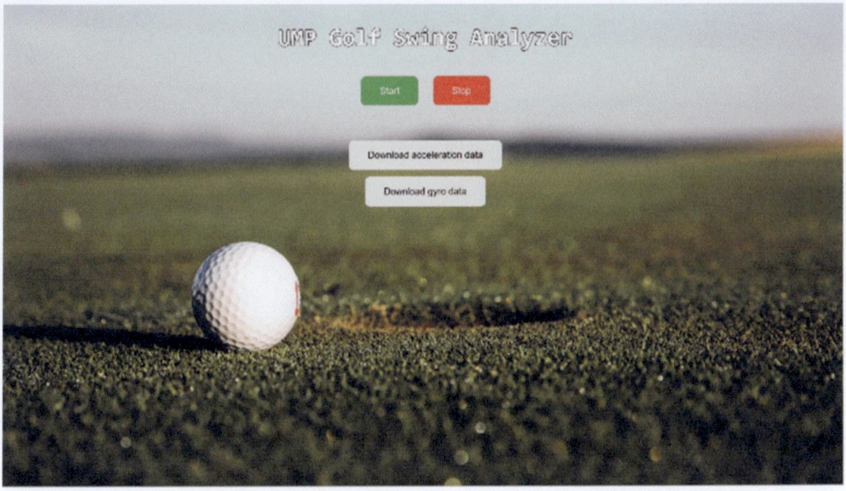

Fig. 7 Webserver interface for user's control

2.3 *Performance Evaluation*

The device is being validated to determine its accuracy and reliability. To achieve this, a comparative analysis is conducted with the MetaMotionS (MMS) device by MbientLab. This commercial IMU uses a Bosch BMI270 6-Axis accelerometer. The technical specifications of BMI270 are shown in Table 1. In this study, the MMS sensitivity was set at 16 g and running at the lowest bandwidth, 12.5 Hz.

Table 1 Specification of BMI270 6-Axis accelerometer

Parameter	Technical data
Digital resolution	Accelerometer (A): 16-bit or 0.06 mg/LSB Gyroscope (G): 16-bit or 0.004 dps/LBS
Programmable measurement range & sensitivity	(A): ±2 g: 16384 LSB/g to (A): ±16 g: 2048 LSB/g (G): ±125 dps: 262.1 LSB/dps to (G): ±2000 dps: 16.4 LSB/dps

3 Results

3.1 Calibration of MPU6050

Specific calibrations are required for the MPU6050 to work optimally, particularly concerning its accelerometer, gyroscope, and bandwidth range. The MPU6050 has predetermined value ranges by default: 2 g for the accelerometer, 250 dps for the gyroscope, and 260 Hz for bandwidth. However, because of these default parameters, the amplitude of the data chart may show significant variance when comparing the device to the MMS. The device's accelerometer, gyroscope, and bandwidth range are changed to 16 g, 1000 dps, and 10 Hz, respectively, in which the MMS serves as the reference point. These modifications can be easily performed by integrating certain specifications into the code that transfers data from the transmitter to the receiver.

3.2 Validation

Two validation procedures were conducted to verify the device. The device and MMS were compared for a simple pendulum movement in the first validation process. The device and the MMS were attached to a shaft of a pendulum setup at the exact location, as illustrated in Fig. 8. The shaft was lifted to a 90°A from the normal vertical axis and released accordingly. The shaft was allowed to move freely for a complete cycle before being stopped. The process was repeated three times, and the outcomes were presented in Figs. 9, 10 and 11.

The acceleration profiles along the horizontal and vertical axis are almost identical, indicating that the device has equal capability to the commercial MMS. The only differences are the magnitude of the accelerations and the data sequence. These slight variations are expected since the device data was determined at 10 Hz, while the MMS sensor used 12.5 Hz.

A distinct difference is found in the acceleration profile along Z-axis. It is observed that the device captured a higher magnitude of data than the MMS. Further investigation found that the MPU6050 sensor was titled; hence the normal to Z-axis is pointing downward. As a result, a higher acceleration along the Z-axis was recorded. To justify the finding, another set of data was obtained by adjusting the placement of the MPU6050 sensor. As indicated in Fig. 12, the acceleration profile changes, and it matches the acceleration profile of the MMS.

The second validation process is the comparison during the actual golf swing. The MMS and the device were attached to the left wrist of a golfer. The golfer is a male with a height of 165 cm and holds a mid-handicap between handicaps 12–15. As shown in Fig. 13, the measured acceleration profile for both devices generally shared the same behavior. Again, as found in the first validation procedure, the bandwidth difference, and the placement of the MPU6050 affected the readings. Another sample was taken for the reliability test, as indicated in Fig. 14; a similar observation is

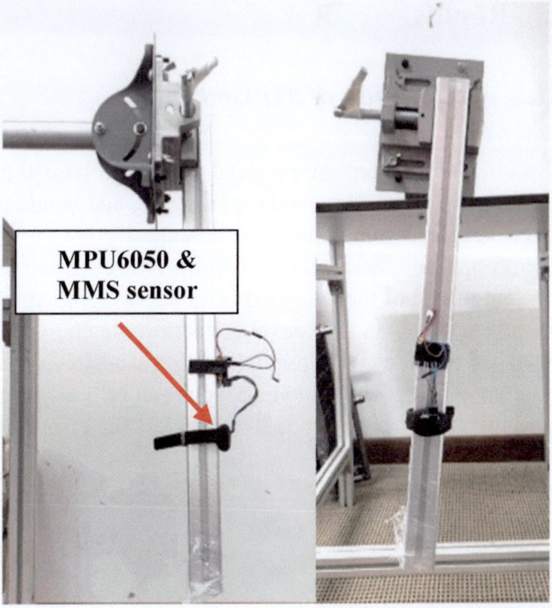

Fig. 8 Pendulum movement setup

obtained, which is the highest variation in the acceleration profile along the Z-axis. As illustrated in Fig. 15, when both devices were attached to the golfer's wrist, due to the golfer's hand's contour and the device's location, the normal of the Z-axis was slightly different, leading to the reading variation. Nonetheless, the outcomes suggest the excellent reliability of the device in determining the swing profile.

Furthermore, to validate the capability of the device to measure swing profile at multiple points, three devices were used as senders and one device as the receiver. The three senders were attached to the golfer's wrist, waist, and head. Based on Fig. 16, the devices can capture the swing profile at those three locations and show a similar swing profile as reported in Reveendran et al. (2023) for wrist and waist. These multipoint data are vital to understanding better the movement of different body parts and the coordination between them.

Furthermore, the ESP-Now communication used in the device offers a more stable communication. It supports maximum distances with low transmission latency (Eridani et al. 2021). As a result, when measuring the swing profile at the three body locations, the device shows a distinct performance in comparison to the MMS. However, the ESP-Now communication in the device consumes a higher power consumption than the BLE. Hence, the selection of the battery capacity and notification on the battery level is crucial.

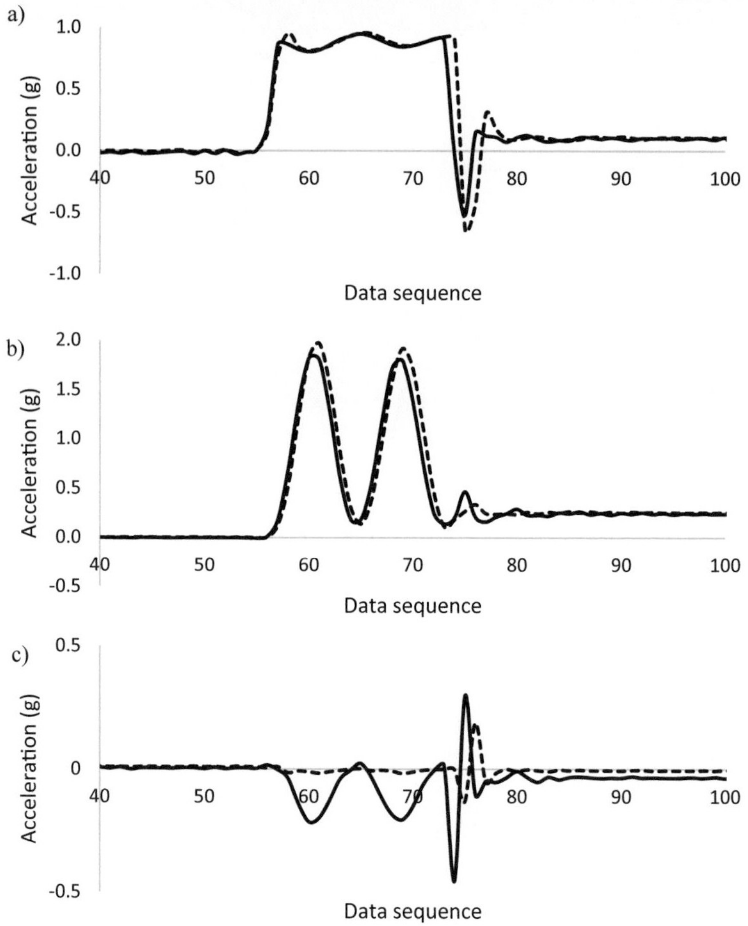

Fig. 9 First sample of pendulum movement analysis. **(a)** Acceleration profile along X-axis, **(b)** acceleration profile along Y-axis, and **(c)** acceleration profile along Z-axis. Solid and dashed lines represent the device and MMS respectively

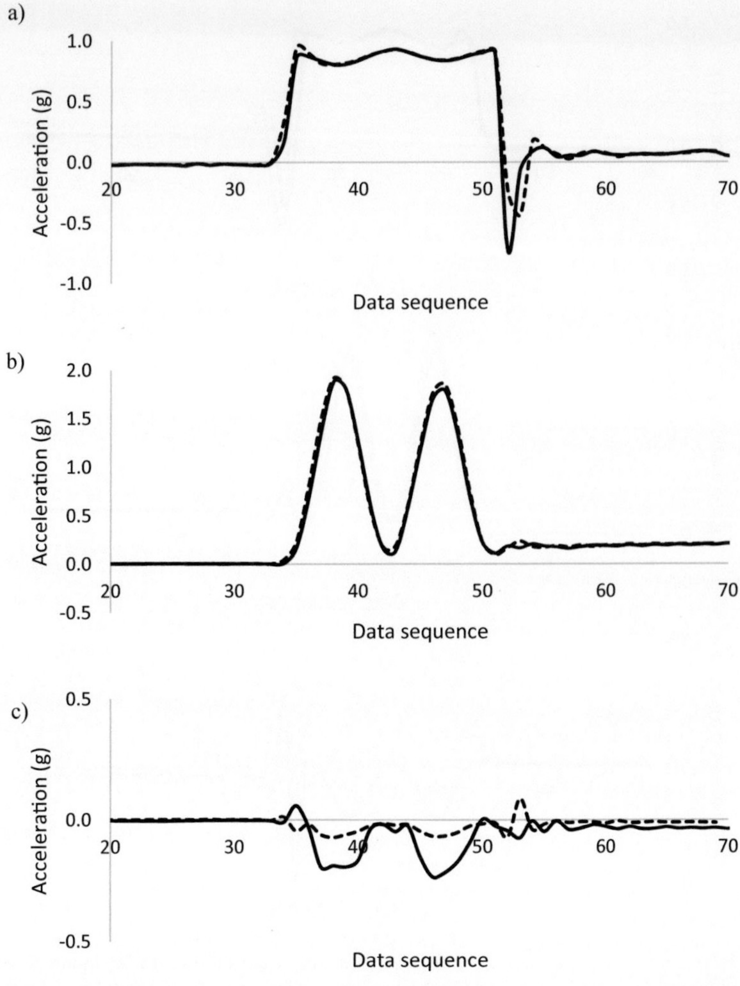

Fig. 10 Second sample of pendulum movement analysis. (**a**) Acceleration profile along X-axis, (**b**) acceleration profile along Y-axis, and (**c**) acceleration profile along Z-axis. Solid and dashed lines represent the device and MMS respectively

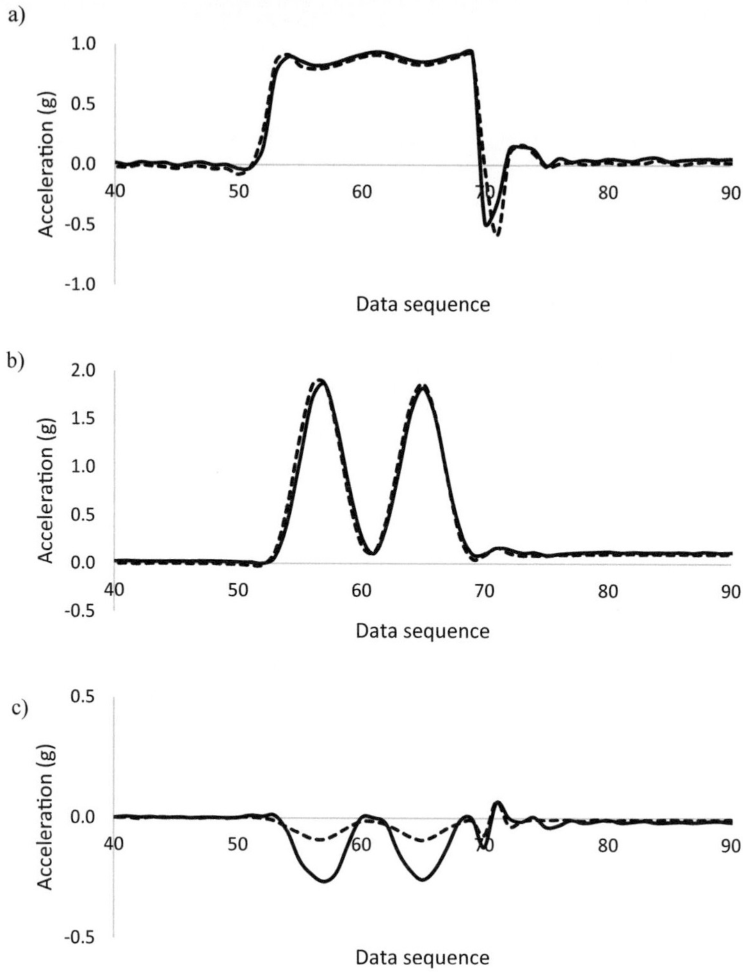

Fig. 11 Third sample of pendulum movement analysis. (**a**) Acceleration profile along X-axis, (**b**) acceleration profile along Y-axis, and (**c**) acceleration profile along Z-axis. Solid and dashed lines represent the device and MMS respectively

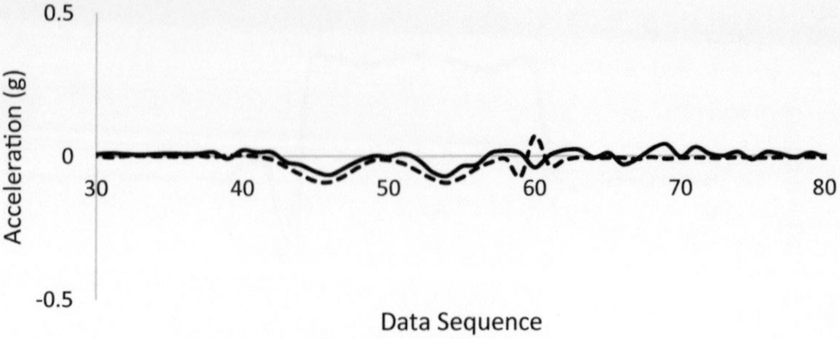

Fig. 12 Acceleration profile along Z-axis with altered MPU6050's position. Solid and dashed lines represent the device and MMS respectively

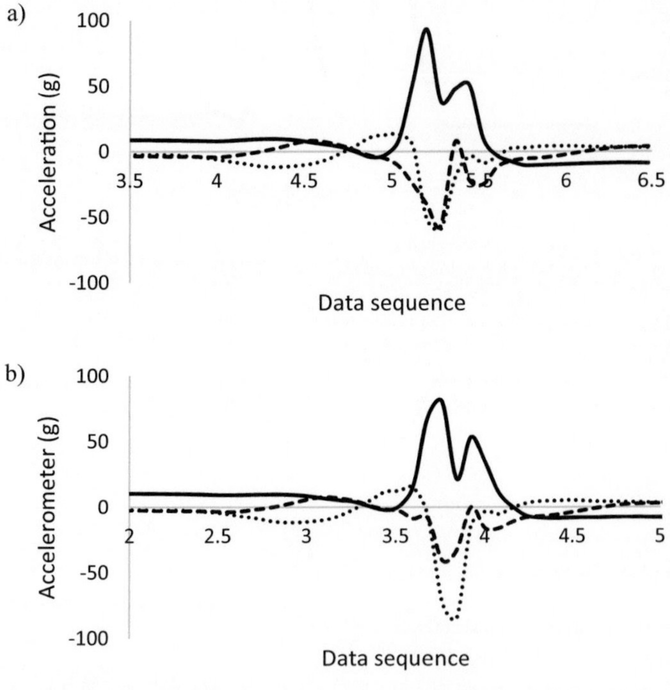

Fig. 13 Acceleration profile during a golf swing. (**a**) Acceleration profile captured by the MMS, and (**b**) acceleration profile captured by the device. Dotted, solid, and dashed lines represent acceleration profile along X, Y, and Z-axis respectively

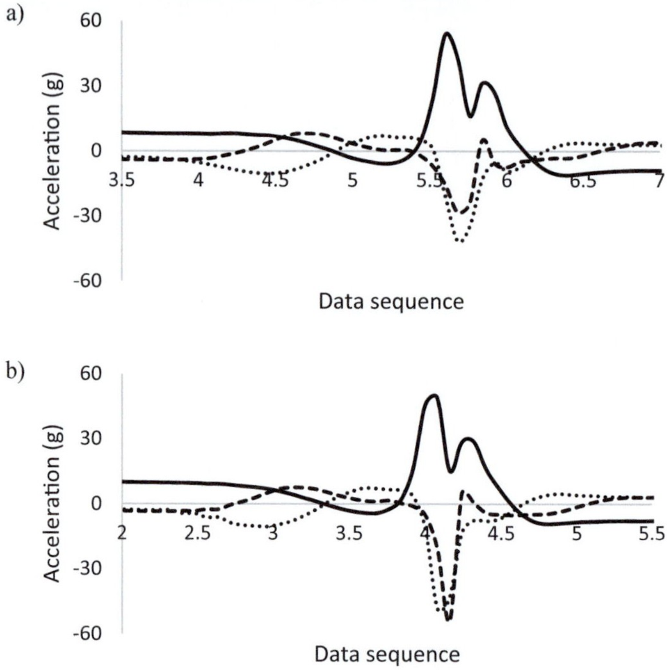

Fig. 14 Acceleration profile during a golf swing. (**a**) Acceleration profile captured by the MMS, and (**b**) acceleration profile captured by the device. Dotted, solid, and dashed lines represent acceleration profile along X, Y, and Z-axis respectively

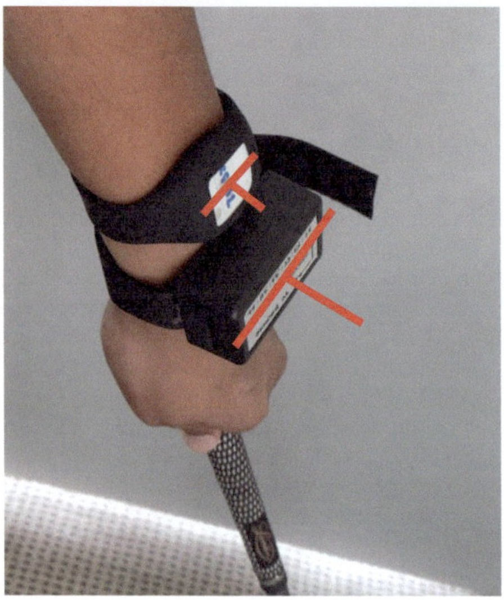

Fig. 15 The location of the device and MMS on the golfer's wrist

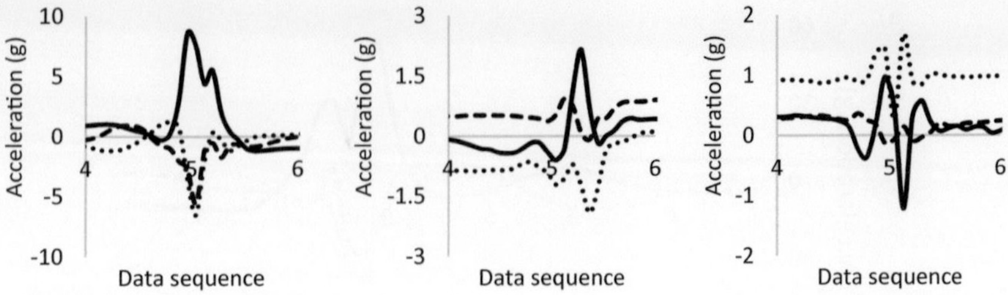

Fig. 16 Acceleration profile during a golf swing captured by the device. (**a**) Acceleration profile captured at wrist, (**b**) acceleration profile captured at head, and (**c**) acceleration profile captured at waist. Dotted, solid, and dashed lines represent acceleration profile along X, Y, and Z-axis respectively

4 Conclusion

This research aims to create a motion tracking device capable of capturing a golfer's swing profile. The device comprises MPU6050, FireBeetle ESP32 microcontroller, and a 3.7v Li-Po battery. The commercial MMS sensor was the reference, and the MPU6050 was set to the nearest equivalent specifications. Based on the findings, the device matches the performance of the MMS, considering the bandwidth difference and the position of the MPU6050. Moreover, multiple devices can be used in which ESP-Now communication was embedded between the FireBeetle ESP32 boards to capture the swing profile of different body parts. Using the device allows golfers to have quantitative feedback on their golf swing, which can be analyzed for improvement.

References

Balzerson D, Banerjee J, McPhee J (2016) A three-dimensional forward dynamic model of the golf swing optimized for ball carry distance. Sports Eng 19(4):237–250

Bourgain M, Rouch P, Rouillon O, Thoreux P, Sauret C (2022) Golf swing biomechanics: a systematic review and methodological recommendations for Kinematics. Sports 10(6):91

Brown C, McNally W, McPhee J (2020) Optimal control of joint torques using direct collocation to maximize ball carry distance in a golf swing. Multibody SysDyn 50(3):323–333

Corke TW, Betzler NF, Wallace ES, Otto SR (2018) A novel system for tracking Iron Golf Clubheads. Proc Inst Mech Eng Part p: J Sports Eng Technol 233(1):59–66

Eridani D, Rochim AF, Cesara FN (2021) Comparative performance study of ESP-now, Wi-Fi, bluetooth protocols based on range, transmission speed, latency, energy usage and barrier resistance. Int Seminar Appl Technol Inf Commun (iSemantic) 322–328

Jiao L, Bie R, Wu H, Wei Y, Ma J, Umek A, Kos A (2018) Golf swing classification with multiple deep convolutional neural networks. Int J Distrib Sens Netw 14(10):155014771880218

Lamb PF, Pataky TC (2018) The role of pelvis-thorax coupling in controlling within-golf club swing speed. J Sports Sci 36(19):2164–2171

Mun F, Suh SW, Park H-J, Choi A (2015) Kinematic relationship between rotation of lumbar spine and hip joints during golf swing in professional golfers. BioMed Eng OnLine 14(1)

Reveendran K, Omar MN, Johari NH, Hassan MH, Aziz A (2023) Investigation on the acceleration of wrist and waist during a golf swing towards the ball trajectory. In: Lecture notes in bioengineering, pp 223–232

Stančin S, Tomažič S (2013) Early improper motion detection in golf swings using wearable motion sensors: the first approach. Sensors 13(6):7505–7521

The Body Mass Index and Health-Related Fitness of Sarawak Primary School Students

Ngien Siong Chin, Boon Sian Teo, and Matheus Yan Yu Chin

Abstract The increasing concern among primary school students is their lack of physical activity (PA) has led to adverse health effects in Malaysia. Though physical education is part of the Malaysian education curriculum, a significant percentage of students continue to maintain inactive lifestyles. The study investigates the BMI and health-related fitness of the Sarawak primary school students. The participants were 1248 primary school students (654 males & 594 females) aged between 10 and 12 years old (2.10 ± 0.82) from urban and rural schools in Sarawak. The SEGAK test battery was utilized and administrated. Based on the BMI, the males' frequency in normal weight BMI increased significantly and slightly increased in overweight BMI, while decreased moderately in underweight and obese BMI from pre to post. For females, their frequency in normal weight and obese BMI slightly increased, but decreased slightly in underweight and overweight BMI from pre to post. Whereas for the age groups, the 10–12 years old group in normal weight BMI increased moderately but decreased slightly in underweight BMI from pre to post. Furthermore, the 10–11 years old group slightly decreased in overweight BMI, whereas increased slightly in obese BMI from pre to post. Next, the 12-year-old group increased slightly in overweight BMI, yet decreased gradually in obese BMI from pre to post. Moreover, the SEGAK test revealed that both genders in superior and excellent fitness levels increased steadily, whereas decreased significantly in fit and poor fitness levels from pre to post. Additionally, participants aged between 10 and 12 years old in superior, excellent, and fit level of fitness increased significantly, but decreased sharply in poor and below average levels of fitness from pre to post. The study showed that intervention programs can foster physical activity participation, leading to beneficial effects on the students' fitness.

Keywords BMI · Health-related fitness · SEGAK · Students · Physical education

N. S. Chin (✉) · B. S. Teo
Department of Physical Education and Health, Institute of Teacher Education Batu Lintang Campus, Kuching, Sarawak, Malaysia
e-mail: ipgm-2785@moe-dl.edu.my

M. Y. Y. Chin
SMK Sungai Maong, Kuching, Sarawak, Malaysia

Z. Mohamed et al. (eds.), *Proceedings of the 2nd International Conference on Innovation and Technology in Sports, ICITS 2023, 27–28 November, Malaysia*, Lecture Notes in Bioengineering, https://doi.org/10.1007/978-981-97-3741-3_18

1 Introduction

Over the years, there has been a consistent pattern of insufficient physical activity and sedentary habits among school students in Malaysia. This is a cause for concern because it is well-established that inadequate physical activity during youth can lead to various health problems later in life. Data from national surveys, including the National Health Morbidity Survey (NHMS) in 2017 and 2019, the Adolescent Healthy Survey (AHS) in 2017, and the South East Asian Nutrition Surveys I (SEANUTS I) in Malaysia, indicated that more than half of Malaysian children and adolescents do not engage in enough physical activity and are categorized as having low levels of physical activity (Institute of Public Health 2020). Additionally, the prevalence of school students' sedentary behaviors has been on the rise, increasing from 47.3% in 2012 to 50.1% in 2017. Moreover, the NHMS 2017 data also revealed school students aged 13–17 years were only physically active for a total of 60 min per day on 5 or more days per week (Institute of Public Health 2017). The current Malaysia Report Card 2022 on the assessment of PA and related contributing factors for children and adolescents in Malaysia has shown overall physical activity obtained a poor grade of D and sedentary behavior obtained a grade C (Institute of Public Health 2017). The results implied that physical activity levels in children and adolescents have not changed, hence the grade remains consistent with NHMS 2016 (Shahril et al. 2023).

The lack of PA among the school students in moderate to vigorous physical activity (MVPA) which increases their sedentary behavior apart from the physical education classes carried out in schools showed that the time allocated for PE was insufficient to meet the MVPA till now. For example, in terms of flexibility assessed through the sit-and-reach test (National Physical Fitness Standard for Malaysian School Student) revealed that Malaysian adolescent's (10–17 years old) levels of flexibility were graded B which is average in comparison to international norms (Ministry of Education Malaysia 2019).

Despite having the National Physical Education Fitness Standard (SEGAK) assessment and intervention guide for physical activities since it was introduced in 2016, the level of fitness among the children has not been satisfactory. Thus, in 2022, the Ministry of Education has introduced the BMI 5-9T program as an additional assessment for BMI besides the yearly National Physical Fitness Standard Test (SEGAK) as part of its policies. The BMI 5-9T also included recommended PA intervention based on the educational Malaysian PE curriculum to reduce the obesity rate of school students from 5 to 9 years of age. Whereas, the schools would need to carry out their own PA interventions respectively in order to improve the BMI and fitness levels of the students.

Obesity is a pressing concern that demands our attention, especially given the increasing number of Malaysian children who are experiencing weight issues. This trend is causing them to adopt sedentary lifestyles, which in turn is having a detrimental impact on their overall health and physical fitness. Furthermore, the process of modernization has exacerbated the severity of this obesity problem, further compromising the health and well-being of children and adolescents.

This study aimed to examine the BMI and health-related fitness of Sarawak primary school students after the implementation of the PA programs in schools to provide better interventions and strategies to improve the fitness levels of the students.

2 Materials and Methods

2.1 *Participants*

A total of 1248 primary school students comprising of 654 males and 594 females aged between 10 and 12 years old (2.10 $\pm$ 0.82) from urban and rural schools in Sarawak participated in the study. The ethical approval was approved by the Institutional Ethics Committee. Informal consent was obtained from the participants and teachers prior to the study. The SEGAK test was conducted and administrated twice by the Physical Education school teachers in accordance with the guidelines and standard operating protocol determined by the Ministry of Education, Malaysia.

The National Physical Fitness Standard Test (SEGAK) battery was utilized to assess the health-related fitness which includes body composition, muscle strength, muscle endurance, flexibility, and cardiovascular endurance. This test is mandatory for all Malaysian students aged 10–17 years of age twice annually, unless they have issues related to their health that prevent them from participating in the test.

The body composition was determined using the Body Mass Index (BMI) test, calculated as weight (kg) divided by height (m) squared. The BMI values adhered to World Health Organization (WHO) standards (Expert and Consultation 2004), with categories as follows: BMI $\leq$ 18.5 for underweight, 18.5–24.9 for normal weight, 25–29.9 for overweight, 30–34.9 for obesity, and BMI $\geq$ 35 for extreme obesity.

There are four assessments in the test battery namely push-up, curl-up, step-up, and the sit-and-reach test. The step-up test gauged cardiovascular endurance, push-up assessed muscle strength, curl-up measured muscle endurance, and the sit-and-reach test evaluated flexibility.

The intervention programs were the physical activities program based on the SEGAK guide planned by the school PE committees respectively which integrate physical activities related to cardiovascular endurance, upper body muscular endurance, abdominal muscular endurance, and flexibility. The gradual progression of the program started after the pre-test in March which started with a basic level of movement exercises followed by intermediate level of exercises of moderate intensity based on the F.I.T.T. principle till the post-test in Ogos according to the Ministry of Education curriculum. All exercises were of low-to-moderate intensity and utilized the students' own-body weight.

The planned physical activities took between 20 and 30 min of either their PE lessons and arranged outside the normal class lessons which included standardized warm-up twice a week for a duration of 20 weeks. The PE teachers would need to integrate the recommended exercises into their 30 min of PE classes by adapting, customizing, and being flexible in allocating the exercises into the PE lesson accordingly. All interventions were conducted by the PE teachers who had a bachelor's degree in PE and were experienced in teaching PE and sports.

The analysis of the data was processed through IBM Statistical Package for the Social Science (SPSS) version 28.0.

3 Results

Table 1 shows the demographic characteristics of the participants. There were 1248 participants involved in the study, with a total of 654 (52.4%) males and 594 (47.6%) females from years 4, 5, and 6. The age groups categories showed that 366 (29.3%) 10 years old, 388 (31.1%) 11 years old and 494 (39.6%). In terms of ethnicity, the majority of the participants were Bumiputera Sarawak with a total of 711 (57.0%), followed by Malay with 502 (40.2%), and Chinese with 35 (2.8%) of the population, respectively. Based on the locality, 520 participants were from urban schools, whereas 728 participants were from rural schools.

Table 2 shows the pre-test and post-test of BMI and SEGAK test. For the Body Mass Index (BMI), results revealed that the frequency of participants increased slightly from pre (53.1%) to post (55.6%) in the normal weight category. Conversely,

Table 1 Demographic characteristics of the participants (N = 1248)

Characteristics	Frequency (*F*)	Percentage (%)	*M* (*SD*)
Gender			1.48 (0.50)
Male	654	52.4	
Female	594	47.6	
Age			2.10 (0.82)
10	366	29.3	
11	388	31.1	
12	494	39.6	
Ethnicity			2.17 (0.97)
Bumiputera Sarawak	711	57.0	
Malay	502	40.2	
Chinese	35	2.8	
Type of school			1.58 (0.49)
Urban	520	41.7	
Rural	728	58.3	

the frequency of the participants declined slightly from pre (8.4%) to post (6.2%) in the underweight category. Besides, the frequency of the participants in the overweight decreased slightly from pre (18.1%) to post (18.0%). The frequency of participants in the obese category also decreased slightly from pre (20.4%) to post (20.2%). For the SEGAK test, the frequency of participants increased gradually from pre (5.1%) to post (9.4%) in superior fitness level. Furthermore, the frequency of participants increased significantly from pre (32.6%) to post (49.8%) in the excellent fitness level. However, the frequency of the participants declined sharply in the fit level of fitness from pre (40.7%) to post (31.2%). The frequency of participants in the poor fitness level decreased gradually from pre (19.6%) to post (9.5%). The frequency of participants in the below-average fitness level decreased significantly from pre (1.9%) to post (0.2%).

Table 3 shows the BMI test on pre and post-test based on gender. For males, results showed that the frequency of participants in the normal weight level increased significantly from pre (45.1%) to post (49.5%), but the frequency of participants in the underweight level declined steadily from pre (9.5%) to post (6.6%). Besides, the frequency of participants in the overweight level increased slightly from pre (18.0%) to post (18.2), while decreased slightly from pre (27.4) to post (25.7%) in the obese level. Whereas, for females, results revealed that the frequency of participants in the normal weight level increased slightly from pre (62.0%) to post (62.3%), while decreased moderately from pre (7.2%) to post (5.7%) in the underweight level. Moreover, the frequency of participants in the overweight level decreased slightly from pre (18.2%) to post (17.8%). Conversely, the frequency of participants in the obese level increased gradually from pre (12.6%) to post (14.1%).

Table 4 shows the BMI test on pre and post-test based on age groups. For the 10-year-old group, the frequency of participants in the normal weight level increased slightly from pre (51.6%) to post (53.3%). Conversely, the frequency of participants

Table 2 Pre-test and post-test of BMI and SEGAK test

Variables	Pre-test		Post-test	
	Frequency (F)	Percentage (%)	Frequency (F)	Percentage (%)
Body mass index (BMI)				
Normal weight (18.5–24.9)	663	53.1	694	55.6
Underweight (≤18.49)	105	8.4	77	6.2
Overweight (25.0–29.9)	226	18.1	225	18.0
Obese (30.0–34.9)	254	20.4	252	20.2
SEGAK test				
Superior	64	5.1	117	9.4
Excellent	407	32.6	621	49.8
Fit	508	40.7	389	31.2
Poor	245	19.6	119	9.5
Below-average	24	1.9	2	0.2

Table 3 BMI test on pre and post-test based on gender

Test	Gender	Level								$\sum n$
		Normal		Underweight		Overweight		Obese		
		n	%	*n*	%	*n*	%	*n*	%	
Pre	Male	295	45.1	62	9.5	118	18.0	179	27.4	1248
	Female	368	62.0	43	7.2	108	18.2	75	12.6	
Post	Male	324	49.5	43	6.6	119	18.2	168	25.7	1248
	Female	370	62.3	34	5.7	106	17.8	84	14.1	

in the underweight level decreased moderately from pre (10.1%) to post (7.4%), and also decreased slightly from pre (20.2%) to post (18.6%) in the overweight level. Besides, the frequency of participants in the obese level increased gradually from pre (18.0%) to post (20.8%). For the 11-year-old group, the frequency of participants in the normal weight level increased significantly from pre (52.8%) to post (56.4%), while decreased substantially from pre (10.1%) to post (6.4%) in the underweight level. The frequency of participants in the overweight level decreased slightly from pre (17.5%) to post (17.0%), but increased slightly from pre (19.6%) to post (20.1%) in the obese level. For the 12-year-old group, the frequency of participants in the normal weight level increased steadily from pre (54.5%) to post (56.7%), whereas decreased slightly from pre (5.9%) to post (5.1%) in the underweight level. The frequency of participants in the overweight level increased slightly from pre (17.0) to post (18.4%), yet decreased considerably from pre (22.7%) to post (19.8%).

Table 5 shows the paired sample t-test on SEGAK test. The mean score of step test was a significant increase from pre (3.37 ± 1.05) to post (3.47 ± 0.94), $\mu = -0.94$, $t(1247) = -2.789$, $p = 0.005$. Besides, the mean score of push-up test was statistically significant from pre (3.58 ± 1.44) to post (4.01 ± 1.29), $\mu = -0.43$, $t(1247) = -10.844$, $p < 0.001$. Similarly, the mean score of curl-up test was a significant increase from pre (4.02 ± 1.22) to post (4.41 ± 1.01), $\mu = -0.39$,

Table 4 BMI test on pre and post-test based on age groups

Test	Age	Level								$\sum n$
		Normal		Underweight		Overweight		Obese		
		n	%	*n*	%	*n*	%	*n*	%	
Pre	10	189	51.6	37	10.1	74	20.2	66	18.0	1248
	11	205	52.8	39	10.1	68	17.5	76	19.6	
	12	269	54.5	29	5.9	84	17.0	112	22.7	
Post	10	195	53.3	27	7.4	68	18.6	76	20.8	1248
	11	219	56.4	25	6.4	66	17.0	78	20.1	
	12	280	56.7	25	5.1	91	18.4	98	19.8	

$t(1247) = -12.015, p < 0.001$. Finally, the mean score of sit and reach test was also statistically significant from pre (2.51 ± 1.04) to post (2.85 ± 1.02), $\mu = -0.34$, $t(1247) = -11.917, p < 0.001$.

Table 6 shows the paired sample t-test for SEGAK test based on gender. For males, results revealed that there was no significant increase on step test from pre (3.25 ± 1.06) to post (3.32 ± 0.96), $t(653) = -1.384, p = 0.167$. Conversely, there was significant increase on push-up test from pre (3.65 ± 1.45) to post (4.01 ± 1.34), $t(653) = -6.545, p < 0.001$. Similarly, there was significant increase on curl-up test from pre (3.98 ± 1.21) to post (4.43 ± 0.99), $t(653) = -6.545, p < 0.001$. Likewise, there was significant increase on sit and reach test from pre (2.43 ± 0.95) to post (2.77 ± 0.96), $t(653) = -10.623$, $p < 0.001$. For females, results revealed that there was significant increase on step test from pre (3.51 ± 1.03) to post (3.63 ± 0.89), $t(593) = -2.634$, $p < 0.001$. Besides, there was significant increase on push-up test from pre (3.52 ± 1.43) to post (4.01 ± 1.24), $t(593) = -8.939$, $p < 0.001$. Furthermore, there was significant increase on curl-up test from pre (4.06 ± 1.22) to post (4.38 ± 1.03), $t(593) = -6.509$, $p < 0.001$. Moreover, there was significant increase on sit and reach test from pre (2.59 ± 1.12) to post (2.94 ± 1.07), $t(593) = -7.760$, $p < 0.001$.

Table 7 shows the paired sample t-test for SEGAK test based on age groups. For the 10 years old group, results revealed that there was significant decrease on step test from pre (3.66 ± 0.94) to post (3.55 ± 0.87), $t(365) = 2.072, p = 0.039$. Besides, there was significant increase on push-up test from pre (3.77 ± 1.49) to post (4.01 ± 1.39), $t(365) = -3.637$, $p < 0.001$. Likewise, there was significant increase on curl-up test from pre (3.98 ± 1.34) to post (4.31 ± 1.12), $t(365) = -5.188, p < 0.001$. Furthermore, there was significant increase on sit and reach test from pre (2.456 ± 1.07) to post (2.83 ± 0.99), $t(365) = -5.903, p < 0.001$. For the 11 years old group, results revealed that there was no significant decrease on step test from pre (3.53 ± 1.06) to post (3.47 ± 1.03), $t(387) = 0.904$, $p = 0.367$. Conversely, there was significant increase on push-up test from pre (3.33 ± 1.43) to post (3.98 ± 1.31), $t(387) = -8.937$, $p < 0.001$. Moreover, there was significant increase on curl-up

Table 5 Paired sample T-test on SEGAK test

Variables	Pre-test		Post-test		*μ (95% CI)*	*t*-test	
	M	*SD*	*M*	*SD*		*t* (df = 1247)	*p*
Step test	3.37	1.05	3.47	0.94	−0.94 (−0.16, −0.03)	−2.789	0.005*
Push-up	3.58	1.44	4.01	1.29	−0.43 (−0.50, −0.35)	−10.844	<0.001*
Curl-up	4.02	1.22	4.41	1.01	−0.39 (−0.45, −0.33)	−12.015	<0.001*
Sit and reach	2.51	1.04	2.85	1.02	−0.34 (−0.40, −0.29)	−11.917	<0.001*

$^{*}p < 0.05$

Table 6 Paired sample T-test of SEGAK test based on gender

Variables	Test	Male				Female			
		M	*SD*	*t* (df = 653)	*p*	*M*	*SD*	*t* (df = 593)	*p*
Step test	Pre	3.25	1.06	−1.384	0.167	3.51	1.03	−2.634	0.009*
	Post	3.32	0.96			3.63	0.89		
Push-up	Pre	3.65	1.45	−6.545	<0.001*	3.52	1.43	−8.939	<0.001*
	Post	4.01	1.34			4.01	1.24		
Curl-up	Pre	3.98	1.21	−10.623	<0.001*	4.06	1.22	−6.509	<0.001*
	Post	4.43	0.99			4.38	1.03		
Sit and reach	Pre	2.43	0.95	−9.151	<0.001*	2.59	1.12	−7.760	<0.001*
	Post	2.77	0.96			2.94	1.07		

* $p < 0.05$

test from pre (3.95 ± 1.25) to post (4.35 ± 1.08), $t(387) = -6.476$, $p < 0.001$. In addition, there was significant increase on sit and reach test from pre (2.46 ± 1.00) to post (2.84 ± 1.07), $t(387) = -6.610$, $p < 0.001$. For the 12 years old group, results revealed that there was significant increase on step test from pre (3.03 ± 1.04) to post (3.40 ± 0.92), $t(493) = -6.807$, $p < 0.001$. Meanwhile, there was significant increase on push-up test from pre (3.65 ± 1.39) to post (4.03 ± 1.19), $t(493) = -6.081$, $p < 0.001$. Besides, there was significant increase on curl-ups test from pre (4.10 ± 1.08) to post (4.53 ± 0.85), $t(493) = -9.081$, $p < 0.001$. Additionally, there was significant increase on sit and reach test from pre (2.51 ± 1.04) to post (2.88 ± 1.00), $t(493) = -8.039$, $p < 0.001$.

4 Discussion

The research seeks to assess the health-related physical fitness levels of Malaysian primary school students concerning gender and age groups following the recent introduction of the BMI 5-9T program to supplement the SEGAK physical activity guidelines that serve as a guide for the schools to implement physical activity interventions aimed at addressing obesity among primary school students in Malaysia. Regarding BMI, there was a slight increase in the normal weight category, but no notable changes in the underweight, overweight, or obese categories which is in line with (Song et al. 2021) study. This may be attributed to insufficient time, and a lack of proper diet implementation in schools, as meaningful changes in exercise and diet typically take time to significantly impact a child's BMI and physical development in health and fitness. Short-term fluctuations in eating habits and activity levels may not immediately manifest in BMI. In contrast, overall fitness levels exhibited a minor improvement, particularly in the superior and excellent fitness categories,

Table 7 Paired sample T-test on SEGAK test based on age groups

Variables	Test	10 years old				11 years old				12 years old			
		M	*SD*	*t* (df = 365)	*p*	*M*	*SD*	*t* (df = 387)	*p*	*M*	*SD*	*t* (df = 493)	*p*
Step test	Pre	3.66	0.94	2.072	0.039*	3.53	1.06	0.904	0.367	3.03	1.04	−6.807	<0.001*
	Post	3.55	0.87			3.47	1.03			3.40	0.92		
Push-up	Pre	3.77	1.49	−3.637	<0.001*	3.33	1.43	−8.937	<0.001*	3.65	1.39	−6.081	<0.001*
	Post	4.01	1.39			3.98	1.31			4.03	1.19		
Curl-up	Pre	3.98	1.34	−5.188	<0.001*	3.95	1.25	−6.476	<0.001*	4.10	1.08	−9.081	<0.001*
	Post	4.31	1.12			4.35	1.08			4.53	0.85		
Sit and reach	Pre	2.56	1.07	−5.903	<0.001*	2.46	1.00	−6.610	<0.001*	2.51	1.04	−8.039	<0.001*
	Post	2.83	0.99			2.84	1.07			2.88	1.00		

* $p < 0.05$

while there was a slight dip in the fit and below-average fitness levels. This could be attributed to physical education and sports teachers implementing regular exercises, aligning with the SEGAK test battery, which includes push-up, sit-up, flexibility, and cardiovascular endurance exercises conducted during PE classes or sports training. Consistently engaging and sustaining in regular exercises may have led to increased fitness levels based on the SEGAK test battery which is in line with (Song et al. 2021) study which mentioned that enhancing physical fitness necessitated the establishment of a consistent exercise routine.

Regarding gender differences, there was a noteworthy decrease in the obese category among males in contrast to females, whereby an increase was observed. This discrepancy could be linked to increased consumption of unhealthy and less nutritious foods among females, often accompanied by sedentary behaviors. The higher prevalence of overweight in females might also be attributed to extended screen time with computers and tablets, along with increased snacking due to heightened adolescent stress, potentially leading to eating disorders and reduced interest in physical activity. In contrast, males were more likely to allocate a significant amount of time to physical activity and regular exercise. The results supported past studies which showed physical activity interventions were able to increase the level of PA among the students though males tend to have a slight increase over females (Shabeshan et al. 2014; Hamlin et al. 2002).

In terms of age groups, there was an alarming rise in obesity prevalence among 12-year-olds. This is of some concern as older age groups appeared to engage in less physical activity, increase snacking, and spend more time on screens, all while prioritizing revision and studying for year-end exams. The current emphasis on academic performance and workload in schools would lessen the students' time in engaging in physical activities which is in line with a study that showed a positive correlation with studying hours with obesity (Calestine et al. 2017).

The SEGAK test results showed a slight increase in students reaching superior and excellent fitness levels, but a decrease in the fit category. Moreover, the percentage of students in the poor and below-average fitness levels decreased in the post-test compared to the pre-test. These findings suggest that any form of physical activity intervention increased students' daily physical activity levels, aligning with the recommended daily guideline of 60 min of moderate-to-vigorous physical activity, ultimately improving their fitness levels (Zenic et al. 2020). Notably, females outperformed males in the superior and excellent fitness levels, challenging the notion of insufficient physical activity among girls, which was reported to be at 85% between 2001 and 2016 (Guthold et al. 2018). Females might have engaged in physical activity through workouts and exercises, often with support from peers and social media networks. Adolescent females tend to rely more on interpersonal reasons, social support, and community engagement for physical activity as they transition from childhood to adolescence.

Additionally, the SEGAK test indicated an increase in students' fitness levels in the superior and excellent categories but a decline in the fit category based on age groups. Furthermore, there was a lower percentage of students in the poor and below-average fitness levels across all age groups in the post-test compared to the pre-test. The physical activities programs were deemed to be effective from school stakeholders, emphasizing creative pedagogical approaches that prioritize physical education and create an environment conducive to student participation in physical activity and exercise. The results supported (Centers for Disease Control and Prevention.: School health guidelines to promote healthy eating and physical activity. 2011; Bermejo-Cantarero et al. 2017) systematic reviews and (Song et al. 2021) study which showed an improvement in the physical fitness of the school students. In addition, (Guthold et al. 2018; Song et al. 2021) recommended any form of physical activity program needs to last approximately 30 min a day and 3 days a week in order to generate some form of positive health effects. These results contradicted (Ng et al. 2020) study which suggested the decrease in moderate physical activity with age was due to environmental changes. However, the increased academic workload, subjects, homework, and study commitments have likely limited older students' time for physical activity, especially primary 6 students preparing for their annual school assessments.

5 Conclusions

The study concluded that any PA interventions can shed a potentially significant impact on physical activity participation, leading to beneficial effects on students' fitness, mental health, and physical well-being. A better approach through strengthening both parental and school involvement to guide parents in adopting effective strategies to promote, foster, and self-regulate students' engagement in physical activity would lead to a healthier balance of activities and quality of life among students.

References

Bermejo-Cantarero A, Álvarez-Bueno C, Martinez-Vizcaino V, García-Hermoso A, Torres-Costoso AI, Sánchez-López M (2017) Association between physical activity, sedentary behavior, and fitness with health related quality of life in healthy children and adolescents: a protocol for a systematic review and meta-analysis. Medicine 96(12):e6407. https://doi.org/10.1097/MD.0000000000006407

Calestine J, Bopp M, Bopp CM, Papalia Z (2017) College student work habits are related to physical activity and fitness. Int J Exerc Sci 10(7):1009–1017

Centers for Disease Control and Prevention (2011) School health guidelines to promote healthy eating and physical activity. MMWR Recommendations and Reports 60(5):1–76

Guthold R, Stevens GA, Riley LM, Bull FC (2018) Worldwide trends in insufficient physical activity from 2001 to 2016: a pooled analysis of 358 population-based surveys with 1·9 million participants. Lancet Glob Health 6(10):e1077–e1086

Hamlin M, Ross J, Sang WH (2002) The effect of 16 weeks of regular physical activity on fitness levels in primary school children. J Educ New Zealand 35(1):45–55

Institute of Public Health (2020) National Health and Morbidity Survey (NHMS) 2019.: vol. I: NCDs—Non-communicable diseases: risk factors and other health problems. Ministry of Health Malaysia

Institute of Public Health (2021) National Health and Morbidity Survey (NHMS) 2017: Adolescent Health Survey 2017. Ministry of Health Malaysia

Ministry of Education Malaysia (2019) Data on National Physical Fitness Standard for Malaysian school student (SEGAK) (Unpublished results)

Ng K, Cooper J, McHale F, Clifford J, Woods C (2020) Barriers and facilitators to changes in adolescent physical activity during COVID-19. BMJ Open Sport Exerc Med 6(1):e000919. https://doi.org/10.1136/bmjsem-2020-000919

Shabeshan R, Wee KSL, Syed Kamaruzaman S (2014) Keberkesanan program intervensi kecergasan terhadap fleksibiliti murid lelaki tingkatan empat di sebuah sekolah di Malaysia. JuKu 2(4):1–9

Shahril MR, Unal TI, Wong JE, Sharif R, Koh D, Lee ST, Poh BK (2023) Results from the Malaysia 2022 report card on physical activity for children and adolescents. J Exerc Sci Fit 21(1):88–94. https://doi.org/10.1016/j.jesf.2022.11.001

Song JH, Song HH, Kim S (2021) Effects of school-based exercise program on obesity and physical fitness of urban youth: a quasi-experiment. Healthcare 9(3):358–368. https://doi.org/10.3390/healthcare9030358

WHO Expert Consultation (2004) Appropriate body mass index for Asian populations and its implication for policy and intervention strategies. Lancet 363(9403):157–163. https://doi.org/10.1016/S0140-6736(03)15268-3

Zenic N, Taiar R, Gilic B, Blazevic M, Maric D, Pojskic H, Sekulic D (2020) Levels and changes of physical activity in adolescents during the covid-19 pandemic: contextualizing urban versus rural living environment. Appl Sci 10(11):1–14

Training Efficiency Application for *Buah Pertama* Seni Silat Cekak Malaysia Using Motion Capture Approach

Aznor Hanah Abdul Halim, M. Hafiz S. Shafawi, Rohani Farook, Mohamad Rizal Abd Rejab, Siti Azura Abuzar, and Mohammad Shahril Salim

Abstract This paper uses the motion capture approach to develop an application of *Buah Pertama* Seni Silat Cekak Malaysia (SSCM) training efficiency. This Silat Cekak method's effectiveness is evaluated by assessing its angular acceleration and velocity. The tracking of 32 markers on the lower body joints of the practitioner in UniMAP's BioMechanical lab is done by employing Motion Capture (MoCap) technology. Three expert practitioners from the SSCM are the subjects for motion capture analysis. The motion capture analysis objective is to recognize potential zones for betterment. Moreover, it could be an asset for other SSCM practitioners in elevating their training methods and overall effectiveness. The facts obtained from this paper may be employed to appraise the abilities of both senior and junior practitioners through application development. The outcome of this paper is an application developed to assist the *Buah Pertama* SSCM training efficiency determination.

Keywords Training efficiency application · *Buah pertama* · Motion capture · Seni silat Cekak Malaysia

A. H. A. Halim (✉) · M. H. S. Shafawi · R. Farook · M. R. A. Rejab · S. A. Abuzar · M. S. Salim
Faculty of Electronic Engineering & Technology (FKTEN), Universiti Malaysia Perlis, Arau, Malaysia
e-mail: aznor@unimap.edu.my

Sports Engineering Research Center (SERC), Universiti Malaysia Perlis, Arau, Malaysia

A. H. A. Halim · R. Farook
Persatuan Seni Silat Cekak Malaysia (PSSCM), Putrajaya, Malaysia

Z. Mohamed et al. (eds.), *Proceedings of the 2nd International Conference on Innovation and Technology in Sports, ICITS 2023, 27–28 November, Malaysia*, Lecture Notes in Bioengineering, https://doi.org/10.1007/978-981-97-3741-3_19

1 Introduction

1.1 Motion Capture

Motion capture is a method for capturing and studying human movement. It has several uses in industries including sports, entertainment, and healthcare. Motion capture is frequently utilized in sports to enhance athletic performance by giving precise details on how the body moves throughout various activities. Motion capture technology has recently become more popular among coaches and athletes to enhance performance and lower injury risks.

Several studies reveal motion capture technology for sports analysis. A review analysis of the physical condition, athletic performance, technical expertise and injury mechanism, prevention, and rehabilitation in sports has been done from 2013 until 2016 (Pueo and Jimenez-Olmedo 2017). There is also an analysis of the acceleration specific to cricketers, for players who field primarily in the infield close to the pitch or outfield close to the boundary (Lockie et al. 2014). A motion capture system has been used to illustrate support leg dynamics during instep kicking (Inoue et al. 2014). Meanwhile, in combat sports, there is research on getting valuable data on joint's kinematic parameters to improve the execution of Silat "Jurus Satu" by using a 3D depth camera motion capture system as a tool (Kaharuddin et al. 2017). Research based on Taekwondo in obtaining information about kinematic indicators that determine the efficiency between two versions of taekwondo roundhouse kick in sport and traditional (Wasik et al. 2022) and the kinematic effects of the target on the velocity of round-house kick (Mustapha et al. 2014) is carried out. There are paper studies on the possibilities of kinematic analysis and the possibilities of their transfer to sport and training practices based on long-term experiences with the SIMI systems (Wąsik et al. 2021). A systematic survey of research publications was conducted in the field of martial arts biomechanics and shows a trend that tends to focus on offensive techniques rather than defensive techniques (Movement and Dlugosza 2016).

1.2 Seni Silat Cekak Malaysia

Seni Silat Cekak Malaysia (SSCM) was founded by Ustaz Haji Hanafi bin Haji Ahmad in 1965 and has actively been practiced until now. Persatuan Seni Silat Cekak Malaysia registered in 1975 had expanded throughout Malaysia with 13 branches (Mustapha et al. 2015; Persatuan Seni Silat Cekak Malaysia 2019). Major universities in Malaysia conduct classes of Seni Silat Cekak Malaysia (SSCM) which has a comprehensive syllabus for students to learn. There are 21 *buah* of Seni Silat Cekak Malaysia to be learned in four levels named *buah asas*, *buah jatuh*, *buah potong* and *tamat* (Malaysia et al. 2022; Halim et al. 2016).

One of the syllabus in SSCM is named *Buah Pertama*, which will use the fend-off technique of *Kaedah A* (Mustapha et al. 2014).

1.3 Motivation

In SSCM, research analysis on hand movement has been conducted by Mustapha et al. (2014), Halim et al. (2016) while in terms of leg movement is still lacking. The aspects of angular acceleration and angular velocity of the leg movement are important to determine the efficiency of the training. SSCM students can only improve their efficiency through normal training without knowing whether their training follows the correct skill or not. They also do not know the exact acceleration and velocity of their leg movement without any help from measurement tools. Therefore, this paper reports an application development by using the motion capture approach to determine training efficiency for *Buah Pertama* Seni Silat Cekak Malaysia (SSCM).

2 Method

2.1 Motion Capture Approach

The analysis in determining the threshold value for performance efficiency is performed at BioMechanic Lab at Universiti Malaysia Perlis. The lab has full features of the motion capture analysis system named Qualisys Tracking Manager. This software will analyze the body movement with six cameras positioned surrounding the subject. Three experts from SSCM who have more than 5 years of experience were selected as subjects. The analysis focuses on *buah asas Buah Pertama* in SSCM. Subjects wore suits with 32 markers for the lower body parts as labeled in Fig. 1 while in Fig. 2 the display in QTM was for data collection.

The experts performed *Buah Pertama* and the data from markers was captured by six Oqus cameras that surround the platform. The markers are labeled in the QTM software to identify the position of the segment and body part of the subject. The movement capture data is then converted to C3D format to be opened in Visual 3D. Visual 3D is used as the biomechanics analysis application that measures the motion and force accumulated to visualize the real subject's movement through simulation of the bone's movement created from the joint measurement of markers on the subject's suit.

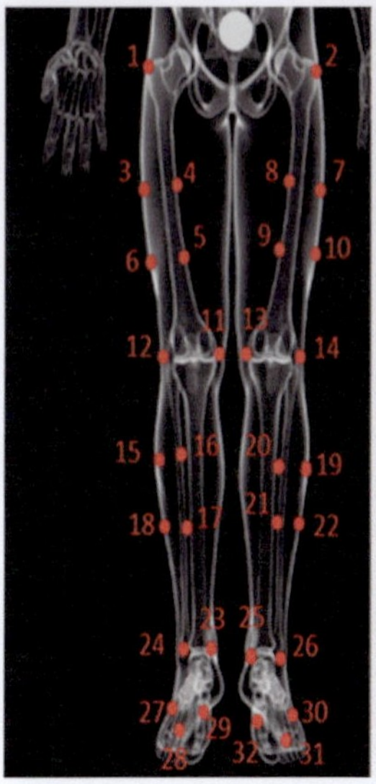

Aspect	Right Leg	Left Leg
Hip	[1] RH	[2] LH
Tigh	[3] RTA [4] RTB [5] RTC [6] RTD	[7] LTA [8] LTB [9] LTC [10] LTD
Knee	[11] RKI [12] RKO	[13] LKI [14] LKO
Shank	[15] RSA [16] RSB [17] RSC [18] RSD	[19] LSA [20] LSB [21] LSC [22] LSD
Ankle	[23] RAI [24] RAO	[25] LAI [26] LAO
Foot	[27] RFO [28] RFF [29] RFI	[30] LFO [31] LFF [32] LFI

Fig. 1 Marker of 32 points and its labels

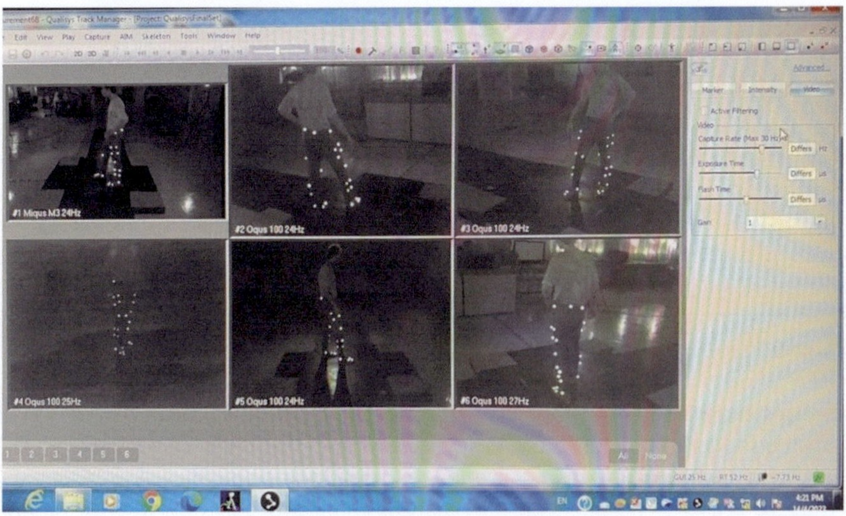

Fig. 2 QTM monitoring during data collection

2.2 *Application Development*

The design of the application involves GUI which provides a simple interface for accessing, visualizing, and analyzing motion capture data. The interface is designed in a user-friendly manner, with easy navigation, and efficient data exploration in

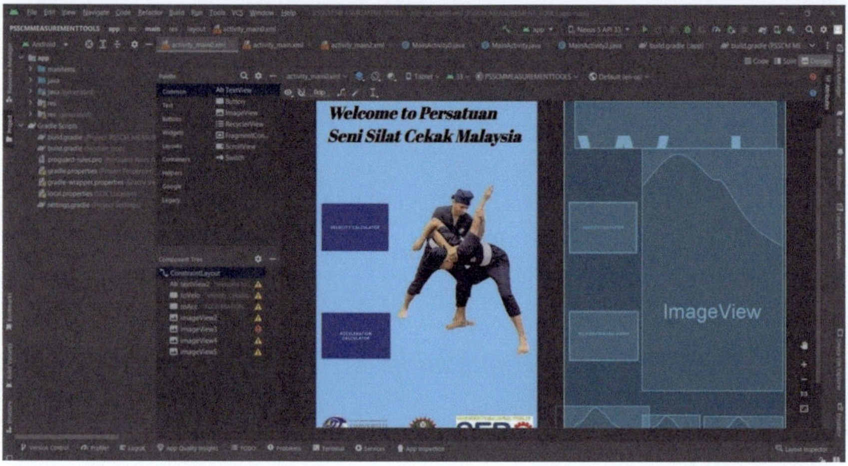

Fig. 3 Android studio application development

mind. In addition, the interface provides comparison tools for analyzing velocity and acceleration patterns and producing meaningful data for performance evaluation. Figure 3 is the GUI design developed during the application development phase using Android Studio. The output of the application gave results indicating the training efficiency based on data input by the user. Three levels of output are stated:

- ***Excellent***—the range of expert levels.
- ***Need more Practice***—the range did not reach the expert level.
- ***Invalid Data***.

3 Result

Two efficiency data used in determining the training performance are Angular Acceleration and Angular Velocity. Six points were selected as data collected from the three experts that later to be the threshold in determining the training efficiency for *Buah Pertama* SSCM are as in Table 1.

Table 1 Data was collected from the three experts using motion capture

Training efficiency points	Expert 1	Expert 2	Expert 3
Angular acceleration right foot (RFT) (rad/s^{-2})	326	290	245
Angular acceleration right shank (RSK) (rad/s^{-2})	205	97	162
Angular acceleration right thigh (RTH) (rad/s^{-2})	116	104	126
Angular velocity right foot (RFT) (rad/s)	5.67	5.0	5.7
Angular velocity right shank (RSK) (rad/s)	3.4	1.7	2.2
Angular velocity right thigh (RTH) (rad/s)	8	5.7	7.8

3.1 Training Efficiency Application

The application GUI of the home layout to determine the performance of *Buah Pertama* SSCM for new practitioners is shown in Fig. 4. Two types of training efficiency, which are Angular Acceleration and Angular Velocity, can be selected on the main screen. Based on the threshold gained from the previous motion capture analysis, this application could determine the new practitioner's performance. The example results on Angular Acceleration are shown in Fig. 5 while for Angular Velocity is shown in Fig. 6. From Figs. 5 to 6, it compares the new practitioner's data with the threshold data side by side and highlights any differences or similarities. This allows the new practitioner to assess their performance and identify areas for improvement. By examining the differences between their data in the application, practitioners gained valuable insights into their training efficiency and technique execution. This visual comparison aids in self-assessment, performance evaluation, and goal setting, contributing to the continuous improvement of the *Buah Pertama* SSCM technique and training efficiency.

Fig. 4 Home layout of the application

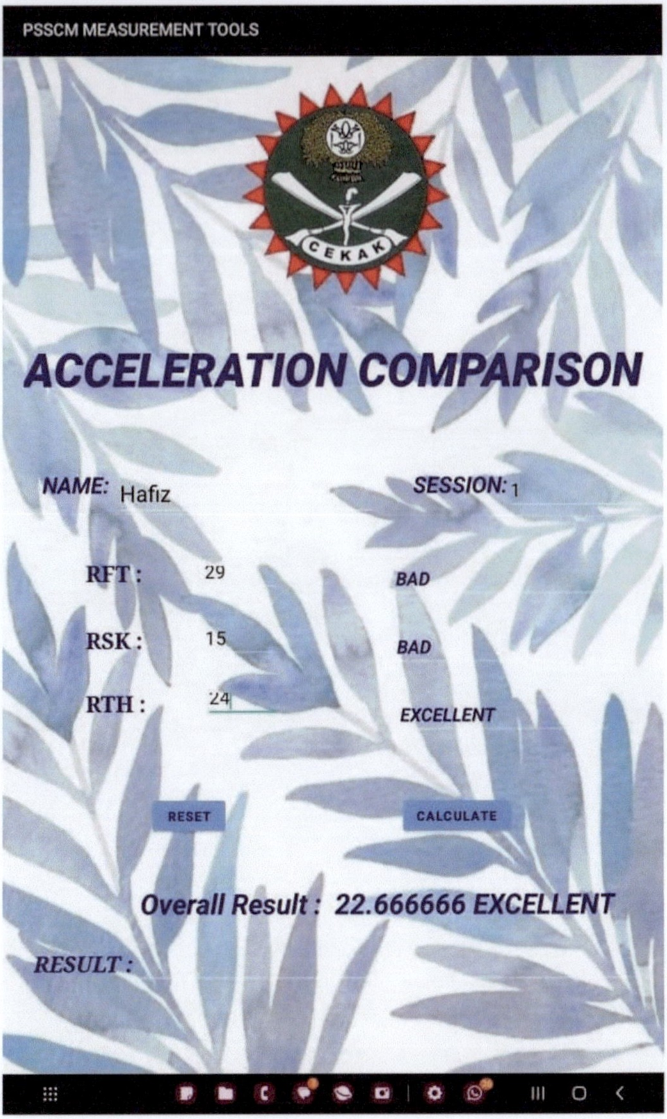

Fig. 5 Results on angular acceleration

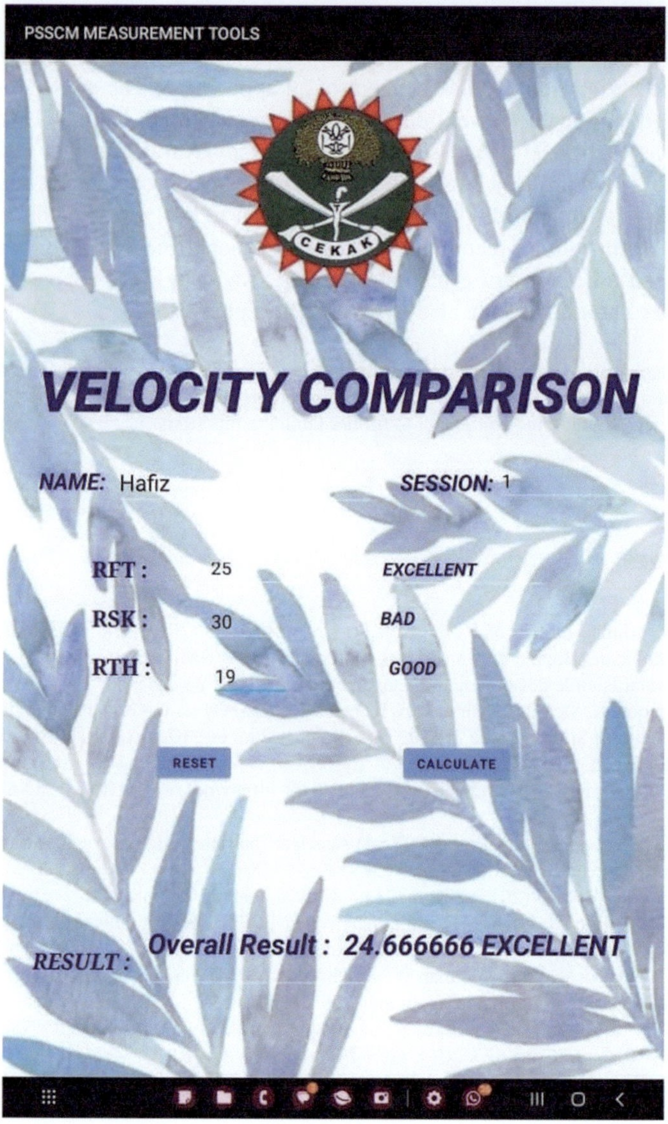

Fig. 6 Results on angular velocity

4 Conclusion

Determining *Buah Pertama* SSCM training efficiency is made possible through the development of the application by using a motion capture approach. Motion capture analysis helped reveal crucial factors influencing training effectiveness in SSCM. Angular acceleration and velocity data analysis by the expert subjects revealed a common pattern and set the benchmark for the tools of training efficiency determination. The improvement of training effectiveness in SSCM has been significantly contributed by this study's development of an application. The application's user-friendly interface enables practitioners to analyze data with the expert. This effective tool enables practitioners to measure themselves easily by tracking their performance regularly and identifying necessary changes. Expert-level proficiency is within reach as practitioners receive assistance via suggestions in optimizing their techniques through the application's recommendations and personalized training guidance. The accuracy and objectivity of evaluating training efficiency were enhanced with the integration of motion capture analysis within the application.

Acknowledgements The authors gratefully acknowledge the Persatuan Seni Silat Cekak Malaysia (PSSCM) and Center of Excellence—Sports Engineering Research Center, Universiti Malaysia Perlis for research funds and equipment to further facilitate this study.

References

Halim AHA, Ahmad N, Farook R, Syuhada A, Omar N, Ali H (2016) The 11th ISDEV international Islamic development management conference (11th IDMAC 2016). In: The 11th ISDEV international Islamic development management conference, pp 1–12

Inoue K, Nunome H, Sterzing T, Shinkai H, Ikegami Y (2014) Dynamics of the support leg in soccer instep kicking. J Sports Sci 32(11):1023–1032. https://doi.org/10.1080/02640414.2014.886126

Kaharuddin MZ et al (2017) Biomechanics analysis of combat sport (Silat) by using motion capture system. IOP Conf Ser Mater Sci Eng 166(1):0–6. https://doi.org/10.1088/1757-899X/165/1/012028

Lockie RG, Callaghan SJ, Jeffriess MD (2014) Acceleration kinematics in cricketers: implications for performance in the field. J Sport Sci Med 13(1):128–136

Movement I, Dlugosza AJ (2013) Kinematic analysis in combative sports

Mustapha G, Mahmud J, Zakaria M, Sulaiman WRW (2015) Biomechanics research on martial arts—The importance of defensive study. Arch Budo 11:187–195

Mustapha G, Zakaria M, Wan Sulaiman WR, Mahmud J (2014) The mechanical aspects of martial arts: total time of execution and Kinematics of Kaedah A. In: Proceedings of the international colloquium on sports science, exercise, engineering and technology 2014 (ICoSSEET 2014), 2014, vol 2014, no ICoSSEET 2014, pp 3–12. https://doi.org/10.1007/978-981-287-107-7

Mustapha G, Wan Sulaiman WR, Abdul Majeed APP, Mohd Yahya NH, Mahmud J (2014) Kinematics and efficacy analysis of the Seni Silat Cekak Malaysia (Kaedah A). Appl Mech Mater 680(October):267–270. www.scientific.net/AMM.680.267

Persatuan Seni Silat Cekak Malaysia (2019) CAWANGAN NEGERI. https://www.silatcekak.org.my/cawangan/. Accessed 01 Nov 2022

Persatuan Seni Silat Cekak Malaysia (2022) Mesyuarat Agung Perwakilan Persatuan Seni Silat Cekak Malaysia kali ke-50

Pueo B, Jimenez-Olmedo JM (2017) Application of motion capture technology for sport performance analysis. Retos 2041(32):241–247. https://doi.org/10.47197/retos.v0i32.56072

Wąsik J, Mosler D, Ortenburger D, Góra T, Cholewa J (2021) Kinematic effects of the target on the velocity of Taekwon-do roundhouse kicks. J Hum Kinet 80(1):61–69. https://doi.org/10.2478/hukin-2021-0103

Wasik J, Mosler D, Gora T, Ortenburger D, Chalimoniuk M, Langfort J (2022) Kinematic differences between traditional and sport version of roundhouse kick executed by male Taekwon-do masters. J Mens Health 18(6). https://doi.org/10.31083/j.jomh1806138

Effectiveness of Cardiovascular Training on Aerobic Capacity of Judo Players

Siti Noor Aisyah Shahabuddin, Adjah Naqkiah Mazlan, Zainal Abidin Zainuddin, Nur Fatihah Raimi, and Berto Apriyano

Abstract The aerobic capacity is a prerequisite for high training and competitive performance in judo. However, there is a lack of emphasis on appropriate cardiovascular training for judokas in the maintenance of their aerobic performance. Therefore, this study was conducted to compare the aerobic capacity of Johor judo athletes before and after four weeks of cardiovascular training. A total of six judo athletes aged between 18 and 20 years old were selected as subjects in this research. A 20 m shuttle run test was conducted to measure aerobic capacity before and after a four-weeks cardiovascular training program. The results of mean and standard deviation for pre-test was 39.9 ± 6.2 and for post-test was 43.0 ± 6.2. The means of VO_{2max} for both groups were compared using a dependent T-test in Statistical Package for Social Science (SPSS). The findings showed that there is a significant difference ($p \leq 0.001$) in the aerobic capacity of judo athletes before and after training cardiovascular. Thus, it can be confirmed that this four-weeks cardiovascular training program is suitable to be implemented to increase the aerobic capacity of judo athletes, as well as a training guide to improve the performance of judo athletes.

Keywords Cardiovascular training · Aerobic capacity · VO_{2max}

A. N. Mazlan (✉) · N. F. Raimi
Sport Innovation Technology Centre, Institute of Human Centered Engineering, Universiti Teknologi Malaysia, 81310 Johor Bahru, Johor, Malaysia
e-mail: adjahnaqkiah@utm.my

S. N. A. Shahabuddin · A. N. Mazlan · Z. A. Zainuddin · N. F. Raimi
School of Education, Faculty of Social Sciences and Humanities, Universiti Teknologi Malaysia, 81310 Johor Bahru, Johor, Malaysia

B. Apriyano
Department of Physical Education, Faculty of Sport Science, Universitas Negeri Padang, Padang, Indonesia

Z. Mohamed et al. (eds.), *Proceedings of the 2nd International Conference on Innovation and Technology in Sports, ICITS 2023, 27–28 November, Malaysia*, Lecture Notes in Bioengineering, https://doi.org/10.1007/978-981-97-3741-3_20

1 Introduction

Judo is a high-intensity sport that takes complex skills and good tactics to win something competition (Burdick 2019). The sport of judo is a system of unarmed combat where the objectives are to knock down the opponent by using limbs and causing the opponent to lose. Judo was established at the end of the ninth century twelve in Japan. Professor Jigoro Kano is the founder of the sport of judo that brought martial arts into human life (Wichers 2018). He also considered that balance is the most important component in judo. This is because if the athlete does not have good balance, it is easy to be thrown by the opponent.

Judo is a combat sport featured in the Olympics, where athletes compete based on their weight categories (Ouergui et al. 2022). Consequently, athletes often need to lose weight to compete in their designated categories. Moreover, judo also incorporates various fitness components to create well-rounded athletes, including cardiovascular endurance. Cardiovascular endurance is the heart's efficiency and capacity to supply oxygen to the muscles during physical activity. Typically, international-level judo athletes participating in matches lasting five minutes (Franchini et al. 2011) can compete in five to seven matches, making sufficient oxygen crucial for sustaining prolonged performance. In addition, assessing an individual's aerobic fitness often involves evaluating their cardiovascular endurance (Brennan 2021). Strong cardiovascular endurance not only makes daily tasks easier to carry out but also lowers the risk of chronic conditions like diabetes and heart problems.

Aerobic fitness level can be measured by taking the heart rate while resting (Jones 2022) and to ensure the heart is at rest, the ideal time to take resting heart rate is after waking up on time morning or at least more than two hours after exercise. Resting pulse rate which is low indicates a good level of aerobic fitness. Cardiovascular endurance is the ability of the heart to supply oxygen to muscles for energy production. Cardiovascular endurance tests are often used to measure aerobic capacity (Jones 2022). Capacity aerobic is the maximum amount of oxygen that can be taken by an individual during exercise at high intensity and is known as VO_{2max} (Bortenschlager 2021). Physiologically, the oxygen that enters the lungs will be sent to the bloodstream and the heart will pump oxygenated blood throughout the body to transfer oxygen to muscle groups to work.

Previous research on aerobic capacity of judo players was conducted based on scale for differences in body size. Here, this study is a review of the assessment of the fitness levels of Judo athletes in Johor to measure their aerobic capacity. Improving aerobic capacity through cardiovascular training has become a necessary goal for every outstanding athlete. Each athlete has physiological differences that result in varying oxygen intake levels among athletes (Park et al. 2016). Fitness and a healthy body are essential preparations for athletes to engage in training. Judo is a well-known self-defense art worldwide that combines physical, mental, and moral development (Protic-Gava et al. 2019). Judo involves overall physical development, including flexibility, power, strength, balance, agility, and coordination. Furthermore, the aerobic capacity of Judo athletes is greater than that of soccer or gymnastics athletes. During matches, athletes need to compete for four minutes. Therefore, every Judo athlete requires a high aerobic capacity to supply oxygen to the muscles to keep them working. Fitness training programs aim to improve the cardiovascular levels of athletes.

Judo is a high-intensity combat sport that requires good physiology and technical excellence to succeed in a match (Moreira et al. 2021). Officially, matches last for four minutes for both men and women. During training or matches, players are often exposed to situations where they must make the right moves. Not only that, but muscle endurance is also required during matches so that athletes can endure and compete against their opponents.

Thus, this study aims to investigate the effectiveness of cardiovascular exercises on the aerobic capacity of Judo athletes from the state of Johor, Malaysia. The study also aims to justify the athletes' aerobic capacity level concerning cardiovascular training.

2 Methodology

This study utilizes a 20 m shuttle run test conducted on Judo athletes to obtain their VO_{2max} scores. The VO_{2max} scores are calculated using the formula proposed by A D Flouris in 2005 based on previous research (Magee et al. 2021) (see Table 1).

The sample represents a subset of elements within the population. In this context, subjects refer to individuals included in the sample. The researcher has selected a sample consisting of Judo athletes based camp in Kluang, Johor. The sample consists of six athletes, comprising 3 male and 3 female athletes aged between 18 and 20 years, who maintain an active lifestyle. The selection of these subjects was based on appropriate characteristics such as age, weight, and height. Additionally, all subjects were required to be free from any injuries or illnesses. As a result, the researcher conducted

Table 1 Estimated VO_{2max} for each 20 m shuttle run

Stage	Shuttle	Distance (m)	VO_{2max} (Ml/Kg/Min)
1	7	140	16.7
2	8	300	23.0
3	8	460	26.2
4	9	640	29.3
5	9	820	32.5
6	10	1020	35.7
7	10	1220	38.8
8	11	1440	42.0
9	11	1660	45.1
10	11	1880	48.3
11	12	2120	51.5

the test according to the sample selection that had been set and based on the study design and procedure (see Fig. 1).

In addition, this research procedure required subjects to fill out an indemnity form beforehand, indicating their consent to participate in this study. Subsequently, the subjects undergo a 20 m shuttle run test (beep test) at the Kluang Stadium running track to measure their VO_{2max}. After that, the subjects will undergo a cardiovascular training program for a duration of 4 weeks. The training frequency is 3 days per week (see Fig. 2).

After the cardiovascular training program is conducted on the subjects, the researcher will repeat the 20 m shuttle run test to measure VO_{2max}. Subsequently, the analysis of the differences in VO_{2max} before and after for Johor Judo athletes will be carried out.

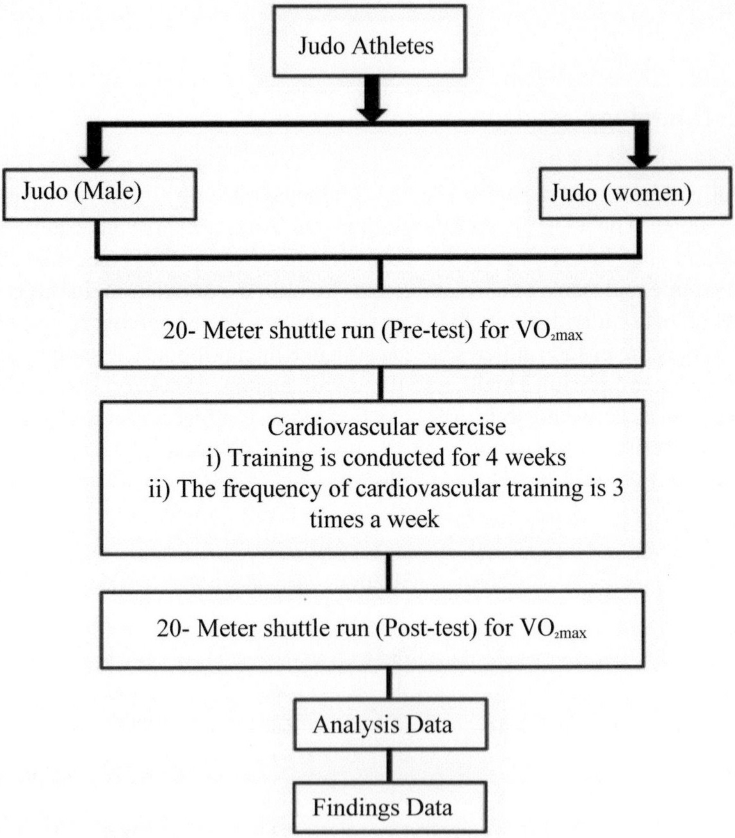

Fig. 1 Research flow chart

	Day 1	Day 2	Day 3
Week 1	• Long Slow Distance 10km	• Fartlek Training	• Run 2.4km
Week 2	• Shuttle run	• Interval run 400m and 800m	• Long Slow Distance 10km
Week 3	• Run 2.4km	• Shuttle run	• Swimming 50m (4 set)
Week 4	• Fartlek Training	• Long Slow Distance 10km	• Interval run 400m and 800m

Fig. 2 Cardiovascular training program (4W-3D)

3 Results and Discussion

The primary objective of this study was to justify the aerobic capacity of Johor judo athletes before and after four weeks of cardiovascular training progress and make a comparison of VO_{2max} readings based on dependent T-test inference using the *SPSS*. The findings are summarized below.

3.1 Effectiveness of Cardiovascular Training on the Aerobic Capacity of Judo Players

Table 2 shows the pre-test and post-test results for measuring the VO_{2max} readings of Johor Judo athletes. The mean and standard deviation for the pre-test was 39.9 ± 6.2, while for the post-test, it was 43.0 ± 6.2. The study's findings indicate that all subjects recorded an improvement in their VO_{2max} scores. The *p*-value obtained for this study is $p \leq 0.001$, and the significance level set is $p \leq 0.05$. This indicates that the cardiovascular training provided significantly improved the aerobic capacity of Johor Judo athletes.

After undergoing cardiovascular training, the 20 m shuttle run test conducted to record VO_{2max} showed significant improvements. Fitness training including cardiovascular exercises can enhance an athlete's VO_{2max} (Bahtra et al. 2020). For individuals with a high VO_{2max}, it can help athletes perform well in competitions. If an

Table 2 The effectiveness of cardiovascular training

Subject	Test	Pre-test	Post-test
S1	20 m shuttle run	48.3	51.5
S2		45.1	48.3
S3		42.0	45.1
S4		29.3	32.5
S5		38.8	42.0
S6		35.7	38.8
	Mean	39.9	43.0
	Standard deviation (SD)	6.2	6.2

athlete has good skills but lacks a high VO_{2max}, they may struggle to compete against international players during matches (Bahtra et al. 2020).

The research findings obtained are parallel because Long-Slow Distance Run was used since it was a continuous training that could increase cardiovascular fitness and planning systematic training. This type of physical activity needs to be a means to an end and must be combined with other means of training, including speed work (Hian and Hoe 2021). This is also same with interval run which is gaining popularity with endurance athletes. Various studies have shown that interval run or training allows for similar or greater endurance, strength, and power performance improvements than traditional endurance training but demands less time and volume (Koral et al. 2018).

Thus, the four-week cardiovascular training program (4W-3D) proved to be effective in improving the aerobic capacity of Johor Judo athletes. The study's findings indicate a significant improvement in the effects of cardiovascular training. Increased VO_{2max} can assist Judo athletes in enduring longer periods without feeling easily fatigued during matches.

3.2 The Comparison VO_{2max} Before and After Cardiovascular Exercise

Table 3 shows the results of the t-test analysis. The analysis using the dependent Samples t-test indicates that the t-value (5) = −150.208, $p \leq 0.001$, is significant. This is because the p-value of ≤ 0.001 is smaller than the set significance level of $p \leq 0.05$. After undergoing cardiovascular training, there is a difference in VO_{2max} readings among Johor Judo athletes.

Based on the analyzed data from the 20 m shuttle run test to obtain VO_{2max} readings before and after undergoing cardiovascular training on the subjects, it can be formulated that there is a significant difference in the aerobic capacity of Judo athletes before and after cardiovascular training. Physical exercise plays a crucial role in maintaining or improving an individual's physical fitness, including cardiovascular

Table 3 The comparison effectiveness of cardiovascular training

	Mean (M)	Standard deviation (SD)	t	df	*p*-value
Pre-test	39.9	6.2	−150.208	5	0.001
Post-test	43.0	6.2			

training (Nugroho et al. 2022). The second finding of the study analysis experienced a significant improvement after undergoing cardiovascular training. This is because the *p*-value is ≤ 0.001, indicating a difference in VO_{2max} readings after the subjects underwent cardiovascular training. Therefore, it can be concluded that Ha1 is accepted and Ho1 is rejected because there is a significant difference in the aerobic capacity of Judo athletes before and after cardiovascular training.

4 Conclusion and Contributions

The results indicate that there is a change in VO_{2max} readings before and after undergoing cardiovascular training. The presented 4W-3D cardiovascular training regimen affects the VO_{2max} readings of each Johor Judo athlete. Maximum oxygen uptake during the 20 m shuttle run test refers to the cardiovascular system's ability to deliver oxygen to the muscles to perform work. Therefore, individuals with high maximum VO_{2max} values have excellent fitness. Thus, this study successfully demonstrates that the 4W-3D cardiovascular training is effective in enhancing the aerobic capacity of Judo athletes and helping improve their performance. Hence, more cardiovascular training should be conducted for Judo athletes to improve their performance, and it is hoped that this training can serve as a guide for all judokas and for future research in diversifying training methods to enhance aerobic capacity.

Acknowledgements Special thanks to Universiti Teknologi Malaysia, Skudai, Johor, Malaysia for research grant scheme Q.J130000.2653.16J92 and Persatuan Judo Negeri Johor team members for your kind support.

References

Bahtra R, Asmawi M, Widiastuti, Dlis F (2020) Improved vo2max: the effectiveness of basic soccer training at a young age. Int J Hum Mov Sports Sci 8(3)

Bortenschlager M (2021) Why aerobic capacity matters and how to measure it

Brennan D (2021) What is cardiovascular endurance

Burdick D (2019) Fifty years of Korean martial arts at Indiana University Bloomington. https://doi.org/10.13140/RG.2.2.27043.71200

Franchini E, Del Vecchio FB, Matsushigue KA, Artioli GG (n.d.) Physiological profiles of Elite Judo Athletes (2011)

Hian TC, Wee EH, Ling OS, Choong TY (2021) The effectiveness of physical activity (long slow distance run) in weight lose for adults. In: International conference on social science & social research
Jones C (2022) What is cardiovascular endurance? (and how to improve it) your heart health will greatly benefit. Share this
Koral J, Oranchuk DJ, Herrera R, Millet GY (2018) Six sessions of sprint interval training improves running performance in trained athletes. J Strength Cond Res 32(3):617–623
Magee MK, White JB, Merrigan JJ, Jones MT (2021) Does the multistage 20-m shuttle run test accurately predict VO2max in NCAA Division I Women Collegiate Field Hockey Athletes?
Moreira DG, Brito CJ, de Almeida Ferreira JJ, Marins JCB, de Durana ALD, Canalejo JC, Revenga JB, Sillero-Quintana M (2021) Lactate concentration is related to skin temperature variation after a specific incremental judo test. J Strength Condition Res 35(8)
Nugroho RA, Mahfud I, Jubaedi A (2022) Pengaruh Latihan Shuttle Run Terhadap VO_{2max} Peserta Ekstrakurikuler Bulutangkis. Sport Sci Educ J 3(1)
Ouergui I, Delleli S, Chtourou H, Formenti D, Bouhlel E, Ardigò LP, Franchini E (2022) The role of competition area and training type on physiological responses and perceived exertion in female Judo Athletes. Int J Environ Res Public Health 19(6)
Park H-Y, Hwang H, Park J, Lee S, Lim K (2016) The effects of altitude/hypoxic training on oxygen delivery capacity of the blood and aerobic exercise capacity in elite athletes-a meta-analysis for exercise nutrition
Protic-Gava B, Drid P, Krkeljas Z (2019) Effects of judo participation on anthropometric characteristics, motor abilities, and posture in young judo athletes. Hum Mov 20(3)
Wichers JBF, Hattink B (2018) Judo as a supportive tool for business and entrepreneurship. Qual Sport 4(2)

Designing the Right Games Strategy: Evidence-Based Technology of Teaching Games for Understanding (TGfU) Model

Diyana Zulaika Abdul Ghani, Zainal Abidin Zainuddin, Asha Hasnimy Mohd Hashim, Halijah Ibrahim, and Ani Mazlina Dewi Mohamed

Abstract Teaching games for understanding (TGfU) model is an approach that implements tactical games to develop and enhance individual skills in sports. Focus X2 is a technology used to evaluate the sports performance through notational analysis. Notational analysis is a procedure that assesses movement pattern and strategy in sports performance. Notational analysis could facilitate learning skills and supported the TGfU model from the empirical scientific feedbacks generated to improve individuals' performance. The aim of this study is to compare the empirical data on professional athletes and school squash players during matches using technology via notational analysis and to apply TGfU teaching model in order to seek the playing pattern for personalized learning development. A total of 17,361 shots were analyzed using the Focus X2 games analysis software. Analysis was based on observation through video recording of the shots performed throughout the tournaments. Analyses were mainly focused on rallies at 16 divided areas of squash court which were evenly divided into front, middle, and back areas. Results indicated that Area 13 (back-sidewall area) is the area with the highest number of shots performed while the lowest number of shots performed was recorded at Area 14 and Area 15 (back-middle areas) with zero percentage. Through these results, although the standard of playing was at a different level, the tactical approaches used were the same. TGfU approaches focus on the tactical aspect first rather than technical had been proven significant where individuals tend to react more by improvising their playing pattern according to the situation using basic techniques. Tactically in squash, playing constantly at the back area of the court and near to the sidewall (Area 13) could ensure the opponent

D. Z. Abdul Ghani (✉)
Sports Innovation & Technology Centre (SITC), Institute of Human Centered Engineering (iHumEN), Universiti Teknologi Malaysia, 81310 Skudai Johor, Malaysia
e-mail: diyanazulaika@utm.my

D. Z. Abdul Ghani · Z. A. Zainuddin · A. H. Mohd Hashim · H. Ibrahim
School of Education, Faculty of Social Sciences and Humanities, Universiti Teknologi Malaysia, 81310 Skudai Johor, Malaysia

A. M. D. Mohamed
Faculty of Education, Universiti Teknologi MARA, Puncak Alam Campus, Shah Alam, Malaysia

Z. Mohamed et al. (eds.), *Proceedings of the 2nd International Conference on Innovation and Technology in Sports, ICITS 2023, 27–28 November, Malaysia*, Lecture Notes in Bioengineering, https://doi.org/10.1007/978-981-97-3741-3_21

to be away from the center of the court and create opportunities to play offensively to pressure the opponent to commit mistakes. Thus, shots were greatly played in this area. Findings via technology signify the effectiveness of TGfU concept in strategizing the playing pattern in squash for individuals and could be used as feedbacks to reflect on individuals' self-learning performance.

Keywords TGfU · Squash · Game strategy · Notational analysis · Technology · Playing pattern

1 Introduction

Playing is synonymous with children. Through play, children employ their creativity to nurture their imaginations in the context of games. Playing also plays a crucial role in fostering the development of a healthy brain, as it involves the interaction of cognitive, physical, and emotional aspects during their engagement with games (Ginsburg 2007). TGfU model is one of the approaches that use the concept of playing to enhance children's learning skills. According to Hopper (2001), TGfU focuses more on the tactical performance rather than technical performance. In physical education, children always involve with specific skills acquisition, especially in specific sports that require special equipment and surrounding.

The TGfU approach, pioneered by Bunker and Thorpe in 1982, offers a distinct perspective on game instruction. This model emphasizes immersing students or athletes in game scenarios that prioritize the importance of tactics, decision-making, and problem-solving. Other variations and terminology include ; 'Game sense', 'play practice', and 'game centered approach' (Webb and Pearson 2008). After that, feedbacks from children and teacher will improve their understanding and skills towards the game. It is believed that tactics and skills intertwine with each other and by applying the tactics first, children will automatically learn the skill as well (Hopper 2001). TGfU enhances the importance of playing the games more than how to play the games.

According to Mandigo et al. (2008), TGfU has four game categories for teaching sport specific purposes;

1. Target games (e.g., Dart, Bowling, or Golf).
2. Net/Wall games (e.g., badminton, netball, or squash).
3. Striking/Fielding games (e.g., Baseball, Softball, or Cricket).
4. Territory games (e.g., Football, Handball, or Basketball).

These categories encompass games with similar structures, aimed at providing broad descriptions to children for each category. The skills and strategies taught can be applied across various sports. For instance, if children grasp the fundamental concept of maintaining possession of an object in a territory-based game (such as utilizing different passing techniques, ball control, supporting teammates, etc.), this understanding can be extended to playing a range of territory-based games whose tactics can be transferred to related sports (such as basketball, soccer, handball, and more). Consequently, this enables them to improve their skills in multiple sports disciplines. As in competitive sports, the model of TGfU has been applied repeatedly through trainings and matches. Their performances were recorded and analyzed using technology to obtain empirical data as a guideline to improve performance. One of the approaches via technology that has been used is notational analysis that was defined as a study of pattern in movement, strategy, and tactics during a game (Hughes and Franks 2004). Figure 1 illustrates the concept of sports performance via technology. Both are intertwined with each other to obtain the optimal data and feedbacks for development and improvement. Notational analysis approach could assist in enhancing the performance via empirical data and feedbacks.

Therefore, notational analysis via technology was applied for this study to compare the empirical data on Malaysian professional athletes and school squash players during matches and how may TGfU model can be applied in personalized learning in order to get the right game strategies in squash. Data obtained can be used as a reference to strategize for children's learning improvement for the future.

2 Methodology

2.1 *Participants*

Data were collected during Malaysia Circuit Squash ($n = 25$), and the National Junior Championship ($n = 25$) tournaments. Participants were Malaysia professional athlete ($n = 5$) and school's squash players ($n = 5$) with mean $\pm$ standard deviation of age was 24.3 ± 6.57 years; weight $= 60 \pm 11.33$ kg; height $= 167.7 \pm 8.58$ m. All athletes were selected based on discussion and suggestion by the Head Coach of Squash Racket Association Malaysia (SRAM) according to the ranking and playing experiences of all Malaysian squash players.

Professional athletes' group were identified as experienced elite players by association and gained World rankings at the least average number of rank positions within 83.2 ± 55.58. On the other hand, school athletes were identified as under Malaysia's junior backup players program.

All the athletes selected were ensured to be right-handed to maintain consistency during analysis. School's ethics committee acknowledged ethical approval for school athletes' participants. All athletes were informed and conscious that their games were being recorded and observed during competition.

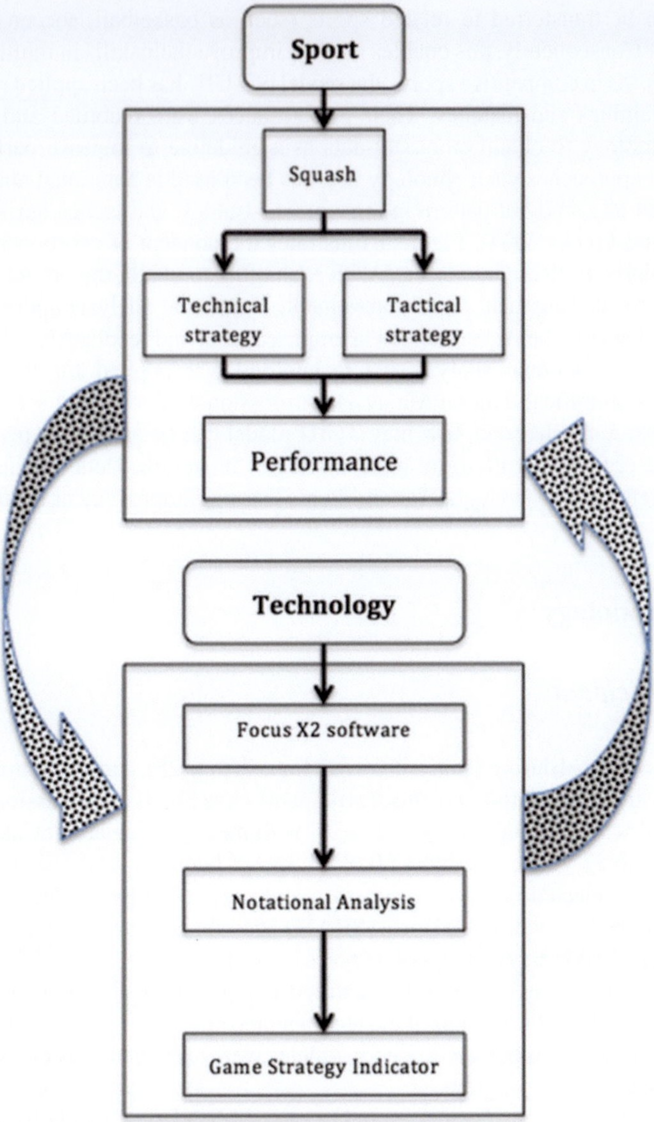

Fig. 1 Conceptual framework of sport specific squash and technology

2.2 *Procedure*

This study involved a notational analysis approach based on the video recording and the number of shots occurring during the rally. Data collection used a video camera, Digital Video Camera Recorder (DCR-PC350E, lens Carl Zeiss Vario-Sonnar T, 720 × 576 pixels, 25 frames per second) which was stationed at the frontal view at the back of the court with an overall view of the court ensured being captured and recorded. The video recording started when players started the serve and ended when the umpire announced the end of the game.

3 Notational Analysis

The technology tool used in this study was Focus X2 elite sports analysis software (www.performanceinnovation.net, UK). The analysis window design was generated based on the playing areas of squash court dimension and the number of shots performed during matches were counted during data analysis. The following variables were coded for each event:

- Area of playing (Area 1, Area 2, Area 3, Area 4, Area 5, Area 6, Area 7, Area 8, Area 9, Area 10, Area 11, Area 12, Area 13, Area 14, Area 15 and Area 16)
- Number of shots performed for each area during rally in matches.

An experienced notational analyst categorized the area of playing as pictured in Fig. 2. The court was divided according to the previous studies done by Hughes et al. (2006), McGarry and Franks (1994, 1995), Murray and Hughes (2001).

Each area mentioned above was encrypted in the editor window. The number of shots executed in each respective area were recorded for statistical analysis.

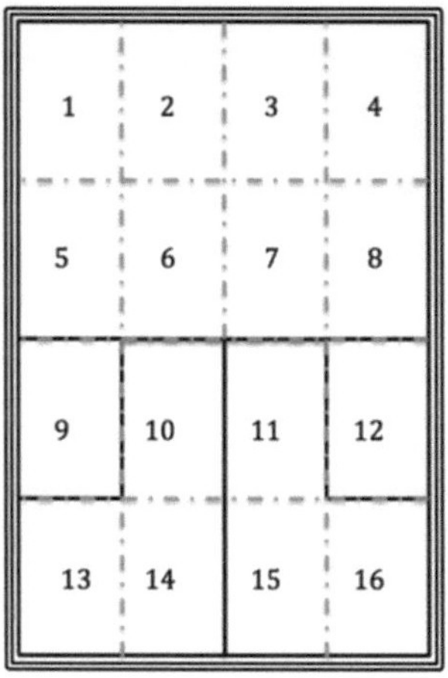

Fig. 2 The division of playing area

3.1 Reliability

The experts for notational data analysts had ten years of playing experience in squash, coupled with seven years of expertise in performance analysis. Data errors are defined as either from the analyst hits the wrong button or incorrectly categorizes an action (James et al. 2002). Thus, to quantify the reliability of the error, an intra-rater reliability approach has been done where the data analyst first viewed half of the total matches and compared with the second time viewed data gathered (the observation has been done twice) using Cohen's Kappa reliability test.

The intra-rater reliability analysis resulted in an overall Kappa value of 0.794 for the area of playing. Therefore, reliability result for the area of playing was indicated as acceptable and reliable to conduct further analysis (Sim and Wright 2005).

4 Focus X2 Software Data Analysis

This study used Focus X2 game analysis software because of its ability to observe and evaluate game performance according to the prior designated features in category panes effectively and efficiently. Focus X2 software also allows the analyst to independently view and control the key aspects of performance which are priorly set up during the beginning of the data analysis process.

Initially, the analyst labeled 16 areas for playing areas. Then, all shots played for all matches were analyzed and recorded using Focus X2 editor window for each playing area. Results attained were exported to Microsoft Excel and SPSS for additional empirical analysis.

5 Results

Professional athletes executed a total of 10,530 shots and Fig. 3. illustrates a full diagram of playing areas distribution performed. Area 13 recorded 2,669 shots indicating the highest compared to other areas (Fig. 3, orange box) and area with second highest shots performed was Area 9 with 2,042 shots (Fig. 3, purple box). Meanwhile, both middle areas at the back area (Area 14 and Area 15) showed the least percentages (Fig. 3, green boxes) of balls played during matches.

On the other hand, school athletes executed a total of 6,831 shots. Figure 4 shows playing areas distribution played by school athletes. Area 13 was recorded as the most frequent area played with 1,478 shots (Fig. 4, orange box) followed by Area 9 with 1,206 shots (Fig. 4, purple box). Similar to professional athletes, school athletes also recorded both Area 14 and Area 15 as the least percentages (Fig. 4, green boxes) of balls played during matches.

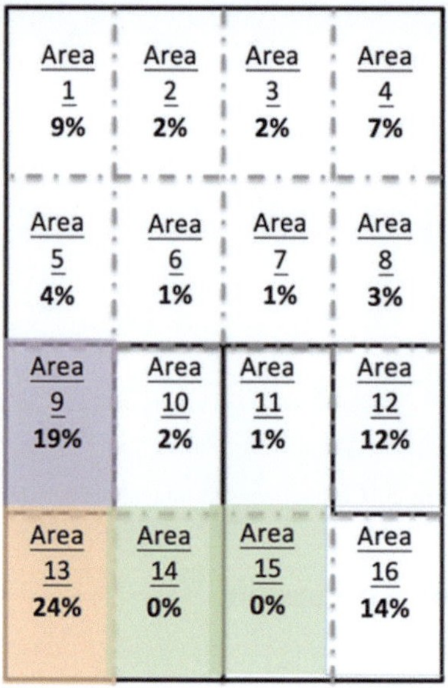

Fig. 3 Distribution of shots performed by professional athletes

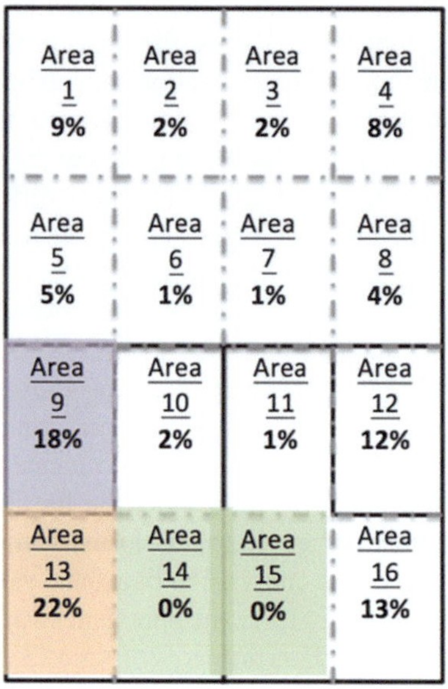

Fig. 4 Distribution of shots performed by school athletes

Figure 5 shows the total shots performed by professional and school athletes. Although the total number of shots performed were different, the pattern of tactical strategies played by both groups is similar in terms of playing area chosen to be played for overall matches.

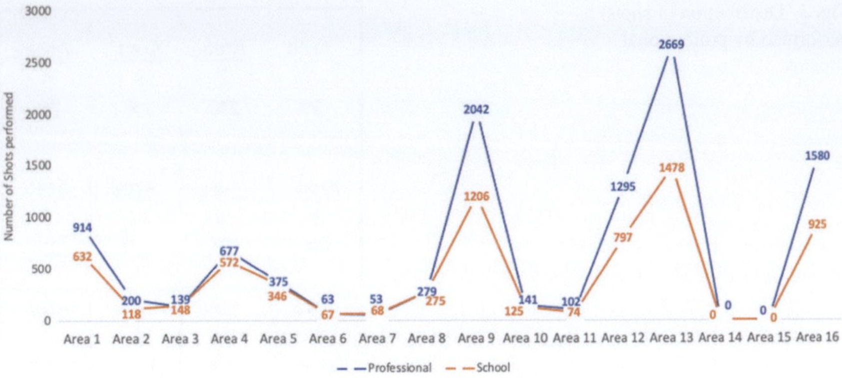

Fig. 5 Comparisons of total shots performed between professional and school athletes

6 Discussion

In general, professional athletes had 61% of shots performed compared to school athletes, with only 39% of shots performed. According to Sim and Wright (2005), professional athletes recorded more game time during competition compared to school athletes thus, the shots performed are more and aligned with the result of this study. This situation may arise from the extensive experience of professional athletes who had been involved in numerous tournaments and possessed knowledge and superior skills compared to the school athletes (Vuckovic et al. 2005). Findings by Hughes et al. (2000) show differences in different standards of playing. They indicated that as the ranking increases, the techniques applied in order to win the rally become more advanced thus prolonging the game duration. This explains the contrast in shot distribution between different standards of playing in this study.

Findings indicated that Area 13 and Area 9 had been recorded as the two most frequent areas played during rally. Although both groups had differences in their playing levels, both groups performed same playing areas throughout matches. Those two playing areas were also located at the back area and near to the side wall and athletes performed shots via backhand skill. Based on (Hughes et al. 2006) and (Vucković et al. 2013), during competition players predominantly play at the back area of the court using backhand skills. In squash, players typically adopt offensive approach to force the opponent away from the center of the court and by strategically directing shots primarily at Area 13 which is at the back of the court, players maneuver the opponents away from the center of the court hence produce distractions and increase the likelihood of winning points.

This basic tactical strategy in squash indirectly implemented TGfU where in the category of wall/net games, it focused on the possession of area. Players either beginners or experts have to ensure that they control the optimal area to win the game and withdraw their opponent away from the optimum area.

On the other hand, both groups also had recorded two areas at the middle and near to the back wall (Area 14 and Area 15) as the lowest percentage of balls played throughout matches. According to the (Vucković et al. 2014), performing shots at the middle areas is a straightforward strategy and gives advantages to the opponent to attack thus players least performing shots at those areas to avoid losing points. Furthermore, due to some circumstances during rally, it could lead to the chances of replaying the rally if the opponents blocked their way and had to ask for Let decision. As a consequence, players rarely performed shots in these areas to avoid disadvantages for their rally and concentrate more on the other benefited areas to gain points (Vucković et al. 2014).

Again, the concept of TGfU was implemented during games by players. Hence, it can be said that TGfU model is important for individuals' learning development where it reflects autonomous tactical and technical strategies during games.

Figure 6 shows the main concept of TGfU, sport specific and technology. TGfu has 4 main categories and squash is in the net/wall games category. Squash consists of technical and tactical strategy where performance is the end product and this performance could be reflected and improved with the aid of technology tool.

Technology procedures produced feedbacks to help increase the performances and from results, showed that players from different standards of playing had applied similar basic tactical strategy throughout the games and these were aligned with TGfU concept where it focuses on game teaching are linked as two essential components; "tactical-to-skill" concept, where the focus is enhancing tactical awareness before skill learning, not tactical understanding instead of skill learning. Supported by the dynamic system theories, the key to learning games as (Chandler 1996; Thorpe

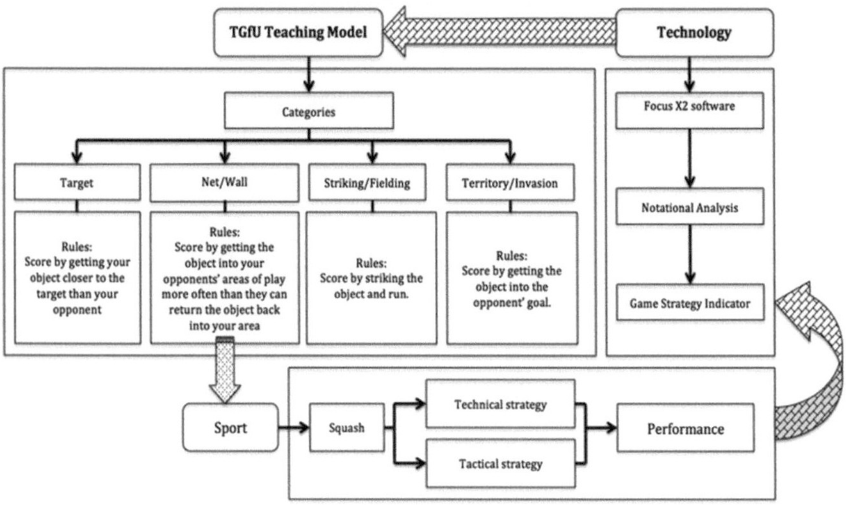

Fig. 6 Operational framework of TGfU teaching model, Sport specific squash and technology

et al. 1986), all supported the TGFU emphasizing games by using the individual self-motivation of games to increase skillfulness.

In TGFU, more focus is given on the tactical learning and lesser on technical requirements could endorse individuals to formulate their own tactical strategies based on the available information they have. Through TGfU, players could acquire tactical strategies based on the integration of the position (center of the court), environment (areas of court) and also movement of the opponents. They created right tactical strategies in order to win the rally. As players generate more significant strategies, they also develop new skills and further refine previously learned skills in order to develop more effective winning strategies.

In the meantime, via technology, it can be seen that the effectiveness of TGfU concept was applied even though the standard of playing for both groups is different but the basic performance shown is similar. Therefore, it can be said that the level of achievement gained through experiences but basic skills developed through games. It is individual self-development through practical application of all tactical and technical strategies simultaneously. These results have shown that technology signifies the importance of applying TGfU concept towards the learning development of individuals and feedbacks from notational analysis helps athletes to understand their own playing pattern during games and can be used as a reference to improve their performance for the future.

7 Conclusion

Teaching games for understanding (TGFU) is an approach to game teaching where the play of a game is implemented before skill refinement. When instructing in games, it proves challenging to disentangle skills from tactics; in actuality, the strategic utilization of skills forms the core of proficient game performance. After analyzing a TGFU concept from a skill to tactical development perspectives, this study used technology to analyze the implementation and practicality of TGfU towards sports specific game teaching. Squash game is the best sport to experiment the importance and effectiveness of TGfU approach where it has many dynamical constraints (space, opponent's motion, etc.) and will challenge individual's capabilities to improve their skills and to show how TGFU and skill focused approaches influence individual's learning. Results obtained signify the effectiveness of TGfU concept in determining and designing right game strategy for young athletes which is similar to professional athletes' game pattern. Nevertheless, young athletes still need more game experience to achieve the same level of performance as professional athletes. This paper establishes that technology strongly substantiates the significance of the TGfU model in game instruction, with its effectiveness being readily apparent. Furthermore, when viewed from the perspective of tactics-to-skills within the TGfU framework, individuals can effectively achieve self-directed learning and development.

Acknowledgements This study was conducted using the Research University Grant (Q.J130000.2631.14J95) of Universiti Teknologi Malaysia (UTM), Malaysia.

References

Chandler T (1996) Teaching games for understanding: reflections and further questions. J Phys Educ Recreat Dance (JOPERD) 67(4):49–51

Ginsburg KR (2007) The importance of play in promoting healthy child development and maintaining strong parent-child. Am Acad Pediatr 182–191. https://doi.org/10.1542/peds.2006-2697

Hopper T (2001) Teaching games for understanding: what does it look like and how does it influence student skill acquisition and game performance? J Teach Phys Educ. http://web.uvic.ca/~thopper/WEB/articles/JTPE/TGFU.htm

Hughes M, Well J, Matthews K (2000) Performance profiles at recreational, county and elite levels of women's squash. J Hum Mov Stud 39:85–104

Hughes MD, Franks IM (2004) Notational analysis of sport: a better systems for improving coaching and performance, 2nd edn., London

Hughes M, Howells M, Hughes M (2006) Using perturbations in elite men's squash to generate performance profiles. In: Rendimiento A (ed) IV Congreso Mundial de Ciencia y Deportes de Raqueta. www.altorendimiento.net

James J, Jones N, Hollely C (2002) Reliability of selected performance analysis systems in football and rugby. In: 4th international conference on methods and techniques in behavioural research, pp 116–118

Mandigo J, Butler J, Hopper T (2008) What is teaching games for understanding? A canadian perspectives. Phys Health Educ J 2(72):14–20

McGarry T, Franks IM (1994) A stochastic approach to predicting competition squash match play. J Sport Sci 12:573–584

McGarry T, Franks IM (1995) Modelling competitive squash performance from quantitative analysis. J Hum Perform 8(2):113–129

Murray S, Hughes M (2001) Tactical performance profiling in elite level senior squash. (P. D, Ed.)Pass.Com performance analysis. Sport Sci Comput (Crowood Press, Wiltshire)

Sim J, Wright CC (2005) The Kappa statistic in reliability studies: use, interpretation, and sample size requirements. J Am Phys Therapy Assoc 85(3):257–268. http://doi.org/15733050

Thorpe R, Bunker D, Almond L (1986) Rethinking games teaching. University of Technology, Loughborough

Vucković G, James N, Hughes M, Murray S, Sporis G, Pers J (2013) The effect of court location and available time on the tactical shot selection of elite squash players. J Sports Sci Med 12:66–73

Vuckovic G, Dezman B, Pers J, Kovacic S (2005) Motion analysis of the international and national rank squash players. In: ISPA 2005. Proceedings of the 4th international symposium on image and signal processing and analysis, pp 334–338. https://doi.org/10.1109/ISPA.2005.195433

Vucković G, James N, Hughes M, Murray S, Milavonic Z, Perš J, Sporis G (2014) A new method for assessing squash tactics using 15 court areas for ball locations. Hum Mov Sci 1–10. https://doi.org/10.1016/j.humov.2014.01.002

Webb PI, Pearson PJ (2008) An integrated approach to Teaching Games of Understanding (TGfU). In: 1st Asia Pacific sport in education conference: Ngunyawaiendi Yerthoappendi Play to Educate, Adelaide. http://ro.uow.edu.au/edupapers/52/

Design and Development of a Punching Analysis Device New Instrument in Boxing

Muhamad Danish Bin Abdul Taib, Nurul Syuhadah Binti Khusaini, Zulkifli Mohamed, Mohd Hanif Mohd Ramli, and Norhelena Aziz

Abstract Boxing is a popular combat sport with a rich history, involving two participants wearing gloves and throwing punches at each other within a ring. The sport has various rules and regulations, including weight classes, round duration, and the use of protective equipment. However, there is a lack of accurate and reliable methods for measuring the maximum strength of punches and the reaction time of boxers, particularly in Malaysia. This study presents an alternative method for measuring punch strength, with a focus on Malaysian boxers' anthropometric performance and strength. The study aimed to overcome the limitations of existing methods, such as the lack of accurate anthropometric data for Malaysian boxers and the absence of readily available punching instruments with IoT capabilities. The punching analysis device developed in this study used Node-Red software for dashboard management, MySQL for the database, and Arduino hardware and Catia for 3D modeling and design. The performance of the punching machine was tested, and the accuracy of the new method was within a margin of error of 5% when compared to other instruments in the market. The study's findings provide valuable insights for coaches, trainers, and athletes, contributing to a better understanding of the mechanics of punches in boxing. The developed method is a reliable and cost-effective way to measure punch strength and reaction time, easily implemented in both training and competition settings. This study's results are essential for improving the training and development of boxers and designing new and effective training equipment for the sport of boxing.

Keyword Boxing · Malaysian anthropometry · Node-RED · MySQL

M. D. B. A. Taib · N. S. B. Khusaini (✉) · Z. Mohamed · M. H. M. Ramli · N. Aziz
School of Mechanical Engineering, College of Engineering, Universiti Teknologi Mara (UiTM) Shah Alam, Shah Alam, Malaysia
e-mail: nurulsyuhadah2089@uitm.edu.my; 2020628352@student.uitm.edu.my

Z. Mohamed et al. (eds.), *Proceedings of the 2nd International Conference on Innovation and Technology in Sports, ICITS 2023, 27–28 November, Malaysia*, Lecture Notes in Bioengineering, https://doi.org/10.1007/978-981-97-3741-3_22

1 Introduction

This research addresses the need for accurate measurement of punching strength and reaction time in boxing, particularly in Thailand and Malaysia. Existing methods like dynamometers and pendulums have limitations, leading to the exploration of new, more effective approaches.

Anthropometric studies have shown correlations between physical attributes and boxing performance. For instance, Baldari's study (Guidetti et al. 2002) on elite amateur boxers in the middleweight class established a significant relationship between physical fitness, muscle strength, and anthropometric measures. Another study by Devender Kumar (2019) in India found a significant correlation between the performance of 70 boxing players and variables like weight, height, BMI, body fat percentage, and leg length.

Inertial sensors have gained popularity for measuring strike quality, frequency, and power output in combat sports. In a systematic review and meta-analysis by Worsey et al. (2019), different types of strikes were found to have varying impact characteristics and speeds. For example, the straight punch in boxing exhibited the highest measured values, averaging 1659 N.

Studies on boxers' reaction time reveal the impact of different formats and resting intervals on performance. Chadli's device (Chadli et al. 2014), developed in 2014, measured reaction time, punch force, and frequency in amateur athletes, aiming to improve intrinsic performance. Çakmakçı's study (Çakmakçı et al. 2019) in 2019 found that the 3×3 min format resulted in significantly faster reaction times compared to the 4×2 min format. Wearable inertial sensors accurately captured punch-out movements but faced challenges in retraction detection, as noted in a study (Kimm and Thiel 2015).

The lack of suitable anthropometric data for Malaysian boxers poses a challenge in designing a new punch analysis tool. Currently, available data does not accurately reflect the physical characteristics of Malaysian boxers, hindering the design and development of an effective tool. Furthermore, no papers or research related to Malaysian boxers' strength have been compared to international standards, emphasizing the need for further research in this area.

This study aims to investigate the punching strength of Malaysian boxers and develop a tool that aligns with their specific anthropometric characteristics. The goal is to enhance the accuracy and effectiveness of punch data tracking, providing valuable insights into their performance in the sport.

2 Methodology

The Functional Decomposition Diagram (FDD) for this study, based on the hardware and software development of a new instrument for IOT-based punching analysis, is shown in Fig. 1

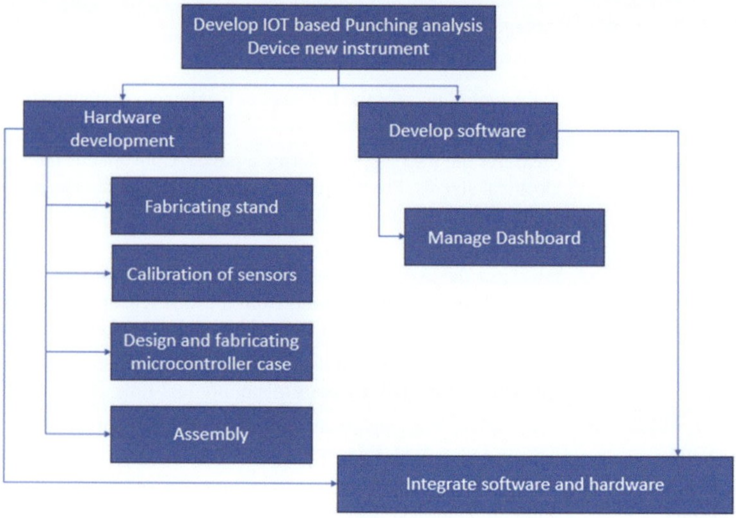

Fig. 1 Functional Decomposition Diagram (FDD) for IoT development

The implemented solution utilizes an ESP32 board as the microcontroller and a 200 kg flat membrane box load cell sensor, as depicted in Fig. 2. To ensure continuous functionality, a rechargeable battery and a USB charging module are employed to power the system. Additionally, an LCD screen is integrated to display the data. The data is transmitted to a Raspberry Pi, functioning as a server running Node-RED software, via the ESP32 board's Wi-Fi module using the MQTT protocol. To enhance data presentation and visualization, a customized website dashboard is created using Node-RED specifically designed for tablet-sized screens. This dashboard incorporates real-time graphs, providing a comprehensive depiction of the data and improving the monitoring experience.

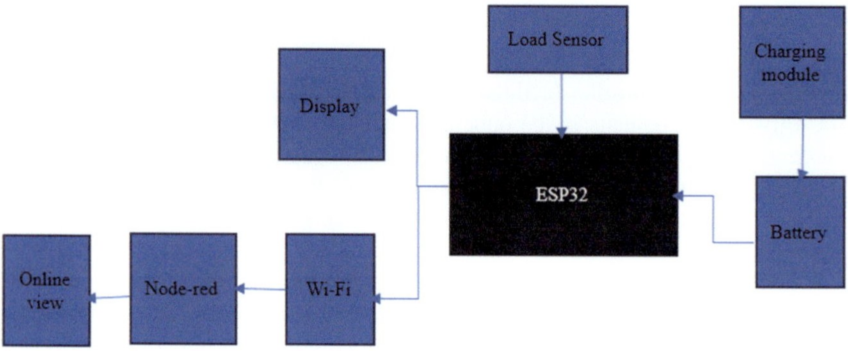

Fig. 2 System block diagram

Table 1 Hardware list of components

No	Component	Description
1	ESP32 microcontroller	Main control unit for the IoT device. Facilitates communication between different components,
2	DSMH-102 load sensor	Measures the force applied during a punch. Converts the force into electrical signals. Provides analog output based on the applied force. Enables detection and quantification of punch force. Max force capacity: 200 kg
3	HX711 sensor module	Interfaces the load sensor with the ESP32 microcontroller. Amplifies and digitizes the analog signals from the load sensor. Converts the signals into a digital format for processing by the microcontroller
4	TFT LCD display	Provides visual output for displaying punch force, maximum force, and battery level. Allows users to monitor real-time data on the display
5	Battery charging module	Manages the charging process of the battery. Regulates the charging voltage and current. Prevents overcharging and over-discharging of the battery
6	Button	Provides a physical interface for user input. Can be used to reset the recorded punch force value. Enables user interaction with the system IoT device
7	Aluminum profile	Provides a sturdy and stable structure for the stand and holding unit. Offers a lightweight and durable material for construction

2.1 Hardware Development

Hardware development encompasses a series of stages including prototyping, testing, and component integration to create a robust IoT device for boxing monitoring. The following outlines the steps involved in hardware development for the IoT device (Table 1).

2.2 System Configuration

The ESP32 microcontroller facilitated seamless communication between hardware and software components in the punching analysis system. Proper wiring connections ensured accurate data transmission from the load sensor to the microcontroller. Using the Arduino IDE and ESP32 libraries, the microcontroller was programmed to detect punches through an external interrupt function, calculating force using the load sensor's calibration factor.

A TFT LCD display, connected to the microcontroller, served as the user interface, showing real-time force readings, and recording the highest force achieved after each punch. A reset button cleared the force value for subsequent punches. The system included a battery charging module for power supply, with its status displayed on the TFT screen.

To complete the system, water flow rate data from a sensor was sent to a Raspberry Pi server via MQTT protocol. The server stored the data in a synchronized database, enabling easy monitoring and analysis. A custom-designed website dashboard provided a user-friendly interface for visualizing the data on tablet-sized devices, enhancing the monitoring experience, and offering insights into water flow patterns.

2.3 *System Integration*

After completing the programming of the IoT components, the next step involved integrating all the individual components into a single circuit for reliable and secure connectivity during field testing. The circuit's schematic design, which visually represents the connections and layout of the hardware, is illustrated in Fig. 3.

3 Results

The development of a punching force meter involves the integration of both hardware and software components. The purpose of this design specification is to outline the key features and functionality of the hardware and software aspects of the punching force meter. This essay will discuss the design specifications for each component separately, highlighting their significance in achieving accurate force measurement, reliable data acquisition, and user-friendly operation.

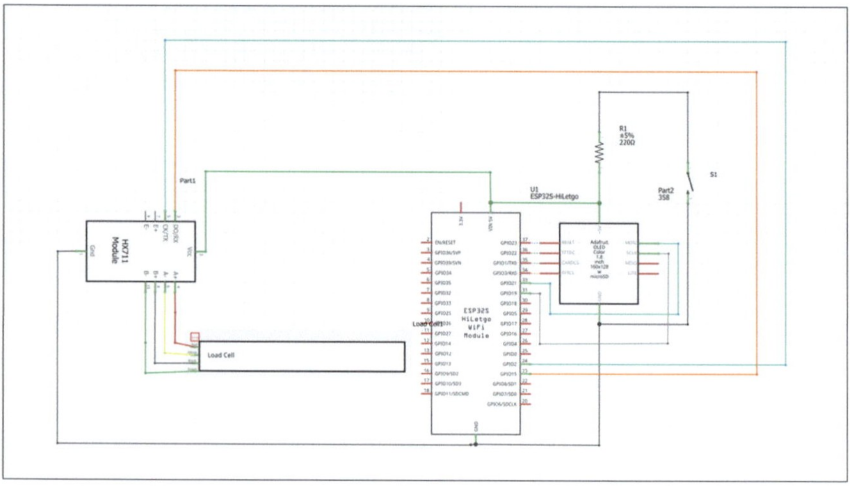

Fig. 3 Schematic diagram of the wiring for the S

3.1 Hardware

The punching force meter's hardware design centers around a boxing pad mounted on an adjustable stand, strategically positioned 75 cm from the participant's front leg. Load sensors are integrated into the pad's structure, carefully placed to capture force accurately. This customized sensor arrangement ensures optimal positioning based on the participant's height and weight, facilitating precise force measurement.

To ensure consistent and accurate force measurement, a meticulous calibration process is employed. The load sensors within the boxing pad undergo calibration using known weights to establish a baseline for accurate readings. This calibration step is crucial in minimizing variations and inaccuracies, ensuring the integrity and reliability of the collected force data (Fig. 4).

The hardware setup features an HX711 amplifier interfaced with an ESP32 microcontroller for real-time data acquisition from load sensors. The ESP32 captures force data instantly, providing participants with immediate feedback to monitor and adjust their punching force for improved performance.

For an enhanced user experience, a TFT LCD display is integrated into the punching force meter, offering real-time visual feedback on force measurements during each punch. This interactive interface facilitates immediate performance evaluation, allowing participants to track their progress effectively.

The punching pad and sensors are mounted on an adjustable stand, customizable based on the participant's height and reach. This flexibility accommodates

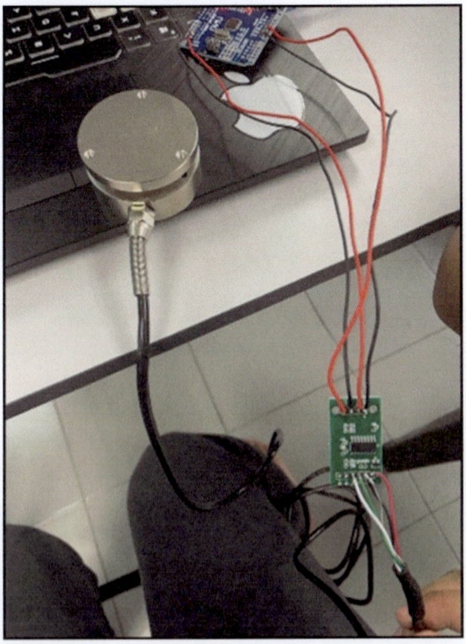

Fig. 4 Calibration of sensor

varying anthropometric characteristics, ensuring optimal sensor placement for accurate force measurement. The adjustable stand's adaptability enhances the punching force meter's usability across a diverse range of participants.

3.2 *Software*

The software design prioritizes a user-friendly experience and seamless operation of the punching force meter. The interface is simple and intuitive, enabling participants to initiate the punching sequence and view real-time force measurements effortlessly. This design minimizes the learning curve for smooth interaction with the system.

Algorithms for data processing and analysis are integrated into the software. The collected force data is processed to calculate the average force for each punch set. This feature allows participants and trainers to assess performance, identify strengths and weaknesses, and make informed decisions for training and technique improvement.

A reset feature, activated by a push-button, resets the system for the next punching sequence. This ensures consistent and precise data acquisition by eliminating residual force readings from previous measurements.

For remote monitoring and data logging, force data captured by the ESP32 is transmitted in real time using the MQTT protocol. This enables remote performance monitoring, data analysis, and feedback from trainers, coaches, or researchers.

The software includes a calibration module for users to calibrate load sensors if needed. This management feature ensures the punching force meter maintains measurement accuracy over time, allowing for periodic calibrations to ensure reliability and consistency of force measurements.

3.3 *Dashboard*

The integration of Node-RED and MySQL in the IoT-based punching analysis system created significant opportunities for data visualization, management, and analysis. This fostered discussions on performance improvement and training strategies for Malaysian boxers.

Real-time performance monitoring was a key discussion point. The Node-RED dashboard allowed coaches and athletes to access immediate data on punching strength and reaction time during training sessions and competitions. This real-time feedback facilitated on-the-spot adjustments to training techniques and strategies, encouraging discussions on areas needing improvement and tactics to capitalize on strengths. This dynamic monitoring contributed to rapid skill development and enhanced competitiveness.

MySQL's historical data tracking capabilities played a pivotal role in discussions on long-term performance analysis. Storing and organizing data over time allowed coaches to assess athletes' progress and performance trends across multiple sessions and events. These historical records prompted discussions on identifying patterns and outliers, enabling tailored training programs to suit individual needs and capitalize on consistent areas of excellence.

Data integrity and reliability were crucial in performance analysis, and MySQL's adherence to the SQL standard ensured consistency and accuracy in data representation. This aspect became a focal point of discussion among sports scientists, coaches, and researchers, emphasizing the credibility and trustworthiness of the recorded data. The assurance of data integrity prompted discussions on the potential wider adoption of similar IoT-based analysis systems in other sports and athletic disciplines.

The integration of Node-RED's real-time data visualization with MySQL's robust data storage fostered collaborative discussions between sports scientists and technology experts. This interdisciplinary collaboration enhanced understanding of the system's technical aspects and its potential impact on training methodologies. Sports scientists provided insights into specific performance metrics like punching strength and reaction time, while technology experts contributed knowledge on data security, scalability, and system optimization.

The discussions expanded beyond individual athlete improvement, as the IoT-based punching analysis system allowed for comparative analysis and benchmarking of Malaysian boxers against international counterparts (Çakmakçı et al. 2019). This comparative approach highlighted global performance standards, motivating further improvements and potential opportunities for international collaboration and the exchange of training methodologies.

The implementation of Node-RED and MySQL in the punching analysis system sparked broader discussions on the implications of IoT technology in sports performance analysis. Researchers and sports technology enthusiasts explored the future of IoT-based sports analytics, potential advancements in wearable technology, and the transformative impact of data-driven decision-making on sports training and performance evaluation (Fig. 5).

The homepage of the dashboard presents a user-friendly interface designed to visualize and analyze the data collected from the IoT-based punching analysis device. The dashboard offers several key components, including a bar chart, line chart, and a table, each serving specific purposes to provide valuable insights to trainers, coaches, and athletes.

(a) Bar Chart: The bar chart on the homepage visually represents punching force data. Each bar corresponds to a specific punch performed by the participant during the testing sessions.
(b) Line Chart: The line chart complements the bar chart by displaying the temporal variation of punching force and reaction time. It shows how the force and reaction time evolve over time as participants perform consecutive sets of punches.
(c) Table: The table presents a tabular format of the recorded data from the punching analysis device. It includes comprehensive details such as punch number, force exerted, and corresponding reaction time.

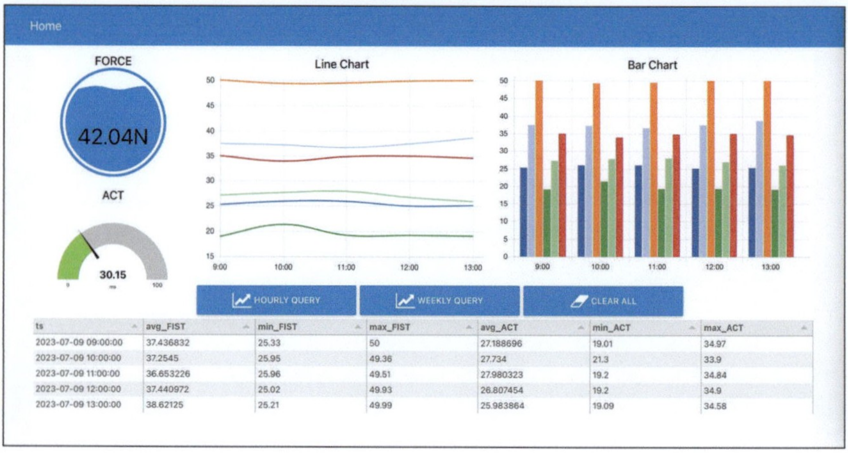

Fig. 5 Dashboard homepage

(d) Force and Reaction Time Display: At the top of the dashboard homepage, there are real-time displays for the current punching force and reaction time as the participants are actively engaging with the punching pad.

The dashboard's user association feature allows participants to input their names, associating recorded punches with specific users. This enables individual performance tracking, data organization, and personalized feedback. Coaches can conduct comparative analysis and monitor long-term progress, fostering data-driven decision-making and improved performance (Fig. 6).

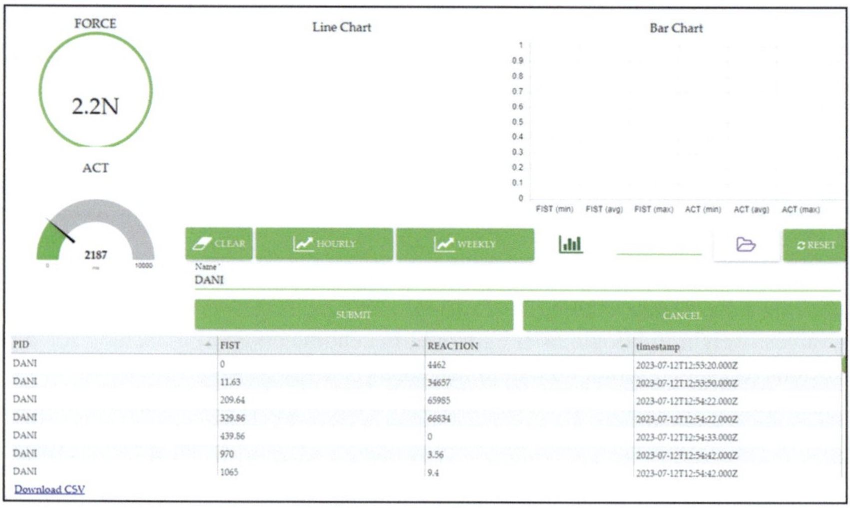

Fig. 6 Dashboard user registration

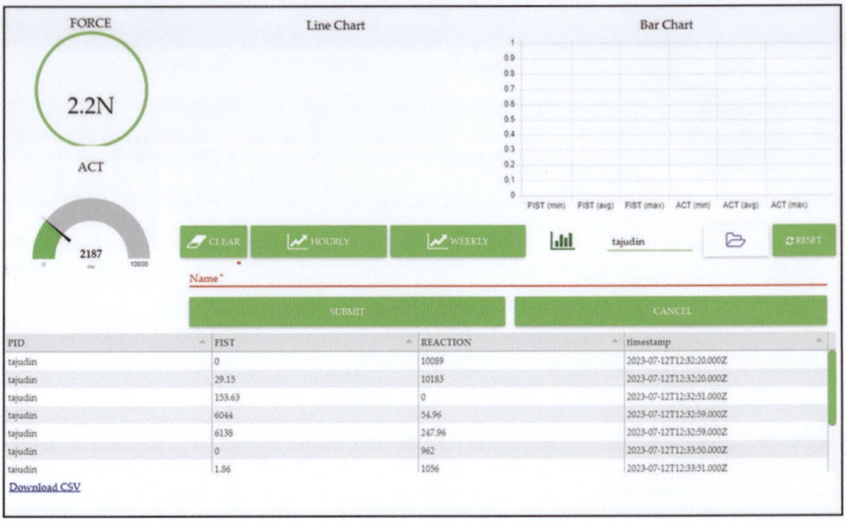

Fig. 7 User data search

Additionally, the dashboard includes a search functionality that allows users to input other participants' names at the top. This feature enables easy access to their progress and historical data of punches. Coaches and trainers can conveniently review and analyze individual performance trends, making it a valuable tool for personalized training strategies and performance evaluation (Fig. 7).

Moreover, the dashboard offers the functionality to download the recorded data as a CSV file. By clicking the corresponding button located at the bottom left, users can easily export the punching data to Excel or any other compatible spreadsheet software. This feature enhances data portability and provides users with the convenience of further analyzing and processing the information outside the dashboard environment (Fig. 8).

3.4 Data

The data presented in the graph represents the relationship between weight, height, punching force, and reaction time of various participants. The graph showcases four variables: weight in kilograms (kg), height in centimeters (cm), punching force in Newtons (N), and reaction time in milliseconds (MS) (Fig. 9).

One noticeable observation from the graph is that weight and height do not exhibit a strong correlation with punching force. Participants with varying weight and height measurements demonstrate a wide range of punching force values. For instance, we can observe that participants with higher weights do not necessarily exert more punching force, nor do taller participants consistently show higher punching force

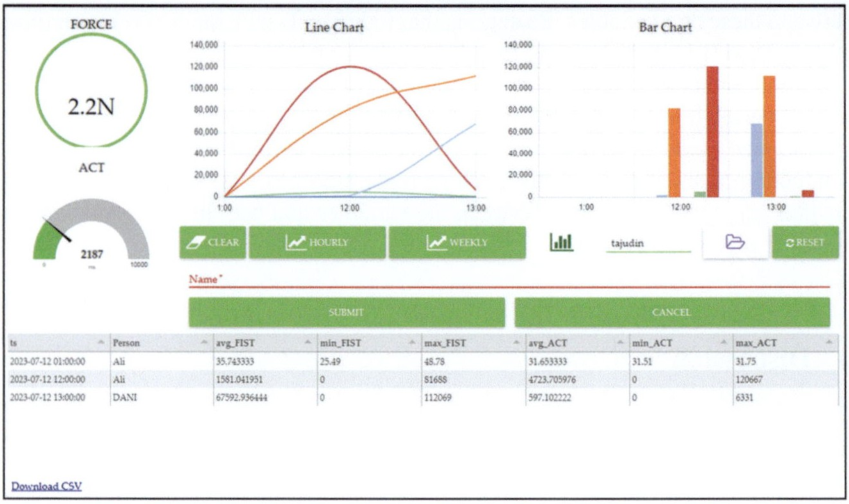

ts	Person	avg_FIST	min_FIST	max_FIST	avg_ACT	min_ACT	max_ACT
2023-07-12 01:00:00	Ali	35.743333	25.49	48.78	31.653333	31.51	31.75
2023-07-12 12:00:00	Ali	1581.041951	0	81688	4723.705976	0	120667
2023-07-12 13:00:00	DANI	67592.936444	0	112069	597.102222	0	6331

Fig. 8 Download CSV data file

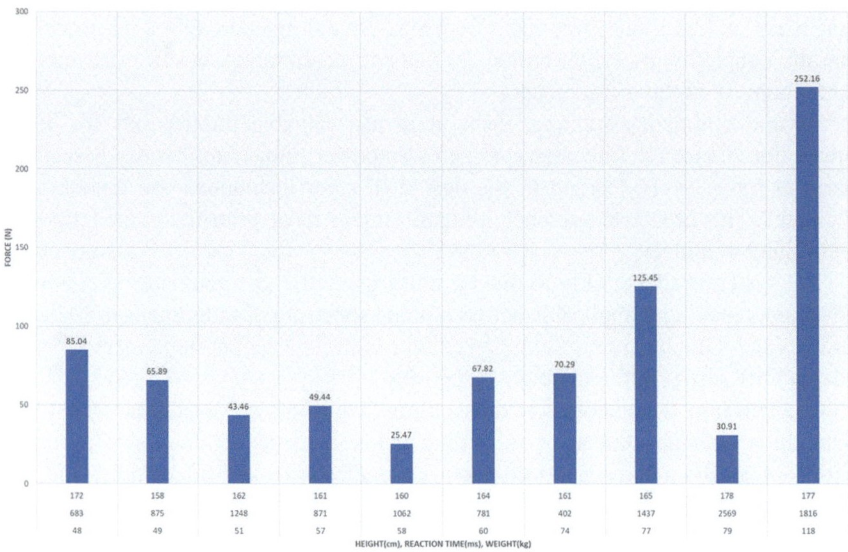

Fig. 9 Graph height, reaction time, weight versus force

values. This suggests that punching force is influenced by other factors beyond weight and height alone.

On the other hand, it appears that reaction time may have some association with punching force. Participants who exert higher punching forces tend to have relatively shorter reaction times. This trend can be observed as a general negative correlation between these two variables. It suggests that individuals with quicker reaction times might generate higher punching forces during the given trial sets.

Further analysis could reveal the presence of outliers, which might affect the overall correlation between the variables. Additionally, a larger and more diverse sample size could strengthen the conclusions drawn from the graph. Moreover, conducting a statistical analysis, such as regression analysis, could provide more insights into the relationships between the variables and help in understanding any potential predictive patterns.

4 Discussion

The punching pad with a load sensor in Malaysia represents a groundbreaking approach to measuring punching force in combat sports. Its strength lies in its personalized sensor placement, strategically adjusting to individual height, reach, and punching technique. This tailored approach ensures precise data collection for a comprehensive analysis of punching force dynamics.

The calibration process is a key element, playing a crucial role in achieving consistent and accurate force measurements. An automated calibration process at startup, coupled with a reset button for manual recalibration, ensures the integrity and reliability of the recorded data.

Beyond optimizing training, the system has valuable implications for injury prevention. Objective force measurements empower trainers to identify potentially harmful force levels, leading to the design of safer techniques and conditioning regimens. This proactive approach mitigates injury risks, prioritizing the long-term well-being of athletes.

The system's adaptability to diverse anthropometric characteristics is a notable advantage, making it applicable across a broad spectrum of individuals in Malaysia and beyond. Graphical representations underscore its concrete implementation and adaptability across different height categories.

In conclusion, the outcomes achieved through this innovative technology have the potential to reshape combat sports training and performance evaluation. The combination of accurate force measurement, personalization, and adaptability offers a robust, data-driven methodology that not only optimizes training but also contributes to evidence-based practices in combat sports and related fields. This forward-looking approach stands poised to shape the future of combat sports training and research globally (Fig. 10).

Fig. 10 Testing with tall height (180)

5 Conclusion

In conclusion, this technical report has explored the design, development, and evaluation of a punching force monitoring system using a punching pad with a load sensor. The results obtained from the conducted experiments demonstrate the system's effectiveness and potential in accurately measuring punching force and providing valuable insights for athletes, trainers, and sports scientists.

Throughout the report, we presented the key components of the system, including the punching pad, adjustable stand, and load sensor. The customized sensor placement based on individual anthropometric characteristics ensures precise force measurements tailored to each participant. The calibration process played a critical role in establishing a baseline for the load sensor, ensuring accurate and consistent data collection.

The experimental results showcased the system's capability to measure and quantify punching force in real time. Participants' punches were accurately captured, allowing for the analysis of maximum force, reaction time, and other relevant parameters. The system's integration with the MQTT protocol facilitated seamless data transfer and real-time monitoring, enhancing the usability and practicality of the system.

The implications of this punching force monitoring system are far-reaching. Athletes and trainers can utilize the obtained data to optimize training programs, identify areas for improvement, and track progress over time. The system's ability to customize sensor placement based on anthropometric characteristics ensures accurate measurements across a diverse range of individuals, making it applicable in various sports and combat disciplines.

Furthermore, the system's potential extends beyond training optimization. The ability to objectively measure punching force can contribute to injury prevention by identifying excessive force that may pose a risk to athletes' well-being. This knowledge enables the implementation of targeted techniques and conditioning programs to mitigate injury risks and ensure long-term athlete health and performance.

In summary, the punching force monitoring system presented in this report offers a reliable and data-driven approach to assess and enhance punching performance. The accuracy of force measurements, customization based on anthropometric characteristics, and real-time monitoring capabilities make it a valuable tool for athletes, trainers, and sports scientists. As technology continues to advance, further research and development in this area hold the potential for even greater insights and applications in combat sports and related fields.

References

Çakmakçı E, Tatlıcı A, Kahraman S, Yılmaz S, Ünsal B, Özkaymakoğlu C (2019) Does once-a-week boxing training improve strength and reaction time? Int J Sport Exerc Train Sci 88–92. https://doi.org/10.18826/useeabd.552086

Chadli S, Ababou N, Ababou A (2014) A new instrument for punch analysis in boxing. Proc. Eng. 72:411–416. 1016/j.proeng.2014.06.073

Devender Kumar ST (2019) Selected anthropometric characterisitics as the predictors of performance in boxing. J Manag Res Anal 6(1):200–207

Guidetti L, Musulin A, Baldari C (2002) Physiological factors in middleweight boxing performance. J Sports Med Phys Fitness 42(3):309–314

Kimm D, Thiel DV (2015) Hand speed measurements in boxing. Proc Eng 112:502–506. https://doi.org/10.1016/j.proeng.2015.07.232

Worsey MTO, Espinosa HG, Shepherd JB, Thiel DV (2019) Inertial sensors for performance analysis in combat sports: a systematic review. Sports 7(1). https://doi.org/10.3390/sports7010028

Innovative Transformer-Based Wireless Power Transfer System for Low Power E-Bicycle

Matthias Foh Thye Tiong, Tian Swee Tan, Kah Meng Leong, Joyce Sin Yin Sia, Kok Yeow You, Sameen Ahmed Malik, Jahanzeb Sheikh, and Kun Ma

Abstract Wireless power transfer is a diverse research area that has started to be implemented in e-bicycle charging applications today. There are a few issues related to the traditional wired charging method such as the risk of electrical hazard and the burdensome maintenance costs. Hence, this study aims to develop a transformer-based wireless power transfer system for e-bicycle wireless charging. The wireless power transfer system operates on the principle of magnetic resonant coupling, ensuring efficient and safe energy transmission. To optimize system performance, an impedance matching approach is introduced, integrating step-up and step-down transformers into the system design. At the end of this study, a prototype has been successfully constructed, exhibiting a 45% transfer efficiency over a separation distance of 10 cm. This achievement signifies a substantial step toward realizing an efficient and sustainable e-bicycle charging facility.

M. F. T. Tiong · T. S. Tan (✉) · K. M. Leong · J. S. Y. Sia · K. Y. You · J. Sheikh · K. Ma
Department of Biomedical Engineering and Health Sciences, Faculty of Electrical Engineering, Universiti Teknologi Malaysia, Johor Bahru, Malaysia
e-mail: tantswee@utm.my

M. F. T. Tiong
e-mail: tfmatthias@graduate.utm.my

T. S. Tan
IJN-UTM Cardiovascular Engineering Centre, Institute of Human Centered Engineering, Universiti Teknologi Malaysia, Johor Bahru, Malaysia

Bioinspired Device and Tissue Engineering Research Group, Universiti Teknologi Malaysia, Johor Bahru, Malaysia

S. A. Malik
Department of Bio-Medical Engineering, Faculty of Electrical Engineering, University of Engineering & Technology Lahore-Narowal Campus, Narowal, Punjab, Pakistan

J. Sheikh
Department of Biomedical Engineering, Sir Syed University of Engineering and Technology, Karachi, Pakistan

Z. Mohamed et al. (eds.), *Proceedings of the 2nd International Conference on Innovation and Technology in Sports, ICITS 2023, 27–28 November, Malaysia*, Lecture Notes in Bioengineering, https://doi.org/10.1007/978-981-97-3741-3_23

Keywords Electric bicycle · Impedance matching · Magnetic resonance · Transformer · Wireless power transfer

1 Introduction

Electric bicycles or e-bicycles are getting more popular as an alternative means of transportation and a recreational activity where cyclists can navigate diverse terrains, including hilly landscapes. According to the market analysis conducted by Allied Market Research, the global e-bicycle market exhibited a value of $40,312 million in 2019 and is projected to reach $120,000 million by 2030 (Akshay 2021). The driving factor behind it is because of the flexibility and environmental friendliness of e-bicycles. With lesser physical effort required and the absence of fuel costs, there is a rise in interest in cycling as a fitness and recreational activity.

In order to promote cycling activity, a sustainable e-bike charging facility needs to be set up. The conventional method to charge those e-bicycles is by physical wire connection, but it has a few disadvantages. It may cause electrical hazards since the cables are exposed to the air (Ram et al. 2020). Furthermore, the natural wear and tear of those wires due to friction and environmental exposure would cause regular inspections, repairs, and eventual replacement (Zhang et al. 2017).

Therefore, an innovative solution such as wireless charging technology is getting more attention as an alternative to this traditional wired charging method (Pellitteri et al. 2019). A wireless power transfer (WPT) system is the transmission of electric power from the power source to electrical loads over free space without the physical connector. It can bring convenience and certain advantages to many applications such as consumer electronic products, medical implants, and electric vehicles charging by eliminating the charging cable.

Started by Nikola Tesla as the pioneer in the 1890s (Zhang and Mi 2016), the WPT system has been studied for over a century. Numerous electromagnetics techniques have been introduced which are briefly explained (Kim et al. 2018) in Table 1. Based on effective transfer distance, near-field techniques such as inductive coupling and capacitive coupling may be not suitable for e-bicycle charging since the transfer efficiency is only optimum in a very short distance. The e-bicycle needs to be placed very close to a charger and in perfect alignment, which is impractical. The far-field technique such as microwave power transfer suffers from high-power loss and utilizes high-frequency operation which may be unsafe for humans (Khan et al. 2020). High-frequency electromagnetic waves can cause tissue heating due to the dielectric properties of human tissue (Campi et al. 2016).

This study focuses on transformer-based magnetic resonance coupling WPT system to transfer energy at a medium distance which is suitable for low-power e-bicycle applications. Transformers are integrated into both transmitter and receiver circuits to achieve impedance matching.

Table 1 Categories of WPT techniques

Category	Near field	Mid field	Far field
Technique	Inductive coupling, capacitive coupling	Magnetic resonance coupling	Microwave
Operating frequency	Hz–kHz	kHz–MHz	GHz
Power transfer distance	mm–cm	cm–m	m–km
Power transfer efficiency	High	Medium	Low
Advantage	High efficiency	Effective in mid-range application	Long distance
Disadvantage	Short range	Higher design complexity	High frequency

2 Background Theory

The general WPT system consists of the power source, transmitter circuit, coils, receiver circuit, and load as shown in Fig. 1. Based on Faraday's law of induction, varying magnetic field are generated by the transmitter coil to induce an electromotive force in the receiver coil (Zhang 2020). It is safe to use on humans since the human body is a non-magnetic medium and does not respond to magnetic fields.

Magnetic resonance coupling has medium transfer efficiency but has a longer power transfer distance compared to inductive coupling (KahMeng et al. 2016). Thus, the usage of this technique is more practical to charge electronic devices such as cell phone, electric vehicle, and implanted medical devices at mid-range distances (Jawad et al. 2017). The resonance is the condition when both the coils are tuned at the same resonant frequency where the highest power transfer efficiency can be achieved (Sah 2013) as shown in Fig. 2. The resonant angular frequency is expressed as in (1) where L and C is the inductance and capacitance.

$$\omega = \frac{1}{\sqrt{LC}} \tag{1}$$

In 2007, Massachusetts Institute of Technology (MIT) researchers proposed a four-coil wireless power transfer system that transmits energy through the resonant coupling between the coils (Kurs et al. 2007). It was a big breakthrough at that time

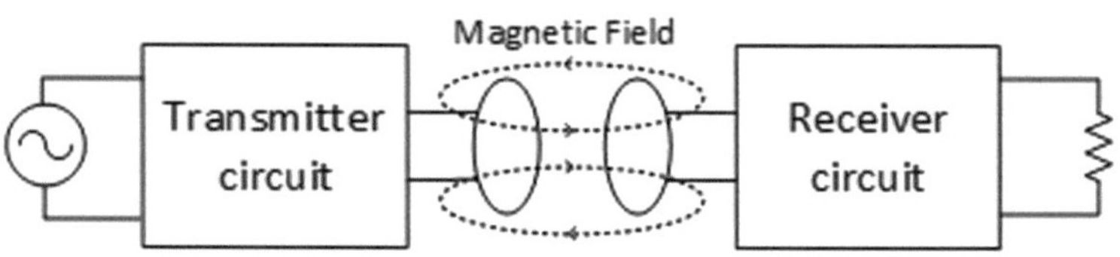

Fig. 1 General WPT system

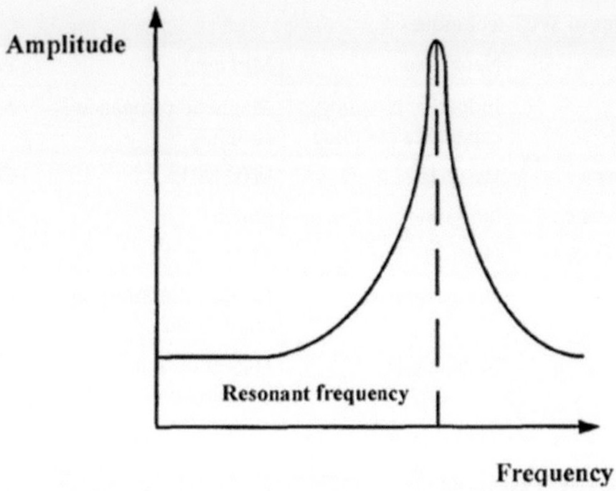

Fig. 2 Graph of amplitude versus frequency (Kurs et al. 2007)

that they were able to light up a 60 W bulb 2 m away. Since then, resonant coupling has become one of the topics of interest in wireless power transfer research.

To achieve a resonance state, impedance matching also known as compensation networks is used to improve the WPT performance. On the primary side, an impedance matching circuit is added to reduce the input apparent power so that more real power can be transferred. While the function of the impedance matching circuit on the secondary side is to minimize the inductance of the secondary coil to optimize the transfer capability (Zhang and Mi 2016).

Initially, there are four basic compensation networks, which are S–S, S-P, P-S, and P-P. The S and P represent the series and parallel connection of the compensating capacitor, C in the circuit respectively while L represents the coil inductance (Zhang et al. 2017) as shown in Fig. 3. On the other hand, more hybrid compensation networks have been proposed recently, such as LCL-LCL (Qu et al. 2019), LCC-SP (Yang et al. 2020), and LCCL-LC (Zhou et al. 2020) networks. Hybrid compensation networks have more compensating components, thus higher design freedom (Yang et al. 2020) compared to basic compensation networks. However, it leads to higher design complexity to tune each compensating component.

Besides the impedance matching circuit, another factor that affects the power transfer efficiency is the coil design. The energy loss due to resistive and radiation loss can be represented by the quality factor, or Q-factor of the coil. A coil with a higher Q-factor can store more energy with lower energy dissipation (Lämmle et al. 2020). It can be represented by the ratio of reactive reactance to resistance as expressed in (2).

$$Q = \frac{\omega L}{R} \tag{2}$$

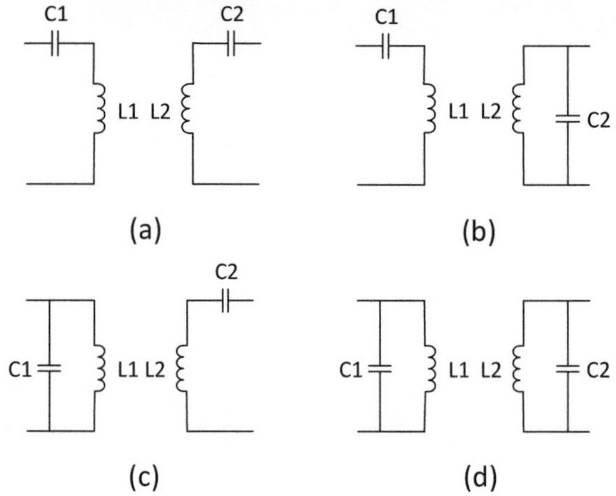

Fig. 3 Four basic compensation networks. **a** S-S network. **b** S-P network. **c** P-S network. **d** P-P network (Zhang et al. 2017)

3 Methodology

Overall, the study started with the fabrication of coils and transformers. The circuit parameter is to be identified by using the coil with the highest Q-factor. Simulation is then carried out, followed by the experiment.

3.1 Transmitter and Receiver Coils

Copper coils with 1.5 mm thickness and 10 cm diameter are fabricated for different numbers of turns, from three to twelve turns as shown in Fig. 4. The parameters of these coils are measured using an LCR meter to determine the coils with the highest Q-factor, which is to be used in this study.

3.2 Transformer

For impedance matching purposes, step-up and step-down transformers are added to the transmitter and receiver circuit respectively. After the parameters of the coil are identified, the transformer gain for both the receiver and transmitter circuit are derived as in (3) and (4) respectively based on (Swee et al. 2017). The subscripts p and s refer to the transmitter and receiver circuit, respectively.

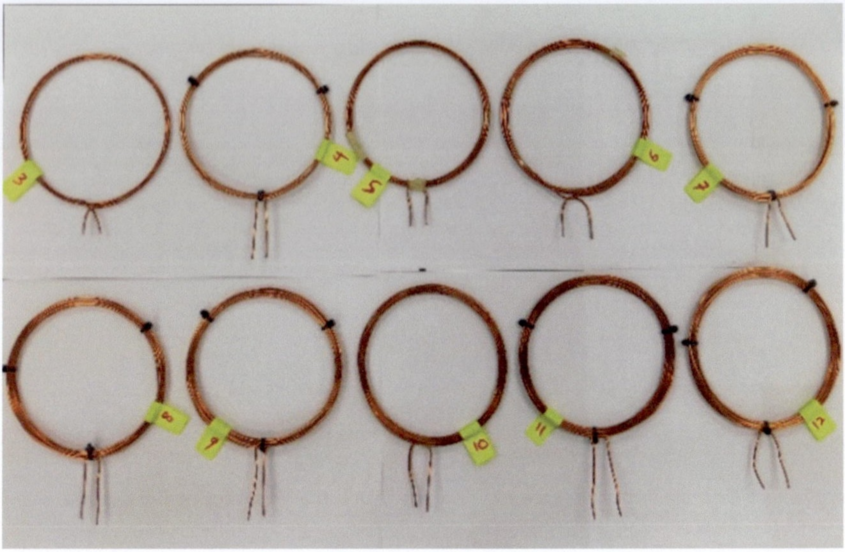

Fig. 4 Fabricated coils different numbers of turns

$$n_s = \sqrt{\frac{L_2 R_1 R_2}{R_l(L_1 R_2 + L_2 R_1)}} \tag{3}$$

$$n_p = n_s \sqrt{\frac{1}{2}\left(\frac{L_1}{L_2}\right)\left(\frac{R_L}{R_s}\right)} \tag{4}$$

The transformer is then fabricated based on (5) as shown in Fig. 5, where N is the number of turns of the coil.

$$n = \frac{N_p}{N_s} = \sqrt{\frac{L_p}{L_s}} \tag{5}$$

3.3 ADS Simulation and Experiment Setup

Before the experiment, Agilent Advanced Design System (ADS) simulator is used to simulate the WPT system for verification. Two types of WPT system are simulated which are the WPT system without impedance matching and with the impedance matching as shown in Figs. 6 and 7. The circuit parameters are summarized in Table 2.

Fig. 5 Fabricated transformers

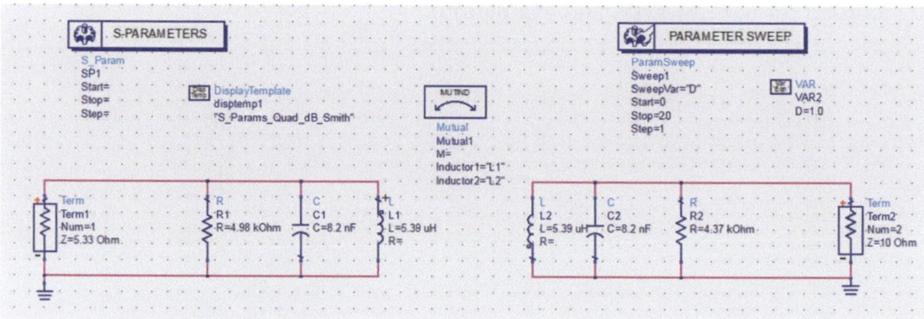

Fig. 6 Schematic diagram of WPT system without impedance matching

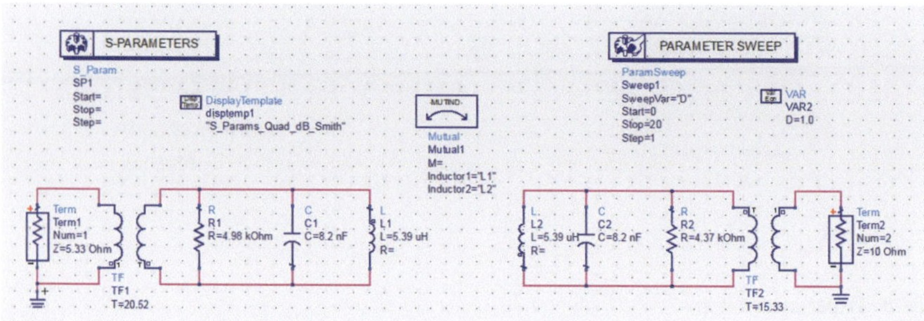

Fig. 7 Schematic diagram of WPT system with impedance matching

Table 2 Circuit parameters

Parameter	Value
Frequency	758 kHz
Coil thickness	1.5 mm
Coil diameter	10 cm
Q_1, Q_2	154.6
L_1, L_2	5.39 μH
C_1, C_2	8.2 nF
R_1	4.98 kΩ
R_2	4.37 kΩ
Z_{in}	5.33 Ω
Z_L	10 Ω
n_p	20.52
n_s	15.33

Then, the system is set up as shown in Fig. 8. The power generator and function generator are used to provide the power source and the frequency signal to the system. The output signal over the load is observed by using an oscilloscope. The result will be recorded for data analysis and evaluation.

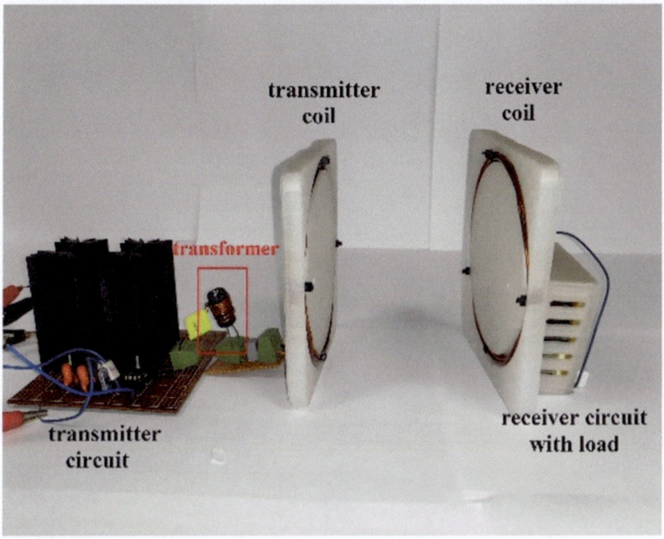

Fig. 8 Experiment setup

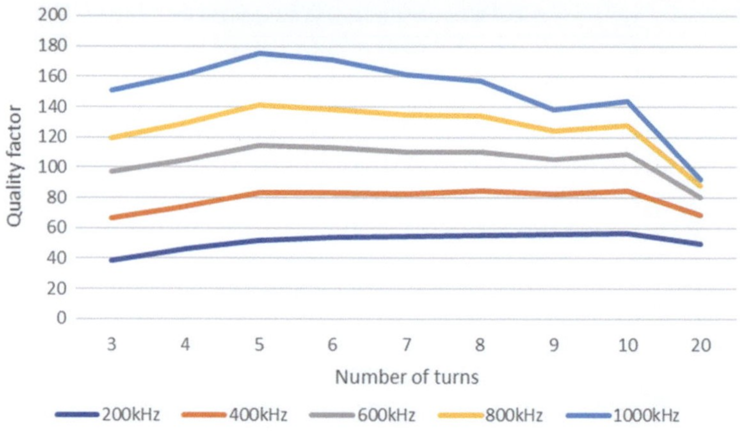

Fig. 9 Q-factor of coil for different number of turns in different frequencies

4 Result and Discussion

4.1 Q-factor

The Q-factor of a 10 cm coil of different numbers of turns is measured using an LCR meter. The result obtained serves as a reference for step-up and step-down transformer fabrication. The coil with five turns has the highest Q-factor as shown in Fig. 9, hence selected for this study. The calculated gain for step-up and step-down transformers are 20.52 and 15.33, respectively.

4.2 Simulation and Experiment Result

The simulation and experiment results are shown in Figs. 10 and 11 for WPT system without and with impedance matching networks, respectively. The distance separation between the transmitter and receiver coils varied from zero to 20 cm to observe its power transfer efficiency. It is notable that the system without an impedance matching system has very high-power transfer efficiency, which is 88% in the experiment. However, the efficiency drops drastically when the coil separation exceeds 2 cm and goes to zero when the distance is over 10 cm. This is impractical for e-bicycle charging applications since it is almost impossible to place the charger near the battery of the e-bicycle. The inductive coupling is usually implemented in applications like electronic consumers such as smartphones, tablets, and laptops which are available in the current market.

For the WPT system with an impedance matching system, the effective power transfer distance is extended up to 15 cm as shown in the ADS simulation result, or

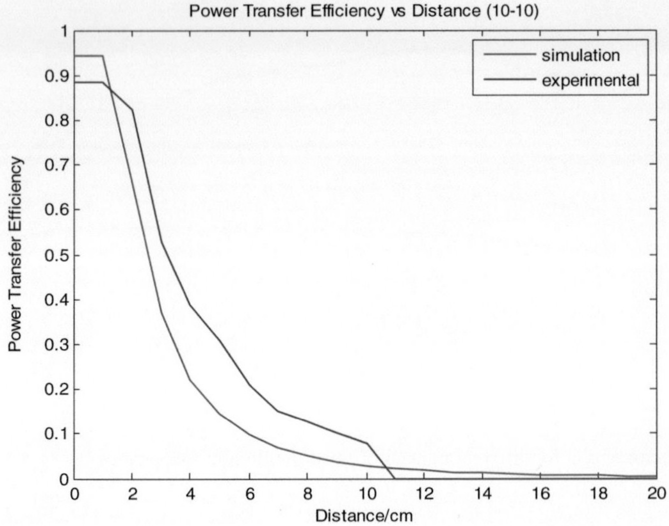

Fig. 10 Simulation result and experimental result for WPT system without impedance matching

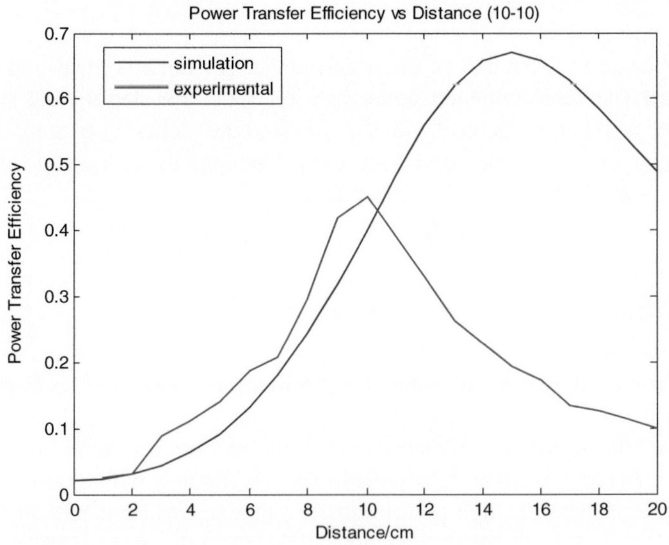

Fig. 11 Simulation result and experimental result for WPT system with impedance matching

10 cm as obtained during the experiment which is five times compared to the system without impedance matching. Although the efficiency may not be as high as before, the system now has more flexibility to the distance variation which could not be avoided in practical situations.

On the other hand, the variation of experiment and simulation results may be due to imperfect impedance matching. The transmitter coil, receiver coil, and transformer are fabricated manually, hence the parameters may slightly be distinct from the calculation being made. The deviated value contributes to the impedance mismatch and leads to lower power transfer efficiency and shorter transfer distance. Besides, there is power loss due to heat in the system. The temperature of the step-up transformer is increasing over time. This would lead to an increase in the resistance of the transformer and affect the system's performance. For future scope, a study of transformer heat reduction can be carried out to further improve the efficiency of the system.

Overall, the transformer-based wireless power transfer system can transfer energy through resonant coupling. Compared with other hybrid compensation topologies such as LCL-LCL, LCC-SP, and LCCL-LC networks, it has a simpler design that considers the gain of the step-up and step-down transformers.

5 Conclusion

Magnetic resonant wireless power transfer is getting more and more attention recently. The non-impedance matching system may have higher efficiency, but it is limited to a very short distance. With impedance matching, the WPT system operates in a resonance state and has higher transfer efficiency over longer distances. A transformer-based wireless power transfer system has been successfully designed in this study. The fabricated system manages to transfer the power up to 45% efficiency over a distance of 10 cm. The variation between simulation and experimental result is because of the imperfect impedance matching due to self-fabrication, and power loss as heat in the transformer. A future study on reducing the heating in transformers is proposed to further improve the system performance. Reasonable efficiency over longer distances is more practical and more desirable for e-bicycle charging applications.

Acknowledgements The authors are grateful to the funder by the Ministry of Higher Education under FRGS, Registration Proposal No: FRGS/1/2020/TK0/UTM/02/105 (5F282) and Universiti Teknologi Malaysia.

References

Akshay SM (2021) Electric bike market size, share, competitive landscape and trend analysis report by product (Pedelecs, Speed Pedelecs, Throttle on Demand, and Scooter & Motorcycle), Drive Mechanism (Hub Motor, Mid-Drive, and Others), and battery type (Lead Acid, Lithium). https://www.alliedmarketresearch.com/electric-bikes-market

Campi T, Cruciani S, Palandrani F, De Santis V, Hirata A, Feliziani M (2016) Wireless power transfer charging system for AIMDs and pacemakers. IEEE Trans Microw Theory Tech 64(2):633–642. https://doi.org/10.1109/TMTT.2015.2511011

Jawad AM, Nordin R, Gharghan SK, Jawad HM, Ismail M (2017) Opportunities and challenges for near-field wireless power transfer: a review. Energies 10(7):1–28. https://doi.org/10.3390/en10071022

KahMeng L, Kin Yun L, TianSwee T, Zahran Khudzari A, Bin Kadiman S (2016) Wireless power transfer system for biomedical devices by using magnetic resonance coupling technique. Res J Appl Sci Eng Technol 12(8):823–827. https://doi.org/10.19026/rjaset.12.2782

Khan SR, Pavuluri SK, Cummins G, Desmulliez MPY (2020) Wireless power transfer techniques for implantable medical devices: a review. Sensors (switzerland) 20(12):1–58. https://doi.org/10.3390/s20123487

Kim D, Abu-Siada A, Sutinjo A (2018) State-of-the-art literature review of WPT: current limitations and solutions on IPT. Electr Power Syst Res 154:493–502. https://doi.org/10.1016/j.epsr.2017.09.018

Kurs A, Karalis A, Moffatt R, Joannopoulos JD, Fisher P, Soljačić M (2007) Wireless power transfer via strongly coupled magnetic resonances. Science (80-) 317(5834):83–86. https://doi.org/10.1126/science.1143254

Lämmle T, Parspour N, Hölz J (2020) Comparison of circular and double-D coil topologies for automotive inductive charging systems. In: 2020 IEEE PELS workshop emerging technologies wireless power transfer WoW 2020, pp 68–73. https://doi.org/10.1109/WoW47795.2020.9291331

Pellitteri F, Campagna N, Castiglia V, Damiano A, Miceli R (2019) Design, implementation and experimental results of an inductive power transfer system for electric bicycle wireless charging. IET Renew Power Gener 14(15):2908–2915. https://doi.org/10.1049/iet-rpg.2020.0056

Qu X et al (2019) Wide design range of constant output current using double-sided LC compensation circuits for inductive-power-transfer applications. IEEE Trans Power Electron 34(3):2364–2374. https://doi.org/10.1109/TPEL.2018.2839769

Ram G et al (2020) Sustainable E-Bike Charging Station that Enables AC, DC and Wireless Charging from Solar Energy†

Sah AK (2013) Design of wireless power transfer system via magnetic resonant coupling at 13.56 MHz. In: Proceedings IOE graduate conference 2013, vol 1, no May, pp 202–210

Swee TT, Meng LK, Malik SA, Hou TJ, Tengku Alang TAI, Yun LK (2017) Contactless power transfer system for low power medical devices. Malaysian J Fundam Appl Sci 13(4–2):438–444. https://doi.org/10.11113/mjfas.v13n4-2.759

Yang J, Jiao C, Zhang X, Yang X, Zhang K (2020) Design of an LCC-SP compensated WPT system and parameters tuning method for misalignment condition. In: 2020 4th international conference HVDC, HVDC 2020, pp 145–150. https://doi.org/10.1109/HVDC50696.2020.9292889

Zhang X (2020) A systematic design flow for wireless power transfer system. In: 2020 IEEE PELS workshop emerging technologies wireless power transfer WoW 2020, vol 2, pp 238–241. https://doi.org/10.1109/WoW47795.2020.9291289

Zhang X, Zhang X, Yao Y, Yang H, Wang Y, Xu D (2017) High-efficiency magnetic coupling resonant wireless power transfer system with class-e amplifier and class-e rectifier. In: 2017 IEEE transportation electrification conference expo, Asia-Pacific, ITEC Asia-Pacific 2017. https://doi.org/10.1109/ITEC-AP.2017.8080933

Zhang W, Mi CC (2016) Compensation topologies of high-power wireless power transfer systems. IEEE Trans Veh Technol 65(6):4768–4778. https://doi.org/10.1109/TVT.2015.2454292

Zhou M, Li S, Chen X, Liu F (2020) A parameter design methodology based on voltage/current stress optimization for LCCL-LC compensated MCR WPT systems. In: 2020 IEEE PELS workshop emerging technologies wireless power transfer WoW 2020, pp 106–111. https://doi.org/10.1109/WoW47795.2020.9291330

Does Physical Reaction Time Training Improves Reaction Time in E-sports?

Muhammad Khairi Muhid, Nur Fatihah Raimi, Aziem Abdul Halik, Adjah Naqkiah Mazlan, and Mohd Nidzam Jawis

Abstract Sitting for extended periods of time while in front of the computer can lead to a sedentary lifestyle, and while physical activity promotes health, there is ongoing debate about its impact on e-sports performance. Considering that exercise is proven to bolster cognitive functions like reaction time, this study investigated physical reaction time training's effect on e-sports players' reaction time. This experimental study involved twenty Universiti Teknologi Malaysia (UTM)'s e-sport athletes. They were divided into two groups which are Training (T) and Control (C) groups. Both groups were required to undergo a pre-test and post-test of reaction time using the Aimlab application. After the pre-test, the T group must undergo physical reaction time training and the C group acted as a control for four weeks, followed by the post-test. Mean differences between pre-test and post-test reaction time scores in both groups were compared using a pair t-test analysis, meanwhile, mean changes of pre-test and post-test between T and C groups were compared using an independent t-test analysis. Findings show that there were significant differences between pre-test and post-test scores of reaction time in the T group (0.43 s $\pm$ 0.02 vs. 0.26 s $\pm$ 0.02, $P < 0.001$) but not in the C group (0.43 s $\pm$ 0.03 vs. 0.44 s $\pm$ 0.04, $P > 0.05$). Additionally, there were also significant differences in the mean changes of pre-test and post-test scores between T and C groups (−0.172 s vs. 0.004 s, $P < 0.001$). As the findings showed that the T group experienced significant decreases in the Aimlab test after the four-week physical reaction time training, this indicates that the training improves reaction time in e-sport players.

M. K. Muhid · N. F. Raimi (✉) · A. A. Halik · A. N. Mazlan
School of Education, Faculty of Social Science and Humanities (FSSH), Universiti Teknologi Malaysia, 81310 Skudai, Johor, Malaysia
e-mail: nurfatihah.raimi@utm.my

N. F. Raimi · A. N. Mazlan
Sports Innovation & Technology Centre (SITC), Institute of Human Centred Engineering (iHumEN), Universiti Teknologi Malaysia, 81310 Skudai, Johor, Malaysia

M. N. Jawis
School of Health Sciences, Health Campus, Universiti Sains Malaysia, 16150 Kubang Kerian Kelantan, Malaysia

Z. Mohamed et al. (eds.), *Proceedings of the 2nd International Conference on Innovation and Technology in Sports, ICITS 2023, 27–28 November, Malaysia*, Lecture Notes in Bioengineering, https://doi.org/10.1007/978-981-97-3741-3_24

Keywords Reaction time · E-sport · Aimlab

1 Introduction

In recent years, the e-sports industry has experienced remarkable expansion, propelling players into the sphere of professional competitive gaming. As e-sports continue to advance, becoming increasingly rigorous and competitive, the emphasis on the well-being of e-sports athletes becomes of utmost importance. E-sports athletes typically dedicate extended hours sitting in front of computers for gaming and honing their skills with specific computer applications. This prolonged sedentary lifestyle can have adverse effects on their health. Some research has suggested that extended periods of inactivity may contribute to conditions such as obesity, cardiovascular problems, chronic diseases, mental health issues, reduced motor skills, impaired cognitive development, and discomfort like eye strain, neck pain, and back pain (Choi et al. 2018; Cunningham et al. 2018; Colm et al. 2020).

Considering the widely established benefits of physical exercise on various aspects of health and overall well-being, previous studies suggested that incorporating exercise could serve as an effective intervention strategy to enhance the performance of e-sports players. This recommendation stems from the potential physiological advantages and cognitive functions enhancement associated with exercise (Donoghue et al. 2018; Dykstra et al. 2021). Enhanced cognitive function, including improved attention, processing speed, and decision-making abilities, can lead to faster reaction times.

Reaction time is defined as the time interval between a stimulus and the response to it (Toth et al. 2020). In various e-sports games such as first-person shooters (FPS), having a swifter reaction time allows players to aim, shoot, or make crucial decisions faster than their rivals, often giving them an edge (Dykstra et al. 2021; Radák 2018). Neglecting to train these aspects in e-sports athletes undoubtedly hampers their performance (Bediou et al. 2018). Hence, it is imperative for athletes to consistently engage in specific training tailored to the video games they play to maintain a satisfactory level of performance.

Each video game offers specialized training exercises aimed at enhancing athlete performance, with a particular focus here on Valorant, a first-person shooter (FPS) game. Valorant centers on firearm-based shooting and character-driven combat. This game provides dedicated tools, such as Aimlab, a training platform created by gamers to sharpen FPS aiming skills as well as to measure and record their accuracy in aiming, tracking targets, and reaction time. This tool is widely known as a specific technical/technology tool to be used by e-sport athletes to train their reaction time.

Be it in e-sport or in physical exercise, the nature of reaction time training is almost similar, which heavily relies on the central nervous system (CNS) in synchronizing visual input with hand movements. On this basis, we assume that physical reaction time training could potentially impact an athlete's performance in e-sports. However, to the best of our knowledge, there is no specific physical reaction time training program that has been created for e-sports athletes to improve their reaction. Hence, this study aimed to investigate physical reaction time training's effect on e-sports players' reaction time.

2 Methodology

2.1 *Sampling*

A purposive sampling technique was chosen in recruiting twenty participants for this study. The participants were e-sport athletes from Universiti Teknologi Malaysia who are engaged in the first-person shooter (FPS) category. They were divided into two groups which were Training (T) and Control (C) groups.

2.2 *Research Procedure*

Upon signing the consent form, all participants were required to undergo a familiarization followed by a pre-test of reaction time measured by Aimlab application in a computer prepared by the researchers at the data collection location. After the pre-test, they were allocated into two groups, either T or C group. The participants in the T group were required to undergo a four-week physical reaction time training meanwhile the participants in the C group acted as control for four weeks. Upon completion of the four weeks, they were required to undergo a post-test of reaction time measured by Aimlab application as in the pre-test. Mean differences in reaction time scores between pre-test and post-test within both groups were compared using a pair t-test analysis in SPSS, meanwhile, mean changes of pre-test and post-test reaction time scores between T and C groups were compared using an independent t-test analysis in SPSS. Figure 1 displays the flow chart of the research procedures.

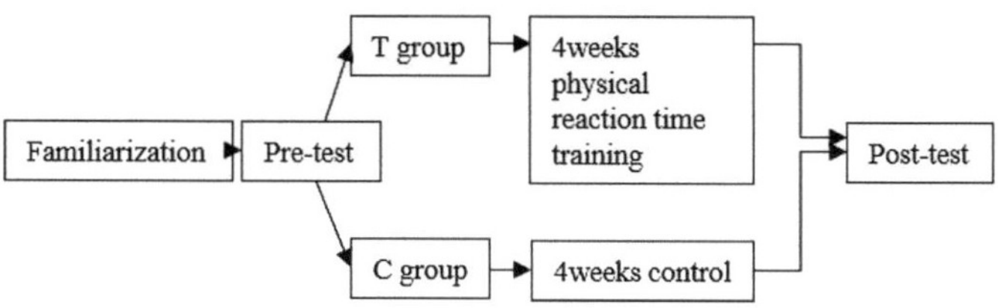

Fig. 1 Flow chart of the research procedure

2.3 *Protocol of Reaction Time Tests Using Aimlab*

Aimlab is a widely used software application and training platform designed to assist individuals in enhancing their aiming precision and reflexes while playing first-person shooter (FPS) video games. It offers a diverse range of training exercises and drills that are specifically designed to improve key skills like accuracy in aiming, tracking targets, and reaction time, all of which are vital in competitive FPS gaming. Aim Lab is a popular choice among e-sports athletes and gamers aiming to refine their aiming abilities and overall performance in games like Counterstrike: Global Offensive, Valorant, Overwatch, and similar titles. It offers a structured and data-driven approach to training, enabling users to monitor their progress and make targeted enhancements to their gaming skills.

Figure 2 displays how the reaction time tests were done in the Aimlab. Blue balls were required to be shot by athletes using the computer mouse. Only one minute was allowed for completion. Athletes were awarded points based on the number of blue balls they hit, the speed at which they dropped them, and the accuracy of their shots. If a shot was missed, points were deducted. Reaction time and accuracy were displayed as shown in Fig. 3. All this information was saved in Aimlab. However, as reaction time was the focus of this study, only the reaction time score was recorded by the researchers to be analyzed.

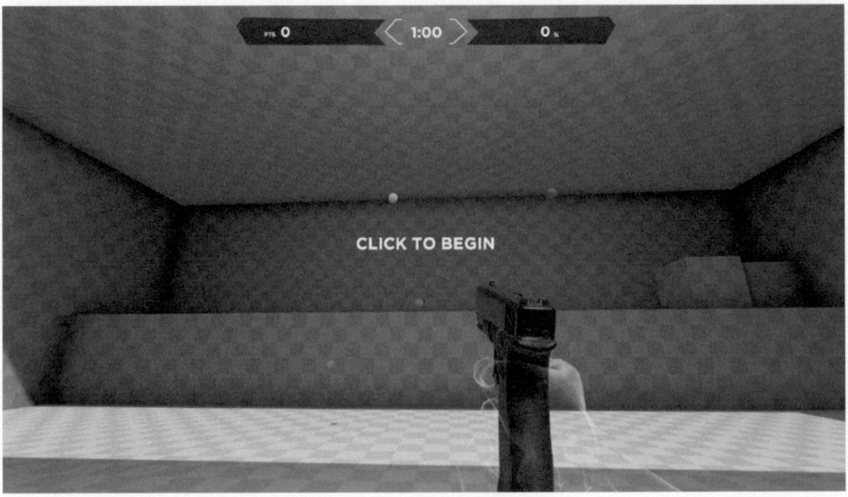

Fig. 2 Reaction time test in Aimlab

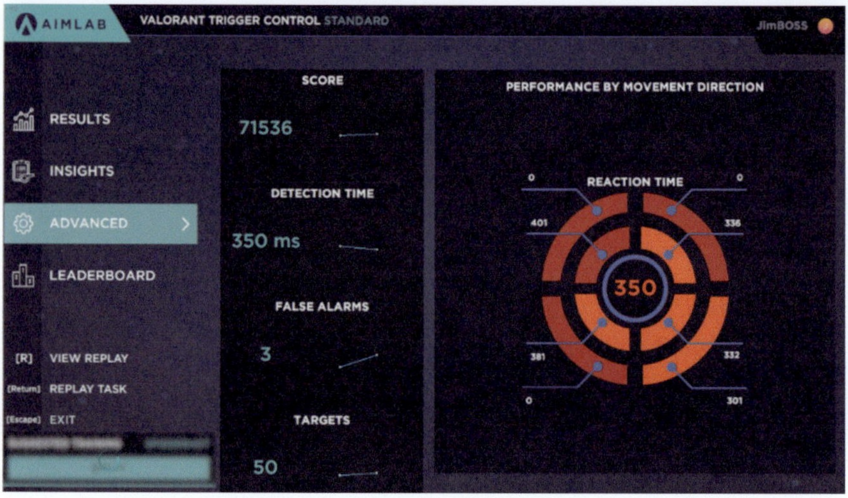

Fig. 3 Scores displayed in Aimlab

2.4 Protocol of Physical Reaction Time Training

In this study, the researchers implemented a four-week physical reaction time training consisting of four types of reaction time exercises which were; (1) Marker Hand Catch, (2) Number Tapping, (3) Card Choice, and (4) Laser Catch. The participants were instructed to perform two types of exercises daily, and they had to adhere to the timetable provided by the researchers to decide which exercises to undertake. The exercises' level of intensity was progressively raised week after week.

Marker Hand Catch. The participants were required to be seated on the ground with three distinct marker colors to their left and another three different marker colors to their right. When the participants heard a color mentioned in the instructional audio, they were tasked with grabbing the marker of the exact color as directed by the audio.

Number Tapping. Participants were instructed to stand in front of a wall containing several numbers written on it. They were then instructed to use the tips of their fingers to touch the number mentioned in the audio instructions.

Card Choice. Two cards were used at one time in this exercise. The researcher first showed the back of a card before displaying the front of the card which had the number on it. When the researcher mentioned a number, the participant had to touch the card using the tips of their fingers.

Laser Catch. The participants were directed to position themselves facing a wall and physically interact with the laser light pointed out by the researcher on the wall.

3 Data Analysis

Mean differences between pre-test and post-test reaction time scores in both groups were compared using a pair t-test analysis, meanwhile, mean changes of pre-test and post-test between T and C groups were compared using an independent t-test analysis in SPSS.

3.1 Comparison Between Mean Scores of Reaction Time in Pre-test and Post-test Within Groups

The paired t-test analysis in Table 1 displays that the mean and standard deviation of reaction time scores for the pre-test were higher than the post-test in the T group (0.43 s ± 0.02 vs. 0.26 s ± 0.02, $P < 0.001$). Meanwhile, the mean and standard deviation of reaction time scores for the pre-test were lower than the post-test in the C group (0.43 s ± 0.03 vs. 0.44 s ± 0.04, $P > 0.05$). As lower scores mean faster reaction time, and vice versa, this indicates that reaction time in the T group became faster after engaging in the four-week physical reaction time training, compared to the reaction time in the C group.

3.2 Comparison of Mean Changes in Reaction Time Between Groups

Table 1 also displays the comparison of mean changes in reaction time between T and C groups which has been analyzed using an independent t-test. The mean changes in reaction time in the T group decreased by 0.172s, meanwhile, the mean changes in reaction time in the C group increased by 0.004s. The analysis showed significant differences in the mean changes of pre-test and post-test scores between T and C groups (−0.172s vs. 0.004s, $P < 0.001$).

4 Discussions

The research findings indicate that the T and C groups displayed contrasting outcomes. The T group, which underwent physical reaction time training for four weeks, demonstrated a significant improvement in reaction time compared to the C group, which did not engage in any reaction time training, physically or using a computer application.

There are a few factors that the researchers believe to contribute to the findings. In the physical reaction training, we included both visual and auditory stimuli in the

Table 1 Comparison between mean scores of reaction time in pre-test and post-test within groups (A) and comparison of mean changes in reaction time between groups (B)

Group	Test		Mean ± SD of reaction time (s)	t	P value		Mean changes (s)	t	P value
T	Pre	(A) Paired t-test	0.43 ± 0.02	27.35	<0.001	(B) Independent t-test	−0.172	−24.62	<0.001
	Post		0.26 ± 0.02						
C	Pre		0.43 ± 0.03	−1.18	0.269		0.004		
	Post		0.44 ± 0.04						

training, mimicking stimuli (visual and auditory) in e-sports. Training these aspects might improve their level of neuromuscular coordination with auditory/visual stimuli given. This assumption is parallel to the reported literatures earlier that success in e-sports depends on players' quick responses to various visual and auditory cues. In addition, the extent of neuromuscular coordination in response to diverse auditory and/or visual stimuli is closely linked to human physical skills (Bickmann et al. 2021; Green et al. 2010).

This finding also aligns with previous literatures, which reported that participating in physical activity can have positive effects on cognitive functions, such as reaction time. These benefits are thought to arise from various processes, including the activation of metabolic brain pathways and mechanisms (Dye et al. 2009; Erickson et al. 2015) and the improvement of brain perfusion, which enhances the delivery of oxygen and nutrients (Netz 2019). It is important to emphasize that both physical and cognitive aspects need to be concurrently addressed to achieve the best results (Maass et al. 2016). Recent reviews also suggest that the most promising approach for efficiently enhancing cognitive reserve involves integrating cognitive tasks with motor tasks, rather than separately training mental and physical functions (Fissler et al. 2013; Herold et al. 2018).

5 Conclusions

In summary, the results of this study have confirmed our initial hypothesis that physical reaction time training can indeed have a positive impact on reaction time in e-sports. This breakthrough represents a valuable contribution to scientific research and has the potential to guide the development of evidence-based guidelines and intervention methods specifically designed for the e-sports community.

Acknowledgements Special thanks to Universiti Teknologi Malaysia, Skudai, Johor, Malaysia for research grant scheme Q.J130000.2653.16J92 and e-sports athletes as participants in this research.

References

Bediou B, Adams D, Mayer R, Tipton E, Green C, Bavelier D (2018) Meta-analysis of action video game impact on perceptual, attentional, and cognitive skills. Psychol Bullet J 144(1):77–110

Bickmann P, Rudolf K, Wechsler K, Tholl C (2021) Comparison of reaction time between eSports players of different genres and sportsmen. Int J eSports Res 1(1):1–16

Choi C, Hums M, Bum CH (2018) Impact of the family environment on juvenile mental health: eSports online game addiction and delinquency. Int J Environ Res Public Health 15(12):2850

Cunningham GB, Fairley S, Ferkins L, Kerwin S, Lock D, Shaw S (2018) ESport: construct specifications and implications for sport management. Sport Manag Rev 21:1–6

Donoghue JF, Balentine J, Schmidt G, Zwibel H (2018) Managing the health of the eSport athlete: an integrated health management model. BMJ Open Sport Exerc Med 5(1):1–6

Dye MW, Green C, Bavelier D (2009) Increasing speed of processing with action video games. Curr Dir Psychol Sci 18(6):321–326
Dykstra R, Koutakis P, Hanson N (2021) Relationship between physical fitness variables and reaction time in eSports gamers. Int J eSports Res 1(1):1–14
Erickson KI, Hillman CH, Kramer AF (2015) Physical activity, brain, and cognition. Curr Opin Behav Sci 4:27–32
Fissler P, Küster O, Schlee W, Kolassa I-T (2013) Novelty interventions to enhance broad cognitive abilities and prevent dementia: synergistic approaches for the facilitation of positive plastic change. Prog Brain Res 207:403–434
Green CS, Pouget A, Bavelier D (2010) Improved probabilistic inference as a general learning mechanism with action video games. Curr Biol 20(17):1573–1579
Herold F, Hamacher D, Schega L, Mueller NG (2018) Thinking while moving or moving while thinking-concepts of motor-cognitive training for cognitive performance enhancement. Front Aging Neurosci 10(228):1–11
Maass A, Düzel S, Brigadski T, Goerke M, Becke A, Sobieray U (2016) Relationships of peripheral IGF-1, VEGF and BDNF levels to exercise-related changes in memory, hippocampal perfusion and volumes in older adults. Neuroimage 131:142–154
Murphy CP, Wakefield A, Birch PDJ, North JS (2020) Esport expertise benefits perceptual-cognitive skill in (traditional) sport. J Expert 3(4):227–237
Netz Y (2019) Is there a preferred mode of exercise for cognition enhancement in older age?-a narrative review. Front Med 6(57):3–10
Radák Z (2018) The physiology of physical training. Academic Press, USA, pp 111–118
Toth AJ, Ramsbottom N, Kowal M, Campbell MJ (2020) Converging evidence supporting the cognitive link between exercise and e-sport performance: a dual systematic review. Brain Sci 10(11):1–34

Preventing Facial Injuries in Sports: A Preliminary Study on Sports-Related Facial Injuries in Malaysia

N. A. H. Abdul Razak and S. Nabil

Abstract Introduction: Facial injury can occur during sports activity. Understanding of epidemiology of these injuries will improve our understanding and could help in preventing such injuries. We aim to explore the epidemiology of sports-related facial injuries. Methodology: This is part of a cross-sectional assessment of facial fractures among patients who have sustained facial soft tissue injury with Ethic Committee approval UKM PPI/111/8/JEP-2021-748. Subjects were recruited over a period of 10 months (from January 2022 until October 2022). Physical and radiographic examination were performed when indicated and findings were recorded. Results: A total of 190 patients sustained facial injuries during the study period. Among those, only 7 were sports related. The mean age was 28.3 ± 18.8 years involving six males and only one female. In each case, the injury was related to five different sports namely boxing, hockey, martial arts, football, and cycling. The majority of patients sustained the injury in the midfacial region, especially the periorbital area. Only one had a facial bone fracture. Discussion: The incidence of sports-related facial injuries is 3.7% in our series. The finding of few females involved in this series is probably due to the sports involvement in Malaysia are predominant by males. More strikingly, most injuries were sustained in the midfacial region especially the periorbital area which is the most exposed part of the face. Conclusion: Sport-related facial injuries are uncommon but when it occurs, it would most likely be involving the midfacial area and can be associated with facial bone fractures. This finding can be used as a targeted measure to prevent sports-related facial injuries.

Keywords Facial injury · Facial fracture · Sports injury · Orbital fracture · Face guards

---

N. A. H. A. Razak · S. Nabil (✉)
Department of Oral & Maxillofacial Surgery, Faculty of Dentistry, Universiti Kebangsaan Malaysia, 50300 Kuala Lumpur, Malaysia
e-mail: syednabil@ukm.edu.my

N. A. H. A. Razak
e-mail: p107099@siswa.ukm.edu.my

Z. Mohamed et al. (eds.), *Proceedings of the 2nd International Conference on Innovation and Technology in Sports, ICITS 2023, 27–28 November, Malaysia*, Lecture Notes in Bioengineering, https://doi.org/10.1007/978-981-97-3741-3_25

1 Introduction

Facial injuries are predominantly influenced by socioeconomic as well as demographic factors. Facial injuries in sports may not be commonly reported like facial injuries due to trauma or falls. Nevertheless, facial injury accounts for less than 30% of sports injuries, while sports injuries are accountable for 11–42% of facial fractures (Hwang 2020). Different sports will pose different risks of injuries and patterns, as all of them will have contact at some point, either with an object or another person.

Although uncommon, those involved in sports medicine should be aware of the risks of facial injuries and associated fractures as well as preventive measures. Injuries can be divided into three, which include soft tissue injury, facial fracture, and dentoalveolar injury. Hwang found that the main cause of sports injuries is due to direct body contact with soft tissue injury of the head and neck as the most common associated injury. He also found that nasal fracture was the most common injury (Hwang 2020).

The purpose of this case series was to review the epidemiology, pattern of injury, and presence of fractures in facial injuries due to sports in order to highlight injuries sustained during sports and associated underlying fractures. Understanding these factors would allow targeted preventive measures in relation to the population at highest risk and the type of injury that is most common.

2 Methodology

This is part of a larger cross-sectional clinical study that was designed to determine the presence of facial fractures among patients who have sustained facial soft tissue injury. This study was conducted in accordance with the Declaration of Helsinki, with the approval of the UKMMC Research Ethic Committee with reference number UKM PPI/111/8/JEP-2021-748. Consent was required from patients for them to be included in this study. The inclusion criteria of the study are any patient above 16 years old or older who sustains any type of facial injuries secondary to sports-related activity and has a radiographic examination done if clinically indicated. Patients will be excluded if they do not provide consent.

The subjects were recruited over a period of 10 months (from January 2022 until October 2022). They were recruited upon presentation to the Emergency Trauma Department (ETD) UKMMC when referrals were made to the Oral and Maxillofacial Surgery (OMFS) Department for the management of facial injuries. Prior to an assessment, subjects were stabilized as per the Advance Trauma Life Saving protocol. Once the subject is stabilized, a thorough history of injury as well as medical history was taken from the subjects or their carer.

Physical examination was conducted as per standard of care. All data were collected and compiled in a proforma by the doctor on duty who oversaw the case. Data collected includes the patient's epidemiological data, such as age, sex, race, patient's past medical history, and mechanism of injury. Following that, radiograph imaging and examination, when indicated, of all subjects taken during their visit was reviewed by the researcher, and any facial bone fractures were observed and recorded based on the anatomical sites. Computed tomography (CT) reports certified by radiologists, when available, were also reviewed, and any fractures reported were also recorded based on anatomical sites.

Statistical analysis was performed using Statistical Package for Social Science (SPSS) Version 26 which was developed by International Business Machines Cooperation (IBM). Descriptive statistics including mean and standard deviation were calculated for selected study variables.

3 Results

There was a total of 190 patients who sustained facial injuries during the study data collection period. Among those, only 7 patients met the inclusion criteria and were included in this analysis. Of the 183 that were excluded were due to: 129 subjects sustained injuries that were not related to sports activity, sixteen subjects were below 16 years old, eight subjects had no radiographic imaging, and 30 subjects did not give their consent or were lost to follow-up. Out of 7 subjects included in this study, 85.7% were males. The mean age of the subjects was 23.8 ± 18.8 years and ranged between 16 and 66 years old. Table 1 describes the summary of the included cases while Table 2 describes the descriptive analysis of the subjects.

3.1 Boxing

Both patients were 17-year-old Malay, fit and healthy males, who were brought to the emergency department as they sustained injury over the lip due to boxing on two different occasions. Both claimed they were wearing protective headgear when sparring. Upon examination, their Glasgow Coma Scale (GCS) was E4V5M6. One sustained laceration wound over the upper lip and the upper labial mucosa demonstrated a laceration and abrasion wound, while the other sustained a laceration wound over the upper lip and labial mucosa. Fortunately, both did not sustain any underlying facial fracture based on radiographic findings. Their wounds were cleaned and sutured and were later reviewed in the outpatient clinic.

3.2 Hockey

The first patient was a 19-year-old Malay, fit and healthy male, who was brought to the emergency department as he sustained a laceration wound and contusion wound

Table 1 Demographic data, site, and type of injury in sports injury

No	Age	Sex	Health status	Type of sports	Facial zone of injury	Soft tissue injury	Fracture of bone	Protective gear	Other injuries
1	17	M	Fit and healthy	Boxing	Upper lip, upper labial mucosa	Laceration, abrasion	No	Yes, helmet	No
2	17	M	Fit and healthy	Boxing	Upper lip, upper labial mucosa	Laceration	No	Yes, helmet	No
3	19	M	Fit and healthy	Hockey	Orbital	Laceration	Yes	No	Traumatic mydriasis
4	16	F	Fit and healthy	Hockey	Upper labial mucosa	Contusion	No	No	Cerebral concussion
5	23	M	Fit and healthy	Martial arts	Orbital	Laceration	No	No	No
6	41	M	Fit and healthy	Football	Orbital	Laceration	No	No	No
7	66	M	Osteoarthritis, right eye blindness	Cycling	Orbital, frontal	Laceration	No	No	No

over the right eye while playing hockey. His injury was sustained due to contact with the hockey stick. Upon examination, his Glasgow Coma Scale (GCS) was E4V5M6. Unfortunately, this patient sustained a lateral orbital wall fracture which required further treatment.

The second patient was a 15-year-old Malay female, fit and healthy, who was presented to the emergency department for a facial injury. Upon examination, her Glasgow Coma Scale (GCS) was E4V5M6, and noted she sustained a contusion wound over the upper labial mucosa which was then treated conservatively. She also sustained a cerebral concussion and was admitted under Neurosurgery for observation. She did not sustain any facial fractures.

3.3 Martial Arts

The patient was a 23-year-old Chinese, fit and healthy male, who attended the emergency department as he sustained an injury over his right eye. Upon examination, his Glasgow Coma Scale (GCS) was E4V5M6, and noted he sustained a laceration wound over his right eye due to contact with a sparring partner's knee. His

Table 2 Descriptive statistics on gender, age, type of sports, presence of fracture, other injuries, and location of injury

Variable	No of patient, N = 7	Total % 100
Gender		
Male	6	85.1
Female	1	14.3
Age		
Mean	28.3 ± 18.8	
Min	15	
Max	66	
<30	5	71.4
>30	2	28.6
Sports type		
Contact	6	85.1
Non-contact	1	14.3
Fracture		
Yes	1	14.3
No	6	85.7
Other injuries		
Yes	2	28.6
No	5	71.4
Location of injury		
Midface	4	57.1
Non-midface	3	42.9

wounds were cleaned and sutured and were later reviewed in the outpatient clinic. He, however, did not sustain any underlying bone fracture.

3.4 Football

The patient was a 41-year-old Malay, fit and healthy male, who presented himself to the emergency department for sustaining injury over the right eye. Upon examination, his Glasgow Coma Scale (GCS) was E4V5M6, and noted he sustained a laceration wound over his right eye. His wounds were cleaned and sutured and were later reviewed in an outpatient clinic. He did not sustain any underlying bone fracture.

3.5 Cyclist

The patient was a 66-year-old Chinese male with underlying hypertension, right knee osteoarthritis, and right eye blindness, presented himself to the emergency department for injury over the right region of the face. Upon examination, his Glasgow Coma Scale (GCS) was E4V5M6 and noted he sustained a laceration wound over his right eye and right frontal after hitting the back of the car. His wounds were cleaned and sutured and were later reviewed in the outpatient clinic. He did not sustain any facial fractures.

4 Discussion

The mechanism of facial injury will usually involve a large amount of force (Truong 2017). The severity of facial injury is expressed by the direction and point of application of force. The pattern of injury is determined by the cross-section area of the object or agent struck. While nature of the injury is determined by the degree of force and the resistance to the force by the face (Pappachan and Alexander 2012). There are three types of facial injury, namely soft tissue injury, facial fracture, and dentoalveolar injury. Soft tissue injury encompasses laceration, contusion, and abrasion injury (Polmann et al. 2020). Facial fractures are classified based on the location of the fracture, i.e., frontal, midface, and mandible. In the event of injury, traditionally soft tissue injuries are observed first. This is followed by the inspection of facial fractures by means of palpating the bony prominence. Injuries should be cleaned, and any open wound should be sutured. Special attention needed to be given to the eye, nose, and mandible as they may interfere with normal function.

In general, there are protective facial equipment used in sports such as mouthguards, goggles, headgears, and helmets. In most sports, using protective headgear is not traditionally worn neither it is mandatory, for example, ball sports, court sports, and water sports (Farrington et al. 2012). It is mandatory in more high-risk sports with potentially life-threatening injuries such as motorsports and boxing (Hino et al. 2017; Tjønndal et al. 2022). Headgear is meant to redistribute the load and reduce the energy of impact forces (Farrington et al. 2012). Full-face helmet was found to be effective in preventing facial injuries compared to a partial helmet (Gobbi et al. 2004). Eye protection, on the other hand, offers protection of the eye from impact and debris with a prevention rate of eye injuries at 90% (Farrington et al. 2012; Mishra and Verma 2012). However, its usage is not popular due to the hindering of the field of vision. Eye protection is also not mandatory in the majority of sports, apart from paintball (Farrington et al. 2012).

Sports-related facial injuries are uncommon with an incidence of 3.7% in our series. In the present cases, all sports injuries involved 6 males who were below 50 years old and healthy individuals apart from one 66-year-old male and one female aged 16. The finding of very few female involvements in this series is probably more due to the sports involvement in Malaysia being predominantly by males. This could also be attributed to the active lifestyle of the male gender with higher involvement in contact sports. On top of that, the male gender was more likely to be exposed to a higher impact by another male during contact due to higher body mass and as a consequence of aggressive play (Bojino et al. 2020). Patients in this study were involved in five different types of sports with different exposure to either body contact or equipment. None of these injuries involved other parts of the body. All but one had a cerebral concussion upon examination. Three patients sustained injuries over the lips, while the other four sustained injuries over the eye, particularly on the right. Although a small number, the distribution seems to follow the pattern by Bolt and Watts where most of the injuries were concentrated over lips and eye region when the causative agent was sports (Bolt and Watts 2004).

The majority of the injuries were sustained at the midfacial region especially the periorbital area which is due to it being the most exposed part of the face. Out of seven patients, three sustained injuries over the lips. Two patients sustained an injury through boxing and claimed that they were both wearing helmets. Tjønndal et al. in their systematic review concluded that headguard in boxing has some protective factors against cuts and fractures (Tjønndal et al. 2022). However, helmet in boxing is usually partially covered with the anterior part of the midface and mandible exposed. While we can stipulate that the helmet may have absorbed most of the impact, some of the impact may have caused enough force to shear the skin that is exposed, therefore causing lacerations over the lips. One study found that the rate of fractures that made visits to ETD was 36.9% (Hojjat et al. 2016).

Our study found one patient who sustained a laceration wound over the eye due to martial arts. It has been found that there is a prevalence of 5.8–8.3% of eye injuries in mixed martial arts, whereas the most prevalent injury involves eyebrow and eyelid laceration (Fliotsos et al. 2021). Prevention strategies should be implemented such as using a padded helmet with a visor, as well as rules of sparring to avoid facial injury should be considered and encouraged. In fact, those who sustained ophthalmic injuries were likely to lose their match (Fliotsos et al. 2021).

Out of four subjects who sustained injuries around the eye, only one had sustained an underlying facial bone fracture. On top of the laceration, this particular patient who had sustained a bone fracture, also sustained contusion over the right eye. Further examination revealed that the patient also sustained traumatic mydriasis over the right eye. This injury was caused by contact with the hockey stick. Hockey is a type of contact ball sport that may predispose patients to accidental injuries from deflected balls and hockey sticks. Orbital fractures can happen as a result of contact with a piece of equipment wider than the orbital socket, which can be explained by the hydraulic theory or the buckling theory (Ohana and Alabiad 2021).

From here, we can observe that bodily contact can cause tissue injury to the orbital region, however, contact with equipment may exacerbate the injury by involving facial bones as well. We can stipulate that equipment transfers a higher energy force to the face. Previous studies have shown that kinetic energy to cause orbital floor fracture was noted to be 1.8–14.7 J (Bullock et al. 1996).

90% of orbital injuries can be prevented with the use of eye protection (Ohana and Alabiad 2021). It is made from polycarbonate lenses which provide 20 times the protection than normal glasses (Ohana and Alabiad 2021). Studies have shown that orbital injury is remarkably increased in those who did not wear a visor (Micieli et al. 2014). Squash mandates their players to wear goggles as the ball can reach up to 90 mph, and there were no reports of ocular injuries (Ohana and Alabiad 2021). Orbital injuries should be prevented as they can cause visual disability; therefore, prevention is very crucial as the eye is an important organ.

There are several limitations of our preliminary study. Firstly, we included all types of sports in our study which gave us inconsistent information with regard to the force of impact. We did not look at the effect of force on the face. The type of force from a human contact differs from the force of equipment contact. In our findings, contact with a solid object without any padding has caused more severe injuries, i.e., orbital fracture as compared to contact with the human force which just causes laceration. Secondly, a small number of subjects in this study might not actually represent the sports injury population in Malaysia as a whole. Future research with a larger population is currently being planned to establish a more comprehensive understanding of the implications of sports injuries to the face. Lastly, we were not able to perform an inferential statistical analysis due to the small sample that sustained facial fractures in this study. Therefore, we were unable to draw conclusions about the factors related to facial injuries in sports based on the sample collected.

5 Conclusion

Sport-related facial injuries are uncommon but when it occurs, it would most likely be involving the midfacial area and can be associated with facial bone fractures. This finding can be used as a targeted measure to prevent sports-related facial injuries. Raising awareness about the frequency and severity of facial injuries is important to encourage the usage of protective facial equipment to reduce complications related to sporting injuries. Given the high prevalence of injury around the eye in our findings, it is important for sports facilities to be well-informed in first aid eye emergencies as well as enforcing the usage of suitable, tailored protective eyewear during sports.

References

Bojino A, Roccia F, Giaccone E, Cocis S (2020) Comprehensive analyses of maxillofacial fractures due to non-professional sports activities in Italy. Dent Traumatol 36(6):632–640. https://doi.org/10.1111/edt.12586

Bolt RW, Watts PG (2004) The relationship between aetiology and distribution of facial lacerations. Inj Extra 35(1):6–11. https://doi.org/10.1016/j.injury.2003.11.010

Bullock JD, Johnson DA, Ballal DR, Bullock RJ, Cox MS (1996) Ocular and orbital trauma from water balloon slingshots: a clinical, epidemiological, experimental, and theoretical study. Trans Am Ophthalmol Soc 94:105–134

Farrington T, Onambele-Pearson G, Taylor RL, Earl P, Winwood K (2012) A review of facial protective equipment use in sport and the impact on injury incidence. Br J Oral Maxillofac Surg 50(3):233–238. https://doi.org/10.1016/j.bjoms.2010.11.020

Fliotsos MJ et al (2021) Prevalence, patterns, and characteristics of eye injuries in professional mixed martial arts. Clin Ophthalmol 15:2759–2766. https://doi.org/10.2147/OPTH.S319025

Gobbi A, Tuy B, Panuncialman I (2004) The incidence of motocross injuries: a 12-year investigation. Knee Surg Sport Traumatol Arthrosc 12(6):574–580. https://doi.org/10.1007/s00167-004-0510-z

Hino S, Takeda A, Kaneko T, Horie N, Shimoyama T (2017) Oral lacerations during motocross: a case report. Dent Traumatol 33(6):482–484. https://doi.org/10.1111/edt.12357

Hojjat H et al (2016) Adding injury to insult: a national analysis of combat sport-related facial injury. Ann Otol Rhinol Laryngol 125(8):652–659. https://doi.org/10.1177/0003489416644617

Hwang K (2020) Field management of facial injuries in sports. J Craniofac Surg 31(2):e179–e182. https://doi.org/10.1097/SCS.0000000000006132

Micieli JA, Zurakowski D, Ahmed IIK (2014) Impact of visors on eye and orbital injuries in the National Hockey League. Can J Ophthalmol 49(3):243–248. https://doi.org/10.1016/j.jcjo.2014.03.008

Mishra A, Verma AK (2012) Sports related ocular injuries. Med J Armed Forces India 68(3):260–266. https://doi.org/10.1016/j.mjafi.2011.12.004

Ohana O, Alabiad C (2021) Ocular related sports injuries. J Craniofac Surg 32(4):1606–1611. https://doi.org/10.1097/SCS.0000000000007618

Pappachan B, Alexander M (2012) Biomechanics of cranio-maxillofacial trauma. J Maxillofac Oral Surg 11(2):224–230. https://doi.org/10.1007/s12663-011-0289-7

Polmann H et al (2020) Prevalence of dentofacial injuries among combat sports practitioners: a systematic review and meta-analysis. Dent Traumatol 36(2):124–140. https://doi.org/10.1111/edt.12508

Tjønndal A, Haudenhuyse R, de Geus B, Buyse L (2022) Concussions, cuts and cracked bones: a systematic literature review on protective headgear and head injury prevention in Olympic boxing. Eur J Sport Sci 22(3):447–459. https://doi.org/10.1080/17461391.2021.1872711

Truong TA (2017) Initial assessment and evaluation of traumatic facial injuries. Semin Plast Surg 31(2):69–72. https://doi.org/10.1055/s-0037-1601370

Machine Learning Insights into Basketball Championship Predictions: An Analytical Comparison

Siti Zainab Ibrahim, Aditya Muhammad Reza, Lew Wei Kean, Nor Azlina Ab. Aziz, and Sharifah Noor Masidayu Sayed Ismail

Abstract Leveraging machine learning techniques to forecast the eventual championship-winning teams through the consideration of diverse sport-related variables has garnered substantial attention in contemporary sports research. In this study, the focus resides in employing machine learning models to forecast National Basketball Association (NBA) champions. The techniques were applied to a selected NBA dataset originated from data.world. Central to this endeavor is the role of feature selection which proves pivotal not only in discerning NBA championships but also in gauging predictive capacities. Five machine learning models encompassing Decision Trees, Random Forests, Logistic Regression, Support Vector Machines (SVMs), and Autoencoders were employed to offer multifaceted insights into the projection of victorious teams and their performance trajectories. Each model exhibited varying degrees of predictive accuracy. The autoencoders model emerged as the pinnacle performer by attaining the highest accuracy levels of 0.888 and 0.810. The findings provided exciting insights into determining the NBA championship and demonstrating the roles of machine learning in predicting championships. This study demonstrates the potential application of machine learning across a wide range of sports arenas. It aspires to foster advancements in predictive analytics thereby enhancing the precision of such forecasts and their broad applicability.

Keywords Predictive analytics · Machine learning model · NBA championship

S. Z. Ibrahim (✉) · A. M. Reza · L. W. Kean
Albukhary International University, Jalan Tun Razak, 05200 Alor Setar, Kedah, Malaysia
e-mail: Zainab.Ibrahim@taylors.edu.my

N. A. Ab. Aziz · S. N. M. S. Ismail
Multimedia University, Jalan Ayer Keroh Lama, 75450 Ayer Keroh, Melaka, Malaysia

Z. Mohamed et al. (eds.), *Proceedings of the 2nd International Conference on Innovation and Technology in Sports, ICITS 2023, 27–28 November, Malaysia*, Lecture Notes in Bioengineering, https://doi.org/10.1007/978-981-97-3741-3_26

1 Introduction

The world of sports has undergone a remarkable transformation with the advent of advanced technology, particularly the integration of machine learning models. In recent years, the application of these models has revolutionized the way sports are analyzed, understood, and even played. Machine learning, a subset of artificial intelligence, involves the utilization of algorithms that enable computers to learn from and make predictions or decisions based on data. This capability has found a profound resonance in the realm of sports, transcending traditional boundaries and unlocking new dimensions of insight.

Sports, whether played professionally or enjoyed recreationally, generate an immense volume of data. From player performance statistics and game outcomes to physiological metrics and fan engagement data, this information-rich environment presents an ideal playground for machine learning. The inherent complexity of sports, characterized by the interplay of numerous variables and dynamic elements, poses intriguing challenges that machine learning models are adept at addressing.

The incorporation of machine learning models in sports extends across a wide spectrum of applications. At a fundamental level, these models enable predictive analytics, offering the ability to forecast game outcomes, player performance, and even injury occurrences. This predictive power has practical implications for sports betting, fantasy leagues, and strategic decision-making by coaches and team managers.

This study aims to predict National Basketball Association (NBA) champions by comparing the performance of machine learning models applied to a comprehensive dataset. The dataset includes team statistics from multiple seasons, allowing this study to analyze performance metrics such as points scored, rebounds, and assists made. The study also aimed to enhance the predictive capabilities of the machine learning models by preprocessing the data and selecting meaningful features for analysis (Kumar et al. 2022). Four supervised machine learning models: Decision Tree, Random Forest, Logistic Regression, and Support Vector Machines (SVMs), along with an unsupervised model called autoencoders were compared. These models provide different approaches to predict champions based on team performance.

To evaluate the models, in addition to the percentage of winning team predicted by each model, accuracy measures were utilized as the primary evaluation metric, measuring the correctness of predictions. Robust validation methods such as train–test split and k-fold cross-validation were employed to ensure the models' generalizability and performance. By using cross-validation techniques, this study may provide useful insights in promoting the generalizability of the models across diverse sports, in particular basketball arenas.

2 Literature Review

In this review, the dynamic landscape of utilizing machine learning models in sports was reviewed. From the methods employed in training these models to their practical applications across various sports domains, the transformative role that data-driven insights play in shaping the modern sporting landscape was delved into. As this section navigates this exciting terrain, it becomes evident that the synergy between machine learning and sports is poised to redefine the way athletic endeavors can be perceived, analyzed, and interacted.

The integration of machine learning models into sports analytics has witnessed rapid advancements, enabling accurate predictions across various aspects of sports. The prediction of game outcomes remains a cornerstone of sports analytics. In the realm of game outcome predictions, Hsu (2021) explored the application of deep learning architectures to forecast soccer match results, highlighting the potential of Convolutional Neural Networks (CNNs) in extracting meaningful features from match images. Rudrapal et al. (2020) employed a novel neural network architecture for predicting soccer match results, showcasing improved accuracy by considering team-level attributes and historical match data.

Advancements in player performance estimation were demonstrated by Gong (2022), who combined player tracking data with recurrent neural networks to predict tennis player performance during matches. Similarly, in cricket, Patil et al. (2020) leveraged machine learning techniques to predict a batsman's runs in different overs, showcasing the potential for data-driven insights in understanding player behavior. Wickramasinghe also has utilized machine learning to forecast cricket player performance by integrating historical data and game conditions (Wickramasinghe 2020), while Ishi et al. devised a model to predict player batting average, leveraging pitch-level data and player characteristics (Ishi et al. 2022).

Transitioning to basketball, predicting basketball game outcomes has been a central focus of machine learning applications (Zhao et al. 2023; Qiyun et al. 2022; Osken and Onay 2022; Yu et al. 2018; Csataljay et al. 2009; Sikka and Rajeswari 2022). Zhao et al. (2023) developed a model based on Graph Convolutional Networks (GCNs) to predict NCAA basketball game outcomes, effectively capturing team interactions and dynamics. Similarly, Qiyun et al. (2022) proposed a hybrid model combining CNNs and Long Short-Term Memory (LSTM) networks to forecast NBA game results, demonstrating competitive accuracy through the incorporation of game sequences and contextual data. Using Artificial Neural Network (ANN), Osken and Onay (2022) achieved a prediction accuracy of 76% of the winning team over a period of five NBA seasons.

The application of machine learning models for sport-related predictions in Malaysia has emerged as a growing field of research (Razali et al. 2023). Football and badminton are widely followed sports in Malaysia, making outcome prediction for both a prime focus. Razali et al. (2023) applied machine learning techniques to predict the results of football using data from the Malaysia Super League. Machine learning models have demonstrated remarkable capabilities in predicting outcomes, player performances, and injuries in the sports domain. From soccer to basketball, the application of these models has enabled data-driven insights to reshape how sports are understood, played, and experienced. While the adoption of machine learning models for sports predictions has demonstrated substantial progress, challenges persist.

Data quality, model interpretability, and the need for domain expertise remain pertinent concerns (Chmait and Westerbeek 2021). The availability of comprehensive and high-quality data, especially for lesser-known sports, is a concern. As the field progresses, it is imperative to address challenges related to data quality, model interpretability, and generalizability to unlock the full potential of machine learning for sports predictions.

3 Methods

3.1 Dataset Collection

Data used for predicting NBA champions was sourced from the official "data.world" website (Data World 2023). This dataset encompasses a wide range of comprehensive team statistics spanning multiple seasons, dated from 1997 until 2022. The raw data set contains 22 features with 725 rows. The dataset chosen offers several advantages for our predictive analysis. It comprises a substantial sample size (Rajput et al. 2023), providing a robust foundation for statistical analysis. Furthermore, it accommodates variations in team performance over time, making it suitable for identifying trends and patterns crucial for championship predictions.

Notably, certain aspects of basketball data, such as player-specific information and detailed game-level data, were excluded from this dataset. While these elements can be valuable in certain analyses, our focus in this study centers on predicting NBA champions based on team-level performance metrics. Consequently, the dataset's exclusion of player-specific and game-level details does not hinder our objectives.

3.2 Feature Selection and Dataset Preprocessing

Key variables considered in this study comprise points scored, rebounds secured, and assists made. Based on the previous study by Csataljay et al. (2009), Migliorati (2021), Jung et al. (2021), these features effectively encapsulate offensive and defensive prowess, enabling the identification of teams with potent scoring or solid defensive strategies. These variables are predicted to directly reflect a team's performance and success throughout a season, offering valuable insights into a team's strength and competitiveness.

Using Phyton libraries, the NBA dataset underwent rigorous preprocessing to render it suitable for predictive analysis. A primary focus of this preprocessing was addressing missing values (Palanivinayagam and Damasevicius 2023) by imputing the mean values for each selected feature. These preprocessing steps were crucial to ensuring data quality and appropriately preparing features for subsequent modeling (Maharana et al. 2022; Garnica-Caparrós et al. 2022).

3.3 Modeling and Evaluation

This study leverages four widely recognized supervised machine learning models: decision tree, random forest, logistic regression, and SVM, alongside one unsupervised machine learning model, autoencoder. These models collectively contribute to the predictive framework for NBA champion forecasting. To evaluate the efficacy of NBA champion prediction models, appropriate evaluation metrics are imperative (Teno et al. 2022). Accuracy, a commonly employed metric, assesses the correctness of predictions by calculating the ratio of correctly predicted outcomes to the total predictions. In this project, accuracy serves as the primary evaluation metric, providing insights into the model's capability to accurately predict championship outcomes within the context of NBA champion prediction (Wang et al. 2022).

In order to validate the NBA champion prediction models and ensure their generalizability, robust validation methods are employed (Horvat et al. 2020). Two key methods utilized in this study are the train–test split and K-fold cross-validation. These methods validate the models' performance on unseen data and ensure their reliability in making predictions beyond the training dataset, contributing to the overall robustness of the predictive framework.

4 Results and Discussions

4.1 Accuracy and Validation

The outcomes presented in Table 1 elucidate the performance of various machine learning models concerning accuracy scores across different splitting ratios. Notably, the autoencoder model emerges as the top performer, consistently surpassing other models with the highest accuracy scores for multiple splitting ratios. Logistic regression also exhibits commendable performance, yielding notably high accuracy scores. Conversely, the decision tree and random forest models display varying levels of accuracy across the different splitting ratios, showcasing their sensitivity to data partitioning. The SVM model, on the other hand, demonstrates stability in its performance, maintaining competitive accuracy scores across the various splits. These findings underscore the critical importance of selecting the appropriate model for a given task, taking into account considerations such as computational complexity and interpretability. Autoencoders, with their ability to capture intricate data patterns, hold significant promise, particularly in scenarios involving complex data structures.

When considering the results of K-fold cross-validation, it becomes evident that model performance exhibits variability. The decision tree and random forest models notably improve its accuracy across multiple folds, indicating its adaptability to different data subsets. In contrast, the accuracy of the logistic regression and SVM models remains relatively stable. The autoencoder model, while initially displaying very high accuracy, experiences a decrease in performance as the number of folds increases. This phenomenon underscores the intricate relationship between model complexity and the number of folds in K-Fold validation.

Overall, the choice of the most suitable model hinges on the dataset as well as the specific requirements of the task at hand. Autoencoders excel in handling complex data patterns, while the logistic regression and SVM models consistently deliver reliable accuracy across various scenarios. The consideration of these findings is pivotal in making informed decisions when selecting the appropriate machine learning model for a given dataset and predictive task.

In Fig. 1, the box plot serves to bolster the robustness of the findings. It presents a visual representation of the accuracy results, employing the Interquartile Range (IQR) within a box, with the median accuracy value denoted by a horizontal line within that box. For the decision tree algorithm, the accuracy values of 0.72 and 0.58 indicate a moderate level of accuracy in predicting NBA champions. This suggests that the model exhibits a reasonable predictive capacity. In contrast, the random forest algorithm attains accuracy values of 0.71 and 0.65, implying a comparatively higher level of accuracy when juxtaposed with the decision tree model. The box plot effectively showcases the dispersion of accuracy values, signifying the model's performance variability.

In the case of logistic regression, the accuracy values are 0.72 and 0.67, revealing a consistent level of accuracy across the two evaluations. The box plot, through its visualization, conveys the extent of accuracy value spread and underscores the model's performance stability. Moving on to the SVM algorithm, it yields accuracy values of 0.72 and 0.63, indicating a moderate level of accuracy with some inherent variability. The box plot visually portrays the distribution of accuracy values, accentuating any outliers or deviations in model performance. In stark contrast, the autoencoders algorithm demonstrates notably higher accuracy values of 0.94 and 0.86, signifying

Table 1 Accuracy results of machine learning models

Algorithms models	Train-Test split				K-fold cross-validation		
	0.2	0.3	0.4	0.5	3	5	10
Decision tree	**0.58**	0.48	0.55	0.51	0.62	**0.72**	0.67
Random forest	0.54	0.57	**0.66**	0.65	0.71	0.71	**0.73**
Logistic regression	0.66	0.68	**0.72**	0.67	**0.72**	**0.72**	**0.72**
SVM	0.63	0.63	0.66	**0.67**	**0.72**	**0.72**	**0.72**
Autoencoders	0.91	0.94	0.94	**0.97**	**0.95**	0.81	0.86

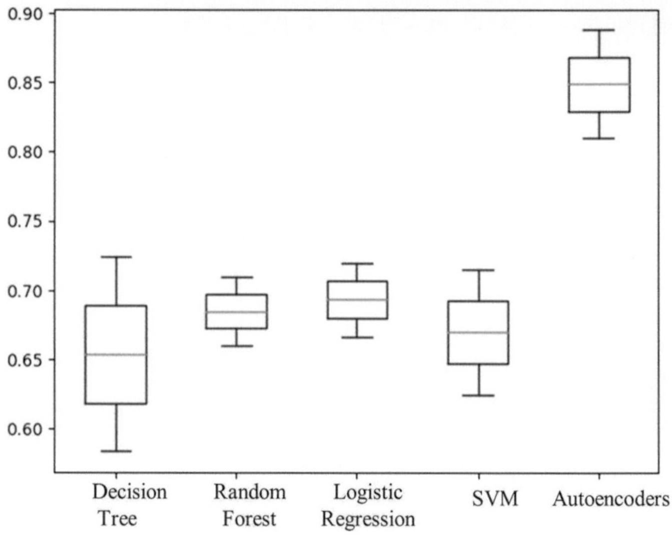

Fig. 1 Performance variability of machine learning models

robust predictive capabilities in identifying NBA champions. The box plot effectively captures the dispersion of accuracy values, elucidating the model's consistent and high-performance attributes.

4.2 Identifying NBA Champion and Predictive Features

Using linear regression, the regression coefficients were calculated to identify the contribution of each selected feature to the prediction. The findings, as presented in Table 2, shed light on the contribution of each feature for the predictive algorithms across two distinct evaluation methods: train–test split and K-fold cross-validation. Notably, the variable "Reb" (Rebounds) emerges as the most consistent factor across all algorithms assessed in both evaluation approaches. Consistency in model performance is a crucial aspect regardless of the chosen algorithm. In this study, irrespective of the specific algorithm employed, the results consistently emphasize the pivotal role of rebounding in influencing the accuracy and efficacy of predictive models. Similarly echoed by Csataljay et al. (2009), Migliorati et al. (2023), the consistent performance of rebounding as a predictive feature emphasizes its potential as a key determinant in championship predictions and reinforces the importance of incorporating it into predictive models.

Incorporating the insights from five distinct machine learning models, our analysis generated a total of ten champion predictions through both the train–test split and k-fold cross-validation methodologies. These predictions collectively serve as a comprehensive basis for evaluating championship contenders. In interpreting the

Table 2 Contributing features for each predictive algorithm

Algorithms models	Train–Test split			K-Fold cross-validation		
	Reb	Pts	Ast	Reb	Pts	Ast
Decision tree	**0.47**	0.30	0.22	0.47	**0.31**	0.24
Random forest	0.36	0.35	**0.28**	0.36	0.35	**0.28**
Logistic regression	0.16	0.13	**0.09**	**0.28**	**0.17**	**0.12**
SVM	0.41	0.40	0.03	**0.52**	**0.49**	**0.20**
Autoencoders	0.52	0.49	0.20	**0.52**	0.49	0.20

Note Reb = Rebounds, Ast = Assist, Pts = Average Points

graphical representation of these predictions, each bar in Fig. 2 corresponds to a specific NBA team. The height of each bar directly mirrors the percentage of predictions attributing championship victory to that particular team. Notably, the bar representing the Miami team stands out prominently, towering above the rest, with an impressive 70% of predictions favoring them as the champions. Conversely, the bars representing the LA Lakers, Golden State, and Chicago teams stand at a considerably lower height, each garnering a mere 10% of predictions. This substantial contrast in bar heights serves as a clear visual indicator of the varying degrees of favorability associated with different NBA teams, as predicted by the ensemble of machine learning models.

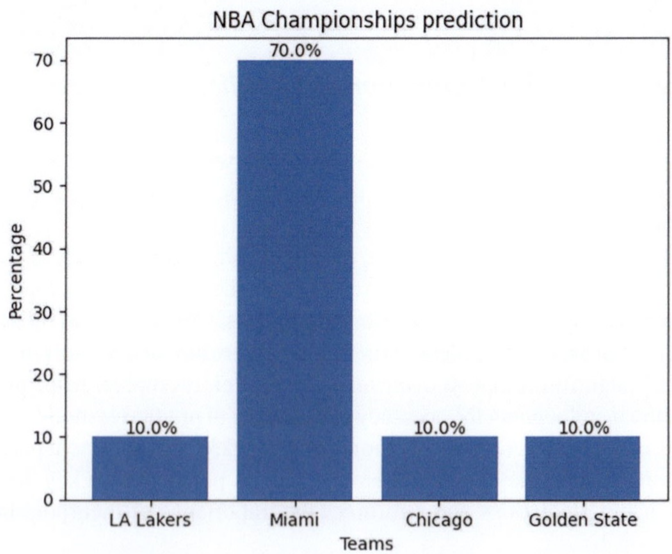

Fig. 2 Predictions of NBA champion

5 Conclusion

This paper has presented a comprehensive NBA championship prediction methodology utilizing machine learning techniques and datasets sourced from the official "data.world" website. The study encompassed critical phases of data preprocessing, meaningful feature selection, and the application of various algorithmic models, both supervised and unsupervised, validated using test–train split ratio and k-fold cross-validation. These models included decision trees, random forests, logistic regression, SVM, and autoencoder, each offering unique insights into championship predictions and team performance. Among the models, the autoencoder model emerged as the top performer, boasting the highest accuracy values of 0.89 and 0.81. This finding shows the possible effectiveness of autoencoders in predicting NBA champions based on the available dataset. Meanwhile, logistic regression and SVM shows stability in the prediction accuracy given the different scenarios present in the dataset.

Future works include blending in other important factors such as player injuries, transfers, coaching changes, and team dynamics as potential determinants, which could influence championship outcomes. Exploring advanced algorithms, including deep learning, holds promise for enhancing prediction accuracy. To ensure robustness of the prediction methods, this study can be further extended to sports arenas where the champions are not dominated by a particular team.

Overall, this paper has contributed valuable insights into the multifaceted features contributing to NBA team championships and demonstrated the applicability of machine learning in sports analytics. By further developing datasets, integrating real-time data, and exploring advanced algorithms, this research can serve as a foundation for refining accuracy and expanding the scope of machine learning applications in the NBA and across various sports arenas.

References

Chmait N, Westerbeek H (2021) Artificial intelligence and machine learning in sport research: An introduction for non-data scientists. Front Sports Act Living 3. https://doi.org/10.3389/fspor.2021.682287

Csataljay G, O'Donoghue P, Hughes M, Dancs H (2009) Performance indicators that distinguish winning and losing teams in basketball. Int J Perform Anal Sport 9(1). https://doi.org/10.1080/24748668.2009.11868464

Data World (2023). https://data.world/etocco/nba-team-stats. Last Accessed June 2023

Garnica-Caparrós M, Memmert D, Wunderlich F (2022) Artificial data in sports forecasting: a simulation framework for analyzing predictive models in sports. IseB 20(3):551–580. https://doi.org/10.1007/s10257-022-00560-9

Gong L (2022) A continuous deep learning system study of tennis player health information and professional input. Comput Intell Neurosci 8599894. https://doi.org/10.1155/2022/8599894

Horvat T, Havas L, Srpak D (2020) The impact of selecting a validation method in machine learning on predicting basketball game outcomes. Symmetry 12(3):431. https://doi.org/10.3390/sym12030431

Hsu YC (2021) Using convolutional neural networks and candlestick representation to predict sports match outcomes. Appl Sci 11(14):6594. https://doi.org/10.3390/app11146594

Ishi M, Patil J, Patil V (2022) An efficient team prediction for one day international matches using a hybrid approach of CS-PSO and machine learning algorithms. Array 14:100144. https://doi.org/10.1016/j.array.2022.100144

Jung W, Watson A, Kuehn S, Korem E, Koltermann K, Sun M, Wang S, Liu Z, Zhou G (2021) LAX-Score: Quantifying team performance in Lacrosse and exploring IMU features towards performance enhancement. In: Proceedings of the ACM on interactive, mobile, wearable and ubiquitous technologies 5(3). https://doi.org/10.1145/3478076

Kumar M, Sharma NB, Handa D (2022) Building predictive model by using data mining and feature selection techniques on academic dataset. Int J Modern Educ Comput Sci 14(4):16–29. https://doi.org/10.5815/ijmecs.2022.04.02

Maharana K, Mondal S, Nemade B (2022) A review: data pre-processing and data augmentation techniques. Glob Transit Proc 3(1):91–99. https://doi.org/10.1016/j.gltp.2022.04.020

Migliorati M (2021) Features selection in NBA outcome prediction through Deep Learning 1–29. http://arxiv.org/abs/2111.09695

Migliorati M, Manisera M, Zuccolotto P (2023) Integration of model-based recursive partitioning with bias reduction estimation: a case study assessing the impact of Oliver's four factors on the probability of winning a basketball game. AstA Adv Stat Anal 107:271–293. https://doi.org/10.1007/s10182-022-00456-6

Osken C, Onay C (2022) Predicting the winning team in basketball: a novel approach. Heliyon 8. https://doi.org/10.1016/j.heliyon.2022.e12189

Palanivinayagam A, Damasevicius R (2023) Effective handling of missing values in datasets for classification using machine learning methods. Information 14(2):92. https://doi.org/10.3390/info14020092

Patil NM, Sequeira BH, Gonsalves NN, Singh AA (2020) Cricket team prediction using machine learning techniques. SSRN. https://doi.org/10.2139/ssrn.3572740

Qiyun Z, Xuyun Z, Hongsheng H, Caizhong L, Yinping L, Rui M (2022) Sports match prediction model for training and exercise using attention-based LSTM network. Digit Commun Netw 8(4):508–515. https://doi.org/10.1016/j.dcan.2021.08.008

Rajput D, Wang WJ, Chen CC (2023) Evaluation of a decided sample size in machine learning applications. BMC Bioinform 24:48. https://doi.org/10.1186/s12859-023-05156-9

Razali N, Mustapha A, Aziz AQAA, Mostafa SA. (2023) Machine learning approach for Malaysia Super League football match outcomes prediction based on Elo rating system. In: Syed Omar SF, Hassan MHA, Casson A, Godfrey A, PP, Abdul Majeed A (eds) Innovation and technology in sports. Lecture Notes in Bioengineering. https://doi.org/10.1007/978-981-99-0297-2_13

Rudrapal D, Boro S, Srivastava J, Singh S (2020) A deep learning approach to predicting soccer match outcomes. In: Behera H, Nayak J, Naik B, Pelusi D (eds) Computational intelligence in data mining. Advances in Intelligent Systems and Computing, vol 990. https://doi.org/10.1007/978-981-13-8676-3_9

Sikka D, Rajeswari D (2022) Basketball win percentage prediction using ensemble-based machine learning. In: The 6th international conference on electronics, communication and aerospace technology. IEEE

Teno GDS, Wang C, Carlsson N, Lambrix P (2022) Predicting season outcomes for the NBA. Commun Comput Inf Sci, 1571 CCIS, 129–142. https://doi.org/10.1007/978-3-031-02044-5_11.

Wang Y, Liu W, Liu X (2022) Explainable AI techniques with application to NBA gameplay prediction. Neurocomputing 483:59–71. https://doi.org/10.1016/j.neucom.2022.01.098

Wickramasinghe I (2020) Naïve Bayes approach to predict the winner of an ODI cricket game. J Sports Analyt 6:75–84

Yu Z, Rennong Y, Guillaume C, Rajiv CS, Rob R (2018) Applying deep bidirectional LSTM and mixture density network for basketball trajectory prediction. Optik 158:266–272. https://doi.org/10.1016/j.ijleo.2017.12.038

Zhao K, Du C, Tan G (2023) Enhancing basketball game outcome prediction through fused graph convolutional networks and random forest algorithm. Entropy (Basel). 25(5):765. https://doi.org/10.3390/e25050765

Development of LED-Based Pacing Running System

Muhammad Adam Hamizan Zainuddin, Muhammad Haniff Gusrial, Nur Aqilah Othman, Hamzah Ahmad, and Norhidayah Mohamad Yatim

Abstract This project aims to enhance running training methodologies by introducing a pacing system integrated with light-emitting diodes (LEDs). Traditionally, athletes manually determine their running pace, but the inherent risk of human errors limits the effectiveness of such approaches. Recognizing the critical role of a healthy pace in influencing an athlete's productivity, endurance, and overall performance, a more precise strategy is imperative. This paper presents the development of an LED-based running pace system, designed to control and monitor the desired pace effectively. The system utilizes MIT App Inventor and an Xtensa microprocessor, implemented on an ESP32 microcontroller development board, chosen for its convenience, user-friendliness, and cost-effectiveness. The accompanying application provides real-time feedback and adjustable features, enabling athletes to progressively enhance their running performance. The testing results affirm the practicality and suitability of the proposed method for real-world applications.

Keywords Running · Pacing system · LED

1 Introduction

Running is an activity that has become widely known not only in the sports arena but also has been incorporated into exercise routine. Running pace is one of the important elements that always become the priority for runners or athletes (professional runners) in maintaining the best running performance. In the professional sports industry, a healthy pace is important for an athlete as it has an impact on their

M. A. H. Zainuddin · M. H. Gusrial · N. A. Othman (✉) · H. Ahmad
Faculty of Electrical and Electronics Engineering Technology, Universiti Malaysia Pahang Al-Sultan Abdullah, Pekan, Pahang, Malaysia
e-mail: nuraqilah@umpsa.edu.my

N. M. Yatim
Faculty of Electronic and Computer Engineering, Universiti Teknikal Malaysia Melaka, Durian Tunggal, Melaka, Malaysia

Z. Mohamed et al. (eds.), *Proceedings of the 2nd International Conference on Innovation and Technology in Sports, ICITS 2023, 27–28 November, Malaysia*, Lecture Notes in Bioengineering, https://doi.org/10.1007/978-981-97-3741-3_27

productivity, endurance, and overall performance. Historically, the running pace is manually counted by the athletes using manual techniques to get their preferred running pace such as timing themselves or watching pacesetters. However, these approaches have their limitations due to the high chances for the human errors to take place, thus a more effective and precise strategy is required (Macomber Kline 2016).

The development of a user-friendly, cost-effective, and portable LED-based pace running system can help empower individuals to take control of their fitness journey, enhancing their overall health and well-being (Church 2020; LeVota 2017). The easy-to-use approaches will be implemented as it will be controlled using an application that is being developed using MIT App Inventor. This feature allows the users especially the athletes to control their desired running pace systematically.

This paper is structured as follows. Section 2 shows the testing process of the system, Sect. 3 defines the overall system's setup, Sect. 4 provides the result of the system, and Sect. 5 concludes the contributions of this paper.

2 System Testing

2.1 WS2815 ARGB LED Testing

The development of the system involved several components and materials. All components needed to function well. WS2815 ARGB LED is used as the main LED strip that produces the outcome for the system. Two different lengths of WS2815 ARGB LED are used for testing the system which involved 5 m and 10 m length strips which corresponded to 300 and 600 numbers of individual LED, respectively. The LED itself is powered by 5 V and 12 V batteries that connect directly to the LED strip (Fig. 1).

The testing was done by including the FastLED library for the coding into Arduino IDE software. The speed of the LED operates in different lengths and supply voltage

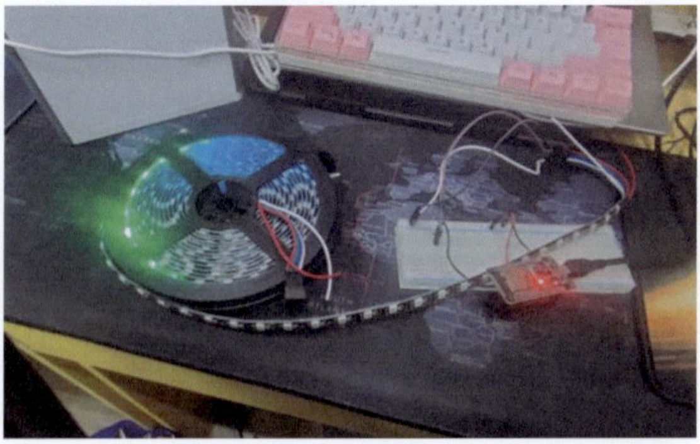

Fig. 1 Setup for WS2815 ARGB LED strip testing

Table 1 Testing using 5 m WS2815 ARGB LED strip

Supply	Test distance	5 m			10 m		
		Time taken (s)	Speed (m/s)	Average speed (m/s)	Time taken (s)	Speed (m/s)	Average speed (m/s)
5 V	Test 1	2.39	2.09	2.08			–
	Test 2	2.39	2.09		–	–	
	Test 3	2.43	2.06				
12 V	Test 1	2.71	1.85	1.92			–
	Test 2	2.85	1.75		–	–	
	Test 3	2.33	2.15				

Table 2 Testing using 10 m WS2815 ARGB LED strip

Supply	Test distance	5 m			10 m		
		Time taken (s)	Speed (m/s)	Average speed (m/s)	Time taken (s)	Speed (m/s)	Average speed (m/s)
5 V	Test 1	4.42	1.13	1.11	5.95	1.68	1.68
	Test 2	4.60	1.09		6.08	1.64	
	Test 3	4.55	1.10		5.85	1.71	
12 V	Test 1	4.45	1.12	1.09	5.65	1.77	1.77
	Test 2	4.49	1.11		5.56	1.80	
	Test 3	4.81	1.04		5.75	1.74	

is measured and calculated as in the Tables 1 and 2.

$$Speed, v = \frac{Distance, d}{TimeTaken, t} \tag{1}$$

Based on the above testing, it shows that the speed of the data transfer between the LEDs is decreasing, while the length of the LED increases. These findings suggest that the increased load on the system's data transfer capabilities slows down the communication process, impacting the speed of the LED effects or animations.

2.2 *Application Development*

A user-friendly application developer called MIT App Inventor is used in building the application. The development process begins by setting up the account and interface with its functionality using a coding-based function. After developing the Loading Screen, the same process is repeated with the development of the Home Page for the application. Several input buttons were included in the Home Page such as ON/OFF, STOP, SPEED, etc. Bluetooth functions were also included in the application to control the system remotely (Raj 2018; ESP32 Bluetooth Low Energy (BLE) on Arduino IDE 2019; ESP32 Bluetooth Classic with Arduino IDE 2019). Figure 2 and 3 show the Loading Screen and Home Page of the developed mobile app in the MIT App Inventor.

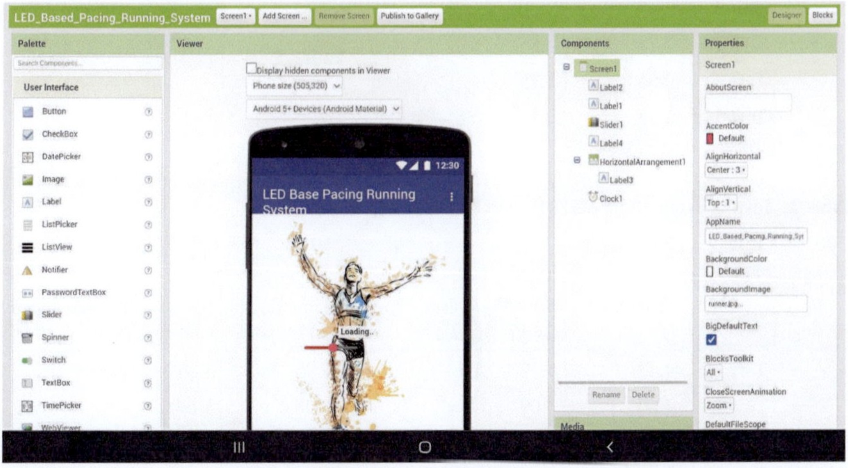

Fig. 2 Loading screen in MIT app inventor

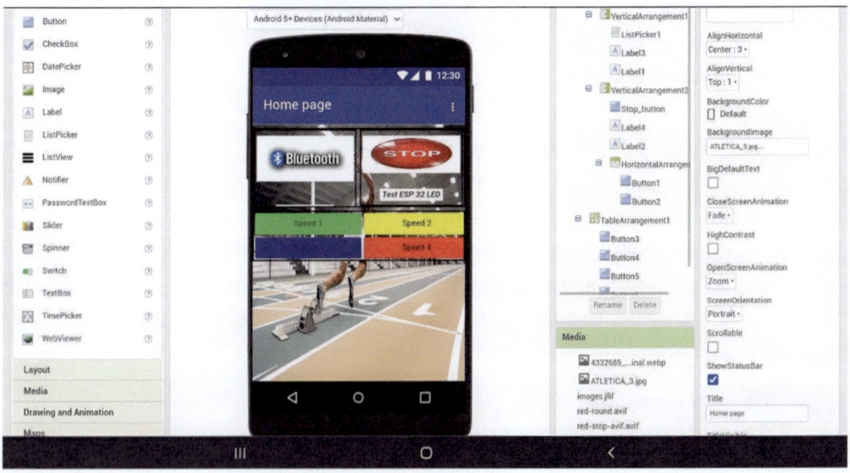

Fig. 3 Home page in MIT app inventor

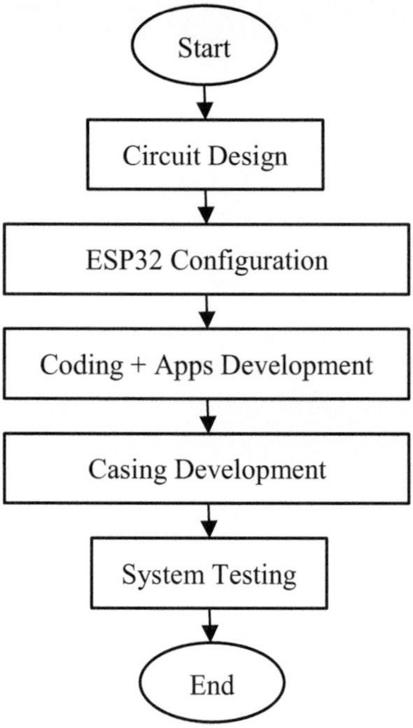

Fig. 4 Flowchart for overall system

3 LED-Based Pacing Running System

3.1 Flowchart

The development of the system begins by designing the circuits involved for the connections between the components. The configuration of the microcontroller which is the ESP32 development board continues to take place and the application with the coding for the system will be developed. After enabling the system to function well, the casing for the system will be developed and the overall system can be tested (Fig. 4).

3.2 Block Diagram

The overall system involved the use of an ESP32 development board as the main microcontroller, a WS2815 ARGB LED strip to show the output, and a mobile app which is the developed application as the main input platform for the system. The system begins by inserting input from the user and the microcontroller will process that input. The microcontroller ensures the input compatibility and its synchronization with the WS2815 ARGB LED strip. Necessary calculations, configurations, or

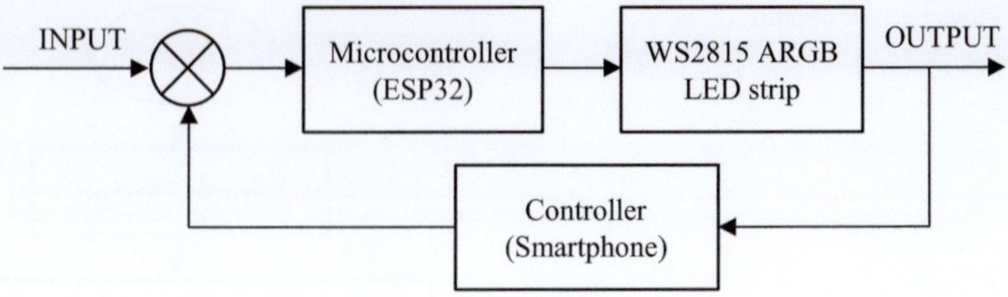

Fig. 5 Block diagram for overall system

transformations, the microcontroller prepares the data for integration with the LED strip. The processed input will generate output that comprises a series of instructions such as color codes, timing patterns, animation sequences, etc., to the desired effects. The smartphone (with the application) that acts as a controller for the system will provide a feedback mechanism and establish Bluetooth connections with the microcontroller. User will receive real-time updates regarding the speed of the LED strip as well as enable the system to be adjusted and controlled (Fig. 5).

3.3 Circuit Design

The overall schematic diagram of the circuit design is shown in Fig. 6. The ESP32 development board has been used for the prototype, together with a buck converter to step down 12 V voltage supply to 3.3 V for the input supply for the ESP32 development board. WS2815 LED strip is rated as a 12 V system.

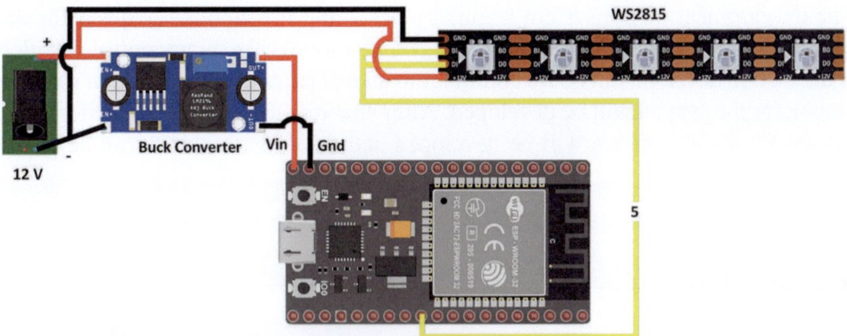

Fig. 6 Schematic diagram of the circuit

Table 3 List of materials and components for the system

Components/Materials
ESP32 development board
WS2815 ARGB LED strip
5V and 12V battery Buck converter
LED diffuser
Container (casing)
Smartphone (for application)

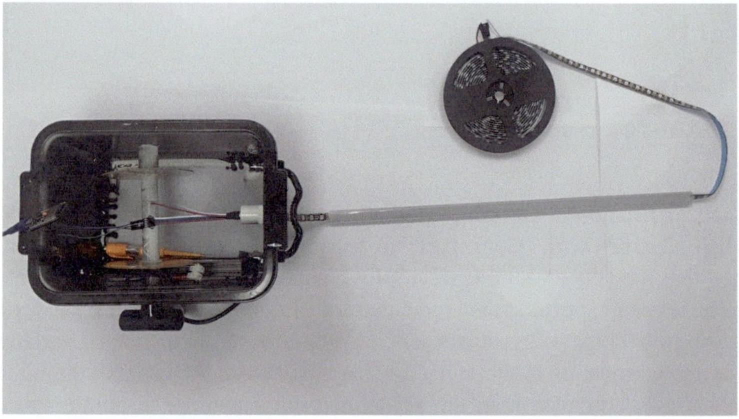

Fig. 7 Overall developed system

3.4 Developed System

The outer casing of the system is made using a black container to make the overall system look neat and tidy. Other related materials and components involved in this project are listed in Table 3 and the initial condition of the system is shown in Fig. 7.

4 Result and Discussion

4.1 Pace Running System

Based on the constructed circuit design, the connections between the circuit with the hardware are tested. At the beginning, the condition of the WS2815 ARGB LED strip is not showing any indications or lights as shown in Fig. 7. Before starting any controlling process, both the microprocessor and the developed application must be

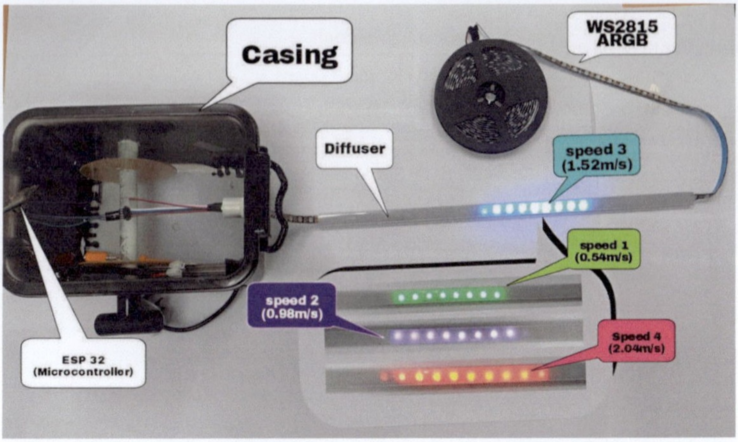

Fig. 8 Final condition of the system

connected via Bluetooth. The double light (blue and red) indicated that the microcontroller ESP32 is ON with "Connected" text displayed while one (1) light (red only) indicated that the ESP32 is OFF with "Not Connected" text displayed as shown in Fig. 9. The text will appear at the Home Page of the application. Using the developed application in MIT App Inventor, users have to choose the displayed variations of speed on the application's interface. By selecting the desired speed, the LED strip illuminates in accordance with the predefined color and speed parameters established during the development phase. Specifically, green, purple, blue, and red lights correspond to speeds of 0.54 m/s, 0.98 m/s, 1.52 m/s, and 2.04 m/s, respectively, as illustrated in Fig. 8. These speeds are adjustable by modifying the time delay function in the code. Currently, the LED lights' speed is set to a value midway between the global average mile pace for men (approximately 2.67 m/s) and the global average mile pace for women (approximately 1.48 m/s) (Alice Barraclough 2023).

4.2 *Application Interface*

When launching the application, the Loading Screen will appear. The Home Page will eventually appear when the loading process is completed. All of the buttons shown on the Home Page can be used to control the overall system. The choosing speed from the Home Page will result in different colors and the speed of the LED movements (Fig. 10).

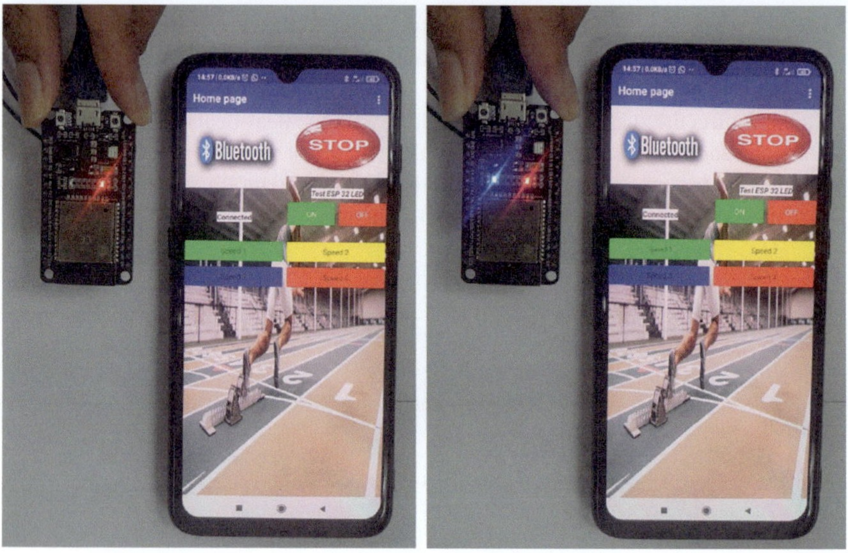

Fig. 9 ESP32 OFF (left) and ON (right) states

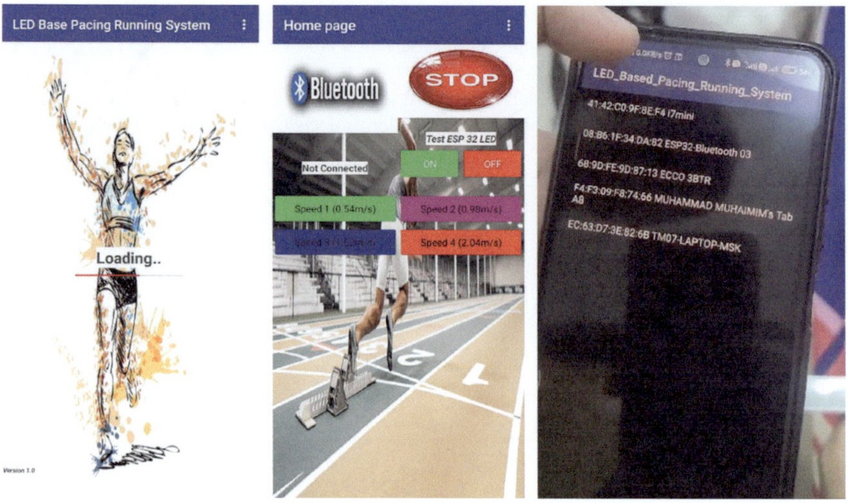

Fig. 10 Loading screen (left), Home page (middle), and Bluetooth connection (right)

5 Conclusion

In conclusion, the development of this system is able to help athletes to ensure a healthy running routine which involves an accurate running pace system by integrating the use of technologies which is smartphones. Integration of LED systems makes the whole development more systematic due to the exact timing of LED ignition to mimic the real running pace. The controllable system makes it a user-friendly as well as an easy-to-use system.

In the future, the system itself can be added with the heartbeats calculator or pulse calculator function that can count the athlete's pulse automatically when running. This will ensure a more healthy routine as the athletes can monitor their health during the training period without 100% relying on the medical check-up.

References

Alice Barraclough (2023) What is the average running speed? Runner's World. https://www.runnersworld.com/uk/training/beginners/a44826741/average-running-speed/

Church B (2020) Flashing lights and smashing world records. Is this the future of athletics? CNN. https://edition.cnn.com/2020/10/09/sport/nike-spikes-athletics-world-record-valencia-spt-intl/index.html

ESP32 Bluetooth Classic with Arduino IDE - Getting Started | Random Nerd Tutorials (2019). https://randomnerdtutorials.com/esp32-bluetooth-classic-arduino-ide/

ESP32 Bluetooth Low Energy (BLE) on Arduino IDE | Random Nerd Tutorials (2019). https://randomnerdtutorials.com/esp32-bluetooth-low-energy-ble-arduino-ide/

LeVota M (2017) LED Rabbit tech enhances training for track athletes. Startland News. https://www.startlandnews.com/2017/05/led-rabbit-tech-enhances-training-for-track-athletes/

Macomber Kline A (2016) Patent Application Publication No.: US 2016 / 0243255 A1, vol 1, no 61

Raj A (2018) ESP32 bluetooth tutorial: how to use serial bluetooth and toggle an LED. circuitdigest.com. https://circuitdigest.com/microcontroller-projects/using-classic-bluetooth-in-esp32-and-toogle-an-led

Innovative 3D-Printed Shoe Insole Design for Youth with Flat Feet

Mohd Firdaus Mahamad Ali, Nurul Izzati Ahmad, and Nadzifah Yaakub

Abstract Flat feet refer to the situation where the curved section of the feet appears flat or mostly flattened to the ground at full. This happened to a large number of youngsters and is a kind of common anatomy irregularity among them. This research aims to identify the issues encountered by these flat-footed youngsters while wearing shoes, identify the related anthropometric data parameters, and develop a design of shoe soles for these youngsters. Thirty numbers of subject persons were studied during this research, which was conducted at Universiti Sultan Zainal Abidin (UniSZA) and Universiti Malaysia Kelantan (UMK). The research was conducted using a questionnaire and analyzed using Excel and previous research data. Few designs and development ideas for the design of the shoe sole were developed in this research. Based on this research, it was found that youngsters with flat feet mostly felt pain and experienced blisters on their feet while wearing the shoes and pain on the heel while walking with the shoes. The new shoe sole design was developed with a 3D printer and took into consideration the most suitable measurement for the height of the ankle and the width of both feet, dedicated for these flat-feet youngsters who are mostly aged 18–24 years old. This research has the potential to benefit the target group and is in line with achieving the Sustainable Development Goal (SDG 3) to ensure a healthy and good lifestyle.

Keywords Flat feet · 3D printed · Insole · Anthropometry

M. F. M. Ali (✉) · N. I. Ahmad
Faculty of Creative Technology and Heritage, Universiti Malaysia Kelantan, Kampus Bachok, 16300 Bachok, Kelantan, Malaysia
e-mail: mfirdaus.ma@umk.edu.my

N. Yaakub
Faculty of Bioresources and Food Industry, Universiti Sultan Zainal Abidin, Kampus Besut, 22200 Besut, Terengganu, Malaysia

Z. Mohamed et al. (eds.), *Proceedings of the 2nd International Conference on Innovation and Technology in Sports, ICITS 2023, 27–28 November, Malaysia*, Lecture Notes in Bioengineering, https://doi.org/10.1007/978-981-97-3741-3_28

1 Introduction

Flat feet, characterized by a lack of natural foot indentation which typically develops by the age of 7–10 (Uden et al. 2017), can result in uneven weight distribution. Flexible flat foot (also known as pes planus or planovalgus) in children, characterized by a lowered medial longitudinal arch, with or without rearfoot eversion (Evans, 2011) is a common reason for seeking orthopedic consultation (Krul et al 2009). However, normative data indicates that 'flat' feet are normal in typically developing children up to eight (Stolzman et al 2015). This condition can arise from genetic factors or internal issues like abnormal muscle tension (hypertonia and hypotonia), weakened foot muscles and ligaments, or obesity (Morrison 2018). External factors like unsupportive footwear or injuries can further exacerbate the problem (Hashim 2019). Flat feet, or pes planus, is a condition where the arches on the inside of the feet are flattened, causing the entire sole of the foot to make contact with the ground. It's relatively common, especially in infants and toddlers, as their arches haven't fully developed. However, in some cases, the condition persists into adulthood. Flat feet, affecting about 20–30% of children and 3–5% of adults, can lead to pain, fatigue, and overpronation, often exacerbated by the lack of supportive footwear available, especially for youth.

Individuals with flat feet often experience symptoms like heel pain, swelling, and an outward-turning foot (overpronation) when walking or standing, leading to shoe wear and potential injuries. Some also display an inward bend of the foot, known as a knock knee (Uden et al. 2017). Leg and foot problems in childhood often worry parents. Rotational issues include in-toeing and out-toeing. Intoeing, most common in infants and young children, is caused by metatarsus adductus, internal tibial torsion, and femoral anteversion. Out-toeing is less common and typically occurs in older children. Out-toeing is caused by external tibial torsion and femoral retroversion. Angular issues include genu varum (bowleg) and genu valgum (knock knee). In cases of pes planus (flatfoot), the foot arch is typically flexible rather than rigid (Caitlyn et al., 2017). Treatments for severe cases include surgery to realign bones, while physiotherapy exercises help alleviate discomfort and improve function. Additionally, wearing arch-supporting footwear is recommended (Halimah 2019). Flat feet can also impact posture, body stability, and pain during physical activities. Pain originating from flat feet can radiate to the ankles, calves, thighs, knees, hips, and even the lower back. This pain is attributed to the increased effort by knee muscles to maintain body stability. Furthermore, flat feet can also be associated with spinal issues. This is attributed to the increased foot load, which may cause individuals, especially adolescents, to lean forward slightly when walking compared to those with a natural foot arch. Additionally, due to the altered foot positioning, known as overpronation, there is premature wear on the soles of shoes (Lee 2016). The prevailing issue highlighted in the study is that current market footwear doesn't cater effectively to those with flat feet. Most shoes are designed with a flat footprint, making them unsuitable for supporting the unique structure of flat feet and potentially leading to injuries. Halimah (2019) emphasized the importance of footwear with appropriate arch support as a viable treatment for flat-feet issues.

Choosing the right footwear is crucial for individuals with flat feet, especially in their formative years. However, many common shoe designs lack adequate support for flat arches, which can exacerbate the condition. Shoes with proper arch support, stability, and cushioning can significantly alleviate discomfort and prevent potential problems associated with flat feet. For youth with flat feet, supportive footwear can make a significant difference in managing symptoms and preventing further issues. Unfortunately, many shoes marketed for children may prioritize fashion over functionality, lacking the necessary support for developing arches.

This research benefits the community by suggesting suitable insoles for adolescents with flat feet, thereby reducing the risk of injuries. It also promotes awareness of ergonomic shoe insole designs for this demographic. Furthermore, this study serves as a valuable resource for future scholars and reviewers, enhancing their grasp of flat feet, ergonomics, and anthropometry. Additionally, it inspires designers to incorporate anthropometric data into their product designs. This research specifically contributes to the development of insoles designed for Malaysian teenagers with flat feet, utilizing locally relevant anthropometric data as it is conducted in Malaysia. These insoles are designed to fit standard shoes, aligning with the common footwear choice among university students. The emphasis is on a contemporary design that incorporates anthropometric measurements, ensuring both comfort and aesthetic appeal for individuals with flat feet. Encouraging youth with flat feet to wear supportive footwear can alleviate discomfort and potentially prevent issues that might arise from misalignment or inadequate support as they grow and remain active. The main objectives of this study are to understanding the footwear challenges faced by teenagers, gathering anthropometric data for sizing, and designing 3D-printed insoles tailored to the needs of those with flat feet.

2 Methodology

2.1 Data Collection

This research was carried out at Universiti Sultan Zainal Abidin (UniSZA) and Universiti Malaysia Kelantan (UMK). This study encompassed 30 respondents, all of whom were between the ages of 18 and 24 and experienced with flat-feet issues.

2.2 Data Analysis

This study used the quantitative and qualitative methods, whereby questionnaires were applied to obtain information from respondents and findings were analyzed using Microsoft Excel. The questionnaire for this study was distributed via Google Forms. The survey consisted of two sections, each of which provided demographic data, such as age, sex, education level, and university type. The second section contains information on the feet types and problems suffered when wearing shoes. The mean, standard deviation, and maximum and minimum values were calculated using descriptive statistics. A 3D modeling tool was created using the Rhinoceros and Solidword applications.

2.3 Design Process

The product design journey commences with conceptual sketching, progresses to evolving these concepts into detailed sketches, constructing prototypes, transitioning into 3D renderings, converting digital files into tangible objects, and ultimately, the refinement phase, resulting in a market-ready product.

3 Results and Discussion

3.1 Problems in Foot Among Respondents

Table 1 presents the survey results regarding foot problems among the respondents. A significant 53.33% of respondents resorted to using additional shoe inserts to provide support for their flat feet, effectively reducing discomfort while wearing shoes. In comparison, 66.66%, or 20 participants, encountered issues such as blisters on their feet, which was a reduction from the previous 90%. Furthermore, 93.33% of respondents reported experiencing foot pain when standing or walking for extended periods. Additionally, 96.66% of participants grappled with heel pain, while the remaining 3.34% did not. The abnormal alignment of the foot bones, typical in individuals with flat feet, often resulted in ankle pain when walking extensively, a concern cited by 93.33% of respondents. Interestingly, the study also revealed that the majority, specifically 86.66% out of 30%, or 26 respondents, did not experience knuckle swelling when wearing shoes for extended durations.

3.2 Analysis of Footprint Size Assessment Findings

Table 2 shows the analysis of footprint size assessment findings among the respondents regarding foot types, it was found that 93.33% (or 28 participants) had a normal foot type, while 6.66% (or 2 respondents) exhibited flat feet. None of the respondents had high arch feet. Regarding the assessment of arch length, 93.33% had an arch

Table 1 Foot problems among respondents

No	Questions	f	%	f	%
1	Do you wear additional shoe linings for foot support?	16	53	14	47
2	Do you experience swelling of the knuckle when wearing the shoes for a long period of time?	4	13	26	87
3	Do you experience pain such as blisters on the sides of the feet when wearing shoes?	27	90	3	10
4	Do you experience foot pain	28	93	2	7
5	Do you experience pain in the heel of the foot?	29	97	3	3
6	Do you experience ankle pain when walking a lot?	28	93	2	7

length of 15 mm, while 6.66% (or 2 participants) with flat feet had an arch length of 20 mm. No respondents with other foot types had an arch length of 20 mm. This study also analyzed the center footprint size among the respondents, with 86.66% (or 26 participants) having a footprint size of 72.2 mm. Additionally, 6.66% had a size of 70.6 mm, and another 6.66% had a footprint center size of 75.5 mm. Regarding the distance between the heels (from left to right), 6.66% had a distance of 75 mm, 83.33% had a distance of 66.5 mm, and 10% had a distance of 55.5 mm.

Table 2 Footprint size assessment

		No. of respondent	Percentage
Foot type	High arch foot	0	0
	Normal foot	28	93.33
	Flat foot	2	6.66
Height foot arch	0 mm	2	6.66
	15 mm	28	93.33
	20 mm	0	0
Center of footprint	72.2 mm	26	86.66
	70.6 mm	2	6.66
	75.5 mm	2	6.66
Length of the heel	75 mm	2	6.66
	66.5 mm	25	83.33
	55.5 mm	3	10
Length of the mid-foot gate size	105 mm	24	80
	100 mm	1	3.33
	99.5 mm	5	16.66

Table 3 Anthropometric data with percentile

Measure	Mean	Standard deviation	Percentile		
			5th	50th	95th
Ankle height	8.57	0.96	6.99	8.57	10.16
Foot breadth R	8.88	0.61	7.87	8.88	9.89
Foot breadth L	8.82	0.61	7.82	8.82	9.81
Foot length R	23.10	1.17	21.17	23.10	25.03
Foot length L	23.11	1.13	21.25	23.11	24.98

3.3 Analysis of Anthropometric Data Parameter

Table 3 shows the anthropometric data parameter with the percentile level. The 95th percentile data for all measurements for teens is utilized on insole because it assures that users with the largest ankle size can easily and comfortably place or insert their feet.

3.4 Analysis of Design Assessment Chart

Table 4 presents the final selection among the three insole designs based on feedback from the 30 respondents. The assessment chart played a crucial role in making the appropriate design decision for adolescents with flat-feet issues. These three design concepts were carefully evaluated and rated by the 30 participants. The design with the highest rating was chosen for shoe production.

The results indicate that Design 2 received the highest overall rating among the three options. This preference can be attributed to Design 2's unique features, including the presence of two additional artificial curves along the edges, distinguishing it from the other designs. Furthermore, Design 2 incorporates the footprint's position into its design, enhancing comfort by featuring artificial arches at the front

Table 4 Design assessment analysis

	Aesthetics	Uses	Technical	Total
	Conformity New shape usability	Youth uses Treating foot problems	Used in shoes artificial arch	
Design 1	150/200 70%	150/200 70%	140/200 75%	440/600 73.3%
Design 2	189/200 94.5%	150/200 75%	172/200 86%	511/200 85.2%
Design 3	170/200 85%	128/200 64%	136/200 68%	434/200 72.3%

and a heel arch. Additionally, Design 2 excels because of its supplementary arches at both the front and back, facilitating the natural arch formation at the center with the support of the artificial arches on the front and back sections of the insoles.

The first design garnered the second highest preference from the respondents. This choice is primarily due to the design's substantial artificial support structure, surpassing the other designs in this aspect. The first design also includes heel support to alleviate pressure on the heel area. It features support structures both at the front and back, contributing to the natural arch formation with the support provided. Conversely, the third design received the lowest preference among respondents. This is because Design 3 only incorporates arch support at the front and back, with the intention of facilitating a natural arch at the center through the arches integrated into the shoe's lining. However, this design lacks central arch support, which affects its overall appeal to respondents.

3.5 *Ideation Sketches*

Based on the analysis of the study that has been obtained from the study findings, here are some ideas put forward by the reviewer.

3.5.1 Mock-Up

Mock-ups were built before models or prototypes were built. Mock-ups that are built do not use real materials. This mock-up emphasizes the size of the height of the ankle, the size of the width of the right and left feet, and the size of the length of the right and left feet so that the additional shoes that are produced are suitable for flat-feet teenagers. Figure 1 shows the mock-up insoles for this study.

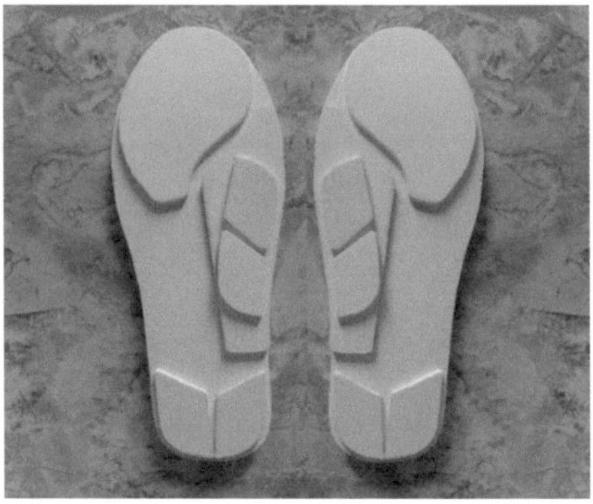

Fig. 1 Mock-up insoles

3.5.2 3D Sketch and Printing

This 3D sketching is done with the Rhinoceros and Solidword applications where this software can be used for their 2D and 3D shapes. This application also uses the resulting 3D digital file to print a print-ready model using a 3D printer. Figure 2 shows the 3D sketch of insoles.

3D printing or additive manufacturing is the process of creating physical three-dimensional objects where materials are applied in layers using additive manufacturing technology that can be controlled using a computer to form objects. Figure 3 shows the final insoles of 3D printing using nylon or polyamide are selected to print shoe liners using 3D printing. This is because nylon has flexible characteristics as well as being strong and durable.

Fig. 2 Insoles 3D sketch

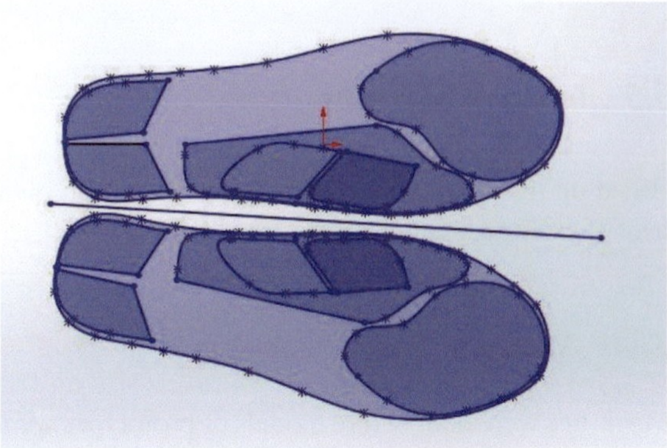

Fig. 3 Insoles 3D printing

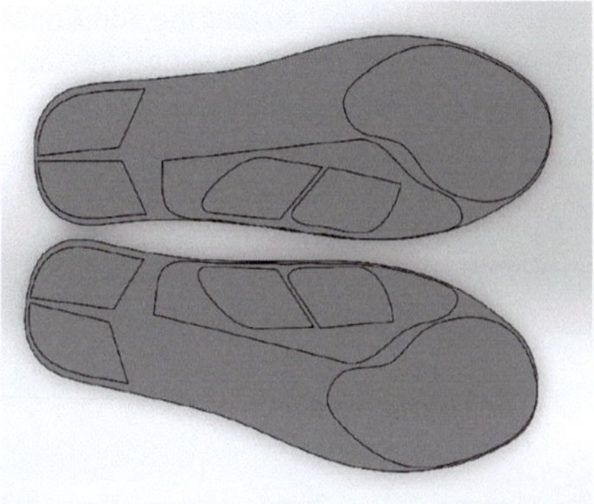

3.5.3 Finishing

Prototypes that are ready to be built need to have appropriate packaging. This packaging work depends on the materials used in the preparation of the prototype project. While doing the packaging, it is necessary to look over some existing products in order to produce component parts according to the working specifications. The purpose of packaging is to add aesthetic, commercial, and durability value to the prototype.

3.6 Differences in Shoe Linings Available in the Market and Those Designed

Figure 4 illustrates the distinctions between the insole shoes found in the market and the designed in this study. Table 5 shows the insole comparison between insole at the market and this study.

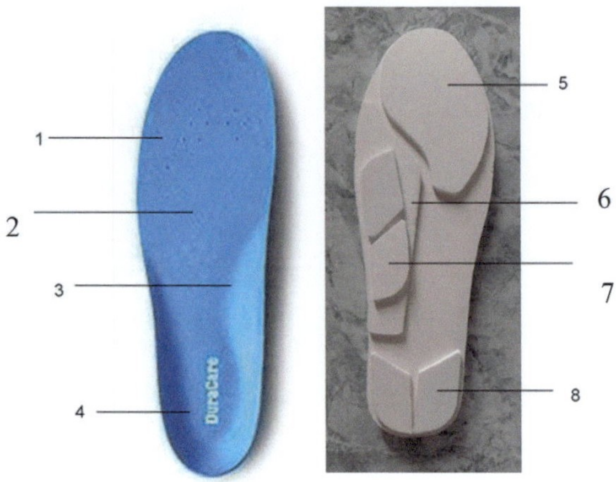

Fig. 4 Shoe linings comparison

Table 5 Shoe linings comparison

	Ready-made insoles in the market		Designed insoles
Details instructions		Details instructions	
1	Air holes	5	Front section support
2	Leg shape curve	6	Leg gate support
3	Edge gate curve	7	Additional gate support
4	Heel curve	8	Heel support

3.7 Observations on Respondents When Using the Mock-Ups

Figure 5 depicts the condition of the respondents' flat feet prior to utilizing the insoles, while Fig. 6 illustrates their condition after using the insoles. This change occurs because elevating the front and back of the shoes creates an arch at the foot's edge, which is naturally formed and supported by the two gate supports at the edge, helping distribute the foot's load.

Figure 7 shows the size of the insoles, where the size of the overall thickness of the insoles is 10 mm and the thickness of the support at the center is 15 mm. The overall length of the case liner is 231.1 mm. The width of the front part is 88.2 for the width of the back also is 66.5, where these data are taken from the results of a survey conducted on teenage money aged 18–24 years from Universiti Malaysia Kelantan and Universiti Sultan Zainal Abidin. These data are also taken from the Institut Keselamatan dan Kesihatan Pekerjaan Negara (NIOSH), where NIOSH has conducted anthropometric reviews across the country on 3,997 subjects.

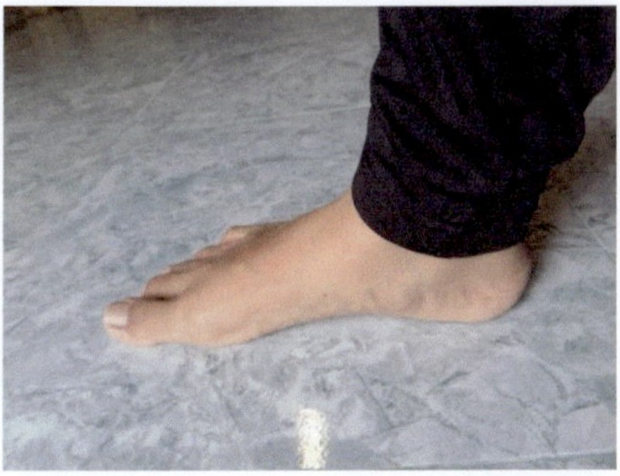

Fig. 5 Flat feet

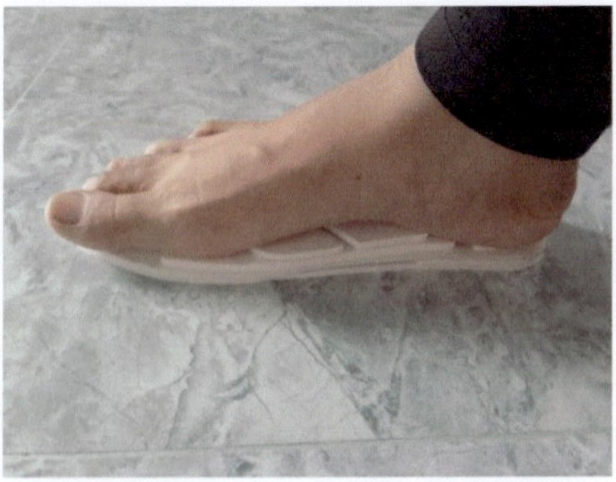

Fig. 6 Flat feet on insoles

Fig. 7 The insole size

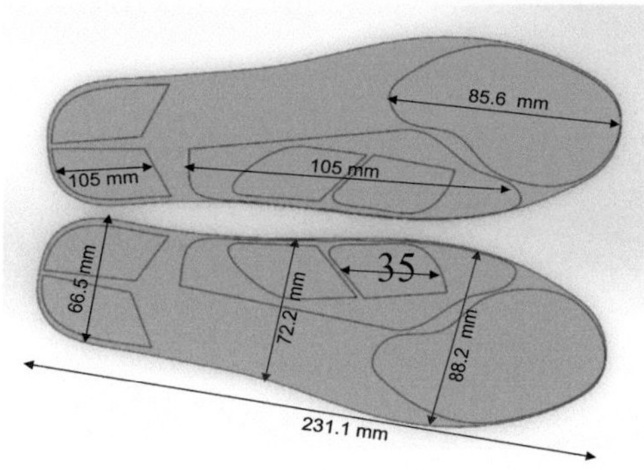

4 Conclusion

The outcomes of this research are anticipated to provide valuable assistance to teenagers dealing with flat feet through the development of 3D insoles. It is envisioned that the production of these insoles can be further extended to cater to specific sizes or genders, thereby enhancing support for individuals with flat feet.

References

Evans AM (2011) The paediatric flat foot and general anthropometry in 140 Australian school children aged 7–10 years. J Foot Ankle Res 4:1–8

Halimah H (2019) Perawatan fisioterapi untuk flat foot pes planus. 28 November 2023. My health. http://www.myhealth.gov.my/perawatan-fisioterapi-untuk-flat-foot-pes-planus/

Krul M, van der Wouden JC, Schellevis FG, van Suijlekom-Smit LWA, Koes BW (2009) Foot problems in children presented to the family physician: a comparison between 1987 and 2001. Fam Pract 26:174–179

Lee D (2016) Risiko tapak kaki jenis rata. 29 December. My metro. https://api.hmetro.com.my/node/164419

Morrison T, Jones S, Causby RS, Thoirs K (2018) Can ultrasound measures of intrinsic foot muscles and plantar soft tissues predict future diabetes-related foot disease? A systematic review. Plos One 13(6):e0199055. https://doi.org/10.1371/journal.pone.0199055

Rerucha CM, Dickison C, Baird DC (2017) Lower extremity abnormalities in children. Am Fam Physician 96(4):226–233

Stolzman S, Irby MB, Callahan AB, Skelton JA (2015) Pes planus and paediatric obesity: a systematic review of the literature. Clin Obes 5(2):52–59. https://doi.org/10.1111/cob.12091

Uden H, Scharfbillig R, Causby R (2017) The typically developing paediatric foot: how flat should it be? A systematic review. J Foot Ankle Res 10(37):1–17

Taekwondo Combat Sports Wireless Scoring System (WSS)

Eliza Law Ong Hie, Khoo Si Xin, Fatin Aliah Phang Abdullah, Izwyn Zulkapri, Nurul Hidayat, Adam Wong Yoong Khang, Kian Sek Tee, and Jaysuman Pusppanathan

Abstract The taekwondo scoring system is a crucial component in accessing and updating the scores accurately during the sports tournament. This paper describes the development of a simple yet portable wireless scoring system (WSS) for taekwondo combative competitions used by juries. WSS uses ESP-WROOM-32 devices and via wireless ESP-NOW protocol. The system consisted of one master device and three slave devices which allowed for score recording, and real-time updates based on input from the juries. The ESP-NOW protocol demonstrated its capability to facilitate efficient data exchange between the master and slave devices. This is to ensure that there are no significant delays in transmitting the data and without any packet loss. The button press detection mechanism on the slave devices was responsive and accurate, and the master device accurately incremented the respective scores according to the transmitted information. The system incorporated an output interface, such as an OLED screen, to display the scores in real time, allowing users

E. L. O. Hie · K. S. Xin · I. Zulkapri · J. Pusppanathan (✉)
Department of Biomedical Engineering, Faculty of Electrical Engineering, Universiti Teknologi Malaysia, 81310 Skudai Johor, Malaysia
e-mail: jaysuman@utm.my

F. A. P. Abdullah
School of Education, Faculty of Social Sciences and Humanity, Universiti Teknologi Malaysia, 81310 Skudai Johor, Malaysia

N. Hidayat
Department of Physics, Faculty of Mathematics and Natural Sciences, Universitas Negeri Malang, Malang, Indonesia

A. W. Y. Khang
Faculty of Electrical Engineering (FKE), Universiti Teknikal Malaysia Melaka (UTeM), Hang Tuah Jaya, 76100 Durian Tunggal, Melaka, Malaysia

K. S. Tee
Department of Mechatronics and Robotics Engineering, Faculty of Electrical Electronic, Universiti Tun Hussein Onn Malaysia, Persiaran Tun Dr. Ismail, 86400 Parit Raja, Malaysia

I. Zulkapri · J. Pusppanathan
Sports Innovation & Technology Centre, Institute of Human Centered Engineering, Universiti Teknologi Malaysia, 81310 Skudai Johor, Malaysia

Z. Mohamed et al. (eds.), *Proceedings of the 2nd International Conference on Innovation and Technology in Sports, ICITS 2023, 27–28 November, Malaysia*, Lecture Notes in Bioengineering, https://doi.org/10.1007/978-981-97-3741-3_29

or competition organizers to conveniently track and monitor the scores throughout the matches. Furthermore, the product is designed to be a cost-effective and user-friendly device, especially catering for small organizations or taekwondo club use. The finishing look is designed to be portable and handheld concept by the juries and coach. Moreover, this paper demonstrates a user acceptance test (UAT) at a real environment testing (taekwondo sparring session) of the WSS to obtain feedbacks by experienced coaches and juries.

Keywords Wireless scoring system · Taekwondo scoring system · ESP-WROOM-32 devices · Combat sports · Sport technology

1 Introduction

With the continuous advancements in technology and the vast knowledge gained in the field of science, numerous devices have been invented to enhance the well-being of human beings. This rapid progress has had a profound impact on various aspects of human activities (Sun 2022). As technology is implemented in many devices, it brings about significant effects in the real world. This parallels the concept of connecting physics principles to real-life applications (Zakaria et al. 2019). Just as physics principles are used to understand and explain the natural world, the implementation of technology in devices allows us to apply scientific knowledge to practical situations, improving our daily lives and shaping the world around us. For example, it has revolutionized the monitoring of environmental quality, the concept of smart homes, and the advancements in medical and healthcare practices. Furthermore, the field of sports science has also experienced significant advancements through the integration of digitalization and intelligent systems to assess athletic performance and training. In the field of organizing international and domestic sports competitions, as well as providing crucial support for athlete training, the use of digital technology and software applications has become indispensable (Can et al. 2011). Additionally, digital technology finds application in real-time monitoring of athletes' vital signs such as blood pressure, heart rate, and body composition testing. It has also been utilized in the scoring systems of various sports including archery, gymnastics, taekwondo, e-sports, etc. The adoption of technology in sports sometimes has been fraught with dispute. This is similar to what has been observed in the past when new technologies are introduced. It can be a challenging process that presents various obstacles. A new scoring system based on information technology (IT) is an innovative way to judge matches fairly. However, there are reliability and expertise problems due to sophisticated differences in human–machine judging criteria (Cho et al. 2021).

As we all acknowledge, taekwondo is one of the modern and famous combat sports from the twentieth century which was introduced to Korea (Cho et al. 2021). The name of taekwondo is derived from the Korean words. The word "Tae" stands for the foot in Korean, "kwon" brings the meaning of fist, and the last word "do" means the art of. Overall, the word "Taekwondo" means the art of foot and fist in Korean

(Combat Sports Medicine 2009). In the context of taekwondo competitions, there are different rule sets and governing bodies that oversee the sport. The two main organizations, the International Taekwon-Do Federation (ITF) and the World Taekwondo Federation (WTF), have distinct competition rules that impact the gameplay and scoring criteria. ITF was established in 1966 with the introduction of rules that prioritized no contact and contact and required athletes to wear protective gear. Matches consisted of two rounds of 2 min each, with a 1-min rest period during eliminations, and three rounds of 2 min each during finals. However, following the disbandment of the ITF by the Korean government, the World Taekwondo (WT), previously known as the WTF, was established in 1973 with the aim of unifying different taekwondo styles. Consequently, the standards set by WT gained widespread acceptance and are now recognized as the official Olympic standards.

There have been various systems employed to facilitate wireless scoring systems in the sports field including infrared systems, Bluetooth, and Wi-Fi. Wireless communication refers to the data transfer between two or more devices without the use of physical cable. Wireless communication systems are relevant to the development of the automatic control of productivity of magnetic separators (Alsayaydeh et al. 2020). By incorporating wireless communication capabilities into magnetic separators, real-time data on magnetic products can be transmitted wirelessly. Therefore, it can potentially enhance the efficiency and effectiveness of the overall system. According to Srivatsa Raghunath, infrared light-based communication has the potential of communication distance between the remote and therefore the display is a smaller amount than 20 feet and alignment access for a beam of IR rays (Mailapalli et al. 2015). IR system has limitation as it is relied on a direct line of sight between the transmitter and receiver and if any obstructions or physical barriers can disrupt the communication. Moreover, Bluetooth is one of the popular short-range wireless communications. However, it tends to consume more power, which makes it less suitable for devices that operate on low power or rely on battery power. Traditional Wi-Fi modules have also been employed in wireless scoring systems but due to their high power consumption and complexity, it is not well-suited for use in taekwondo wireless scoring systems.

Beside the introduction of these wireless communication systems, there are some existing products for the wireless scoring system used in taekwondo. One of the most popular products is developed by Daedo which is a company dedicated to martial arts. In normal, Daedo wireless scoring system consists of two main components which are the electronic sensor and the scoring system software. The electronic sensor is embedded in the protective gear such as the headgear and chest guard. By detecting the kicks and impacts using the sensor, the scoring system software can process the data that is received from the sensor and calculate the scores for each match. Nevertheless, like any technology, the Daedo wireless taekwondo scoring system also has limitations. One of the significant limitations is the possibility of technical malfunction which can occasionally lead to the wrong scoring result. Although some efforts are made to minimize the impacts, some technical issues still can arise and affect the accuracy of the scoring system. Furthermore, it also has the limitation for the dependency on the sensor battery power. If the batteries run out of power during the competition, it can also disrupt the scoring result.

2 Background Problem

Based on the previous research on this topic, the existing product is designed for professional competitions, especially in the Olympics, thus making it financially impractical for smaller organizations. Additionally, the complexity of the scoring system software required the individuals with specific training, knowledge, and a high personal skill to conduct the taekwondo match. However, the traditional manual scoring system applied in the match requires a lot of manpower and has some limitations in capturing and recording scores. The manual scoring system will often lead to interruption and delays in the competition's flow. This is caused by the judges needing plenty of time to record the scores using a paper-based system. Several commercial-grade scoring systems have been developed to automize the judging process during tournaments. However, these systems are too costly for small organizations, clubs, and academies to purchase.

In the scoring system for the current taekwondo competitions, there are at least three juries required to agree on the judgment of the scores according to the subjective criteria. The judgment is based on the perception of juries about the tactics and techniques that were performed by the taekwondo athletes. Therefore, the scoring system of taekwondo depends on the judges' decisions (Zakaria et al. 2019). However, to ensure and increase the precision and the accuracy of the competition, a wireless scoring system is designed to encounter the problem. Although there is a cable scoring system developed nowadays, it is less portable and limits the movement of judges during competitions.

Secondly, some of the scoring systems with modern technology are using touch screens. However, the touch screen is very touch-sensitive, unintuitive, and easy to mistouch, and it is not suitable for combative competitions which require juries to concentrate on the players instead of looking at the scoring devices. The cost for a touchscreen-based system is much higher compared to others.

This project aims to address the challenges associated with the taekwondo wireless scoring system by developing a master and slave system using ESP-WROOM-32 via ESP-NOW protocol. One of the significant objectives is to provide a more affordable and user-friendly wireless scoring system tailored for campus and small organizations. In addition, the wireless scoring system is designed to be portable and handheld to ensure convenience for the juries. Simplifying the operations would also eliminate the need for highly skilled personnel and enable juries to handle the system independently.

3 Methodology

The methodology of this project is aimed to design a wireless portable scoring system for the use in taekwondo competitions by using the ESP-WROOM-32 which is the SoC (System on Chip) microcontroller that has the integrated Wi-Fi dual Bluetooth module as the main component (Alsayaydeh et al. 2020). The judge scoring system implemented in this project utilizes four ESP32 devices, with one device serving as the master and the other three devices functioning as slaves. Based on Fig. 1 as shown, three of the slave devices will be handheld and controlled by the juries at the corner of the sparring area while the master device will be controlled by the center jury.

The Arduino IDE is required to write the whole coding for this system and upload to the ESP-WROOM-32 boards accordingly. The wireless communication between the master and the slave devices is established via the ESP-NOW protocol. The master device is equipped with two OLED screens to display the scores for the blue and red teams, respectively. Each of the slave devices is designed with ten buttons of which five buttons are for the blue team and other five buttons for the red teams. Moreover, each button is assigned for different score values from 1 to 5 marks. The

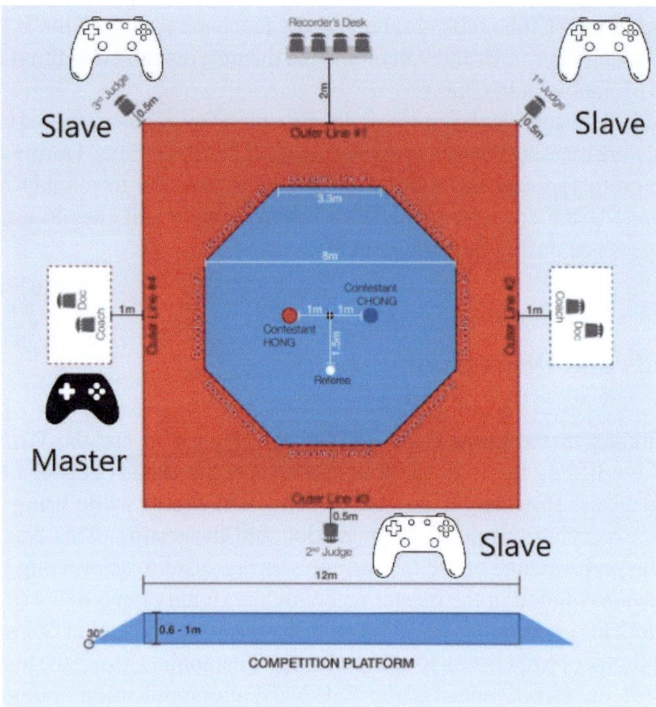

Fig. 1 Overview of the wireless scoring system and competition ring setup

flowchart for the master device in this system can be reviewed based on Fig. 2, while the master–slave device is based on Fig. 3 as shown below.

The developed hardware which consists of one master and three slave devices with embedded ESP32 boards was successfully developed. The prototype for the master device is shown in Fig. 4, while the prototype of the slave device is shown in Fig. 5. The inner view, which is the designed PCB board and embedded electronic components of the master and slave devices, can be viewed through Fig. 6.

As shown in Figs. 4, 5, and 6, the system consists of multiple main components, including the ESP32 boards, push buttons, OLED screens, and LEDs. The following Table 1 shows the main electronic components used with functions, respectively.

4 Test Run for WSS

In order to evaluate the performance and to identify the potential challenges, a comprehensive User Acceptance Test (UAT) is conducted. The main objective of UAT is to evaluate and to gather and obtain feedback from users. To perform UAT, juries, taekwondo players, and coaches are specially invited to carry out the test run and UAT. This is done with the collaboration of Universiti Teknologi Malaysia (UTM) Taekwondo Club. Table 2 tabulates the feedback results of the UAT.

The following Figs. 7, 8, and 9 demonstrate the progress and condition during the UAT in UTM Taekwondo Club.

As a result of the user acceptance test, the functionality, robustness, and usability of the devices were meeting the users' requirements and expectations. During combative events, the prototype successfully demonstrated its capacity to reliably record and update scores. Users considered the system to be simple and easy to use, giving a smooth experience during the competitions.

5 Results and Discussion

In the beginning of the project, the devices are planned to use the Wi-Fi module function of the ESP32 board to transmit and receive the signals between the master and slave devices. However, a significant delay was found while using the Wi-Fi module after carrying out the implementation and the testing. This delay directly impacted the performance of the taekwondo scoring system, as pressing buttons on the slave device resulted in the master receiving the signal after a delay of one and a half seconds. This condition is not allowed in this project. Thus, after doing research, ESP-NOW protocol was chosen to replace the Wi-Fi module. During the testing phase of the project, the effectiveness of the ESP-NOW communication system has been evaluated. The data transfer between the devices is stable and there are no noticeable delays or packet loss for communication.

Fig. 2 Flowchart for WSS master device

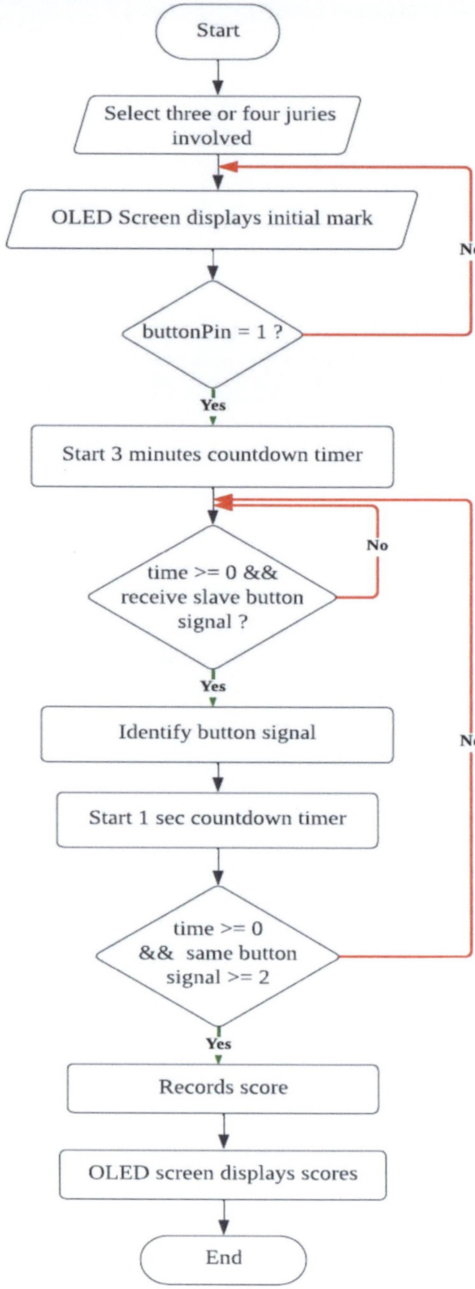

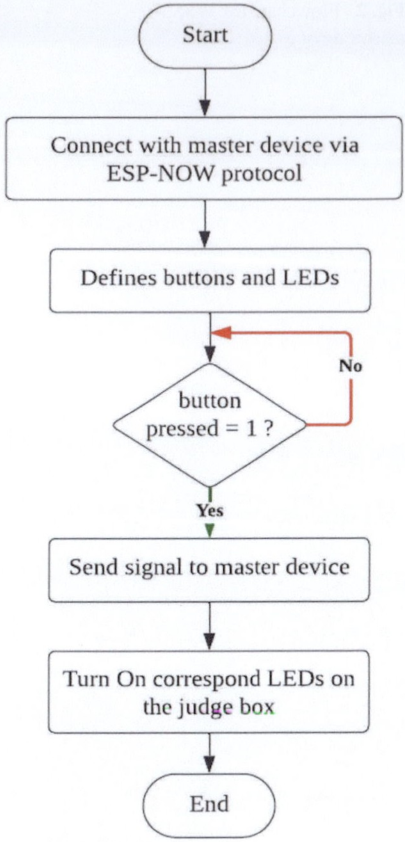

Fig. 3 Flowchart for WSS slave device

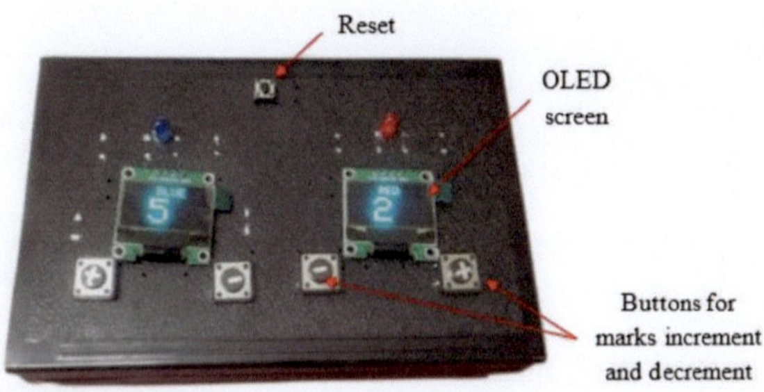

Fig. 4 Prototype of the master device

Based on existing research and practical experience, the ESP32 microcontroller has proven to be capable of reliable communication over a significant range when using the ESP-NOW protocol. According to the research and specific experiment published in the Journal of Universal Excellence (Dipukumar et al. 2008), the communication range between ESP32 boards in an open field environment can

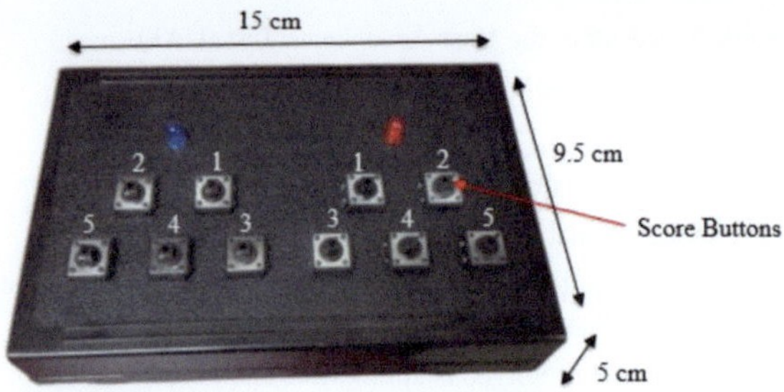

Fig. 5 Prototype of the slave device

Fig. 6 The inner view of the devices

Table 1 Main electronic components used with functions, respectively

Main components	Functions
ESP-WROOM-32	Acts as the main controller embedded in the master and slave device
Push button	Used as the inputs for the user to press
OLED screen	Visually displays the scores for the blue and red teams
LEDs	Function as indicators

demonstrate stable signal transmission up to 190 m. This finding provides empirical evidence that the ESP32 can maintain reliable communication over a substantial distance. In conclusion, the testing can conclude that there are no limitations in signal transmission within the required range for the scoring system.

As a result, the implementation of the ESP-NOW protocol in the ESP32 is successful in making the button press detection on the slave devices accurate and

Table 2 Result of the User Acceptance Test (UAT)

Acceptance requirements	Feedback and result
The button press detection is reliable and accurate	Pass
The score calculation is accurate	Pass
The signal detection and scores update on master are accurate and real time without significant delay	Pass
The size of the prototype is appropriate and portable	Pass

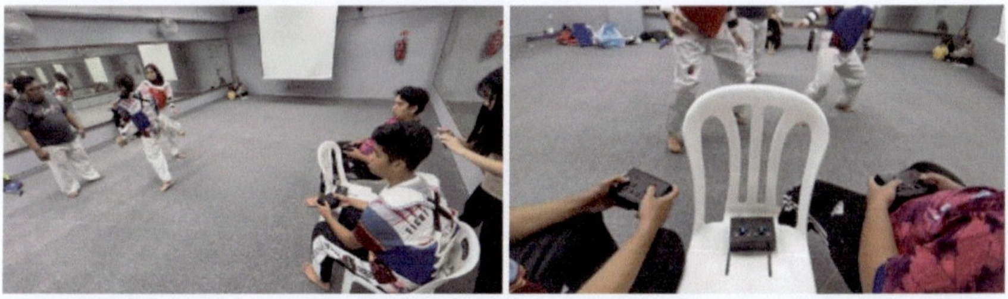

Fig. 7 User testing in UTM Taekwondo club

Fig. 8 The judge shows the score display to the player

seamlessly sending the corresponding data to the master device. To ensure the reliability and accuracy of the button press detection mechanism, the user acceptance test was carried out and the test result shows it is reliable and accurate.

Apart from that, comprehensive testing was conducted to evaluate the functionality of the master device. The master device demonstrated seamless reception of

Fig. 9 Discussion and feedback section with the users

button press data from the slave devices, operating without any noticeable delays. It effectively performed score calculations for the players in the taekwondo match, ensuring real-time updates and displaying them on the OLED screen. The firmware programmed within the master device accurately incremented the respective scores based on the received button press data. Through rigorous testing with various button combinations, we consistently observed correct score updates. To enhance visibility and facilitate score monitoring during competitions, an OLED screen was employed as a suitable output interface. This feature provides users and competition organizers with convenient, real-time access to monitor and track the scores.

Furthermore, the master device incorporated a two-minute timer feature specifically designed for taekwondo combat events. This timer ensures that scores are only calculated when the match is in progress. Besides, there was another 1-s countdown timer for the button signal detection that was programmed in the master device. According to the rules of taekwondo matches, there must be a majority of judges who agree with the award or deduct, then the scores will only be counted. Thus, if one of the buttons is pressed then the 1-s countdown will start. Then, within this 1 s, if the master receives another 1 or 2 button signals from the slave devices which means there are more than or equal to two juries agreeing on the award of the mark, then only the system will calculate the mark and show on the OLED screen.

In conclusion, the primary objective of the project was to develop a taekwondo wireless scoring system by using the ESP-WROOM-32 boards via the ESP-NOW protocol. The system consists of one master and three slave devices, allowing for precise recording and real-time updating of scores based on button presses. Moreover, the objective is to provide a wireless scoring system that is both cost-effective and user-friendly, specifically catering to campus and small organizations. The final product of the project is also designed to be portable and handheld in order to prioritize convenience for the judges. By simplifying the operations, we also aimed to eliminate the requirement for highly skilled personnel and enable the judges to conduct the system independently. Through the implementation and rigorous testing, the project successfully achieved its goals. The reliability of the wireless communication performance using the ESP-NOW protocol was confirmed during the testing phase, showcasing its capability to facilitate seamless and efficient data exchange between the master and slave devices.

Acknowledgements The authors would like to acknowledge the Taekwondo Club of Universiti Teknologi Malaysia and the community, instructors, trainers, and athletes for their support for this work.

References

Alsayaydeh JAJ, Indra WA, Khang AWY, Hossain AKMZ, Shkarupylo V, Pusppanathan J (2020) The experimental studies of the automatic control methods of magnetic separators performance by magnetic product. ARPN J Eng Appl Sci 15

Can H, Lu M, Gan L (2011) The research on application of information technology in sports stadiums. Phys Procedia 22

Cho D, Lee H, Lee W, Kang S (2021) Detecting anomalous kicks in Taekwondo with spatial and temporal features. IEEE Access 9

Combat Sports Medicine (2009) Combat Sports Medicine. https://doi.org/10.1007/978-1-84800-354-5

Dipukumar M, More R, Jahagirdar SVS (2008) A Survey on Electronic multipurpose scoreboard. Int Res J Eng Technol

Mailapalli DR, Benton J, Woodward TW (2015) Biomechanics of the Taekwondo axe kick: a review. J Human Sport Exerc 10

Sun F et al (2022) A flexible lightweight triboelectric nanogenerator for protector and scoring system in Taekwondo competition monitoring. Electronics (Switzerland) 11

Zakaria NH, Phang FA, Pusppanathan J (2019) Physics on the go: a mobile computer-based physics laboratory for learning forces and motion. Int J Emerg Technol Learn 14

A Basic Optical Lidar-Based FPV Drone Sport Configuration

Saumik Dash, Ahmad Ridhwan Wahap, Fatin Aliah Phang Abdullah, Wan Mohd Farhan Bin Wan Mohd Fuaad, Syed Faris Syed Omar, and Jaysuman Pusppanathan

Abstract This study introduces a basic approach to designing, implementing, and assessing a specialized quadcopter with altitude-hold stabilization, specifically designed for first-person view (FPV) drone racing pilot training. The quadcopter is designed to tackle the challenges of flying and maintaining flight altitude during air hover by utilizing the features of the F405 Flight Controller (FC) stack and the Mateksys Optical Flow 3901-L0X Lidar Sensor. The methodology involved thoroughly selecting components, integrating hardware, and configuring software, specifically emphasizing tweaking the Proportional–Integral–Derivative (PID) parameters for altitude control. Exceptional results were observed in the experimental flight test, which aimed to simulate conditions encountered in FPV racing. The altitude-hold stabilized quadcopter repeatedly demonstrated precise altitude control, strong stability when undergoing rapid changes in altitude, and a high level of response to pilot inputs. The present study contributes to the progress of FPV drone racing

S. Dash · J. Pusppanathan (✉)
Department of Biomedical Engineering, Faculty of Electrical Engineering, Universiti Teknologi Malaysia, 81310 Skudai Johor, Malaysia
e-mail: jaysuman@utm.my

A. R. Wahap
Department of Control and Mechatronic, Faculty of Electrical Engineering, Universiti Teknologi Malaysia, 81310 Skudai Johor, Malaysia

F. A. P. Abdullah
School of Education, Faculty of Social Sciences and Humanity, Universiti Teknologi Malaysia, 81310 Skudai Johor, Malaysia

W. M. F. B. Wan Mohd Fuaad
Malaysian Research Accelerator for Technology and Innovation (MRANTI), HIVE 5 (Enterprise 4), Taman Teknologi MRANTI, Bukit Jalil, 57000 Kuala Lumpur, Malaysia

S. F. S. Omar
National Sports Institute of Malaysia, Bukit Jalil, 57000 Kuala Lumpur, Federal Territory of Kuala Lumpur, Malaysia

J. Pusppanathan
Sports Innovation & Technology Centre (SITC), Institute of Human Centered Engineering (iHumEN), Universiti Teknologi Malaysia, 81310 Skudai Johor, Malaysia

Z. Mohamed et al. (eds.), *Proceedings of the 2nd International Conference on Innovation and Technology in Sports, ICITS 2023, 27–28 November, Malaysia*, Lecture Notes in Bioengineering, https://doi.org/10.1007/978-981-97-3741-3_30

technology and facilitates newbie FPV racers into the sport. Integrating the F405 FC stack and Mateksys Optical Flow 3901-L0X Lidar Sensor exemplifies the possibility of the hardware and software configurations. Moreover, this study provides insights into the broader range of applications for altitude control and stabilization algorithms, demonstrating the versatility and accuracy of these systems in practical situations.

Keywords FPV drone · Altitude stabilization · Pilot training · INAV · Lidar sensor

1 Introduction

Unmanned Aerial Vehicles (UAVs) have significantly increased in popularity and practicality in the last decade. They have been adopted in various fields, including surveillance, agricultural, and leisure activities (Tezza et al. 2021). Among these applications, quadcopters in the context of First-Person View (FPV) drone racing have emerged as an exhilarating and emerging sport or recreational activity. FPV drone racing necessitates considerable pilot expertise and accuracy, frequently compelling competitors to maneuver their drones around complex tracks at exceptionally fast velocities, thereby assessing their dexterity and reaction time (Barin et al. 2017). This sport elicits intense excitement and is a leisurely activity and an important platform for developing skilled drone pilots. In this context, the acquisition of flight skills can significantly impact the outcome of competitions, determining success or failure.

Maintaining stable flight and exercising exact control over the quadcopter's altitude is crucial for success in FPV drone racing (Tezza et al. 2021). For entry-level beginners, maintaining a consistent flight altitude above the ground is of utmost importance when navigating the diverse topography of a racing circuit. The precise adjustment of a quadcopter's altitude is necessary to effectively respond to swift variations in elevation, enabling it to navigate obstacles with dexterity and efficiency. The attainment of such a high degree of precision and control, which is crucial for achieving success in competitive contexts, continues to pose a significant challenge for drone pilots and designers.

To meet up these basic requirements of FPV drone racing, engineers and enthusiastist communities and individuals have consistently investigated diverse technological advancements to improve the quadcopter performance. Many research studies have explored altitude control and stabilization methodologies to design technologies that ensure a consistent and stable flying encounter, even in the most demanding racing situations.

Even projects such as the AI-based altitude hold and path planning for drones are under development for finding optimum PID tuning or control algorithms to stabilize drone flight and make flights easier for pilots. On the other hand, FPV drones can be somewhat stabilized using a simple Lidar sensor with a typical flight controller board. The functionality of this sensor is identical to that of an optical mouse, enabling the quadcopter to sustain its location without relying on GPS. The device exhibits a functional radius of up to 2 m, employing a VL53L0X Lidar rangefinder with a PMW3901 optic flow module. The sensor is linked to the flight controller over a UART interface and is exclusively compatible with INAV firmware (Matek Optical Flow 2021, 2023).

2 Background Study

An essential element of FPV drone racing involves meticulously managing the quadcopter's vertical position. Precise altitude control is of utmost importance in effectively navigating the racecourse, as it empowers the pilot to execute maneuvers, evade obstructions, and optimize velocity. The absence of dependable altitude stabilization poses significant challenges and frustrations in drone racing, particularly for inexperienced pilots. Sixty percent of novice pilots prefer a stabilized quadcopter and begin training with one for their first flight (Tezza et al. 2021). This initial learning stage is also essential for asset and pilot safety. High-speed drones can be dangerous for pilots, and drone performance depends on pilot experience (Pfeiffer and Scaramuzza 2021). Hence, beginner pilots are most likely to crash the drone, putting drone assets and pilots in an unsafe environment.

2.1 *Flight Mode for Drone Maneuvering*

There are three widely used flight modes among race pilots to address various stabilization parameters which are:

(i) Angle mode: The roll and pitch of the drone are limited to a maximum angle in angle mode, so even if you push the sticks to their maximum position, the drone will not flip or roll beyond the user-defined angle (usually 45 degrees). The drone will return to its level orientation when the pilot releases the sticks (Acro Mode 2021).

(ii) Horizon mode: This flight mode is similar to angle mode, except that the drone will flip and roll when the sticks are fully deflected. Performing acrobatic-style rolls and flips in this mode allows the drone to return to its normal position (Acro Mode 2021).

(iii) Acro (Rate) mode: There is no aid in leveling in acro mode, also known as manual mode and rate mode, and the drone's role or pitch will remain at the same angle of orientation when the pilot lets off of the sticks. Constant manual adjustments are required in this mode to maintain drone control and avoid crashing into the ground (Acro Mode 2021).

Though these flight modes are regularly used, they don't provide any assistance or stability control for beginner pilots.

2.2 Altitude Hold Using Lidar

In the context of the above-mentioned challenge, this project aims to construct an altitude-hold stabilized quadcopter specifically intended to train individuals in FPV drone racing. The objective is to build a drone that provides superior altitude control which is essential for newbies to First-Person View (FPV) racers. This research project utilizes state-of-the-art technology to tackle the altitude control issue, specifically the F405 Flight Controller (FC) stack and the Mateksys optical flow and Lidar sensor as in Fig. 1 (Optical Flow 2019).

Active remote sensing is what Lidar does best. To measure objects on the ground, an active system must provide energy, such as light. A laser is rapidly fired in a Lidar system, emitting light. This ray of light eventually reaches the ground, where structures and foliage reflect it. The Lidar sensor then records the reflected light. The Lidar system calculates how long it takes for light to reach the ground and return. However, this measurement only considers the parameters under the sensor's Field of Vision (FOV), as shown in Fig. 2. The distance traveled is determined by the elapsed time. Elevation gain is calculated by transforming journey distance.

Hovering a drone requires it to stay at a fixed altitude above the ground, regardless of the terrain or how quickly the height changes. This is important for safety and getting around. So, another goal of this study is to try and improve the setup in various flight conditions and environments so that the stabilization works along with the available flight modes. The following section will further explain the basic setup for Mateksys Lidar and optical flow sensor with a flight controller. This serves as a quick guideline for newbies to set up their own FPV drone.

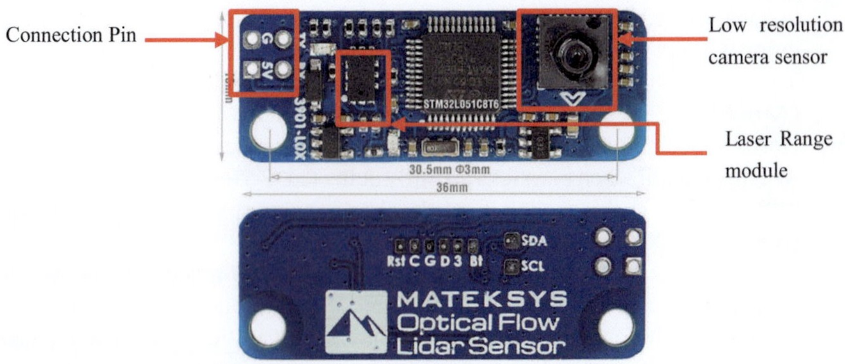

Fig. 1 Mateksys Lidar and optical flow sensor (Optical Flow 2019)

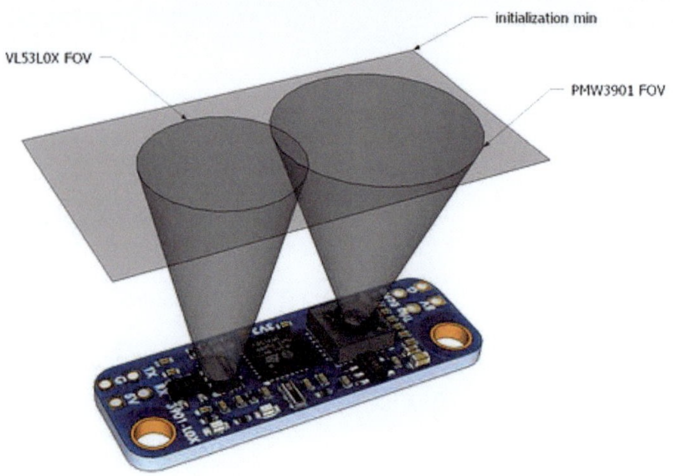

Fig. 2 Mateksys Lidar and optical flow sensor (Matek Optical Flow 2021, 2023)

3 Methodology

Before digging into specific details, it is imperative to present an overview of the development process for the altitude-hold stabilized quadcopter. The flowchart shown below delineates the fundamental phases of development (Fig. 3).

3.1 Component Selection

Choosing the altitude-hold stabilized quadcopter components is vital in its overall development. Every element is selected after thoroughly evaluating its compatibility, performance, and appropriateness in meeting the project's requirements. The basic component selection for a FPV quadcopter is tabulated in Table 1.

3.2 Hardware Setup

The hardware development phase of the altitude-hold stabilized quadcopter encompassed a sequence of procedures with the chosen components' (Table 1) effective integration and operational capability. The selected quadcopter frame, which possesses both lightweight and durable qualities, is utilized as the fundamental structure for assembling the hardware components. The carbon fiber frame effectively accommodates the chosen high-performance brushless motors, a stable and harmonious integration.

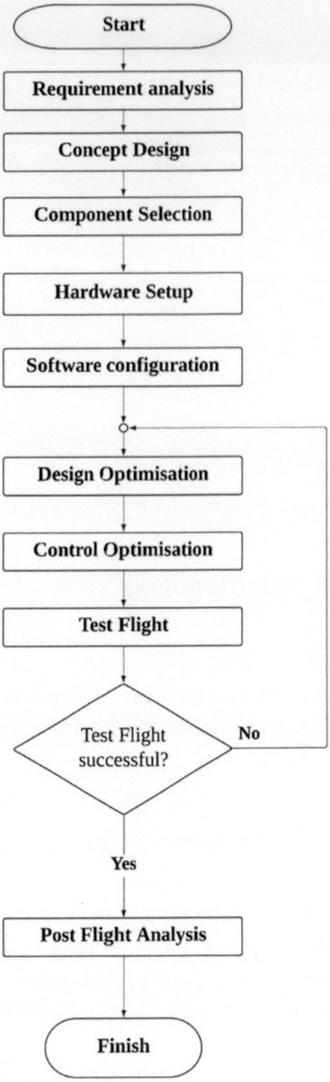

Fig. 3 Project development flow chart

The integration of the F405 Flight Controller Stack, which includes the flight controller, onboard power distribution, and electronic speed controllers (ESCs), is assembled on the quadcopter's frame. Meticulous consideration is given to the wiring and power distribution to ensure all elements' consistent and dependable functioning. Integrating the Mateksys Lidar sensor with the flight controller facilitated the quadcopter's acquisition of real-time altitude data. The sensor is soldered to a UART Serial Communication port on the flight controller. The wiring diagram is shown in Figs. 4 and 5.

Table 1 Main electronic components used with functions, respectively

Component	Quantity	Description
F405 V3 BLS 50A 4in1 Flight Stack	01	F405 fc ESC stack aims to prioritize affordability without compromising functionality. The device is equipped with a 50A Blheli_S Electronic Speed Controller (ESC) and has a barometer sensor (SpeedyBee 2022)
Optical flow 3901-L0X Lidar Sensor	01	The Mateksys Lidar sensor is a crucial choice, effectively fulfilling the fundamental need for height measurement. Using laser-based technology, this sensor provides accurate and up-to-date altitude information, essential for ensuring a constant altitude during first-person view (FPV) racing. Integrating the F405 FC stack guarantees a smooth and uninterrupted flow of communication and data interchange
Brushless DC motor	04	Four high-performance brushless motors with KV1950 were selected based on their capacity to respond quickly and generate sufficient power, which is crucial for swift adjustments in altitude and agile maneuvering
Frame	01	The frame selection is paramount in determining the quadcopter's agility and durability. A quadcopter frame that is both lightweight and durable was chosen to offer a stable foundation for the various components (Urdea 1009)
Li-Po battery	01	A 4s-6s Li-Po battery is a power source suitable for motors with a 1950 kV rating

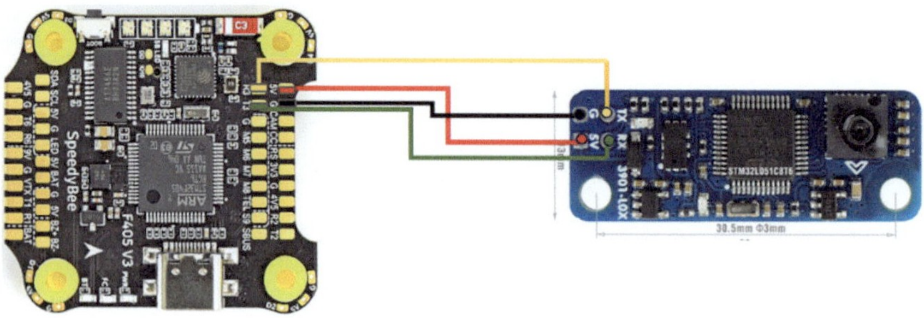

Fig. 4 Flight controller and Lidar sensor wiring diagram

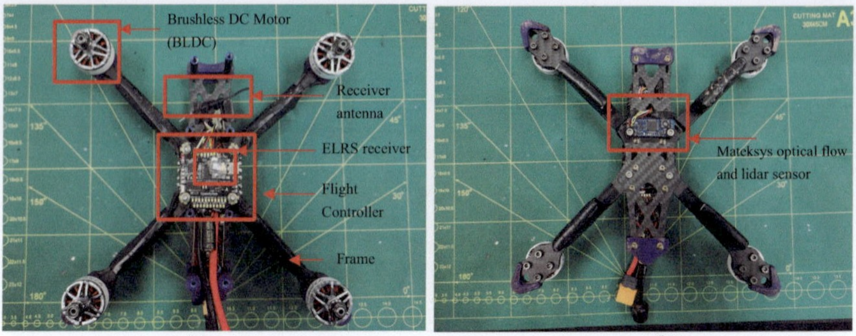

(a) (b)

Fig. 5 Detailed view **a** Top view and **b** Bottom view

A landing gear with a minimum of 2 cm clearance from the ground (see Fig. 6) is required for optical flow initialization by the Lidar sensor (Optical Flow 2019). This also serves as a safety measure to minimize the risk of damage to the quadcopter and its surrounding environment that may arise from unexpected collisions.

The hardware development phase was deemed successful, culminating in the creation of a completely constructed and operational altitude-hold stabilized quadcopter. Establishing this hardware configuration served as a foundation for the following phases of the project, including the software setup and PID tuning. These elements collectively enhance the quadcopter's capacity to effectively sustain altitude control and stability, essential prerequisites for FPV drone race training.

Fig. 6 Isometric view

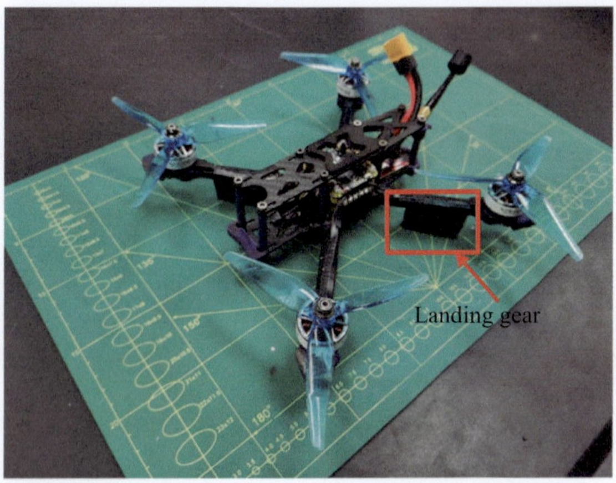

3.3 Software Configuration

The critical step of the altitude-hold stabilized quadcopter involved the software setup, primarily focusing on installing navigation-enabled flight control software (INAV). The INAV software is an open-source flight controller configurator catering to fixed-wing and multi-rotor aircraft. Its features make it a suitable option for our quadcopter endeavor.

The initial stage of the software configuration process entailed the installation of the open-source INAV firmware (iNavFlight 2021) onto the F405 Flight Controller Stack using a flashing procedure. The INAV system provides a user-friendly interface and a wide range of configuration options, making it highly suitable for customizing the behavior of the quadcopter to meet specific altitude control and stabilization needs (Fig. 7).

On the INAV configurator, we can set up the initial settings such as receiver, flight modes, motor orientations, and ports. To set up the Lidar sensor, we need to set up the respective serial port to get the values correctly. The Lidar is connected to serial port 3 in the hardware setup, as shown in Fig. 4. To ensure the functionality of optic flow, it is essential to have the following sensors in operation: Gyroscope, Accelerometer, Barometer, Optical Flow, and Sonar. We must configure the port to MSP (Multiwii Serial Protocol) for the "Rangefinder" and "optical flow" drop-down options to set it up as shown in Fig. 8.

After configuring the port, the sonar and flow indicator should light up, as shown in Fig. 9. The red indicator in Fig. 9 means the circuit connection is correct, but the sensor is not working due to a low power supply. Once the battery is connected, the

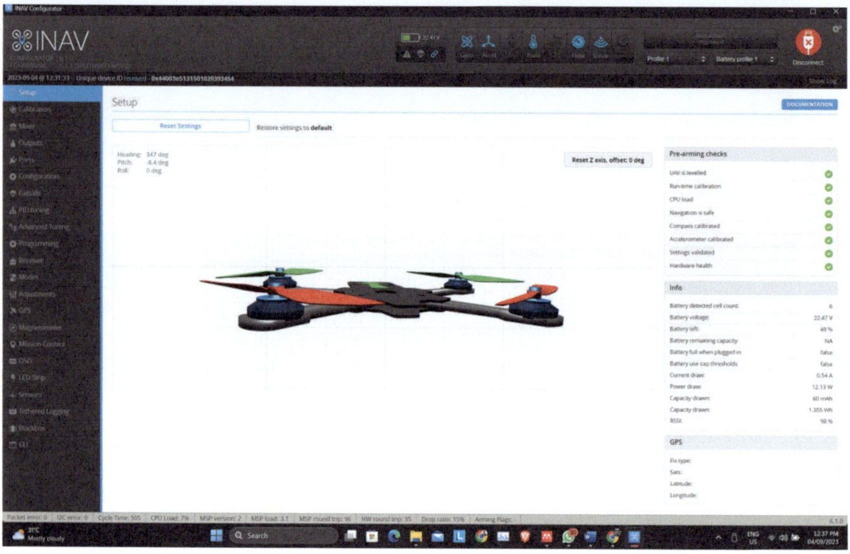

Fig. 7 INAV configurator overview

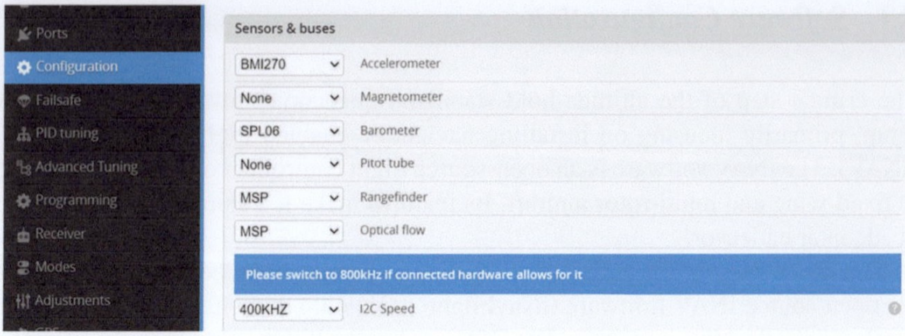

Fig. 8 Sensor connection configuration

flow and sonar indicator will turn blue, as shown in Fig. 10, which means that now the sensor is reading surrounding data and passing it to the flight controller.

Once the configuration is done, the data response can be seen if the debug port is set for the sensor, as there is no dedicated GUI for the Lidar sensor. Moving the drone up and down and side to side will show respective graphs in the debug port (see Fig. 11).

A remote radio controller channel must be configured to turn ON or OFF for the altitude hold by assigning to a designated switch on the radio controller. In accordance with Fig. 12, we must set both "surface" and "Nav althold" in the same radio channel, for this instance, Channel 8 switch is used (Fig. 13).

Fig. 9 Idle sensor

Fig. 10 Active sensor

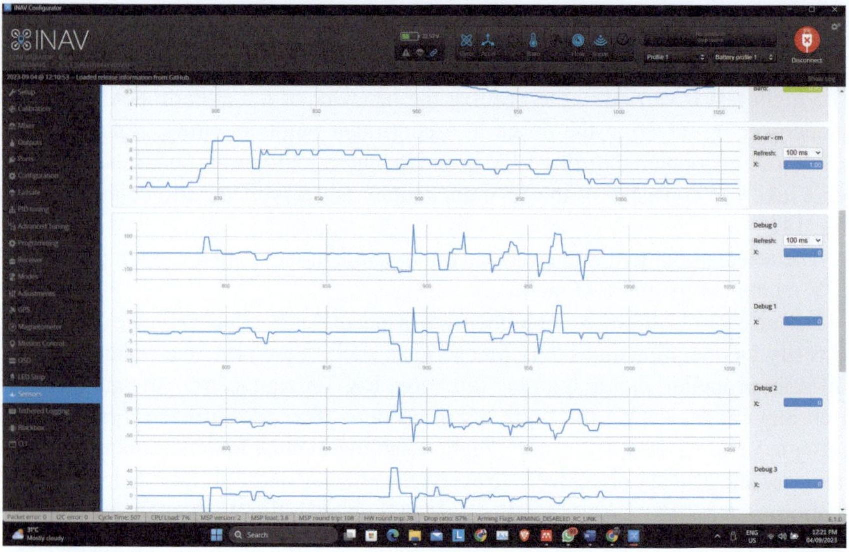

Fig. 11 Sensor activated (Sensor reading)

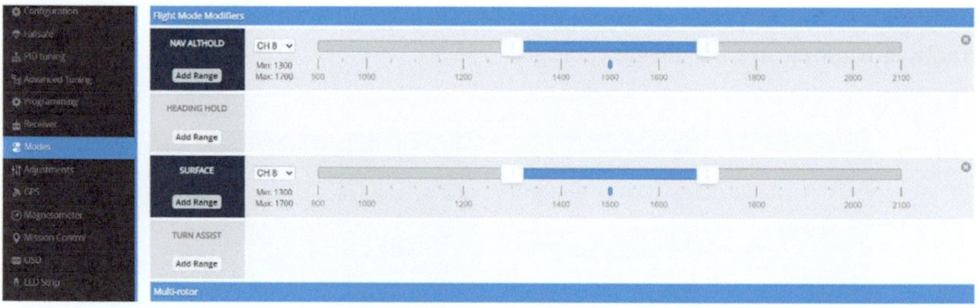

Fig. 12 Sensor activation switch setup (INAV)

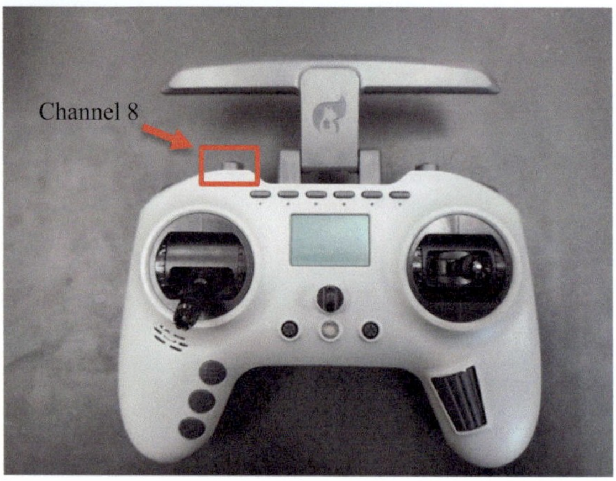

Fig. 13 Sensor activation switch setup (Radio)

4 Testing Procedure

The flight test is carried out to observe the behavior of the altitude-hold stabilized quadcopter by hovering the drone in a controlled outdoor setting, replicating the conditions encountered during FPV drone racing in real-world scenarios. The main tools involved consisted of the FPV quadcopter and a typical radio controller as in Fig. 14.

The selected outdoor site emulates the dynamism and varied topography of real FPV racing competitions (Fig. 15).

Before each test flight, a comprehensive inspection is conducted on the quadcopter to verify its operation and safety precautions, such as the presence of landing gear, fail-safe, propeller, motor spinning direction, RX-TX signal strength, and others to limit any danger during the flight test procedure. The pilot utilized a radio transmitter to regulate the quadcopter's flight, enabling manual intervention as required. However, the primary objective is to assess the efficacy of the altitude-hold and stabilization system. The system's robustness and adaptability were evaluated through experimental flights done under different situations. The quadcopter's flight behavior is captured using video recording. The data is afterward examined to assess the quadcopter's capacity to uphold altitude precision, stability in the face of rapid altitude fluctuations, and responsiveness to pilot instructions. These factors are essential for practical training in FPV drone racing.

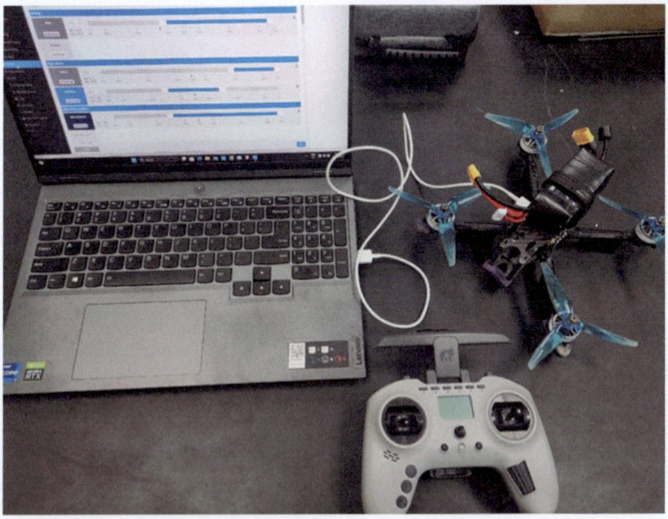

Fig. 14 Experimental setup (Equipment)

Fig. 15 Flight test hovering

5 Results and Discussion

The experimental findings illustrate the effective creation of the altitude-hold stabilized quadcopter for FPV drone race training, utilizing the Mateksys Lidar sensor. The quadcopter demonstrated notable performance in outdoor test flights, thereby confirming the effectiveness of the altitude control and stabilization technology.

The test flights were conducted based on three main conditions:

(i) Different flight modes (angle, horizon, and acro modes)
(ii) Altitude-hold sensor ON or OFF before arming the drone.
(iii) Altitude-hold sensor turned ON during flight or staying ON during flight or turned OFF during flight.

The pilot turned ON or OFF the sensor by switching switch channel-8 on the radio controller as shown in Figs. 12 and 13 earlier. Please note that this designated switch channel can be customized to any preferred channel on the radio controller. The summarized test flight procedure and observation results are as follows.

The recorded video shows that the Lidar sensor works properly in the specified settings but fails in others as summarized in Table 2. While the Lidar is turned off, the pilot has absolute control over the quadcopter in all three flight modes. When it is turned on before arming and stays on during the flight, for angle mode, the altitude-hold works perfectly. The movement is smooth, and the throttle power will not increase if the trainee pilot increases the throttle stick.

Horizon mode shows similar results but restricts some movement, like flip or other advanced maneuvers, which are key elements of horizon mode. In this setup, acro mode performs the poorest. Even though it can hover and hold altitude when stationary, it fails if we try to move around. This resulted from the sensor's position and the acro mode's control algorithm not being compatible. When it is off before arming and turned on during the flight, then for all flight modes, it goes up rapidly,

Table 2 Summarized flight test observation

No	Flight test condition			Altitude-hold result	Observation result
	Flight mode	Altitude-hold sensor condition before arming the drone	Altitude-hold sensor condition during flight		
1	Angle	Off	Off	Not applicable	The pilot has full control
2		On	On	Successful	Fully Stabilized Flight
3		Off	On	Successful	Fly up automatically but stabilize after a moment
4		On	Off	Fail	Fly up rapidly and uncontrollably afterward
5	Horizon	Off	Off	Not applicable	The pilot has full control
6		On	On	Successful	Stable flight but critical maneuver restricted
7		Off	On	Successful	Fly up automatically but stabilize after a moment. After stabilization, the result is the same as no. 6 remarks
8		On	Off	Fail	Fly up rapidly and uncontrollably afterward
9	Acro	Off	Off	Not applicable	The pilot has full control
10		On	On	Successful	It can hover at a preset altitude, but movement is heavily restricted
11		Off	On	Successful	Fly up automatically but stabilize after a moment. After stabilization, the result is the same as no 10 remarks
12		On	Off	Fail	Fly up rapidly and uncontrollably afterward

and then, after some time, it stabilizes itself in the preset altitude. It's more like an overshoot, which can be minimized by further PID tuning. It performed similarly to the previous case for the rest of the flight time.

For the last test condition, where the altitude-hold function is set to On before arming the drone and turned Off during flight hovering, the quadcopter shoots up uncontrollably for all the flight modes. Following post-flight review and troubleshooting, it was determined that the reason is due to the motor vibration during the transition from Lidar-stabilized function to standard flying mode. Motor vibration can affect the Lidar performance (Ghazali and Rahiman 2022) as the Mateksys Optical flow and Lidar sensor don't work alone. The flight controller combines the barometer, accelerometer, and Lidar reading data to estimate the final altitude. Thus, by tuning each sensor and developing a new control algorithm, the drone's performance can be elevated in future research (Flores and Flores 2021; Zakaria et al. 2019; Alsayaydeh et al. 2020).

6 Conclusion

This research has demonstrated the implementation of the F405 flight controller stack, Mateksys optical flow, and Lidar sensor into the altitude-hold stabilized quadcopter exemplifying the prospective utilization of hardware and software configurations of the drone.

The altitude-hold stabilized quadcopter signifies a notable accomplishment within the FPV drone race training domain. The system's effective creation and testing have the potential to pave the way for future progress in drone technology, offering advantages to experienced FPV racers as well as individuals new to the pursuit of mastering the thrilling activity of FPV drone racing.

This study represents an initial advancement in creating safer, more accurate, and more readily available first-person view (FPV) racing training tools for entry-level newbies. Moreover, it lays the foundation for subsequent advancements in this particular domain.

Acknowledgements The authors would like to acknowledge the support from the FPV drone community of Malaysia, drone enthusiasts, and all open-source software developer teams. This project is based on students' project assignment as an introduction to drone technology. Funding support by UTM Encouragement Research (UTMER Q.J130000.3851.20J51)

References

Acro Mode, Self-Level Mode, Angle Mode & Horizon Mode Differences (2021) https://dronesumo.com/acro-mode-self-level-mode-angle-mode-horizon-mode/

Alsayaydeh JAJ, Indra WA, Khang AWY, Hossain AKMZ, Shkarupylo V, Pusppanathan J (2020) The experimental studies of the automatic control methods of magnetic separators performance by magnetic product. ARPN J Eng Appl Sci 15

Barin A, Dolgov I, Toups ZO (2011) Understanding dangerous play: a grounded theory analysis of high-performance drone racing crashes. In: CHI PLAY 2017—Proceedings of the annual symposium on computer-human interaction in play, pp 485–496. Association for Computing Machinery, Inc. https://doi.org/10.1145/3116595.3116611. Can H, Lu M, Gan L (2011) The research on application of information technology in sports stadiums. Physics Procedia, vol 22

Flores A, Flores G (2021) Development and implementation of a new flight mode called Cartesian flight mode for easy piloting of muli-rotor UAVs. In: 2021 international conference on unmanned

aircraft systems, ICUAS 2021 237–242. Institute of Electrical and Electronics Engineers Inc.. https://doi.org/10.1109/ICUAS51884.2021.9476827
Ghazali MHM, Rahiman W (2022) Vibration-based fault detection in drone using artificial intelligence. IEEE Sens J 22:8439–8448
iNavFlight (2021) GitHub—iNavFlight/inav: INAV: navigation-enabled flight control software. GitHub. https://github.com/iNavFlight/inav
Matek Optical Flow (2021) Lidar Sensor 3901-L0X | Flying Tech. https://www.flyingtech.co.uk/electronics/matek-optical-flow-lidar-sensor-3901-l0x
Matek Optical Flow (2023) Lidar Sensor 3901-L0X 3d model. https://3dmdb.com/en/3d-model/matek-optical-flow-lidar-sensor-3901-l0x/9520016/
Optical Flow (2019) 3901-L0X—Matek Systems. http://www.mateksys.com/?portfolio=3901-l0x#tab-id-2
Pfeiffer C, Scaramuzza D (2021) Expertise affects drone racing performance
SpeedyBee (2022) F405 V3 BLS 50A 30x30 FC&ESC Stack—Speedy Bee. https://www.speedybee.com/speedybee-f405-v3-bls-50a-30x30-fc-esc-stack/
Urdea M (2021) Stress and vibration analysis of a drone. In: IOP conference series: materials science and engineering, vol 1009. IOP Publishing Ltd.
Tezza D, Laesker D, Andujar M (2021) The learning experience of becoming a FPV drone pilot. In: HRI '21 : companion of the 2021 ACM/IEEE international conference on human-robot interaction
Zakaria NH, Phang FA, Pusppanathan J (2019) Physics on the go: a mobile computer-based physics laboratory for learning forces and motion. Int J Emerg Technol Learn 14

Optimizing Volunteer Scheduling for Sporting Events: A Round-Robin Algorithm Approach

Choong Ting Quan, Norfadila Mahrom, Ahmad Faizal Annuar, Suhizaz Sudin, and Rafikha Aliana A. Raof

Abstract Volunteers play a crucial role in the operation of sporting events. When there is a large number of volunteers required, more information regarding their personal details, preferred tasks, and skills is required to be stored. This information must be stored in a proper place to prevent tedious problems to occur. There are a few problems encountered, one of them is managing the volunteers in the sporting event because the manual scheduling process consumes a lot of time and is generally ineffective. Then, the inadequate number of skilled volunteers for certain sports occurred. Thus, the objectives are to develop a web-based Sports Volunteer Scheduling System with the implementation of round-robin scheduling algorithms and to analyze the accuracy, actual and average waiting time, and turnaround time of the algorithm in scheduling volunteers for the sports event. The results obtained show that the round-robin scheduling algorithm achieved 100% accuracy in allocating volunteers for the event. It also managed to assign the sports and venues to all volunteers in the database based on their skill and task preference. When the number of volunteers increases, the performance metrices to analyze the round-robin scheduling algorithms also increase. As a result, the web-based Sports Volunteer Scheduling System with the implementation of round-robin scheduling algorithm can equally assign volunteers who share the same task preference and skill to a specific sport and venue.

C. T. Quan · N. Mahrom (✉) · S. Sudin · R. A. A. Raof
Faculty of Electronic Engineering and Technology, University Malaysia Perlis (UniMAP), Pauh Putra, 02600 Arau, Malaysia
e-mail: norfadila@unimap.edu.my

N. Mahrom · S. Sudin · R. A. A. Raof
Sports Engineering Research Centre (SERC), Centre of Excellence, University Malaysia Perlis, Pauh Putra, 02600 Arau, Malaysia

Advanced Computing, Centre of Excellence, University Malaysia Perlis, Pauh Putra, 02600 Arau, Malaysia

A. F. Annuar
Faculty of Mechanical Engineering and Technology, University Malaysia Perlis, Pauh Putra, 02600 Arau, Malaysia

Advanced Manufacturing System Research Group, Faculty of Mechanical Engineering and Technology, University Malaysia Perlis, Pauh Putra, 02600 Arau, Malaysia

Z. Mohamed et al. (eds.), *Proceedings of the 2nd International Conference on Innovation and Technology in Sports, ICITS 2023, 27–28 November, Malaysia*, Lecture Notes in Bioengineering, https://doi.org/10.1007/978-981-97-3741-3_31

Keywords Round-robin sports scheduling · Sports event · Sports timetabling

1 Introduction

A volunteer is a person who willingly offers their time to perform tasks or services assigned to them. A scheduler is a system, mostly made of software applications that help with scheduling processes. Previously, managing the volunteers in the sporting event was done manually such as using a system like an Excel spreadsheet or a Google Form to register volunteers as well as using pen and paper for recording, storing, and checking the volunteer records. The abundance use of spreadsheets, Google Form submissions and hardcopy records that lack the necessary volunteer data are troublesome, tedious, and prone to error in data organization. As a result, a lot of time and resources will go to waste due to inefficiencies.

Therefore, this project aimed to develop a web-based sports volunteer scheduling system by implementing the round-robin scheduling algorithm. The accuracy, actual and average waiting time, and turnaround time of the algorithm in scheduling volunteers for the sports event were also analyzed in this project. The scheduling system was expected to schedule and record all activities with a minimal amount of time and effort. This system is targeted to give the opportunity to all UniMAP students, whether they are athletes or not, the chance to volunteer and participate in sporting events (Taks et al. 2015). Moreover, students can enhance their knowledge and skill by looking to make a shift and take on tasks they are not familiar with. For example, the student gains new knowledge and skill by practicing and trying newly assigned tasks. This enables the students to get out of their comfort zone. They also get to establish connections for the task that could benefit their future leadership development (Dickson et al. 2022).

All the volunteer data and details such as sports events, venues, and many more can be stored in the database. The volunteers can only access the system online, whereas the sports coordinator can access both the database and the system online whenever they want and from any location. The sports volunteer coordinator is able to run the scheduler and allocate volunteers to various sports and venues based on their skills and task preferences. This will give the volunteer coordinator flexibility in organizing volunteers and less time will be spent in managing the scheduling process (Waters 2021).

On the other hand, the volunteers can view the assigned schedule in the system to obtain information such as which sport, venue, date, and time they were assigned by the system. This system concentrates on finding solutions to address issues with scheduling numbers of volunteers who share the same skill and task preference for specific sports but lack the necessary number of skilled volunteers for other sports.

2 Background Study

Currently, the University Malaysia Perlis (UniMAP) Sports Centre allocates volunteers for sporting events based on the information obtained via Google Forms, spreadsheets, and hardcopy document records. There is no platform designated for the volunteers to register and check the schedule assigned to them, as well as to view the sports schedule.

Previously, only students who are athletes themselves were given the chance to volunteer in sporting events. This is because these students have a better understanding of sporting events compared to those students who do not have any athletic experience. However, even though some students are athletes, they might not be appropriate for certain sports events. For example, a student-athlete excels and has experience in football events but becomes a volunteer in a basketball event. This shows the possibility that the student will face new challenges as a volunteer.

There are two types of scheduling algorithms called preemptive and non-preemptive. The non-preemptive scheduling algorithm refers to a process that will remain active once it enters the running state until it has completed the execution. The preemptive scheduling algorithm refers to a high-priority process that will be executed first when it enters the system. The existing low-priority process will be stopped and pre-empted for executing the high-priority process (Thakur 2019). In this project, a scheduling algorithm named round-robin was implemented and its performance in terms of accuracy, actual and average waiting time, and turnaround time were analyzed.

The project development of this project is based on the Software Development Life Cycle (SDLC) Agile model. The SDLC Agile model combines incremental and iterative process models with a focus on user satisfaction and adaptability of processes through the quick delivery of functional software. In order to deliver specific features for a release, the project is divided into smaller incremental builds using the agile methodology. The traditional SDLC models such as the waterfall model are based on predictive strategy, whereas the Agile model is based on an adaptive software development model. The Agile software development model assisted developers in creating projects that quickly alter to changing requests over time (SDLC 2023).

3 Methodology

3.1 Project Specifications

In the planning stage of SDLC, the objectives and problem statements of the volunteer scheduling system for the sporting event are determined. The scope of this project is to develop a system that can schedule volunteers for different sports at different venues based on their skill and task preference. A web-based development platform integrated with the database is used for this project. The target users are the sports volunteer coordinator and volunteers in the UniMAP Sports Centre. The planned sports are badminton, basketball, and football. On the other hand, the planned venues are Kubang Gajah Multipurpose Hall, Universiti Malaysia Perlis Sports Centre, and Perlis Youth and Sports Complex. Each sport will have two sessions which are conducted on different days and times.

Next, the target users that are the sport's event coordinator and volunteers are determined. The volunteers are the students in UniMAP. There are five job scopes available for the volunteers to choose: crowd management, technical, sports medicine, event planning, and sports event promoter. Each of the sports events such as badminton, basketball, and football will have volunteers for these five job scopes. The sports coordinator can access the system to run the scheduler for scheduling algorithms and assigning schedules to the volunteers. The scheduler process is implemented by using round-robin scheduling algorithms. The fundamentals of the round-robin algorithm and the scheduler process must be understood in order to develop a functional web-based system (Raof et al. 2018). It is based on the allocated volunteers as the input to generate the performance metrics and later stored in the database. The sports coordinator also can register the volunteer or update their details in case the volunteer is unable to perform the mentioned function and view the volunteer details as well as the schedule of the sports event assigned.

Before the volunteer can access the system, they have to register their information in the database via a registration form from the system. The information to be stored in the database will contain the name of the student, student matric number, faculty, gender, age, email, phone number, hostel name, skill, medical history, task preference, password, and retype password. Additional information is added to the database when the sports coordinator runs the schedulers such as the performance metrics of the scheduling algorithms, sports, and assigned schedules. The sports table in the database is defined by the sports coordinator and will contain the sport's events, venue, date and time, skill capacities, and task preference.

The volunteers can view and update their personal information. The volunteer will have the Accept/Reject the Assigned Schedule feature for response to the assigned schedule. They are also able to view the sports schedule and decide whether to accept or reject the assigned schedule.

The system is developed by using Visual Studio Code with Python programming language. For front-end web development and frameworks such as HTML, CSS, and JavaScript are used. For the back-end development, Python programming language and Django framework are used. PostgreSQL as the database software application is used for integrating with web development.

3.2 Use Case Diagram

The use case diagram as shown in Fig. 1 describes the interactions between various user types, known as actors.

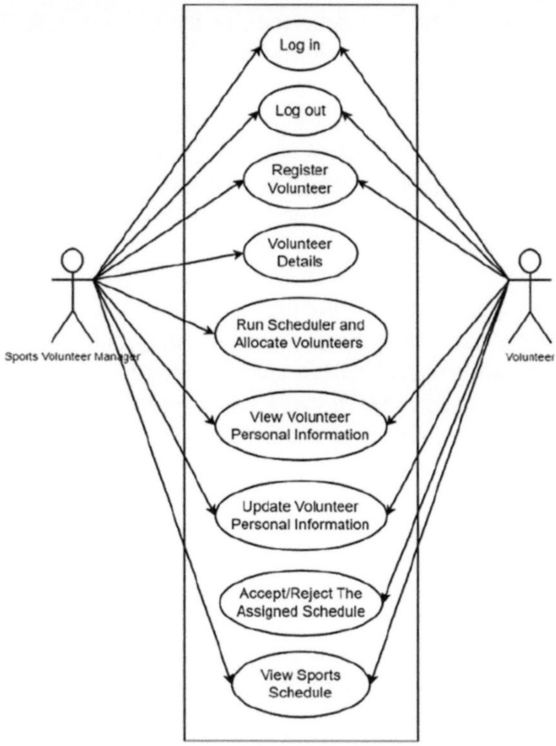

Fig. 1 Use case diagram

For this project, the sports coordinator and the volunteer are the two actors in the use case diagram (Irwin et al. 2019). The web-based sports volunteer scheduling system allows both users to log in and log out of the system. However, only the sports coordinator has features like adding, updating, deleting volunteer details, running the scheduler, and allocating volunteers. On the other hand, the volunteer has the feature to accept or reject the assigned schedule. Both users can register volunteers, view volunteer personal information, update their details, and view sports schedules.

3.3 Volunteers Scheduler

There are two important actions that can be performed by the sports coordinator in the system. The first part is to allocate or relocate the volunteers to a specific sport and venue, and the second part is to run the scheduler. If the coordinator chooses to allocate the volunteers, then the system will validate whether the coordinator has allocated the volunteers once. If the coordinator has allocated the volunteers once, then the system will display an error message as a response. Otherwise, the system allocates the volunteers and displays the result of allocated and unallocated volunteers as well as the time generated for the assigned schedule for the coordinator. If the coordinator clicks to relocate the volunteers, then the system will validate whether the coordinator has allocated the volunteers before. If the coordinator has not allocated the volunteers before, then the system will display an error message together with the previously allocated and unallocated volunteers as well as the time generated to the coordinator. Otherwise, the system relocates the volunteers and displays the latest result of allocated and unallocated volunteers as well as the time generated to the coordinator.

Upon the coordinator's decision to initiate the scheduler, the system utilizes the allocated volunteers and their registration times to generate the schedule. This is to ensure the fulfillment of the specific number of volunteers required for each job scope. The performance metrics are generated concurrently during scheduling to analyze the effectiveness of round-robin algorithm. The resulting metrics are displayed to the coordinator and also added to the database. Other information such as allocated volunteers, unallocated volunteers, and time generated will also be displayed to the coordinator.

There are several performance metrics equations involved in scheduling the volunteers, which are accuracy, actual waiting time, actual turnaround time, average waiting time, and average turnaround time (Singhal 2018; ElDahshan et al. 2017; Alhaidari et al. 2019). The accuracy of the round-robin algorithm in scheduling the volunteers is computed by dividing the number of scheduled volunteers by the total assigned schedules to all volunteers in the allocated_volunteers list. Equation (1) is used to calculate the accuracy.

$$accuracy = \frac{number\_of\_scheduled\_volunteers}{total\_assigned\_schedules} \tag{1}$$

In order to obtain the total turnaround time, the sum of the turnaround time is computed. The turnaround time is computed by getting the difference between the number of scheduled volunteers and the index of the volunteer's task preference as shown in Eqs. (2) and (3).

$$\begin{aligned} turnaround\_time = {} & num\_scheduled\_volunteers \\ & - index(task\_preference) \end{aligned} \tag{2}$$

$$total\_turnaround\_time + = turnaround\_time \tag{3}$$

The actual waiting time is calculated as shown in Eq. (5). The actual waiting time is computed by dividing the total waiting time by the number of scheduled volunteers. The total waiting time is computed by summing the difference between the turnaround time with the length of the allocated volunteers and plus one as shown in Eq. (4).

$$total\_waiting\_time + = turnaround\_time$$

$$- \, len(allocated\_volunteers) + 1 \tag{4}$$

$$actual\_waiting\_time = \frac{total\_waiting\_time}{number\_of\_scheduled\_volunteers} \tag{5}$$

The actual turnaround time is calculated as shown in Eq. (6). The actual turnaround time is computed by dividing the total turnaround time by the number of scheduled volunteers.

$$actual\_turnaround\_time = \frac{total\_turnaround\_time}{number\_of\_scheduled\_volunteers} \tag{6}$$

The average waiting time is calculated as shown in Eqs. (7). The average waiting time is computed by dividing the total waiting time by the total number of allocated volunteers.

$$average\_waiting\_time = \frac{total\_waiting\_time}{total\_allocated\_volunteers} \tag{7}$$

The average turnaround time is calculated as shown in Eqs. (8). The average turnaround time is computed by dividing the total turnaround time by the total number of allocated volunteers.

$$average\_turnaround\_time = \frac{total\_turnaround\_time}{total\_allocated\_volunteers} \tag{8}$$

4 Results and Discussion

The web-based Sports Volunteer Scheduling System has been successfully developed in this project. Figure 2 shows the login page for both volunteers and sports coordinators. Any new volunteer must click on *Register New Volunteer* to register as a volunteer in the system.

Whenever a volunteer is allocated to specific sports and venues, this information will be displayed on the *Notification Board* as shown in Fig. 3. The volunteer must respond to the assigned schedule by clicking on the Accept or Reject buttons. This action will be recorded in the volunteer response table in the database. If the volunteer accepts the assigned schedule, the sports coordinator will not take any action. However, if the volunteer chooses to reject, the sports coordinator will remove the response from the database as well as the *Notification Board*.

The system allows the sports coordinator to view the generated sports schedule. Figure 4 shows the schedule of volunteers assigned to each session of a specific game. For a specific game session, the skill and task preference of each volunteer

Fig. 2 Login page of the sports volunteer scheduling system

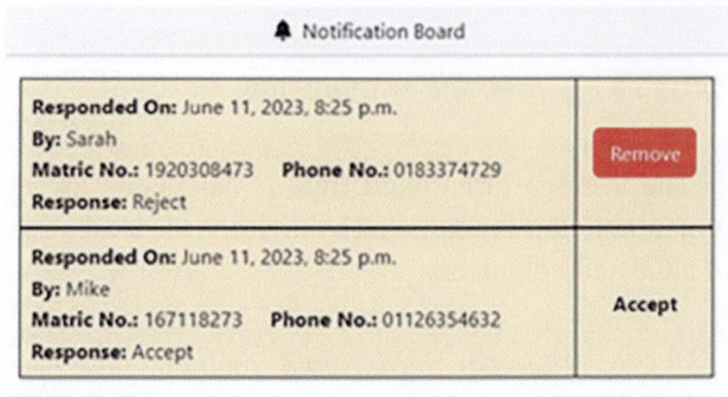

Fig. 3 Volunteer responses in the notification board

are also displayed to ease the sports coordinator in managing the volunteers during the running session.

The system has been tested to allocate and schedule different numbers of volunteers as shown in Tables 1 and 2. The data set of volunteers used for testing is obtained from the UniMAP Sports Centre. The data consists of volunteers for sports events that have been organized by the UniMAP Sports Centre. It is observed that the scheduler has achieved an accuracy value of 1.00 (100%) which indicates it managed to assign the sports and venues to all volunteers in the database based on their skill and task preference. The actual waiting time indicates that the volunteers must wait for a specific amount of time in the queue before being executed.

The difference in the average waiting time obtained from different sets of volunteers to schedule is due to one unit of time quantum, in which the volunteers had to wait in the queue for their turn for allocation. The same goes with the average turnaround time where the round-robin algorithm needs to consider both allocated and unallocated volunteers. Overall, the results show an increasing trend in the waiting time and turnaround time as the number of volunteers increased.

As the volunteer count increases, the allocation and scheduling duration proportionately increases. Several factors influence the efficacy of the round-robin performance. Foremost among the factors is context switching overhead. Employing the

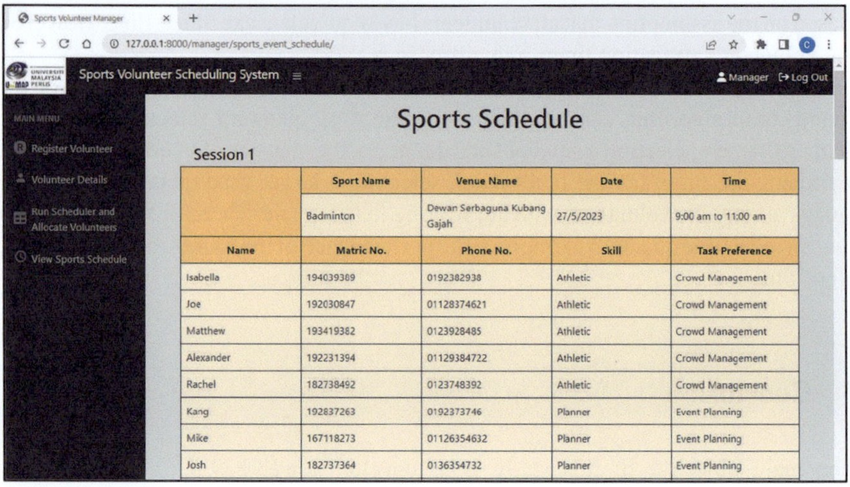

Fig. 4 Generated sports schedule

Table 1 Comparison between 250, 500, and 1000 volunteers

Parameters	250	500	1000
Accuracy	1.00	1.00	1.00
Actual waiting time (s)	58.24	120.79	214.73
Actual turnaround time (s)	147.94	297.99	586.24
Average waiting time (s)	161.68	340.13	577.40
Average turnaround time (s)	410.68	839.13	1576.40

Table 2 Comparison between 1500, 2500, and 3500 volunteers

Parameters	1500	2500	3500
Accuracy	1.00	1.00	1.00
Actual waiting time (s)	314.99	522.22	711.66
Actual turnaround time (s)	878.67	1468.81	2064.71
Average waiting time (s)	837.68	1378.67	1840.35
Average turnaround time (s)	2336.68	3877.67	5339.35

round-robin algorithm invariably results in frequent context switching among volunteers. Each is provided with a fixed time quantum, and upon its expiration, the current volunteer is relegated to the back-end of the queue, allowing the subsequent volunteer to commence. This frequent interchange poses a potential decrement in system performance.

The next factor is different volunteer execution times. Theoretically, the round-robin algorithm handles all volunteers evenly by allocating them the time slices.

Based on the assumption that all volunteers have equivalent execution time. However, the volunteer's execution times may differ in reality.

The third factor is ineffective time quantum selection. The performance of the round-robin algorithm can be affected by the time quantum selection (Adekunle et al. 2014). By selecting a smaller time quantum, the chances of overhead in context switching are high. On the other hand, selecting a larger time quantum can cause delays among the volunteers who are waiting in the queue. Therefore, it is crucial to find an ideal time quantum to improve the performance of the round-robin scheduling algorithm.

5 Conclusion

The main objective of this study is to develop a web-based Sports Volunteer Scheduling System with the implementation of round-robin scheduling algorithm. Python programming language with Django framework was used to develop the system. This system was integrated with the PostgreSQL database application for storing, manipulating, and viewing the data. This system has been successfully launched on the local host.

The second objective of this study is to analyze the accuracy of the round-robin algorithms in scheduling volunteers for sports events. The allocated volunteers were used as the input for running the scheduler and generating the performance metrics of the scheduling algorithms. As obtained from the result, the round-robin scheduling algorithms have the same accuracy of 100% for volunteer allocation and volunteer re-allocation.

The third objective of this study is to analyze the actual and average waiting time and turnaround time of the round-robin scheduling algorithms in scheduling volunteers for the sports event. After the re-allocation of the volunteers, the obtained actual and average waiting time and turnaround time were the same as the volunteer allocation. Finally, during observation, as the number of volunteers increases, the performance of the round-robin scheduling algorithm was simultaneously increased.

References

Adekunle YA, Ogunwobi ZO, Jerry AS, Efuwape BT, Ebiesuwa S, Ainam JP (2014) A comparative study of scheduling algorithms for multiprogramming in real-time systems. Int J Innov Sci Res 12(1):180–185

Alhaidari F, Balharith T, Al-Yahyan E (2019) Comparative analysis for task scheduling algorithms on cloud computing. In: 2019 International conference on computer and information sciences (ICCIS), pp 1–6

Dickson TJ, Darcy S (2022, November) Surviving and thriving from the volunteer involvement in mega-sport events. Org.au website. https://volunteeringstrategy.org.au/wp-content/uploads/2022/11/VRP_Surviving-and-Thriving-from-the-Volunteer-Involvement-in-Mega-Sport-Events.pdf. Accessed 07 Jan 2023

ElDahshan K, Elkader AA, Ghazy N (2017) Round robin based scheduling algorithms, a comparative study. Autom Control Syst Eng J 17(2):42. ISSN 1687-4811. https://doi.org/10.13140/RG.2.2.23618.68804

Irwin NJ, Bennett A, Carlos K, Kiswani J, Harris CR, Dascalu S, Harris FC (2019) ARIA 3.0: a modern approach to web-based music festival registration systems. In: Proceedings of the ISCA 28th international conference on software engineering and data engineering (SEDE 2019). EPIC series in computing, vol 64, September 30–October 2, San Diego, CA, pp 215–224

Raof RAA, Sudin S, Mahrom N, Rosli ANC (2018) Sport tournament automated scheduling system. MATEC Web Conf 150:05027. https://doi.org/10.1051/matecconf/201815005027

SDLC—Agile Model (n.d.) Tutorialspoint.com website. https://www.tutorialspoint.com/sdlc/sdlc_agile_model.htm. Accessed 07 Jan 2023

Singhal A (2018, October 19). Round robin. Gatevidyalay.com; Gate Vidyalay. https://www.gatevidyalay.com/round-robin-round-robin-scheduling-examples/

Taks M, Chalip L, Green BC (2015) Impacts and strategic outcomes from non-mega sport events for local communities. Eur Sport Manag Q 15(1):1–6. https://doi.org/10.1080/16184742.2014.995116

Thakur D (2019, January 8) Operating system scheduling algorithms. Computer Notes website. https://ecomputernotes.com/fundamental/disk-operating-system/scheduling-algorithms. Accessed 19 Dec 2022

Waters A (2021, April 28) Volunteer scheduling: how to boost efficiency using tech. Galaxydigital.com. https://www.galaxydigital.com/blog/volunteer-scheduling

Green Technology in Sports: The Role of Waste Management in Sustainable Stadiums

Ahmad Noor Syukri Zainal Abidin, Mohamad Rusydi Mohamad Yasin, Syukran Hakim, Mohd Amir Shahlan Mohd Aspar, and Noor Zaihah Jamal

Abstract The present review study explores the growing importance of environmentally sustainable technology in the field of sports. The sports industry has experienced an increase in the utilization of environmentally friendly technologies due to a growing focus on sustainability and environmental awareness. These technologies cover the utilization of renewable energy sources, sustainable materials, and effective waste management systems in sports facilities. Massive sporting events, which draw in substantial crowds, have a significant ecological footprint, encompassing the production of waste, utilization of energy, and release of carbon dioxide. The implementation of efficient waste management practices plays a crucial role in mitigating the environmental impact of such events and safeguarding the long-term viability of sports facilities. This study delves into the several strategies utilized in waste management, including recycling, composting, and waste-to-energy conversions. The incorporation of digital and smart technology has also been emphasized, exemplified by advancements such as intelligent waste bins and digital monitoring platforms. Various case studies conducted in different regions, such as Kazakhstan, Bandung, and Kerman Graduate University, offer valuable insights into the pragmatic application of these technologies. Notwithstanding the progress made, there exist areas of research that remain unexplored, particularly in the full examination of the implementation of green technology within sports facilities. Potential avenues for future research encompass a more comprehensive investigation into the sociological dimensions of waste management within the realm of sports, as well as an

---

A. N. S. Z. Abidin (✉) · M. R. M. Yasin
Faculty of Manufacturing and Mechatronic Engineering Technology, Universiti Malaysia Pahang Al-Sultan Abdullah, 26600 Pekan, Pahang, Malaysia
e-mail: ansyukri@miros.gov.my

A. N. S. Z. Abidin
Malaysian Institute of Road Safety Research (MIROS), Institut Penyelidikan Keselamatan Jalan Raya Malaysia (MIROS), Lot 125-135, Jalan TKS 1, Taman Kajang Sentral, 43000 Kajang, Selangor, Malaysia

S. Hakim · M. A. S. M. Aspar · N. Z. Jamal
Faculty of Electrical and Electronic Engineering Technology, Universiti Malaysia Pahang Al-Sultan Abdullah, 26600 Pekan, Pahang, Malaysia

Z. Mohamed et al. (eds.), *Proceedings of the 2nd International Conference on Innovation and Technology in Sports, ICITS 2023, 27–28 November, Malaysia*, Lecture Notes in Bioengineering, https://doi.org/10.1007/978-981-97-3741-3_32

examination of the influence of legislation and regulation in fostering ecologically sustainable behaviors.

Keywords Green technology · Waste management · Sustainable stadiums · Circular economy · Policy and regulation

1 Introduction to the Importance of Green Technology in Sports

The advent of eco-friendly technology in the domain of sports has been a significant development in recent years, marked by an increasing emphasis on the concepts of sustainability and environmental consciousness (Tanveer et al. 2022; Padhy and Achary 2022; Farooq et al. 2022; Murthy and Ramakrishna 2022; Negrete-Cardoso et al. 2021; Jamaludin and Yassin 2013; Jamaludin et al. 2018a, 2017; Abidin et al. 2022; Nasir et al. 2015; Khan et al. 2021; Rosli 2021; Atalla et al. 2020; Lee et al. 2021; Sauve and Acker 2020). In the contemporary context, the phrase "green technology" refers to the utilization of environmentally friendly methods, substances, and sources of energy in the realm of sports facilities and events (Khan et al. 2021; Rosli 2021; Jamaludin et al. 2017; Atalla et al. 2020; Lee et al. 2021; Sauve and Acker 2020; Hasya et al. 2022; Babu et al. 2021; Alsibai et al. 2018; Wang 2020; Abdullah et al. 2022; Rosli et al. 2021; Md Kamil et al. 2022; Yu et al. 2022; Baharuddin et al. 2022; Ramli et al. 1060; Sufian et al. 2021; Hamzah et al. 2022; Ghavam and Vahidi 2022). The utilization of renewable energy sources, efficient waste management systems, and sustainable materials in the construction and functioning of sports facilities has been seen (Ghavam and Vahidi 2022; Baharuddin et al. 2023; Babu et al. 2023, 2022; Atalla et al. 2022; Farag et al. 2019; Hoareau et al. 2021; Nasir et al. 2013; Dinglasan et al. 2021; Jamaludin et al. 2018b; Abdildin et al. 2021; Rosli et al. 2019; Ali et al. 2022; Binti Kamarudin et al. 2021; Zhang et al. 2021). The utilization of green technology in the domain of sports holds considerable relevance and should not be undervalued.

Sports events of significant magnitude that attract a multitude of individuals have a notable environmental impact. This encompasses the generation of waste, utilization of energy, and release of carbon emissions. The integration of environmentally friendly technologies in sports facilities and events holds significant promise for reducing their ecological footprint, actively supporting global initiatives to address climate change, and setting a good example for other industries (Zhang et al. 2021). The effective management of waste plays a crucial role in safeguarding the long-term viability and sustainability of sporting arenas. When a large number of folks congregate to see various activities, a significant amount of waste is generated, including food waste, packaging, and miscellaneous items. In the event that there are inadequate waste management systems in place, the trash stated above has the capacity to contribute to environmental contamination and the exhaustion of natural resources, as evidenced by various scientific references (Tanveer et al. 2022; Yu et al. 2022; Hamzah et al. 2022).

Furthermore, the efficient management of waste at sports stadiums heavily relies on the effective deployment of diverse methods. This includes the exploitation of recyclable materials, the implementation of recycling efforts, and the promotion of waste reduction among event attendees. In addition, a number of stadiums have effectively implemented innovative approaches, such as the adoption of composting methods for organic waste management or the conversion of waste materials into viable sources of energy (Padhy and Achary 2022; Murthy and Ramakrishna 2022; Negrete-Cardoso et al. 2021). The implementation of effective waste management systems not only serves to alleviate the environmental ramifications linked to sports stadiums but also contributes to their economic sustainability. By employing waste reduction measures, stadiums have the capacity to diminish their expenditures associated with garbage disposal and potentially generate supplementary revenue by selling recyclable materials or harnessing energy obtained from waste (Tanveer et al. 2022; Padhy and Achary 2022; Hamzah et al. 2022).

The objective of this analysis is to provide a comprehensive assessment of the current state of green technology and waste management in the field of sports, by critically examine the latest trends and advancements in the specific field under consideration, examine the challenges and constraints that impede the effective implementation of these breakthroughs, and identify new directions for future research and innovation, with a particular focus on sustainable stadiums.

2 Green Technology and Waste Management: Current Trends and Future Directions

2.1 Overview of Green Technology in Sports

The field of green technology in sports is undergoing substantial advancement, marked by continuous improvement in several areas focused on addressing the environmental impacts linked with sports events and facilities. The implementation of sustainable energy sources, such as solar and wind power, is employed to provide electricity to stadiums and other infrastructure facilities. To provide an example, several stadiums have adopted the practice of installing solar panels on their rooftops in order to generate electricity, thereby reducing their reliance on fossil fuels (Zhang et al. 2021). Moreover, there has been a significant emphasis on the significance of water conservation in the architectural planning of sports stadiums, underscoring the necessity for effective management of water resources across diverse sectors, such as stadium design (Wang 2020). This aligns with the overarching goals of green technology, which aims to foster sustainability, mitigate CO_2 emissions, and minimize the production of solid waste (Padhy and Achary 2022).

The integration of renewable energy sources is accompanied by the application of sustainable materials in the field of sports, encompassing the incorporation of such materials in the design and functioning of sports facilities. This involves the use of recycled or sustainably sourced materials, as well as the integration of architectural features designed to reduce energy use, such as the inclusion of natural lighting and ventilation (Zhang et al. 2021). Furthermore, the incorporation of environmentally friendly technology in the context of sports involves the application of digital and intelligent technologies to improve operational efficiency and reduce resource consumption. One strategy for addressing the issue of paper waste is the adoption of digital ticketing systems, while another approach involves the use of intelligent systems for managing energy and waste. The utilization of Industry 4.0 technologies in order to promote supply chain sustainability has been recognized as a significant factor in facilitating intelligent waste management within supply chains (Zhang et al. 2021). Within the wider scope of waste management, various research themes have been recognized as significant areas of emphasis in recent years. These themes include electronic waste, the transition toward a circular economy, plastic waste, bio-based waste management, lifecycle assessment, ecological impacts, and the management of construction and demolition waste (Tanveer et al. 2022). This observation highlights the increasing significance and pertinence of sustainable practices and green technologies across several industries, including the field of sports.

2.2 *Current Trends in Waste Management*

In the last few decades, the sports business has witnessed significant advancements in waste management, marked by a growing emphasis on sustainability and waste reduction. One notable development that has arisen is the widespread acceptance and implementation of circular economy principles, which strive to minimize waste through the optimization of material reuse and recycling (Tanveer et al. 2022). The potential of green monomers derived from renewable sources has also been discussed, emphasizing the role of green technology in promoting sustainability, reducing CO_2 emissions, and decreasing solid waste generation (Padhy and Achary 2022). This aligns with the broader objectives of waste management in sports, where the focus is on using sustainable materials and practices. For instance, stadiums have been adopting innovative strategies such as composting organic waste or transforming waste materials into usable energy sources. The importance of recycling is further underscored by the strong linkage between the circular economy and sustainable production, waste management, and recycling (Negrete-Cardoso et al. 2021).

E-waste management is another area that has received significant attention, with a call for global best practices. The need for a common legal framework globally, the importance of Extended Producer Responsibility (EPR), and the transformation of the informal sector have been highlighted as crucial aspects of effective E-waste management (Murthy and Ramakrishna 2022). Furthermore, the relationship between renewable energy consumption, economic growth, and environmental quality has been explored, revealing that fossil fuels have a negative impact on the environment and health. In contrast, renewable energy sources, such as those derived from waste, can reduce health and pollution-related expenses (Khan et al. 2021). Moreover, while there have been relevant implementations of programs related to waste management, such as the BGC program, there remains a need for community-level waste management activities to achieve significant behavior change (Hamzah et al. 2022).

A notable advancement in waste management within the domain of sports pertains to the increasing adoption of digital and smart technology for the purpose of enhancing waste management procedures. Multiple sources have reported the deployment of intelligent waste bins that possess the ability to sort and compress waste, as well as the utilization of digital platforms for waste monitoring and reporting (Tanveer et al. 2022; Padhy and Achary 2022; Farooq et al. 2022; Murthy and Ramakrishna 2022; Negrete-Cardoso et al. 2021; Khan et al. 2021; Yu et al. 2022; Hamzah et al. 2022). The significance of incorporating smart technology in waste management is further highlighted by the emphasis placed on the use of industry 4.0 technologies to promote supply chain sustainability (Zhang et al. 2021).

3 Opportunities and Innovations in Green Technology and Waste Management

The use of environmentally friendly technologies and effective waste management strategies in sports facilities and events may present many technological challenges. One of the main challenges pertains to the significant initial costs linked to the implementation of ecologically sustainable technology, including renewable energy systems and contemporary waste management equipment. While the utilization of these technologies has the capacity to yield long-term economic advantages, the initial financial investment might present a significant barrier for many enterprises (Farooq et al. 2022; Sauve and Acker 2020). Another technological challenge is in the inadequacy of suitable technology for specific applications. For example, there is a wide array of technologies available for managing widely encountered waste kinds. However, the options for handling more specialized waste streams, such as hazardous waste or electronic waste, may be quite restricted (Farooq et al. 2022; Sauve and Acker 2020). Inevitably, there may arise challenges pertaining to the maintenance and operation of eco-friendly technologies. The effective functioning of these technologies may require individuals to possess particular knowledge and abilities, which may not be easily available in all organizations (Farooq et al. 2022; Sauve and Acker 2020).

3.1 Opportunities for Green Technology in Sports

Despite the inherent restrictions, the domain of sports presents enough opportunities for the effective integration and implementation of ecologically sustainable technologies. The increasing awareness and understanding of environmental issues have prompted a growing demand from many stakeholders, such as supporters, athletes, and other individuals engaged in sports, for sports organizations to proactively tackle sustainability matters. This occurrence creates a substantial incentive for enterprises to invest resources toward the adoption and deployment of environmentally sustainable technologies (Zhang et al. 2021). Moreover, the ongoing advancements in technology are augmenting the feasibility and cost-effectiveness of incorporating eco-friendly solutions into sports facilities. To provide an example, there has been a notable decrease in the price of solar panels in recent times, leading to improved cost-effectiveness for stadiums and other establishments to autonomously generate renewable energy (Zhang et al. 2021). Furthermore, the use of ecologically sustainable technologies might provide a competitive advantage for sports companies. Organizations that demonstrate a persistent commitment to sustainability possess the capacity to enhance their reputation and attract backing from sponsors, partners, and individuals who prioritize the responsible management of the environment (Padhy and Achary 2022).

3.2 Innovation in Waste Management

The implementation of innovative waste management procedures has provided sports organizations with new opportunities to reduce their environmental impact. A significant advancement in this particular domain pertains to the shift toward a circular economy, which is distinguished by the minimization of waste and the optimization of resource reutilization and recycling (Tanveer et al. 2022; Padhy and Achary 2022; Yu et al. 2022). Numerous sports organizations are presently employing innovative waste management strategies, such as the adoption of composting methods for organic waste, the conversion of waste into energy, and the incorporation of recyclable materials for packaging food and beverages (Tanveer et al. 2022; Padhy and Achary 2022; Yu et al. 2022). Moreover, the application of digital technologies is being utilized to boost waste management procedures. An example of this can be seen in the implementation of smart waste bins, which are equipped with the ability to independently classify and compact garbage. Furthermore, the utilization of digital platforms can be employed for the goal of waste monitoring and reporting, as evidenced by multiple studies (Tanveer et al. 2022; Padhy and Achary 2022; Yu et al. 2022).

4 Case Studies of Green Technology and Waste Management in Different Regions

4.1 Case Study: Green Technology Development in Kazakhstan

Kazakhstan, a nation abundant in natural resources, has been making significant progress in the advancement and adoption of environmentally friendly technologies. The significance of sustainable development has been acknowledged by the government, leading to investments in green technologies with the aim of diversifying the economy and mitigating the country's environmental footprint (Hoareau et al. 2021; Abdildin et al. 2021). The sports sector has witnessed the implementation of green technology in various domains. The nation has made significant investments in the development of energy-efficient sporting facilities, the utilization of renewable energy sources, and the implementation of efficient waste management systems. These initiatives have not only resulted in a decrease in the environmental effect of sports events but have also played a role in fostering the growth of the green technology industry inside the country (Abdildin et al. 2021), as shown in Fig. 1.

Furthermore, based on the case study, the emergence of environmentally friendly technology in the field of sports has been a noteworthy advancement in recent times,

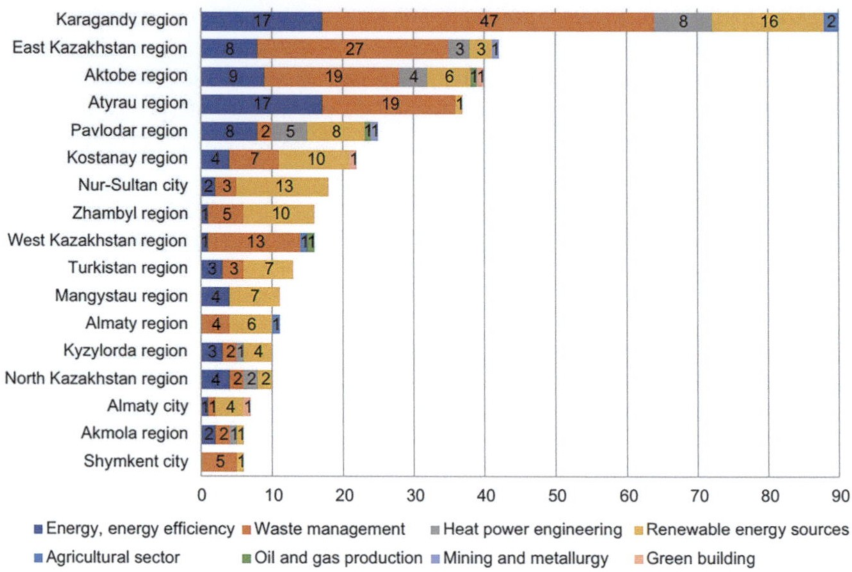

Fig. 1 The utilization of green technologies across the 17 administrative-territorial entities of Kazakhstan in various areas of the economy (Abdildin et al. 2021)

characterized by a growing focus on the principles of sustainability and environmental awareness. Nevertheless, when examining firms from various sectors and areas, it becomes evident that there is a limited level of awareness and interest in adopting environmentally friendly technologies (Yu et al. 2022; Baharuddin et al. 2022, 2023; Ramli et al. 1060; Sufian et al. 2021; Hamzah et al. 2022; Ghavam and Vahidi 2022; Babu et al. 2023; Atalla et al. 2022; Farag et al. 2019; Hoareau et al. 2021; Nasir et al. 2013; Dinglasan et al. 2021; Jamaludin et al. 2018b; Abdildin et al. 2021). The absence of enthusiasm can be ascribed to various factors:

(a) The market for green technologies is inadequately established.
(b) The lack of a universally accepted approach, consistent language, and comprehensive classification system for green technologies is evident.
(c) There is a lack of comprehensive statistical data available at both national and regional levels about the use of green technologies.
(d) There is a lack of comprehensive incentive support measures available to enterprises undergoing the shift toward adopting environmentally sustainable technologies.
(e) Enterprises encounter difficulties in quantifying the financial advantages resulting from the use of environmentally friendly technologies, which gives rise to doubt regarding their concrete economic implications.
(f) There exists a significant deficiency of adequately skilled individuals in the field of green technology, encompassing professionals such as ecologists, economists, engineers, and technicians.
(g) The market for green equipment services and supply is currently in an underdeveloped state.
(h) The lack of a cohesive platform for many parties within the green technology market, encompassing investors, technology providers, enterprises, and both public and private entities.

4.2 Case Study: Waste Management in Bandung Green and Clean Program

The Bandung Green and Clean Program (BGC) is an environmental project implemented in the city of Bandung, Indonesia, with the primary objective of enhancing waste management practices and fostering the principles of environmental sustainability. The initiative encompasses a range of stakeholders, including the local government, community groups, and corporations (Hamzah et al. 2022). One of the fundamental elements of the program entails the advocacy for trash reduction and recycling within sporting facilities. This includes the execution of recycling initiatives, the utilization of materials that can be recycled, and the encouragement of waste reduction among individuals attending the event. The program has effectively achieved its objective of decreasing waste production during sports events and has served as a commendable model for other cities in the surrounding area (Hamzah et al. 2022). The BGC program, which has been in existence for a duration of four years, serves as a testament to the possibilities of waste management within the realm of sports. The success of the subject under consideration is assessed using five fundamental criteria, namely relevance, effectiveness, impact, efficiency, and sustainability, as shown in Table 1.

The success of the BGC program offers significant insights on the integration of green technology in the realm of sports. Waste management encompasses more than just waste reduction, as it encompasses the cultivation of community engagement, the assurance of economic viability, and the promotion of sustainability. Sports stadiums and facilities have the potential to derive inspiration from the BGC program, thereby

Table 1 Key aspects of the BGC program and their significant to sustainable stadium (Hamzah et al. 2022)

Aspect	Description	Significant to sustainable stadium
Relevance	• Alignment with objectives and policies • Collaborative design with six partners • High community participation in planning • Activities in line with objectives and targets	A relevant program ensures that the strategies and activities of a sports stadium align with its sustainability goals. It emphasizes the importance of collaboration and community involvement, ensuring that the stadium's initiatives resonate with its audience
Effectiveness	• Change in community behavior toward waste management • Increased waste segregation • On track to achieve waste reduction objectives	An effective stadium waste management program will lead to tangible results, such as reduced waste and increased recycling. This ensures that the stadium's operations have a minimal environmental footprint and promote sustainable practices
Impact	• Social: environmental awareness and collaboration • Economic: benefits from composting, recycling, and trash banks • Environmental: cleaner and greener neighborhoods	The impact of a program determines the broader benefits a sustainable stadium brings to its community. Beyond the confines of the stadium, it can foster environmental consciousness, boost local economies, and enhance the quality of life in surrounding areas
Efficiency	• Cost-effectiveness • Expansion despite budget constraints	Efficiency in a stadium's sustainability initiatives ensures that even with limited resources, significant positive changes can be achieved. It underscores the importance of strategic planning and resource allocation in stadium operations
Sustainability	• Continuation of positive changes post-program • Dependency on local stakeholder support • Community funding	Sustainability ensures that the green initiatives of a stadium are long-lasting and not just short-term projects. A sustainable stadium will have systems in place that ensure its operations remain eco-friendly, even as external factors change

modifying their approaches to develop environmentally friendly and sustainable venues.

4.3 Case Study: Waste Management in Kerman Graduate University of Technology

The Kerman Graduate University of Technology in Iran has successfully built a comprehensive waste management system with the aim of mitigating the environmental consequences associated with its operational activities. The system encompasses the practice of waste segregation at its origin, the recycling of products that are capable of being recycled, and the composting of organic waste (Ghavam and Vahidi 2022). The institution has included sustainable technology within its infrastructure, such as the utilization of energy-efficient lighting and heating systems, as well as the integration of solar panels to generate electricity. The implementation of these efforts has not only resulted in a decrease in the university's environmental footprint but has also served as a tangible demonstration of the successful integration of sustainable technology and efficient waste management practices within an educational setting (Ghavam and Vahidi 2022).

Stadiums, being expansive public facilities, possess a substantial ecological impact. The aspect of sustainability extends beyond the selection of construction materials and the energy efficiency of lighting systems. The waste management strategy of a stadium is a vital element in ensuring its sustainability. The magnitude of waste produced by a large number of attendees at a singular occasion can be considerable, and the manner in which this waste is handled can have a significant influence on the natural surroundings. In order to gain insight into the waste composition that a stadium may encounter, it is instructive to examine a smaller scale case study, such as Kerman Graduate University of Technology. The waste profile of a university may exhibit variations compared to that of a stadium, particularly during significant gatherings, as shown in Table 2 (Ghavam and Vahidi 2022).

The data presented in the table indicates that organic waste constitutes a substantial proportion of the overall garbage, with nylon, cardboard, and glass following suit

Table 2 Physical analysis of Kerman Graduate University of Technology waste (Ghavam and Vahidi 2022)

Waste type	%
Organic	29.50
Paper and cardboard	22.40
Plastics (including nylon and PET)	33.10
Metals	1.70
Glass	10.70
Textiles	2.70
Total	100

in descending order. This data has the ability to provide stadium management with valuable insights into the waste streams that may arise, particularly during events featuring food kiosks and merchandise sales (Wang 2020; Abdullah et al. 2022; Rosli et al. 2021; Md Kamil et al. 2022; Yu et al. 2022; Baharuddin et al. 2022; Ramli et al. 1060; Sufian et al. 2021; Hamzah et al. 2022; Ghavam and Vahidi 2022). In order to achieve a stadium that is environmentally sustainable, it is vital to possess a comprehensive comprehension of this breakdown, such as the following:

- The composting of organic waste offers a viable solution to mitigate the need for landfill capacity while simultaneously generating valuable compost suitable for landscaping or local agricultural use.
- The recycling of nylon and plastics has the potential to decrease the need for new materials and the energy consumption associated with their production.
- Efficient recycling of glass and metals, when segregated at the point of origin, can effectively contribute to the red.
- Efficient recycling of glass and metals, when segregated at the point of origin, can effectively contribute to the reduction of the stadium's environmental impact.

5 Research Gaps and Future Research Directions

Despite the notable progress made in the fields of green technology and waste management in the context of sports, there remain several areas of inquiry that have not received sufficient attention. One of the main shortcomings pertains to the lack of extensive scholarly investigation about the implementation of environmentally sustainable technology in sports venues, specifically in relation to the management of waste. Despite the existence of several studies on different aspects of green technology and waste management, there is a lack of comprehensive study that encompasses all dimensions of these subjects. To adequately address the current deficiencies in research, it is crucial to propose a range of prospective solutions and delineate areas for future investigation. An area that warrants additional investigation is to the conduct of comprehensive research on the utilization of environmentally friendly technologies and waste management strategies in sports facilities. These studies possess the capacity to examine the technological, economic, social, and regulatory aspects of these issues, therefore providing valuable insights for sports organizations, legislators, and other pertinent stakeholders. A further suggested action involves conducting additional research on the sociological and behavioral aspects of waste management in the context of sports. This may involve conducting surveys or interviews with diverse individuals, including spectators, athletes, staff members, and other pertinent stakeholders. The aim of conducting these surveys and interviews is to gain a deeper understanding of individuals' viewpoints on waste and sustainability and to identify possible strategies for promoting behavioral changes (Khan et al. 2021; Lee et al. 2021). There is an urgent need for additional research to explore the impact of policy and regulation on the promotion and development of environmentally sustainable technology and waste management techniques in the field of sports. This may involve doing comparative studies on different policy and regulatory strategies and evaluating their impact on the adoption of environmentally sustainable technologies and waste management practices (Murthy and Ramakrishna 2022; Ainur Munira Rosli 2021).

Acknowledgements The authors are very grateful to the Ministry of Higher Education (MoHE) Malaysia and the Universiti Malaysia Pahang Al-Sultan Abdullah for their great financial support, which made this study possible. This study was made possible by grants from the Fundamental Research Scheme (FGRS) FRGS/1/2022/TK10/UMP/02/57 and the Universiti Malaysia Pahang Al-Sultan Abdullah Fundamental Research Grant RDU220312.

References

Abdildin YG, Nurkenov SA, Kerimray A (2021) Analysis of green technology development in Kazakhstan. Int J Energy Econ Policy 11:269–279

Abdullah NA, Abdullah FF, Sufian AH, Abidin AN, Jamaludin AS, Razali MN (2022) Effect of degradation by temperature onto nitrile rubber elastomer mechanical properties. Mater Today: Proc 48:1941–1946

Abidin AN, Azmi AH, Kassim KA, Jamaludin AS, Razali MN (2022) A review on automotive tires significant characteristic identification for general consumers. In: Proceedings of the 2nd energy security and chemical engineering congress, pp 375–385

Ali MA, Sivarao, Abdullah Z, Izamshah R, Kassim MS, Jamaludin AS (2022) Multi-response optimization of machining simulation approach using grey relational analysis (Lecture notes in mechanical engineering), pp 154–157

Alsibai MH, Hamran NNN, Nasir A, Abdullah AA (2018) A monitoring system for EPW safe use. J Telecommun, Electron Comput Eng (JTEC) 10(1–2):131–135

Atalla Y, Hashim Y, Ghafar ANA, Jabbar WA (2020) Temperature characteristics of finfet based on channel fin width and working voltage. Int J Electr Comput Eng (IJECE) 10:5650

Atalla Y, Hashim Y, Ghafar ANA (2022) The impact of channel fin width on electrical characteristics of Si-FinFET. Int J Electr Comput Eng (IJECE) 12:201

Babu D, Nasir A, Jamaludin AS, Rosle MH (2021) Holding, grasping and sensing of prosthetic robot arm like a real human hand, a journey beyond limits: an extensive review. In: Human-centered technology for a better tomorrow, pp 485–504

Babu D, Nasir A, Farag M, Sidik MH, Rejab SB (2022) Development of prosthetic robotic arm with patient monitoring system for disabled children; preliminary results. In: 2022 9th International conference on electrical and electronics engineering (ICEEE)

Babu D, Nasir A, Ravindran, Farag M, Jabbar WA (2023) 3D printed prosthetic robot arm with grasping detection system for children. Int J Adv Sci, Eng Inf Technol 13:226

Baharuddin NJ, Abdul Manaf AR, Jamaludin AS (2022) Study on die shoulder patterning method (DSPM) to minimise springback of U-bending. Int J Automot Mech Eng 19:9509–9518

Baharuddin NJ, Manaf ARA, Jamaludin AS (2023) Study of springback behavior on U-bending part using die shoulder patterning method (DSPM). AIP Conf Proc 2544(1):040020

Binti Kamarudin SN, Lee H, Ishak I, Manaf AR, Jamaludin AS, Shaharudin MA, Zawawi MZ (2021) Rapid direct continuous method for hot embossing of glass microlens array combined with CO_2 laser irradiation and external preheating/cooling (Lecture notes in mechanical engineering), pp 669–675

Dinglasan JJ, Luyon JP, Abiog II DE (2021) DLSU-D green innovations on waste management: adoption of greener lifestyles in new normal. J Sustain Perspect 1

Farag M, Azlan NZ, Alsibai MH, Ghafar AN (2019) Slippage detection for grasping force control of robotic hand using force sensing resistors. In: Proceedings of the 2019 5th international conference on computer and technology applications

Farooq M, Cheng J, Khan NU, Saufi RA, Kanwal N, Bazkiaei HA (2022) Sustainable waste management companies with innovative smart solutions: a systematic review and conceptual model. Sustainability 14:13146

Ghavam A, Vahidi H (2022) Waste management to improve the condition in Kerman Graduate University of Technology. J Adv Environ Health Res 10:25–38

Hamzah AH, Sohifah S, Nurhasanah N (2022) The implementation of waste management in Bandung Green and clean program based on multistakeholder. J-MAS (Jurnal Manajemen dan Sains) 7:97

Hasya N, Razali MNM, Jamaludin AS, Saffe SNBM, Hadi MA (2022) Micromilling performance analysis—cold air assisted aluminium machining behaviour observation. In: Innovative manufacturing, mechatronics and materials forum 2022 (iM3F 2022), Online conference, Pekan, Malaysia, pp 124–130

Hoareau CE, Ahmad N, Nuid M, Rubiyatno, Khoi DN, Kristanti RA (2021) Sustainable technology in developed countries: waste municipal management. Ind Domest Waste Manag 1:48–55

Jamaludin AS, Yassin A (2013) Analysis of laser sintered materials using finite element method. Sains Malaysiana 42(12):1727–1733

Jamaludin AS, Hosokawa A, Furumoto T, Koyano T, Hashimoto Y (2017) Evaluation of the minimum quantity lubrication in orthogonal cutting with the application of finite element method. Int J Mech Mechatron Eng IJMME-IJENS 17(01):104–109

Jamaludin AS, Hosokawa A, Furumoto T, Koyano T, Hashimoto Y (2018) Study on the effectiveness of extreme cold mist MQL system on turning process of stainless steel Aisi 316. IOP Conf Ser: Mater Sci Eng 319:012054

Jamaludin AS, Akira H, Furumoto T, Koyano T, Hashimoto Y (2018) High precision estimation on physical behavior for cutting with various tool rake angle by finite element method (Lecture notes in mechanical engineering), pp 715–723

Khan SA, Godil DI, Quddoos MU, Yu Z, Akhtar MH, Liang Z (2021) Investigating the nexus between energy, economic growth, and environmental quality: a road map for the sustainable development. Sustain Dev 29:835–846

Lee H, Binti Kamarudin SN, Ishak I, Manaf AR, Jamaludin AS, Shaharudin MA, Zawawi MZ (2021) Feasibility study of wafer scale laser assisted thermal imprinting of glass nanostructures (Lecture notes in mechanical engineering), pp 917–923

Md Kamil NH, Jamaludin AS, Mhd Razali MN, Ghaffar ANA (2022) Temperature and heat flow analysis in a drying chamber through finite element method (Lecture notes in mechanical engineering), pp 309–316

Murthy V, Ramakrishna S (2022) A review on global E-waste management: urban mining towards a sustainable future and circular economy. Sustainability 14:647

Nasir A, Akagi T, Dohta S, Ono A, Masago Y (2013) Development of low-cost wearable servo valve using buckled tube driven by servo motor. Appl Mech Mater 393:532–537

Nasir A, Akagi T, Dohta S, Ono A (2015) Analysis of low-cost wearable servo valve using buckled tubes for optimal arrangement of tubes. In: 2015 IEEE international conference on advanced intelligent mechatronics (AIM)

Negrete-Cardoso M, Rosano-Ortega G, Álvarez-Aros EL, Tavera-Cortés ME, Vega-Lebrún CA, Sánchez-Ruíz FJ (2021) Circular economy strategy and waste management: a bibliometric analysis in its contribution to sustainable development, towards a post covid-19 era

Padhy M, Achary PG (2022) Polymer composites in green technology: a review. Int J Comput Commun Technol 8–27

Ramli MR, Razak NA, Ismail I, Jamaludin AS, Manaf AR (2022) Effect of dimple size onto wear rate of mild steel AISI 1060 surface (Lecture notes in mechanical engineering), pp 99–102

Rosli AM, Jamaludin AS, Razali MN, Sani AS, Hamzah SB, Osman MS (2019) Modelling of fuzzy inference system for micro milling—a preliminary study through FEM (Lecture Notes in Mechanical Engineering), pp 445–456

Rosli AM, Jamaludin AS, Razali MNM (2021) Recent study on hard to machine material—micromilling process. Evergreen 8:445–453

Rosli AM, Jamil N, Jamaludin AS, Razali MN, Yusoff AR (2021) Tool wear observation during unconventional low speed machining using low cost micromilling (Lecture notes in mechanical engineering), pp 589–597

Sauve G, Van Acker K (2020) The environmental impacts of municipal solid waste landfills in Europe: a life cycle assessment of proper reference cases to support decision making. J Environ Manage 261:110216

Sufian AH, Xun TZ, Abidin AN, Jamaludin AS, Razali MN (2021) Study on tire tread design effect onto tire-road contact behavior through FEM (Lecture notes in mechanical engineering), pp 893–902

Tanveer M, Khan SA, Umar M, Yu Z, Sajid MJ, Haq IU (2022) Waste management and green technology: future trends in circular economy leading towards environmental sustainability. Environ Sci Pollut Res 29:80161–80178

Wang, L.: The design of the membrane structure of sports stadiums has a saving effect on water resources. Journal of Coastal Research. 104, (2020).

Yu Z, Waqas M, Tabish M, Tanveer M, Haq IU, Khan SA (2022) Sustainable supply chain management and green technologies: a bibliometric review of literature. Environ Sci Pollut Res 29:58454–58470

Zhang A, Venkatesh VG, Wang JX, Mani V, Wan M, Qu T (2021) Drivers of industry 4.0-enabled smart waste management in supply chain operations: a circular economy perspective in China. Prod Plan Control 34:870–886

Harnessing Solar Energy: A Green Technology Approach in Sports Facility Management

Devin Babu, Syukran Hakim, Ainur Munira Rosli, Mohamad Rusydi Mohamad Yasin, and Muhammad Hisyam Rosle

Abstract The present review study examines the growing importance of solar energy within the realm of sports facility management. In light of the worldwide focus on environmental sustainability, solar energy has emerged as a crucial renewable resource, particularly due to the widespread use of photovoltaic (PV) systems. The integration of solar technology in sports facilities can yield significant advantages, such as decreased energy expenses and a diminished environmental impact. This is particularly relevant due to the substantial energy requirements associated with these facilities. This change aligns with the wider cultural trend toward sustainability, improving the reputation of sporting facilities and gaining support from stakeholders that prioritize environmental concerns. Nevertheless, the process of integration necessitates meticulous planning and design, taking into account the initial capital outlay and guaranteeing the attainment of maximum solar absorption. The evaluation additionally emphasizes the significance of digital technology in the administration of green energy and the wider utilization of solar energy in many areas such as building, industrial transport, and urban planning. Case studies offer valuable insights into the effective integration of environmentally friendly technologies within sporting facilities. The study finishes by highlighting areas of research that require further investigation and outlining potential future approaches. It emphasizes the necessity of conducting targeted studies on the utilization of solar energy in sports facilities, exploring its compatibility with electric vehicles, examining the possibilities of green hydrogen, and exploring the convergence of solar energy with digital technology.

D. Babu · S. Hakim · A. M. Rosli (✉) · M. R. M. Yasin
Faculty of Electrical and Electronic Engineering Technology, Universiti Malaysia Pahang Al-Sultan Abdullah, 26600 Pekan, Pahang, Malaysia
e-mail: ainurmunira@gmail.com

D. Babu · S. Hakim · A. M. Rosli · M. R. M. Yasin · M. H. Rosle
Faculty of Manufacturing and Mechatronic Engineering Technology, Universiti Malaysia Pahang Al-Sultan Abdullah, 26600 Pekan, Pahang, Malaysia

M. H. Rosle
Faculty of Electrical Engineering Technology, Universiti Malaysia Perlis, 02600 Arau, Perlis, Malaysia

Z. Mohamed et al. (eds.), *Proceedings of the 2nd International Conference on Innovation and Technology in Sports, ICITS 2023, 27–28 November, Malaysia*, Lecture Notes in Bioengineering, https://doi.org/10.1007/978-981-97-3741-3_33

Keywords Solar energy · Sports facility management · Green technology · Digital technology · Green hydrogen

1 Introduction to the Importance of Solar Energy in Sports Facility Management

The global focus on environmental sustainability has expedited the shift toward renewable energy sources (Ekström et al. 2021; Dikaiakos et al. 2023; Farag et al. 2019; Dadzie and Djokoto 2022; Lee et al. 2021; Abdullah et al. 2022; Babu et al. 2023; Hasya et al. 2022; Wrana et al. 2022; Alsibai et al. 2018; Binti Kamarudin et al. 2021; Jamaludin et al. 2018a; Mohommad and Kim 2022; Ramli et al. 1060; Md Kamil et al. 2022; Chen and Kuo 2022). Solar energy is widely recognized as a prominent option due to its attributes of being a sustainable, environmentally friendly, and plentiful form of power (Mohommad and Kim 2022; Ramli et al. 1060; Md Kamil et al. 2022; Chen and Kuo 2022; Babu et al. 2022, 2021; Ali et al. 2022; Buturache and Stancu 2021; Borysiak and Brych 2021; Abidin et al. 2022; Nasir et al. 2015; Rosli 2021; Baharuddin et al. 2023; Ibrahim et al. 2023). The aforementioned transition is observable through the growing utilization of photovoltaic (PV) systems, which possess the capacity to substantially diminish reliance on non-renewable energy sources and mitigate the release of greenhouse gases (Dadzie and Djokoto 2022; Ibrahim et al. 2023). The incorporation of solar energy and other sustainable technology holds significant importance within the domain of sports facility management (Ibrahim et al. 2023; Rosli et al. 2019, 2021; Nasir et al. 2013; Xu et al. 2023; Tseng et al. 2022; Atalla et al. 2020; Jamaludin and Yassin 2013; Pedrazzi et al. 2022; Sufian et al. 2021; Baharuddin et al. 2022; Jamaludin et al. 2018b; Dik et al. 2022). Sports facilities, encompassing stadiums and training centers, commonly exhibit considerable energy demands for essential functions such as illumination, heating, and air conditioning (Dik et al. 2022; Atalla et al. 2022; Jamaludin et al. 2017; Wei et al. 2020; Wahid et al. 2022; Ghalambaz et al. 2023). By implementing solar energy solutions, these establishments have the potential to significantly reduce their energy costs and environmental impact (Ekström et al. 2021; Mohommad and Kim 2022). In addition to the physical advantages, the incorporation of solar energy in sports facilities aligns with the wider societal trend toward sustainability. These activities not only provide a positive impact on environmental conservation but also enhance the reputation of the sports facility and its associated businesses (Mohommad and Kim 2022). The adoption of a proactive stance toward sustainability and societal responsibility has the potential to attract support from environmentally conscious stakeholders and the general public (Ekström et al. 2021; Mohommad and Kim 2022).

Nevertheless, the process of incorporating solar energy into sporting facilities is not without challenges. The successful implementation of photovoltaic (PV) systems necessitates careful and thorough planning and design (Xu et al. 2023). This is crucial in order to strategically position the PV systems to maximize solar absorption while also ensuring that the essential functions of the facility are not compromised. Although the initial financial commitment may be significant, the potential for long-term cost savings on energy expenditures and the possibility of generating additional income by selling excess energy to the power grid render this undertaking a valuable pursuit (Dadzie and Djokoto 2022; Dik et al. 2022). In addition, the management framework for facilitating the adoption of environmentally friendly energy services is undergoing transformation due to the emergence of digital technology. This technological advancement has the potential to harmonize the processes of generating, distributing, and utilizing green energy (Borysiak and Brych 2021). The utilization of solar energy transcends the realm of sporting facilities. An illustrative example may be observed in the conversion of data centers into environmentally sustainable hubs through the utilization of solar energy, which has demonstrated notable benefits in terms of both ecological preservation and economic viability (Dikaiakos et al. 2023). Moreover, the emergence of electric vehicles, which are capable of integrating with the power grid, offers the potential for mutually beneficial collaborations between the transportation and renewable energy industries (Dik et al. 2022).

The purpose of this review is to provide a comprehensive overview of the existing research on solar energy in the context of managing sports facilities. This paper aims to explore the advantages and difficulties associated with the integration of solar energy, as well as the significant contribution of digital technology in promoting environmentally friendly energy services. Additionally, it will examine the various applications of solar energy in different sectors and discuss the potential future developments in this field, such as the interaction between electric vehicles and the employment opportunities within the green energy industry.

2 The Role of Green Technology in Sports Facilities

2.1 Green Technology and Its Impact on Sports Facilities

Green technology, also known as clean technology, refers to the utilization of scientific and technological progress to create products and services that are environmentally sustainable. Within the domain of sports facilities, the incorporation of green technology assumes a crucial function in reducing energy usage, mitigating carbon emissions, and enhancing overall ecological sustainability (Ekström et al. 2021; Dadzie and Djokoto 2022).

There exists a wide array of green technology options that may be implemented in sports facilities, encompassing various aspects such as energy-efficient lighting, sophisticated waste management methods, and renewable energy systems. Among these options, solar energy systems, in particular, have been identified as very influential and capable of bringing about significant changes (Ekström et al. 2021; Xu et al. 2023). The utilization of photovoltaic (PV) systems for capturing solar energy not only diminishes reliance on fossil fuels but also significantly mitigates the release of

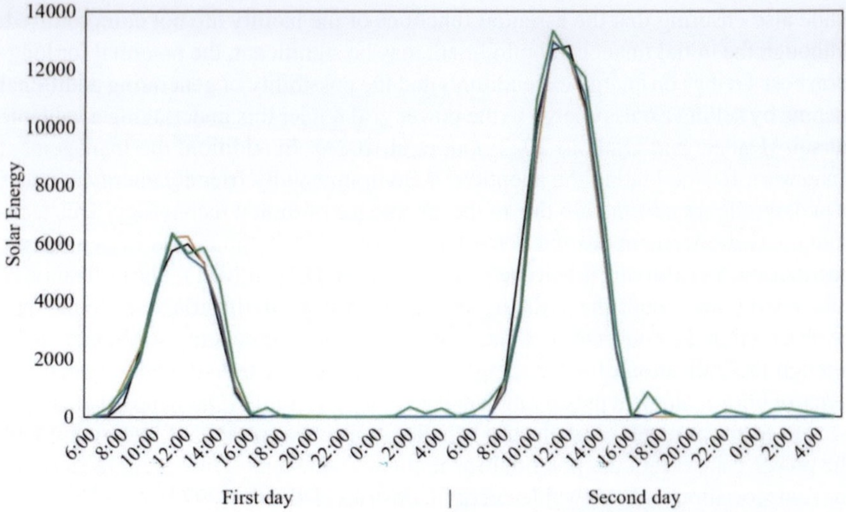

Fig. 1 Forecasted excessive solar energy in just 48 h by a photovoltaic system (Buturache and Stancu 2021)

greenhouse gas emissions (Dadzie and Djokoto 2022; Ibrahim et al. 2023). In addition, it is worth noting that solar systems frequently generate an excess of energy, which can be redirected to the electrical grid. This surplus energy serves as a supplementary source of income for the facility, as shown in Fig. 1 (Ekström et al. 2021; Buturache and Stancu 2021).

The use of environmentally friendly technology in sports facilities, such as the conversion of medium-sized data centers into sustainable hubs through the utilization of solar energy, has exhibited notable benefits in terms of both the environment and finances (Dikaiakos et al. 2023). Furthermore, the investigation of innovative energy solutions, such as the creation of green hydrogen for industrial transportation, highlights the potential of renewable energy in several industries (Pedrazzi et al. 2022). The impact of green technology on sports facilities extends beyond environmental and economic benefits. Additionally, it promotes social advantages. For example, the implementation of sustainable practices contributes to the reputation of the sports facility and its associated entities, aligning with the principles of environmentally aware stakeholders (Ekström et al. 2021; Chen and Kuo 2022). Implementing proactive actions not only enhances the reputation but also aligns with the prevailing cultural trend toward sustainability.

2.2 *The Integration of Solar Energy in Sports Facilities*

The utilization of solar energy in sports facilities generally involves the implementation of photovoltaic (PV) systems, which are highly proficient in converting sunlight into electricity that can be utilized. These systems consist of several components, such as solar panels, inverters, and, in more modern configurations, energy storage solutions like batteries, which enable the consumption of energy even when sunlight is not available (Dadzie and Djokoto 2022; Ibrahim et al. 2023). Furthermore, establishing a smooth and uninterrupted integration with the electrical grid is of utmost importance (Dadzie and Djokoto 2022; Dik et al. 2022).

The strategic approach is necessary for the thorough design and installation of photovoltaic (PV) systems in sports facilities. This entails the task of aligning the solar panels in an optimal orientation to effectively catch the maximum amount of sunshine. Additionally, it requires ensuring that the installation space is predominantly free from shading and being able to meet the electrical requirements of the facility (Farag et al. 2019; Buturache and Stancu 2021). The optimization of energy capture is heavily influenced by various factors, including the tilt and orientation of solar panels. Following a successful installation, photovoltaic (PV) systems have the capacity to meet a significant portion of a facility's electrical needs. This reduces reliance on the grid and thus reduces energy costs (Farag et al. 2019; Xu et al. 2023). The incorporation of energy storage technologies, such as batteries, enhances the overall efficiency of the system by storing excess solar energy, as shown in Fig. 2. This stored energy can then be utilized during periods of reduced sunlight or increased electricity demand (Farag et al. 2019; Ibrahim et al. 2023).

In addition to the practical advantages of energy generation and cost reduction, the integration of solar energy in sports facilities also fulfills an educational objective. Real-time displays that exhibit energy generation and use have the potential to play a crucial role in educating facility consumers and visitors. These activities not only highlight the benefits of solar energy but also emphasize the wider discourse on

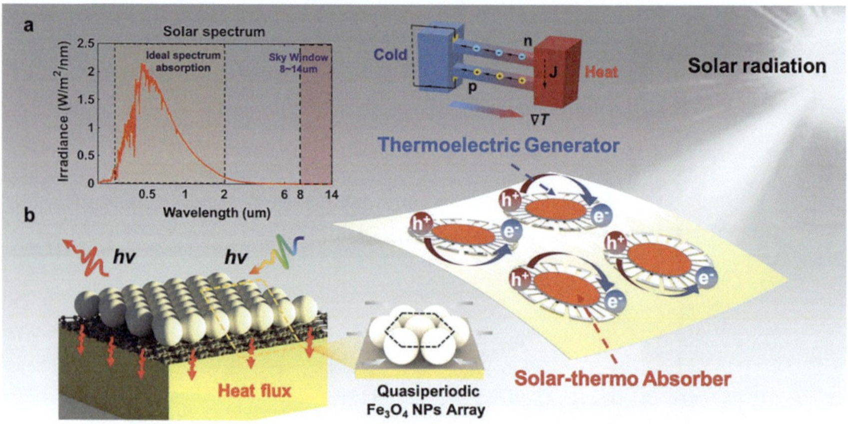

Fig. 2 The quasiperiodic scalable selective solar absorber converts low-grade sun energy into thermoelectric energy. Optimal solar absorption occurs between 0.3 and 2 μm, from ultraviolet to near-infrared (Xu et al. 2023)

energy preservation and environmental sustainability (Farag et al. 2019; Chen and Kuo 2022).

2.3 *Case Studies of Green Technology Implementation in Sports Facilities*

The use of environmentally friendly technology in sports facilities has been demonstrated through multiple case studies, highlighting its potential advantages and feasible implementations. A noteworthy study was undertaken on a university campus, utilizing a sophisticated system that integrated cloud administration with a wireless sensor network to effectively monitor real-time green energy (Tseng et al. 2022). This novel methodology enabled the monitoring and control of energy usage in various campus facilities, including sports amenities, in real time. The result demonstrated a significant decrease in energy usage, highlighting the effectiveness of these technologies (Tseng et al. 2022).

In an alternative setting, a research study showcased the utilization of hydrogen-powered forklifts operating on green energy sources within a fruit and vegetable market located in Italy (Pedrazzi et al. 2022). Although this particular environment deviates from the conventional sports facility, the knowledge and understanding gained from it are of great significance. The study emphasized the benefits of employing green hydrogen, produced by a photovoltaic system with a capacity of 454.3 kWp, for the purpose of energizing forklifts, as shown in Fig. 3. The implementation of this sustainable energy solution resulted in a notable reduction in carbon dioxide emissions and substantial cost savings in energy compared to traditional forklifts that rely on fossil fuels (Pedrazzi et al. 2022).

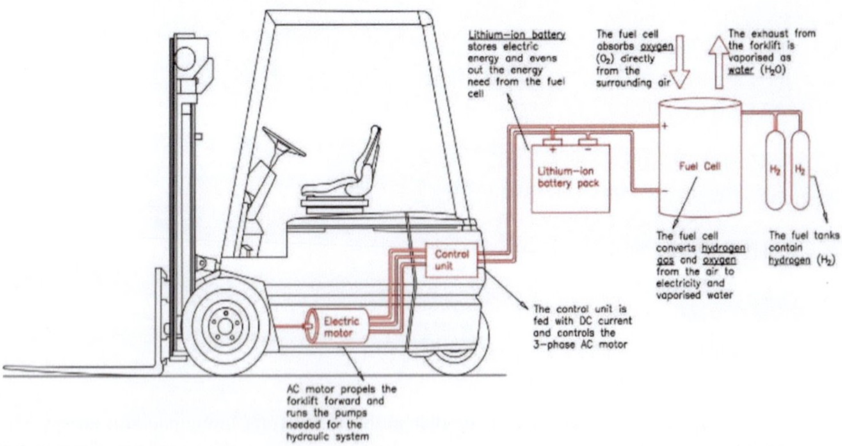

Fig. 3 Fuel cell forklift scheme (Pedrazzi et al. 2022)

Additionally, the research conducted on the Italian market placed significant emphasis on the potential of fuel cells as a highly promising technology for the production of clean energy. The effectiveness and sustainability of photovoltaic systems were demonstrated by the observation that each sub-installation has the capacity to generate enough green hydrogen to power a forklift for an entire year (Pedrazzi et al. 2022). The presented case studies provide evidence of the feasibility and benefits of integrating green technology into sports facilities. The aforementioned sources provide significant precedents and valuable insights for sports facility directors who are considering the incorporation of sustainable technologies (Mohommad and Kim 2022; Tseng et al. 2022; Pedrazzi et al. 2022).

3 The Application of Solar Energy in Various Sectors

3.1 Solar Energy in Building and Construction

The utilization of solar energy within the building and construction industry is experiencing significant growth due to the worldwide focus on sustainable and energy-efficient architectural practices. Considering the significant contribution of buildings to global energy consumption, the integration of solar energy systems can effectively alleviate this environmental impact (Xu et al. 2023). Commonly, these systems involve the placement of photovoltaic (PV) panels on the roofs or exteriors of buildings, which enables the direct conversion of sunlight into electrical energy. The electricity generated can be used immediately, stored in batteries for future use, or fed back into the grid. In addition to photovoltaic (PV) systems, solar-thermal technology presents an alternative approach for the utilization of solar energy. These systems demonstrate proficiency in harnessing solar energy to provide water heating, serving both residential needs and wider uses in building heating (Xu et al. 2023). Solar-thermal systems have exhibited efficient energy conversion, indicating its promise in future solar co-harvesting systems (Xu et al. 2023).

The incorporation of solar energy into buildings not only reduces energy costs and carbon emissions but also promotes energy self-sufficiency and has the potential to increase property values. In addition, the integration of solar energy is in perfect harmony with renowned green construction standards and certifications such as LEED and BREEAM. The alignment of the building not only serves to decrease its environmental impact but also enhances the reputation of the structure and its occupants (Binti Kamarudin et al. 2021). Furthermore, a prior investigation on Solar Thermal Energy Storage (STES) has revealed a notable increase in scholarly inquiry within this field, with prominent nations such as China, the United States, India, Spain, and Germany at the forefront (Ghalambaz et al. 2023). The increase in solar energy utilization in the building sector highlights the worldwide dedication to harnessing this renewable energy source, hence stressing its significance in modern construction techniques.

3.2 Solar Energy in Industrial Transport

The utilization of solar energy in the realm of industrial transportation is a burgeoning and rapidly growing field. This involves the utilization of solar energy for either the purpose of charging electric vehicles or producing environmentally friendly hydrogen fuel for automobiles driven by fuel cells. It is worth mentioning that electric vehicles, when combined with renewable energy sources, have the capacity to function as a means of transportation as well as a versatile energy carrier. This enables them to establish connections with the power grid, buildings, and various other systems (Dik et al. 2022).

The feasibility of utilizing solar-powered green hydrogen in industrial transportation is emphasized in a persuasive case study conducted by earlier researchers (Pedrazzi et al. 2022). The focus of this study was on a fruit and vegetable market in Italy, where forklifts were powered by green hydrogen. This green hydrogen was produced by an electrolysis process driven by solar energy. The results of the study indicated that this particular strategy significantly reduced carbon dioxide emissions and energy costs when compared to conventional forklifts that rely on fossil fuels (Pedrazzi et al. 2022). Additionally, the research presented a 454.3 kWp photovoltaic system as a solution to meet the energy demands. It emphasized the need of intelligent management in enhancing the distribution of commodities and promoting sustainability. Remarkably, it was found that each sub-installation has the capacity to produce enough green hydrogen to power a forklift for a whole year (Pedrazzi et al. 2022).

In addition, it has been observed that the utilization of solar power and energy storage for the generation of green hydrogen has demonstrated that lithium-ion capacitors possess a reduced internal resistance throughout their operational duration in comparison to conventional lithium-ion batteries. This characteristic contributes to the optimization of the energy storage procedure, hence boosting its overall efficiency (Ibrahim et al. 2023). This case study highlights the potential of solar energy to bring about transformation in reducing the environmental impact of industrial transportation, as well as improving operational efficiency and cost-effectiveness. These pioneering efforts provide useful information for several companies and sectors that are considering the incorporation of solar energy into their transportation systems.

3.3 Solar Energy in Urban Planning and Environmental

The significance of solar energy in urban planning and environmental design is apparent in the pursuit of creating sustainable and resilient urban landscapes. This pertains to the integration of solar energy systems into urban infrastructure, which includes buildings, parks, and transportation networks, as well as the deliberate planning of urban layouts to maximize sun exposure (Dikaiakos et al. 2023; Chen and Kuo 2022).

As an example, the conversion of a data center of moderate size into an environmentally friendly facility powered by solar energy not only demonstrated the benefits in terms of the environment and finances but also emphasized the possibilities of incorporating operational data through the Internet of Things (IoT) sensor devices (Dikaiakos et al. 2023). The prevalence of such integrations is on the rise, as solar-powered street lights and public charging stations for electric vehicles are becoming common fixtures in urban areas worldwide. In addition, it is worth noting that several public amenities, including educational institutions, medical facilities, and recreational venues, have adopted the utilization of solar energy. This transition has resulted in significant decreases in both energy costs and ecological impacts (Dikaiakos et al. 2023; Chen and Kuo 2022). Furthermore, solar energy is being included into urban planning and architectural blueprints to enhance energy efficiency and promote sustainability. This involves the strategic orientation and arrangement of buildings to maximize solar exposure, the utilization of solar-reflective materials to limit the phenomenon of urban heat islands, and the integration of green spaces to offset carbon emissions (Dikaiakos et al. 2023; Chen and Kuo 2022). It is worth mentioning that the incorporation of solar panels in urban environments has been observed to yield insulation and cooling benefits. Additionally, the presence of green spaces has been discovered to impact air and ground temperatures (Tseng et al. 2022).

In addition, the investigation of solar-thermal technology for the purpose of energy acquisition has revealed the possibility of advanced solar co-harvesting systems, thereby highlighting the importance of solar energy in the context of urban planning (Xu et al. 2023). The integration of solar energy applications into urban planning and environmental design plays a crucial role in the development of "solar cities." These cities serve as prime examples of not only sustainable and resilient communities but also showcase qualities of livability and vibrancy. The aforementioned efforts highlight the significant impact that solar energy may have on urban environments and the improvement of overall well-being (Dikaiakos et al. 2023; Chen and Kuo 2022; Xu et al. 2023; Tseng et al. 2022).

4 Harnessing Solar Energy in Sports Facility Management

4.1 Solar-Powered Electric Vehicles: A Sustainable Approach in Sports Facility Operations

Electric vehicles (EVs) have emerged as a disruptive option in the field of sports facility management, providing a sustainable means of transportation that operates on renewable energy sources, particularly solar energy. The transportation industry plays a substantial role in global greenhouse gas emissions, making it imperative to transition toward electric vehicles powered by renewable energy sources. This transition represents a crucial step in reducing carbon footprints (Dik et al. 2022). Solar energy, which is captured and utilized via photovoltaic systems, has the capability to directly power electric vehicles (EVs). The achievement of this objective can be facilitated through the utilization of solar panels positioned at charging stations, or alternatively, through the indirect means of the grid receiving its electricity from solar or other renewable energy sources, thereby providing the necessary charge (Dik et al. 2022). This technique not only mitigates the environmental impact of electric vehicles (EVs), but also provides economic benefits. According to research findings (Dik et al. 2022), solar energy generally demonstrates more cost-effectiveness compared to electricity generated from fossil fuels.

Moreover, the revolutionary vehicle-to-grid (V2G) technology enhances the potential of electric vehicles (EVs) in the context of sports facilities. The utilization of this technology enables electric vehicles (EVs) to transfer electrical power back to the grid in instances of increased demand or reduced supply. The batteries of electric vehicles (EVs) serve as decentralized energy storage units, thereby enhancing the stability and resilience of the electrical grid (Dik et al. 2022). Moreover, the utilization of neural networks in the neural-based predictive strategy enhances the adaptability and resilience of renewable technologies, specifically in the context of forecasting the power production of photovoltaic (PV) systems. This approach effectively incorporates solar radiation and weather data. The utilization of this predictive capability has the potential to enhance the optimization of the charging and discharging cycles of electric vehicles (EVs) within sports facilities, hence ensuring the effective management of energy resources (Buturache and Stancu 2021). The integration of solar energy into the management of sports facilities, specifically for the purpose of powering electric vehicles (EVs), serves as a prime example of the intersection between sustainability and innovation. This development is crucial in establishing a more environmentally friendly future for sports operations (Buturache and Stancu 2021; Dik et al. 2022).

4.2 *Solar Energy's Role in Job Creation and Sports Facility Management*

The utilization of solar energy, as a component of the wider sustainable energy movement, presents significant opportunities for employment generation, particularly in the field of sports facility management. The solar energy industry, due to its reliance on manual labor, requires a wide range of skills that encompass multiple occupations. The individuals encompassed in this category consist of engineers and technicians who bear the responsibility of installing and maintaining solar systems, researchers who are at the forefront of developing groundbreaking solar technology, and sales and advisory experts who actively promote the adoption of solar solutions (Mohommad and Kim 2022).

Moreover, the widespread usage of solar energy has a significant impact on other sectors. The integration of solar systems into sporting facilities has the potential to drive growth in the building industry. Likewise, the manufacturing sector has the potential to benefit from the production of solar equipment, while the education and training sectors can thrive by meeting the needs of the growing demand for proficient solar specialists (Mohommad and Kim 2022). Solar-thermal technology for energy harvesting has emerged as a significant development in the field of solar energy. The technology under consideration, known for its efficient energy conversion capabilities, has promise for the development of advanced solar co-harvesting systems, particularly in the context of sports facilities (Xu et al. 2023).

The solar sector has the potential to greatly contribute to economic growth by creating job opportunities, particularly in places that are facing high unemployment rates or undergoing a transition from traditional energy industries. Furthermore, this transition is in perfect harmony with the Sustainable Development Goals, specifically highlighting the importance of Affordable and Clean Energy (Goal 7) and Decent Work and Economic Growth (Goal 8) (Mohommad and Kim 2022). In the field of sports facility management, the use of solar energy not only guarantees enhanced operational efficiency but also establishes a pathway toward a sustainable future with ample employment opportunities (Mohommad and Kim 2022).

5 Gaps and Future Research Directions

The burgeoning field of solar energy, especially its integration in sports facilities, has seen significant advancements. However, there remain gaps that necessitate further exploration to fully harness its potential in sports facility management.

i. Solar Energy Implementation in Sports Facilities

- While there is a wealth of research on solar energy across sectors, a focused exploration on its application within sports facilities is scant. This includes understanding its cost-effectiveness, long-term sustainability, and the integration of IoT for adept energy management (Tseng et al. 2022).
- Comprehensive studies should delve into the technical, economic, and social facets of solar integration in sports arenas, aiming to curate best practice guidelines for facility managers (Ekström et al. 2021; Dadzie and Djokoto 2022).

ii. Solar Energy's Role in Electric Vehicles and Job Creation

- The synergy between solar energy, electric vehicles, and its potential to spur green job opportunities is evident. However, a deeper understanding of the mechanisms, impacts, and potential barriers is essential (Buturache and Stancu 2021; Dik et al. 2022).
- Research could encompass the design of solar-powered EV charging stations, the evolution of vehicle-to-grid technologies, and the socio-economic ramifications of green energy initiatives (Buturache and Stancu 2021; Dik et al. 2022).

iii. Green Hydrogen in Industrial Transport

- The prospect of green hydrogen, especially in industrial transport, is still nascent and demands rigorous exploration. Validating the findings of extant studies and probing the feasibility of green hydrogen across contexts is crucial (Pedrazzi et al. 2022).
- Potential areas of study include the design of solar-powered electrolysis systems, the efficacy of hydrogen fuel cells in vehicles, and the comparative economic and environmental impacts of green hydrogen vis-à-vis other fuels (Pedrazzi et al. 2022).

iv. Solar Energy and Digital Technological Convergence

- The confluence of solar energy and digital technology is in a state of flux, with rapid technological strides necessitating continuous research (Xu et al. 2023).
- Future studies might focus on harnessing AI and machine learning for precise solar energy forecasting, crafting digital platforms tailored for green energy services, and the seamless integration of solar energy into smart grids (Mohommad and Kim 2022; Ibrahim et al. 2023).

By bridging these research gaps, we can augment our comprehension of solar energy's potential and challenges, paving the way for sustainable and resilient energy systems, especially in the realm of sports facility management.

Acknowledgements The authors are very grateful to the Ministry of Higher Education (MoHE) Malaysia and the Universiti Malaysia Pahang Al-Sultan Abdullah for their great financial support, which made this study possible. This study was made possible by grants from the Fundamental Research Scheme (FGRS) FRGS/1/2022/TK10/UMP/02/57 and the Universiti Malaysia Pahang Al-Sultan Abdullah Fundamental Research Grant RDU220312.

References

Abdullah NA, Abdullah FF, Sufian AH, Abidin AN, Jamaludin AS, Razali MN (2022) Effect of degradation by temperature onto nitrile rubber elastomer mechanical properties. Mater Today: Proc 48:1941–1946

Abidin AN, Azmi AH, Kassim KA, Jamaludin AS, Razali MN (2022) A review on automotive tires significant characteristic identification for general consumers. In: Proceedings of the 2nd energy security and chemical engineering congress, pp 375–385

Ali MA, Sivarao, Abdullah Z, Izamshah R, Kassim MS, Jamaludin AS (2022) Multi-response optimization of machining simulation approach using grey relational analysis (Lecture notes in mechanical engineering), pp 154–157

Alsibai MH, Hamran NNN, Nasir A, Abdullah AA (2018) A monitoring system for EPW safe use. J Telecommun, Electron Comput Eng (JTEC) 10(1–2):131–135

Atalla Y, Hashim Y, Ghafar ANA, Jabbar WA (2020) Temperature characteristics of FinFet based on channel fin width and working voltage. Int J Electr Comput Eng (IJECE) 10:5650

Atalla Y, Hashim Y, Ghafar ANA (2022) The impact of channel fin width on electrical characteristics of Si-FinFET. Int J Electr Comput Eng (IJECE) 12:201
Babu D, Nasir A, Jamaludin AS, Rosle MH (2021) Holding, grasping and sensing of prosthetic robot arm like a real human hand, a journey beyond limits: an extensive review. Human-centered technology for a better tomorrow, pp 485–504
Babu D, Nasir A, Farag M, Sidik MH, Rejab SB (2022) Development of prosthetic robotic arm with patient monitoring system for disabled children; preliminary results. In: 2022 9th International conference on electrical and electronics engineering (ICEEE)
Babu D, Nasir A, Ravindran, Farag M, Jabbar WA (2023) 3D printed prosthetic robot arm with grasping detection system for children. Int J Adv Sci, Eng Inf Technol 13:226
Baharuddin NJ, Abdul Manaf AR, Jamaludin AS (2022) Study on die shoulder patterning method (DSPM) to minimise springback of U-bending. Int J Automot Mech Eng 19:9509–9518
Baharuddin NJ, Manaf ARA, Jamaludin AS (2023) Study of springback behavior on U-bending part using die shoulder patterning method (DSPM). AIP Conf Proc 2544(1):040020
Binti Kamarudin SN, Lee H, Ishak I, Manaf AR, Jamaludin AS, Shaharudin MA, Zawawi MZ (2021) Rapid direct continuous method for hot embossing of Glass microlens array combined with CO_2 laser irradiation and external preheating/cooling (Lecture notes in mechanical engineering), pp 669–675
Borysiak O, Brych V (2021) Methodological approach to assessing the management model of promoting green energy services in the context of development smart energy grids. In: Financial and credit activity problems of theory and practice, vol 4, pp 302–309
Buturache A-N, Stancu S (2021) Solar energy production forecast using standard recurrent neural networks, long short-term memory, and gated recurrent unit. Eng Econ 32:313–324
Chen H-S, Kuo H-Y (2022) Green energy and water resource management: a case study of fishery and solar power symbiosis in Taiwan. Water 14:1299
Dadzie J, Djokoto S (2022) Investigating drivers stimulating demand for green renovation of existing buildings and systems. J Sustain Dev 15:88
Dik A, Omer S, Boukhanouf R (2022) Electric vehicles: V2G for rapid, safe, and green EV penetration. Energies 15:803
Dikaiakos MD, Chatzigeorgiou NG, Tryfonos A, Andreou A, Loulloudes N, Pallis G, Georgiou GE (2023) A cyber-physical management system for medium-scale solar-powered data centers. Concurr Comput: Pract Exp 35
Ekström T, Burke S, Wiktorsson M, Hassanie S, Harderup L-E, Arfvidsson J (2021) Evaluating the impact of data quality on the accuracy of the predicted energy performance for a fixed building design using probabilistic energy performance simulations and uncertainty analysis. Energy Build 249:111205
Farag M, Azlan NZ, Alsibai MH, Ghafar AN (2019) Slippage detection for grasping force control of robotic hand using force sensing resistors. In: Proceedings of the 2019 5th international conference on computer and technology applications
Ghalambaz M, Sheremet M, Fauzi MA, Fteiti M, Younis O (2023) A scientometrics review of solar thermal energy storage (STES) during the past forty years. J Energy Storage 66:107266
Hasya N, Razali MNM, Jamaludin AS, Saffe SNBM, Hadi MA (2022) Micromilling performance analysis—cold air assisted aluminium machining behaviour observation. In: Innovative manufacturing, mechatronics & materials forum 2022 (iM3F 2022), Online conference, Pekan, Malaysia, pp 124–130
Ibrahim T, Kerekes T, Sera D, Mohammadshahi SS, Stroe D-I (2023) Sizing of hybrid supercapacitors and lithium-ion batteries for green hydrogen production from PV in the Australian climate. Energies 16:2122
Jamaludin AS, Yassin A (2013) Analysis of laser sintered materials using finite element method. Sains Malaysiana 42(12):1727–1733
Jamaludin AS, Hosokawa A, Furumoto T, Koyano T, Hashimoto Y (2017) Evaluation of the minimum quantity lubrication in orthogonal cutting with the application of finite element method. Int J Mech Mechatron Eng IJMME-IJENS 17(01):104–109

Jamaludin AS, Hosokawa A, Furumoto T, Koyano T, Hashimoto Y (2018a) Study on the effectiveness of extreme cold mist MQL system on turning process of stainless steel Aisi 316. IOP Conf Ser: Mater Sci Eng 319:012054

Jamaludin AS, Akira H, Furumoto T, Koyano T, Hashimoto Y (2018b) High precision estimation on physical behavior for cutting with various tool rake angle by finite element method (Lecture notes in mechanical engineering), pp 715–723

Lee H, Binti Kamarudin SN, Ishak I, Manaf AR, Jamaludin AS, Shaharudin MA, Zawawi MZ (2021) Feasibility study of wafer scale laser assisted thermal imprinting of glass nanostructures (Lecture notes in mechanical engineering), pp 917–923

Md Kamil NH, Jamaludin AS, Mhd Razali MN, Ghaffar ANA (2022) Temperature and heat flow analysis in a drying chamber through finite element method (Lecture notes in mechanical engineering), pp 309–316

Mohommad A, Kim J (2022) Jobs impact of green energy (IMF working papers), p 1

Nasir A, Akagi T, Dohta S, Ono A, Masago Y (2013) Development of low-cost wearable servo valve using buckled tube driven by servo motor. Appl Mech Mater 393:532–537

Nasir A, Akagi T, Dohta S, Ono A (2015) Analysis of low-cost wearable servo valve using buckled tubes for optimal arrangement of tubes. In: 2015 IEEE international conference on advanced intelligent mechatronics (AIM)

Pedrazzi S, Morini E, Nasti M, Pizzileo S, Allesina G (2022) Green hydrogen powered forklifts in industrial transport: case study of an Italian fruit and vegetable market. Int J Heat Technol 40:145–150

Ramli MR, Razak NA, Ismail I, Jamaludin AS, Manaf AR (2022) Effect of dimple size onto wear rate of mild steel AISI 1060 surface (Lecture notes in mechanical engineering), pp 99–102

Rosli AM, Jamaludin AS, Razali MN, Sani AS, Hamzah SB, Osman MS (2019) Modelling of fuzzy inference system for micro milling—a preliminary study through FEM (Lecture notes in mechanical engineering), pp 445–456

Rosli AM, Jamaludin AS, Razali MNM (2021) Recent study on hard to machine material—micromilling process. Evergreen 8:445–453

Rosli AM, Jamil N, Jamaludin AS, Razali MN, Yusoff AR (2021) Tool wear observation during unconventional low speed machining using low cost micromilling (Lecture notes in mechanical engineering), pp 589–597

Sufian AH, Xun TZ, Abidin AN, Jamaludin AS, Razali MN (2021) Study on tire tread design effect onto tire-road contact behavior through FEM (Lecture notes in mechanical engineering), pp 893–902

Tseng K-H, Chung M-Y, Chen L-H, Wei M-Y (2022) Applying an integrated system of cloud management and wireless sensing network to green smart environments—green energy monitoring on campus. Sensors 22:6521

Wahid KAA, Fahmi I, Puteh Z, Saniman MNF, Rusulan K, Azhar K, Muhammad WMW (2022) Self-desalination seawater jig based on solar thermal energy. J Adv Res Fluid Mech Therm Sci 91:92–101

Wei Q, DeBlock RH, Butts DM, Choi C, Dunn B (2020) Pseudocapacitive vanadium-based materials toward high-rate sodium-ion storage. Energy Environ Mater 3:221–234

Wrana J, Struzik W, Gleń P (2022) Natural energy stored in groundwater deposits as a new way of obtaining green energy for urban planners, architects and environmentalists. Energies 15:4716

Xu Z, Li Y, Gao G, Xie F, Ju R, Yu S, Liu K, Li J, Wang W, Li W, Li T, Qiu C-W (2023) Scalable selective absorber with quasiperiodic nanostructure for low-grade solar energy harvesting. APL Photonics 8

Advancement of IoT-Based Smart Systems for Real-Time Sports Performance Analysis: Potential and Challenges

Abdul Nasir, Devin Babu, M. H. Muhammad Sidik, Norain Binti Abdullah, and Waheb A. Jabbar

Abstract Internet of Things (IoT) technology in the field of sports has brought about a significant transformation in the analysis of athletic performance, as it facilitates the seamless collection and processing of data in real time. Nevertheless, enduring obstacles encompass issues pertaining to data security, privacy, and the imperative for resilient real-time data processing capabilities. Mainly, data transfer rate is very important when it is related to real-time application. However the emergence of 5G, edge computing, machine learning, and artificial intelligence offers substantial prospects for augmenting the functionalities of these systems. This enhancement would result in more precise and prompt understandings of athletes' performances, improvised training plan for specific objectives, and high potentially able to decrease risk of common injury. However, there are limitations even though these kinds of technologies have been implemented. On the other hand, cyber physical attacks threat is another aspect that must be taken into consideration in developing IoT devices. This article would also help the readers and researchers to understand the

A. Nasir · D. Babu
Faculty of Electrical and Electronic Engineering Technology, Universiti Malaysia Pahang Al-Sultan Abdullah, 26600 Pekan, Pahang, Malaysia
e-mail: abdnasir@umpsa.edu.my

M. H. Muhammad Sidik (✉)
Faculty of Mechanical and Automotive Engineering Technology, Universiti Malaysia Pahang Al-Sultan Abdullah, 26600 Pekan, Pahang, Malaysia
e-mail: hanafisidik@umpsa.edu.my

N. B. Abdullah
Faculty of Manufacturing and Mechatronic Engineering Technology, Universiti Malaysia Pahang Al-Sultan Abdullah, 26600 Pekan, Pahang, Malaysia

Bio-Coke Research Institute, Kindai University, Osaka, Japan

W. A. Jabbar
School of Engineering and Built Environment, Birmingham City University, Birmingham, West Midlands, England, UK

Z. Mohamed et al. (eds.), *Proceedings of the 2nd International Conference on Innovation and Technology in Sports, ICITS 2023, 27–28 November, Malaysia*, Lecture Notes in Bioengineering, https://doi.org/10.1007/978-981-97-3741-3_34

importance of data processing, smart systems, security and privacy and future trends in IoT application for sports performance analysis.

Keywords Internet of Things · Sports performance analysis · Real-time data processing · Smart systems · Data security

1 Importance of IoT in Sports Performance Analysis

The Internet of Things (IoT) has brought a significant transformation in various industries such as manufacturing, agriculture, medical and healthcare, and even in sports (Pires and Santos 2018; Rejeb et al. 2022; Jain et al. 2021; Abdulmalek et al. 2022). This modern industrial evolution is supported by utilization of IoT due to its ability to facilitate real-time data collection and analysis (Yu and Wang 2020; Jamil et al. 2021; Babu et al. 2021). Internet of Things (IoT) devices, such as wearable sensors and smart equipment, have the capability to gather a wide range of data points, encompassing metrics such as an athlete's heart rate, speed, movement patterns, and other relevant information. The aforementioned data can subsequently be subjected to analysis in order to derive valuable insights pertaining to an athlete's performance, pinpoint areas that require enhancement, and potentially forecast and preempt potential injuries (Silvers-Granelli et al. 2021). The incorporation of Internet of Things (IoT) technology in sports performance analysis has resulted in the emergence of intelligent systems capable of promptly processing and evaluating data (Rosli et al. 2021; Nižetić et al. 2020). These systems have the capability to offer real-time feedback to athletes and coaches, enabling them to make necessary adjustments during training sessions or competitions. An intelligent system has the capability to analyze the data obtained from a runner's wearable device, subsequently offering immediate feedback regarding their running technique (Clermont et al. 2019). This feedback serves the purpose of enhancing efficiency and mitigating the likelihood of sustaining injuries.

Moreover, the integration of Internet of Things (IoT) technology in smart systems has the potential to make significant contributions to the expansive domain of sports science (Baharuddin et al. 2022; Schlegel and Hill 2020). This is achieved through the provision of extensive datasets that can be utilized for research purposes. The utilization of these datasets facilitates the examination of patterns, the identification of optimal strategies, and the formulation of novel approaches to training (Khan et al. 2022). Nevertheless, the integration of Internet of Things (IoT) technology in the realm of sports performance analysis presents certain obstacles. Significant considerations encompass data security, privacy, and the imperative for robust, real-time data processing capabilities (Khoshkhah et al. 2022; Rosli et al. 2019).

The objective of this review is to offer a comprehensive examination of the present condition of Internet of Things (IoT)-based intelligent systems for the analysis of real-time sports performance. The review will additionally put forth potential remedies for these deficiencies, offering a strategic plan for future investigation and advancement.

2 IoT and Real-Time Data Processing

2.1 *The Role of IoT in Real-Time Data Collection*

The Internet of Things (IoT) has brought about a paradigm shift in the field of data collection by facilitating the seamless acquisition of real-time and continuous data from diverse sources. In sports performance analysis, Internet of Things (IoT)-based devices, such as wearable sensors, smart equipment, and video surveillance systems, have the capability to gather a diverse range of data points. These data points encompass an athlete's physiological parameters, movement patterns, and environmental conditions (Zhou et al. 2021). Zhou et al. examined the application of Internet of Things (IoT) technology in surveillance systems, emphasizing the significance of real-time data acquisition for efficient monitoring and decision-making processes (Zhou et al. 2021). In sport science, the collection of real-time data can offer prompt insights into an athlete's performance, enabling on-the-spot adjustments to be made during training sessions or competitive events. Lukas et al. introduced SPORTSENSE, a video-based event retrieval system. This system can visualize the analysis results in real time, store the analysis results persistently for offline activities, and leverage the stored analysis results for intuitive sketch-based video retrieval as in Fig. 1 (Probst et al. 2018).

On the other hand, Jamali and Giman emphasized the utilization of the Internet of Things (IoT) for the purpose of collecting real-time data for flood monitoring (Jamali and Giman 2021). While the principles of real-time data collection and monitoring may not have a direct correlation to sports performance analysis, they possess similarities that allow for their application within the sports domain, especially to monitor the environmental changes that might affect the athlete performance and health (Jamaludin et al. 2018a; Jamali and Giman 2021; Abdullah et al. 2022; Donnelly et al. 2016; Nowak et al. 2022). Sven et al. conclude that climate change is associated with the increases in health-related risks for athletes such as UV exposure, heat stress, UV exposure, exposure to air pollutants, spread of vectors, and natural reservoirs as well as other risks such as extreme weather events (Schneider and Mücke 2021). Thus, monitoring systems that are focusing on feeding affirmative information about surrounding environmental changes during sports activity are crucially needed (Hasya et al. 2022; Torres-Ronda et al. 2022; Md Kamil et al. 2022; Alexandersen et al. 2023). For example, Mainura et al. proposed an improved fuzzy

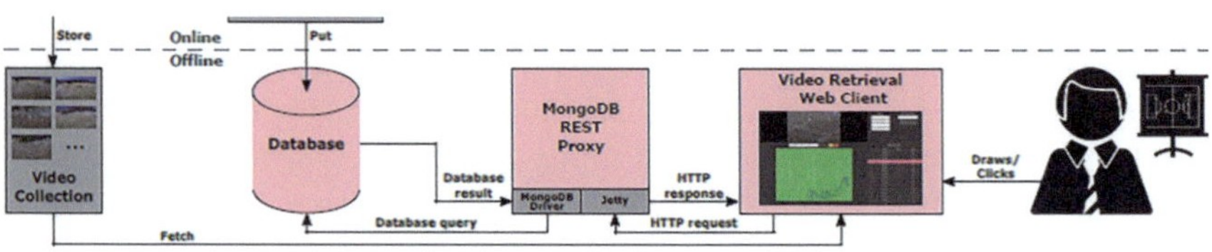

Fig. 1 Architecture of SPORTSENSE system (Probst et al. 2018)

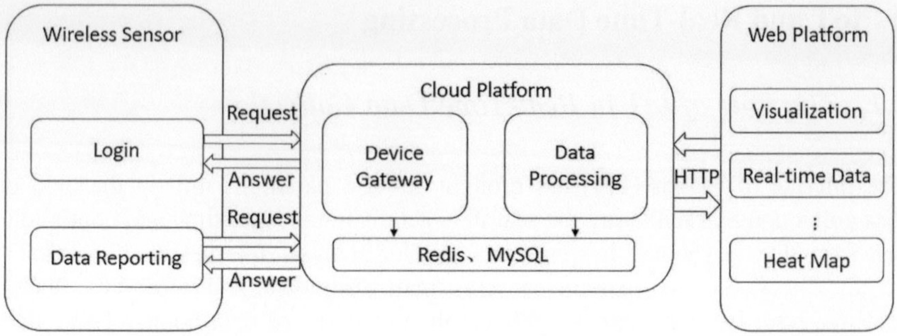

Fig. 2 Architecture of outdoor air quality monitoring system for sports athlete (Li and Xu 2021)

logic-based controller for humidity monitoring control in sports halls. Then, Juan Li et al. as in Fig. 2 developed a real-time online monitoring system for outdoor air quality for sports athletes (Li and Xu 2021).

2.2 Real-Time Data Processing and Decision-Making

After the collection of data in real time, it is necessary to undertake processing and analysis in order to derive significant insights. Real-time data processing is a crucial aspect in this context. Real-time data processing refers to the practice of analyzing data immediately after its collection, enabling prompt decision-making (Khan et al. 2022; Abidin et al. 2022; Sudharsan et al. 2021). In the study on sustainable smart cities, utilization of real-time routing and underscoring the significance of real-time data processing in facilitating efficient decision-making (Rehman et al. 2021; Rosli et al. 2021) found that a mobility support 5G architecture with real-time routing able to decrease the loss of data against network disconnectivity and increase the reliability. Figure 3 shows the success ratio when data rate is increased from 30 to 150 bits per second. Even though data rate is decreased, but it is still in acceptable range (Sachs et al. 2018).

In the field of sports performance analysis, the utilization of real-time data processing has the potential to offer instantaneous feedback regarding an athlete's performance (Silva et al. 2020; Yuan et al. 2020). This enables coaches and athletes to make well-informed decisions in the midst of training sessions or competitive events. Qingfeng proposed a computer big data-based technology real-time signal transmission system for sports competitions (Shi 2023). The proposed system is able to remove noise from signal transmission and also capable of detecting weak signals submerged in the background of strong color noise, which is superior to the traditional time domain detection method.

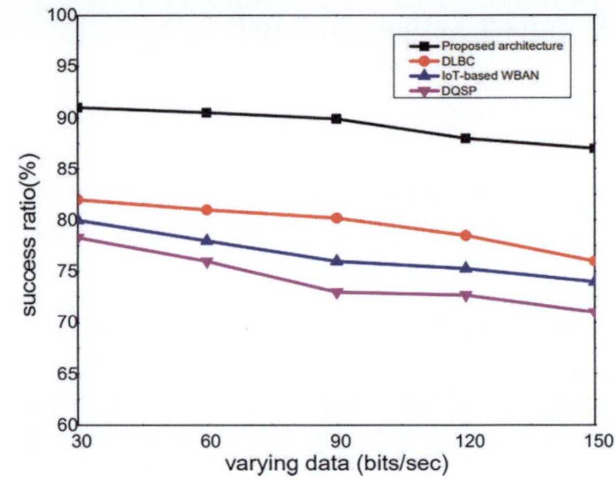

Fig. 3 Analysis of the success ratio under a varying data rate of 30 to 150 bits/s (Sachs et al. 2018)

2.3 *Challenges in Real-Time Data Processing*

Although real-time data processing in IoT-based systems offers numerous benefits, it is not devoid of challenges. These factors encompass the necessity for strong and dependable data processing capabilities, the management of substantial quantities of data, and the assurance of data security and privacy (Gurcan and Berigel 2018). The necessity for strong and dependable data processing capabilities is of utmost importance in order to guarantee accurate and timely processing of the collected data. The need for robust computational capabilities poses a significant challenge for IoT devices, given their inherent limitations in terms of processing power.

One additional challenge pertains to the management of substantial quantities of data. IoT devices have the capacity to accumulate substantial volumes of data, potentially surpassing the data processing capabilities of the system. Efficient data management strategies are necessary in order to ensure the processing of only pertinent data, thereby mitigating the system's workload. The preservation of data security and privacy poses a considerable obstacle. Due to the ability to gather and analyze data in real-time, IoT devices possess a susceptibility to being targeted by cyber-attacks. The implementation of robust security measures is necessary in order to safeguard data and guarantee the preservation of privacy (Khoshkhah et al. 2022; Ting et al. 2022).

3 Smart Systems and IoT

3.1 The Concept of Smart Systems in IoT

Smart systems in the context of the Internet of Things (IoT) pertain to systems that possess the capability to independently process and analyze data, make informed decisions based on the analyzed data, and execute actions without requiring human intervention. These systems utilize cutting-edge technologies such as machine learning, artificial intelligence, and cloud computing to offer intelligent and automated solutions (Rehman et al. 2021; Elkhoukhi et al. 2022). Sudharsan et al. investigated on real-time offensive words spotting network for consumer IoT devices (Sudharsan et al. 2021) and found that is capable of efficiently detecting multiple keywords in real-time. This system showcases the potential of intelligent systems to effectively process and analyze data in real time, enabling informed decision-making based on the analyzed information.

In a similar vein, Elkhoukhi explored the application of intelligent systems in the context of real-time occupancy detection within smart buildings (Elkhoukhi et al. 2022). The system showcases the potential of intelligent systems to effectively handle and analyze substantial amounts of data in real time, enabling them to derive insights and make informed decisions based on the analyzed data.

3.2 Application of Smart Systems in Sports Performance Analysis

The utilization of intelligent systems has demonstrated notable applications in the analysis of sports performance. The ability to extract and evaluate data from diverse sources, including wearable sensors, smart equipment, and video surveillance systems, enables the generation of valuable insights regarding an athlete's performance. This data analysis can identify areas that require improvement and even forecast and mitigate potential injuries (Kofahi et al. 2022; Ali et al. 2022; Liao et al. 2020). Kofahi et al. developed IoT-enabled intelligent system for Mixed Martial Arts (MMA) athletes that has the capacity to continuously monitor and measure the athlete's body temperature, strike force, and the number of strikes that the athlete has received from the opponent, providing promising real-time data analysis and decision-making capabilities (Kofahi et al. 2022). Figure 4 show the location of sensors on protection equipment and this study showcases the potential of smart systems to effectively process and analyze health-related information in real time, thereby facilitating informed decision-making.

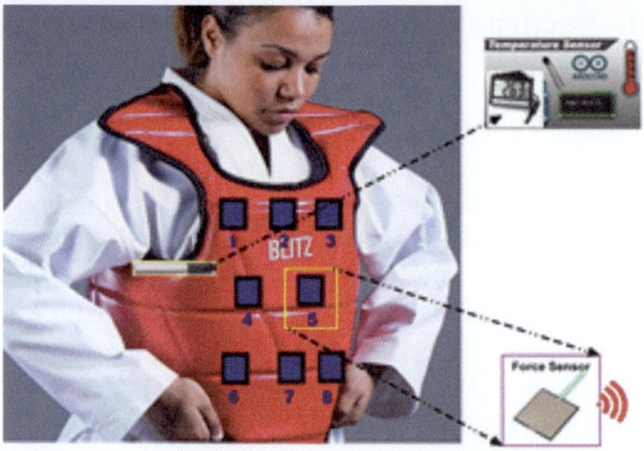

Fig. 4 Implementation of integrated sensors on the proposed system (Shi 2023)

3.3 Limitations and Challenges of Current Smart Systems

Notwithstanding the inherent benefits, contemporary intelligent systems encounter various constraints and obstacles. The aforementioned factors encompass the necessity for resilient and dependable data processing capabilities, management of substantial quantities of data, assurance of data security and privacy and requirement for compatibility between various IoT devices and systems (Liao et al. 2020; Jamaludin et al. 2018b). The necessity for robust and dependable data processing capabilities is of utmost importance in order to guarantee accurate and timely processing of the collected data (Tramarin et al. 2019). The need for robust computational capabilities poses a significant challenge for IoT devices, given their inherent limitations in terms of processing power.

Dealing with substantial quantities of data presents an additional obstacle. IoT devices have the capacity to accumulate extensive volumes of data, potentially exceeding the data processing capabilities of the system. Efficient data management strategies are necessary to ensure the processing of only pertinent data, thereby mitigating the system's workload. The maintenance of data security and privacy presents a noteworthy challenge. Given their ability to gather and analyze data in real-time, IoT devices possess a susceptibility to cyber-attacks. The implementation of robust security measures is necessary in order to safeguard data and guarantee privacy (Tawalbeh et al. 2020; Ramli et al. 1060; Laghari et al. 2021). One additional challenge in the realm of IoT pertains to achieving interoperability across diverse devices and systems. The seamless communication and collaboration of IoT devices and systems are of utmost importance in facilitating the efficient operation of smart systems, given the growing proliferation of such devices and systems.

4 Security and Privacy in IoT-Based Systems

4.1 The Importance of Security in IoT Systems

The preservation of security holds significant importance in IoT systems, primarily due to the handling of sensitive data within these systems. In the field of sports performance analysis, IoT devices are utilized to gather a diverse range of data points (Laghari et al. 2021; Jamaludin et al. 2017; Manzoor et al. 2023). These data points encompass an athlete's physiological parameters, movement patterns, and the prevailing environmental conditions. If this data was to be accessed by unauthorized individuals, it has the potential to result in significant breaches of privacy or be exploited for nefarious intentions. In their research on IoT-based smart climate agriculture systems, Ting et al. emphasize the importance of safeguarding the security of IoT systems by encompassing not only the protection of the data they manage, but also the preservation of the system's integrity (Ting et al. 2022). A security breach has the potential to impede the operational integrity of the system, resulting in the generation of erroneous data and analysis, and potentially jeopardizing the well-being of the athletes.

4.2 Current Techniques for Ensuring Security and Privacy

Various methodologies are presently employed to safeguard the security and privacy of IoT systems. The aforementioned elements encompass encryption, authentication, and the utilization of secure communication protocols. In a study on the generation of daily modal split estimation and OD matrices using IoT data in application of secure communication protocols shows that the importance of secure communication protocols in guaranteeing the integrity of data exchanged between IoT devices and the system is underscored (Khoshkhah et al. 2022). Meanwhile, Xinyu et al. propose a cancelable biometric modality by utilizing a high-density surface electromyogram (HD-sEMG) encoded by hand gesture password, for user authentication. The proposed system will overcome a noncancelability of traditional biometrics (e.g., fingerprint) for user authentication that might increase the privacy disclosure risks once the biometric template is exposed. Based on the analysis result, the proposed system with 95% variance preserved, the security of the proposed HD-sEMG biometrics is equivalent to a 140-bit key with a total of 2^{140} expected attempts of brutal attack required to intrude the system. The security strength of HD-sEMG biometrics against brutal attack is presented in Table 1.

Table 1 Security strength against cyber attack

Preserved variance with PCA-based dependence removal	Entrophy E
100%	1318 bits
99%	313 bits
98%	238 bits
97%	198 bits
96%	164 bits
95%	140 bits

4.3 Challenges and Limitations in Ensuring Security and Privacy

Despite the existing methodologies, the task of safeguarding the security and privacy of IoT systems is accompanied by a multitude of obstacles. Several factors must be considered in the context of ensuring the security of devices, such as the imperative of employing strong encryption methods, the potential vulnerability to physical assaults, and the necessity of safeguarding data storage (Manzoor et al. 2023). The implementation of strong encryption methods is imperative in order to safeguard data integrity and confidentiality throughout the processes of transmission and storage. Nevertheless, the computational resources necessary for ensuring strong encryption pose a significant obstacle for IoT devices, which generally possess constrained processing capabilities.

One additional risk involves the potential for cyber and physical attacks on the devices (Shah and Sengupta 2020). IoT devices are frequently deployed in real-world environments, making them susceptible to physical attacks. This has the potential to result in the compromise of the device and unauthorized access to the collected data (Shah and Sengupta 2020; Lee et al. 2021; Bennett et al. 2018; Baharuddin et al. 2023; Li et al. 2019). The challenge of secure data storage is also evident. To mitigate the risk of unauthorized access, it is imperative to ensure secure storage of the substantial amounts of data amassed by IoT devices. The utilization of robust data storage solutions is imperative in addressing the challenge posed by the constrained storage capacities of IoT devices. Cyber and physical attacks can be detected based on energy auditing as in Fig. 5 (Bennett et al. 2018).

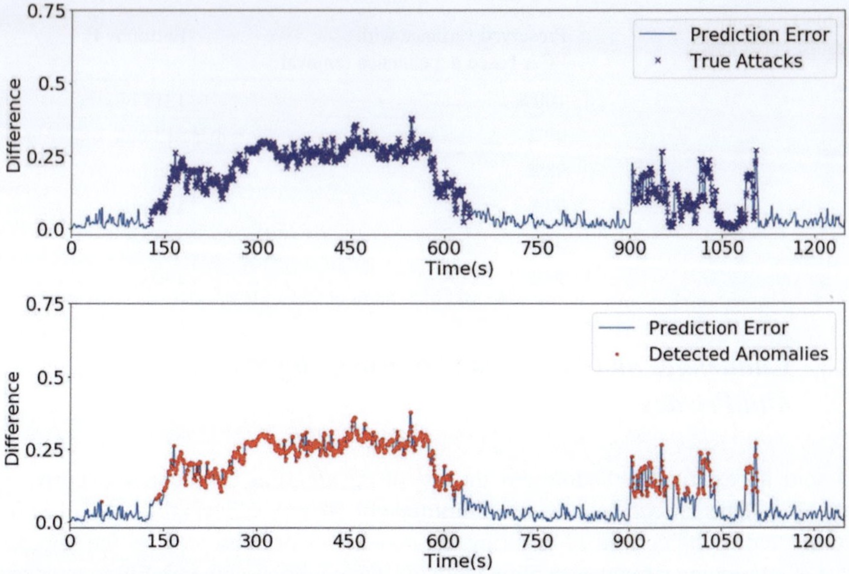

Fig. 5 Power prediction error with true attacks and detected anomalies. Cyber-attacks (between 900 and 1100 s) and physical attacks (between 150 and 600 s) (Li et al. 2019)

5 Future Trends and Opportunities

5.1 The Role of 5G and Edge Computing in IoT (Reference Jain et al. 2021)

The emergence of 5G and edge computing is poised to bring about a transformative impact on the IoT, particularly in sports performance analysis. The implementation of 5G technology has the potential to greatly augment the real-time data collection and processing capabilities of IoT devices, owing to its ability to facilitate high-speed data transfer (Li et al. 2019; Sufian et al. 2021; Zhao et al. 2019). This phenomenon has the potential to result in enhanced precision and promptness in the comprehension of an athlete's performance (Rehman et al. 2021). In contrast, edge computing refers to the practice of performing data processing tasks in close proximity to the source of data generation within a network. The implementation of this approach has the potential to decrease the latency that is typically encountered when transmitting data to a centralized location for processing. As a result, the real-time capabilities of IoT devices can be further improved (Rehman et al. 2021).

5.2 The Potential of Machine Learning and AI in IoT

The utilization of machine learning and artificial intelligence (AI) exhibits considerable promise within the realm of Internet of Things (IoT), specifically in the domain of sports performance analysis for predicting results, athletes profiling and performances, injuries diagnosis and even can be extended to predict advertising behaviors (Du et al. 2021; Worsey et al. 2021; Goud et al. 2019; Binti Kamarudin et al. 2021; Bunker and Thabtah 2019; Chhabra et al. 2023; Ghazali et al. 2022; Jamaludin and Yassin 2013 Naglah et al. 2018; Tedesco et al. 2022; Barnes and Wang 2023). The aforementioned technologies possess the capability to analyze the extensive quantities of data gathered by Internet of Things (IoT) devices in order to detect patterns, generate predictions, and offer insights that may not be readily discernible through conventional methods of analysis (Qian et al. 2022). In optimization of neural networks for time-series analysis in IoT systems in potential of machine learning and artificial intelligence, found that accuracy is improved compared without using any classifiers (Qian et al. 2022).

5.3 Opportunities for Future Research and Development

The progressions in technological domains such as 5G, edge computing, machine learning, and Artificial Intelligence (AI) offer substantial prospects for prospective investigations and advancements in IoT-based intelligent systems designed for the analysis of sports performance (Glebova et al. 2023). These encompass the advancement of more resilient and dependable data processing capabilities, the investigation of novel data collection methodologies, and the enhancement of more secure and privacy-centric systems. Moreover, there exists the potential to investigate the incorporation of additional nascent technologies, such as blockchain and quantum computing, within the realm of the IoT (Li and Zhu 2023; Khaund 2020; Yadav et al. 2023). The implementation of these technologies has the potential to significantly augment the security, privacy, and data processing functionalities of IoT systems. In addition, progress and achievement findings in sports science are high potentially can be implemented for public usage such as for wearable devices and healthcare monitoring.

Acknowledgements The author wishes to express profound gratitude to the Universiti Malaysia Pahang Al-Sultan Abdullah for their invaluable support and financial assistance under internal research grant scheme, RDU230328.

References

Abdullah NA, Abdullah FF, Sufian AH, Abidin AN, Jamaludin AS, Razali MN (2022) Effect of degradation by temperature onto nitrile rubber elastomer mechanical properties. Mater Today: Proc 48:1941–1946

Abdulmalek S, Nasir A, Jabbar WA, Almuhaya MA, Bairagi AK, Khan M-M, Kee S-H (2022) IOT-based healthcare-monitoring system towards improving quality of life: a review. Healthcare 10:1993

Abidin AN, Azmi AH, Kassim KA, Jamaludin AS, Razali MN (2022) A review on automotive tires significant characteristic identification for general consumers. In: Proceedings of the 2nd energy security and chemical engineering congress, pp 375–385

Alexandersen A, Pettersen SD, Johansen D (2023) Quantifying athlete wellness: investigating the predictive potential of subjective wellness reports through a player monitoring system

Ali MA, Sivarao, Abdullah Z, Izamshah R, Kassim MS, Jamaludin AS (2022) Multi-response optimization of machining simulation approach using grey relational analysis. Lecture Notes in Mechanical Engineering, pp 154–157

Babu D, Nasir A, Jamaludin AS, Rosle MH (2021) Holding, grasping and sensing of prosthetic robot arm like a real human hand, a journey beyond limits: an extensive review. In: Human-centered technology for a better tomorrow, pp 485–504

Baharuddin NJ, Abdul Manaf AR, Jamaludin AS (2022) Study on die shoulder patterning method (DSPM) to minimise springback of U-bending. Int J Autom Mech Eng 19:9509–9518

Baharuddin NJ, Manaf AR, Jamaludin AS (2023) Study of springback behavior on U-bending part using die shoulder patterning method (DSPM). Adv Mater Sci Manuf Eng

Barnes SJ, Wang W (2023) Understanding consumer advertising via audio analytics of sports videos. In: 5th international conference on artificial intelligence and information communication (ICAIIC 2023), pp 33–36

Bennett CL, Brady E, Branham SM (2018) Interdependence as a frame for assistive technology research and design. In: Proceedings of the 20th international ACM SIGACCESS conference on computers and accessibility

Binti Kamarudin SN, Lee H, Ishak I, Manaf AR, Jamaludin AS, Shaharudin MA, Zawawi MZ (2021) Rapid direct continuous method for hot embossing of glass microlens array combined with CO_2 laser irradiation and external preheating/cooling. Lecture Notes in Mechanical Engineering, pp 669–675

Bunker RP, Thabtah F (2019) A machine learning framework for sport result prediction. Appl Comput Inform 15(1):27–33

Chhabra G, Kumar S, Gupta S, Nagpal P (2023) Machine learning for individual performance analysis and sports analytics. In: Artificial intelligence to analyze psychophysiological human lifestyle, pp 113–127

Clermont CA, Duffett-Leger L, Hettinga BA, Ferber R (2019) Runners' perspectives on 'smart' wearable technology and its use for preventing injury. Int J Human-Comput Interact 36:31–40

Donnelly AA, MacIntyre TE, O'Sullivan N, Warrington G, Harrison AJ, Igou ER, Jones M, Gidlow C, Brick N, Lahart I, Cloak R, Lane AM (2016) Environmental influences on Elite sport athletes well being: from gold, silver, and bronze to blue green and gold. Front Psychol 7

Du X, Fuqian X, Hu J, Wang Z, Yang D (2021) Uprising E-sports industry: machine learning/AI improve in-game performance using deep reinforcement learning. In: Proceedings of the 2021 international conference on machine learning and intelligent systems engineering (MLISE 2021), pp 547–552

Elkhoukhi H, Bakhouya M, El Ouadghiri D, Hanifi M (2022) Using stream data processing for real-time occupancy detection in smart buildings. Sensors 22:2371

Ghazali NF, Shahar N, As'ari MA (2022) Badminton strokes recognition using inertial sensor and machine learning approach. In: 2022 2nd international conference on intelligent cybernetics and technology applications (ICICyTA 2022), pp 1–5

Glebova E, Gerke A, Book R (2023) The transformational role of technology in sports events. Sports Manag Uncertain Environ 169–187

Goud PSHV, Roopa YM, Padmaja B (2019) Player performance analysis in sports: with fusion of machine learning and wearable technology. In: International conference on computing methodologies and communication (ICCMC), pp 600–603

Gurcan F, Berigel M (2018) Real-time processing of big data streams: lifecycle, tools, tasks, and challenges. In: ISMSIT 2018—2nd international symposium on multidisciplinary studies and innovative technologies, proceedings

Hasya N, Razali MNM, Jamaludin AS, Saffe SNBM, Hadi MA (2022) Micromilling performance analysis—cold air assisted aluminium machining behaviour observation. In: Innovative manufacturing, mechatronics & materials forum 2022 (iM3F 2022), online conference, Pekan, Malaysia, pp 124–130

Jain V, Mishra AK, Ojha MK (2021) The impact of internet of things in manufacturing industry. Lecture Notes in Mechanical Engineering, pp 517–527

Jamali A, Giman JP (2021) Performance analysis of IOT based flood monitoring framework in sub-urban. In: 2021 IEEE 7th international conference on smart instrumentation, measurement and applications (ICSIMA 2021), pp 99–104

Jamaludin AS, Yassin A (2013) Analysis of laser sintered materials using finite element method. Sains Malaysiana 42(12):1727–1733

Jamaludin AS, Hosokawa A, Furumoto T, Koyano T, Hashimoto Y (2017) Evaluation of the minimum quantity lubrication in orthogonal cutting with the application of finite element method. Int J Mech Mechatron Eng IJMME-IJENS 17(01):104–109

Jamaludin AS, Hosokawa A, Furumoto T, Koyano T, Hashimoto Y (2018) Study on the effectiveness of extreme cold mist MQL system on turning process of stainless steel Aisi 316. IOP Conf Series: Mater Sci Eng 319:012054

Jamaludin AS, Akira H, Furumoto T, Koyano T, Hashimoto Y (2018) High precision estimation on physical behavior for cutting with various tool rake angle by finite element method. Lecture Notes in Mechanical Engineering, pp 715–723

Jamil H, Umer T, Ceken C, Al-Turjman F (2021) Decision based model for real-time IOT analysis using big data and machine learning. Wirel Pers Commun 121:2947–2959

Khan MG, Huda NU, Zaman UK (2022) Smart warehouse management system: architecture, real-time implementation and prototype design. Machines 10:150

Khaund S (2020) Blockchain: from Fintech to the future of sport. 21st century sports 191–203

Khoshkhah K, Pourmoradnasseri M, Hadachi A, Tera H, Mass J, Keshi E, Wu S (2022) Real-time system for daily modal split estimation and OD matrices generation using IOT data: a case study of tartu city. Sensors 22:3030

Kofahi NA, Al-Khatib RM, Alomari A, Mansi TA (2022) A smart real-time IOT-based system for monitoring health of athletes. Int J Comput Digit Syst 12:141–148

Laghari AA, Wu K, Laghari RA, Ali M, Khan AA (2021) Retracted article: a review and state of art of internet of things (IOT). Archiv Comput Methods Eng 29:1395–1413

Lee H, Binti Kamarudin SN, Ishak I, Manaf AR, Jamaludin AS, Shaharudin MA, Zawawi MZ (2021) Feasibility Study of wafer scale laser assisted thermal imprinting of glass nanostructures. Lecture Notes in Mechanical Engineering, pp 917–923

Li N, Zhu X (2023) Design and application of blockchain and IoT-enabled sports injury rehabilitation monitoring system using neural network. Soft Comput 27:11815–11832

Li F, Shi Y, Shinde A, Ye J, Song W (2019) Enhanced cyber-physical security in internet of things through energy auditing. IEEE Internet Things J 6:5224–5231

Li J, Xu J (2021) Outdoor air quality real-time monitoring system for sports athletes. In: IEEE advanced information technology, electronic and automation control conference

Liao B, Ali Y, Nazir S, He L, Khan HU (2020) Security analysis of IOT devices by using mobile computing: a systematic literature review. IEEE Access 8:120331–120350

Manzoor SI, Jain S, Singh Y, Singh H (2023) Federated learning based privacy ensured sensor communication in IOT networks: a taxonomy, threats and attacks. IEEE Access 11:42248–42275

Md Kamil NH, Jamaludin AS, Mhd Razali MN, Abd. Ghaffar AN (2022) Temperature and heat flow analysis in a drying chamber through finite element method. Lecture Notes in Mechanical Engineering, pp 309–316

Naglah A, Khalifa F, Mahmoud A, Ghazal M, Jones P, Murray T, Elmaghraby AS, El-baz A (2018) Athlete-customized injury prediction using training load statistical records and machine learning. In: 2018 IEEE international symposium on signal processing and information technology (ISSPIT)

Nižetić S, Šolić P, López-de-Ipiña González-de-Artaza D, Patrono L (2020) Internet of things (IOT): Opportunities, issues and challenges towards a smart and sustainable future. J Clean Prod 274:122877

Nowak AS, Kennelley GE, Krabak BJ, Roberts WO, Tenforde KM, Tenforde AS (2022) Endurance athletes and climate change. J Climate Change Health 6:100118

Pires M, Santos V (2018) Assessing the impact of internet of everything technologies in football. Adv Intell Syst Comput 745:375–388

Probst L, Rauschenbach F, Schuldt H, Seidenschwarz P, Rumo M (2018) Integrated real-time data stream analysis and sketch-based video retrieval in team sports. In: 2018 IEEE international conference on big data (Big Data)

Qian C, Einhaus L, Schiele G (2022) ElasticAI-creator: optimizing neural networks for time-series-analysis for on-device machine learning in IoT systems. In: Proceedings of the 20th ACM conference on embedded networked sensor systems (SenSys 2022), pp 941–946

Ramli MR, Razak NA, Ismail I, Jamaludin AS, Manaf AR (2022) Effect of dimple size onto wear rate of mild steel AISI 1060 Surface. Lecture Notes in Mechanical Engineering, pp 99–102

Rehman A, Haseeb K, Saba T, Lloret J, Ahmed Z (2021) Mobility support 5G architecture with real-time routing for sustainable smart cities. Sustainability 13:9092

Rejeb A, Rejeb K, Abdollahi A, Al-Turjman F, Treiblmaier H (2022) The interplay between the internet of things and agriculture: a bibliometric analysis and research agenda. Internet of Things 19:100580

Rosli AM, Jamaludin AS, Razali MNM (2021) Recent study on hard to machine material—micromilling process. Evergreen 8:445–453

Rosli AM, Jamaludin AS, Razali MN, Sani AS, Hamzah SB, Osman MS (2019) Modelling of fuzzy inference system for micro milling—a preliminary study through FEM. Lecture Notes in Mechanical Engineering, pp 445–456

Rosli AM, Jamil N, Jamaludin AS, Razali MN, Yusoff AR (2021) Tool wear observation during unconventional low speed machining using low cost micromilling. Lecture Notes in Mechanical Engineering, pp 589–597

Sachs J, Wikstrom G, Dudda T, Baldemair R, Kittichokechai K (2018) 5G radio network design for ultra-reliable low-latency communication. IEEE Netw 32:24–31

Schlegel MU, Hill C (2020) The reach of sports technologies. Lecture Notes in Mechanical Engineering, pp 91–110

Schneider S, Mücke HG (2021) Sport and climate change—how will climate change affect sport? German J Exerc Sport Res

Shah Y, Sengupta S (2020) A survey on classification of cyber-attacks on IoT and IIoT devices. In: Ubiquitous computing, electronics and mobile communication conference

Shi Q (2023) Research on real-time signal transmission system of competitive sports competition under computer big data technology. In: 2023 IEEE international conference on control, electronics and computer technology (ICCECT)

Silva AF, Conte D, Clemente FM (2020) Decision-making in youth team-sports players: a systematic review. Int J Environ Res Public Health 17:3803

Silvers-Granelli HJ, Cohen M, Espregueira-Mendes J, Mandelbaum B (2021) Hamstring muscle injury in the athlete: state of the art. J ISAKOS 6:170–181

Sudharsan B, Malik S, Corcoran P, Patel P, Breslin JG, Ali MI (2021) OWSNet: towards real-time offensive words spotting network for consumer IOT devices. In: 2021 IEEE 7th world forum on Internet of Things (WF-IoT)

Sufian AH, Xun TZ, Abidin AN, Jamaludin AS, Razali MN (2021) Study on tire tread design effect onto tire-road contact behavior through FEM. Lecture Notes in Mechanical Engineering, pp 893–902

Tawalbeh L, Muheidat F, Tawalbeh M, Quwaider M (2020) IOT privacy and security: challenges and solutions. Appl Sci 10:4102

Tedesco S, Scheurer S, Brown KN, Hennessy L, O'Flynn B (2022) A survey on the use of artificial intelligence for injury prediction in sports. In: 2022 IEEE international workshop on sports technology research (STAR 2022)—Proceedings, pp 127–131

Ting L, Khan M, Sharma A, Ansari MD (2022) A secure framework for IOT-based smart climate agriculture system: toward Blockchain and edge computing. J Intell Syst 31:221–236

Torres-Ronda L, Beanland E, Whitehead S, Sweeting A, Clubb J (2022) Tracking systems in team sports: a narrative review of applications of the data and sport specific analysis. Sports Med—Open 8

Tramarin F, Mok AK, Han S (2019) Real-time and reliable industrial control over wireless lans: algorithms, protocols, and future directions. Proc IEEE 107:1027–1052

Worsey MTO, Espinosa HG, Shepherd JB, Thiel DV (2021) One size doesn't fit all: supervised machine learning classification in athlete-monitoring. IEEE Sens Lett 5(3)

Yadav J, Misra M, Rana NP, Singh K, Goundar S (2023) Blockchain's potential to rescue sports: a social media perspective. In: Distributed computing to blockchain architecture, technology and applications, pp 405–414

Yu T, Wang X (2020) Real-time data analytics in Internet of Things systems. In: Handbook of real-time computing, 1–28

Yuan C, Yang Y, Liu Y (2020) Sports decision-making model based on data mining and neural network. Neural Comput Appl 33:3911–3924

Zhao L, Matsuo IBM, Zhou Y, Lee WJ (2019) Design of an industrial IoT-based monitoring system for power substations. IEEE Trans Ind Appl 55(6):5666–5674

Zhou X, Xu X, Liang W, Zeng Z, Yan Z (2021) Deep-learning-enhanced multitarget detection for end–edge–cloud surveillance in smart IOT. IEEE Internet Things J 8:12588–12596

Analysis of Normal and Standing Upper SU 5 Category Para-Shuttle Badminton Performance Parameters Using Dartfish

V. V. Prasanth and G. Nallavan

Abstract Badminton is one of the famous sports that is played in 176 member nations around the world, with five continental confederations organized by the BWF. With the help of video analysis, we can find the strengths and weaknesses of the player, which will be very helpful in improving their performance. Also, the game's statistics are analyzed with the help of the data that has been tagged in the form of an Excel sheet. The errors they make are identified with the help of video data by tagging events and visualizing the data, which is also an inherent feature of Dartfish. This work focuses on the comparative analysis of normal and standing Upper SU 5 category para-badminton players with the help of video. This analysis will help the para-badminton players improve their performance by taking insights from the generated report. The normal match was downloaded from YouTube with better quality, and the para-badminton match was shot in real time during the Chief Minister's Trophy that was held in 2023 at Tamil Nadu Physical Education and Sports University, Chennai is used to analyze. At the end of this work, the data that we took from both matches is compared, and statistical data is provided. Using this type of analysis, the para-player's ability when compared to the normal person's ability can be measured, and the coach can improve their performance with the help of the video montage that is shared using the cloud platform. This can be accessed by both the players and coaches for their betterment in upcoming games.

Keywords Badminton video analysis · Para-Badminton analysis · Video analysis

V. V. Prasanth (✉) · G. Nallavan
Department of Sports Technology, Tamilnadu Physical Education and Sports University, Chennai 600 127, India
e-mail: prasanth199277@gmail.com

Z. Mohamed et al. (eds.), *Proceedings of the 2nd International Conference on Innovation and Technology in Sports, ICITS 2023, 27–28 November, Malaysia*, Lecture Notes in Bioengineering, https://doi.org/10.1007/978-981-97-3741-3_35

1 Introduction

Analysis of player performance using a video helps us to identify and get a lot more data, which is required to identify the opportunities where additional training and coaching are required to overcome the skill set that the player is lacking in achieving the desired result (Abdullahi and Coetzee 2017). In this work, the matches that happened for normal and para-badminton players are analyzed with the help of Dartfish, which will provide us with a lot of insights through visualizing the data using the report section in the same software. Along with this, the montages of each event can be segregated player-wise and shared through the cloud platform. The montages can be viewed by the coaches and players to get an idea of their performance, and they can work on their improvement to achieve their goals.

To achieve better results in analysis from the analyst's point of view, the highest number of analysts who use the tool in the Olympics is Dartfish. The same is to be used to identify the key areas for improvement and the better performance that was achieved in that game or match. In Dartfish, we now have all the facilities to create a tagging panel, import video, tag it, create montages, and create Excel data about the events using the event that we wanted to identify with the help of the tagging panel, which is customizable for our convenience. Dartfish also provides us with the feature of extracting insights and data visualization using the Excel data that is generated during the tagging process. We can also use Excel, Tableau, SPSS, and other tools to visualize the data that we extract from the match video. In this work, we have used the inbuild report panel to create the match data and provide insights and data visualization for understanding the events that occurred during the match.

2 Methods

2.1 Participants

In this work, we took one para-badminton and one normal match to find the variance among these two matches, which were randomly picked for analysis purposes and occurred as a state-level event that occurred in Tamil Nadu. At first, the para-badminton match of singles from Chief Minister's Trophy 2023 was taken, and analysis was done with the help of Dartfish. Also, we took a normal badminton match, which is held during the All India Junior Ranking Badminton Tournament and is randomly selected. The selected matches are analyzed with the help of Dartfish using the tagging panel that was created with the number of players, field map, and other important events that are part of the game and their probabilities. So, we may have a clear picture of how the normal and para-badminton matches are getting varied in various events that occur during the match (Gawin and Seidler 2015).

2.2 Procedures

In the initial stage of the analysis, the rules of the game and events that play a major role in badminton are to be understood clearly by the analyst. Once this phase is completed, we can go ahead and create a tagging panel with the list of events that are used in the analysis work that we are carrying out throughout. For reference purposes, here is an image of the tagging panel below. Also attached is the setup panel for creating this tagging panel (Prasanth and Nallavan 2022). In the initial stage, we created a group box with service, receive, foul, and point using the keywords for both teams, Team A and Team B. Also, the scoring panel is fixed outside the group box, using which we can increase or decrease the score for both teams, which is included with the property sheet. Once the above activities are done, the events are purely based on the probability that they may or may not happen (Salim et al. 2010). Based on that, we design the outcomes of service outcome, receive outcome, foul outcome, and point outcome, which are considered to be the important events in badminton, as we discussed above. After this, the badminton court image is loaded to locate exactly in which location of the action is performed in that match for a clear understanding by the analyst, coach, and player (Barreira et al. 2016; Cabello Manrique and González-Badillo 2003). Using this, the analyst can segregate the areas where they have missed and where they have the opportunity to score a point. The next phase is that we have to tag the person who is involved in that particular event with the player name using the player group box. Along with this, the match setup panel is also created with the value button and continuous value button in a group, match details, and team manager. Then the downloaded match will be imported, and analysis will be started using the other tools based on their needs (Figs. 1 and 2).

2.3 Data Analyses

The data that is analyzed using Dartfish is visualized in the below images, which also show the list of events, zone maps, and other details of the match in detail below (Figs. 3, 4, 5, 6, 7 and 8).

3 Results

To analyze the matches in Dartfish randomly, we took two matches, one from the All India Junior Ranking Badminton Tournament and the other from the Chief Minister's Trophy for para-badminton players. In this work, we have analyzed events like service, receive (this includes the receive that was made even in terms of rally, along with receive after service), fault, and point, which are the key events that have to be noted. In the singles match that happened at the All India level, the player who

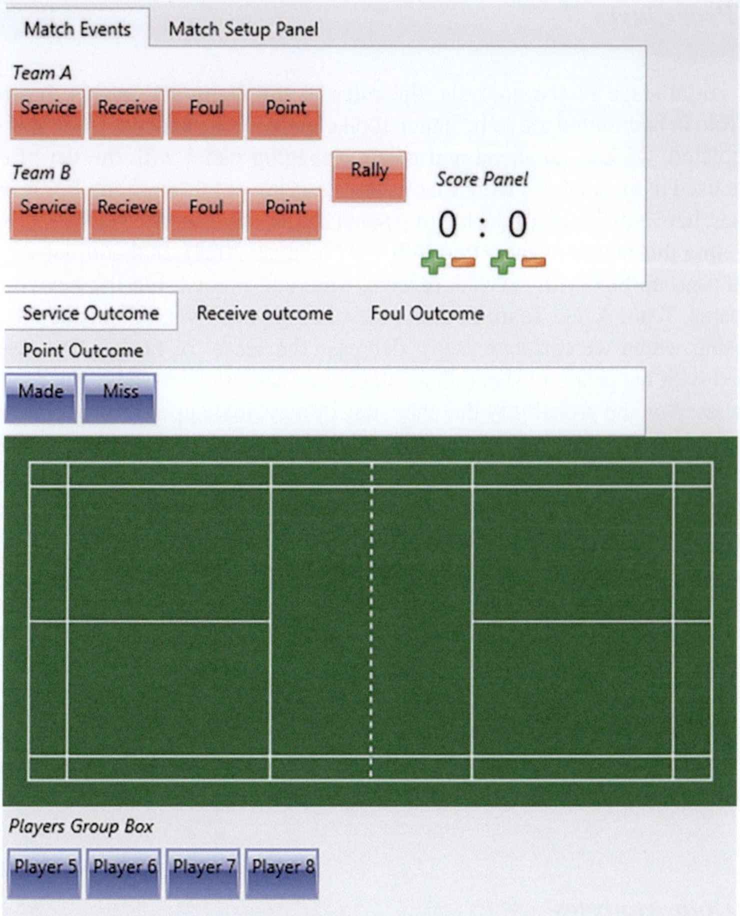

Fig. 1 Tagging panel for shuttle-badminton analysis. This panel is used to analyze the events of the game and the possible outcomes, including where and who performed the activity

reaches 21 points is declared to be the winner of the game, or else the player who leads by two points first is the winner of the best of three games. In this normal match, A team's player Syush Shetty took 21 points, and B team's Darshan Pujari took 14 points in the first game. Totally, 34 services happened in the first game, out of which 20 were from Syush Shetty and 14 from Darshan Pujari. The total received by Syush Shetty is 122, out of which 4 times he missed receiving the shuttle and 118 times he received it successfully, and Darshan Pujari's total received is 127, out of which 7 times he missed the shuttle and 120 times he received the shuttle successfully (Table 1).

In the second game, Syush Shetty took 21 points and Darshan Pujari took 16 points. Totally, 37 services happened in the first game, out of which 20 were from Syush Shetty and 16 from Darshan Pujari. The total received by Syush Shetty is

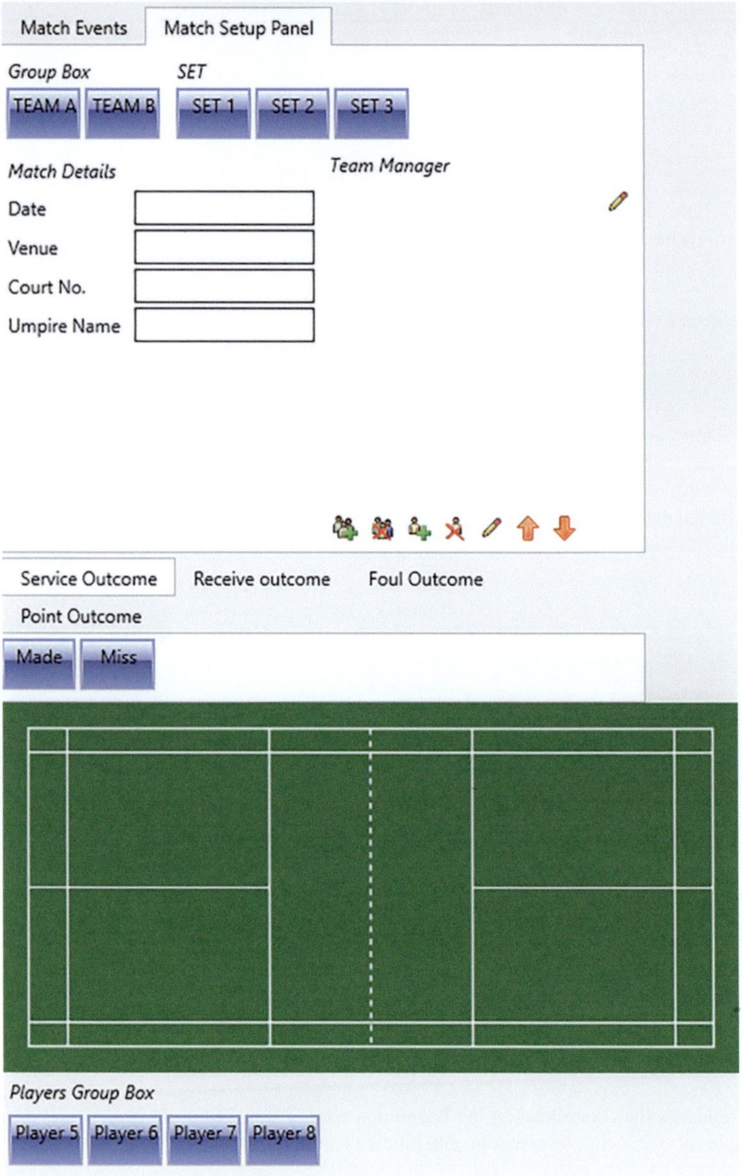

Fig. 2 Match setup panel for tagging shuttle badminton

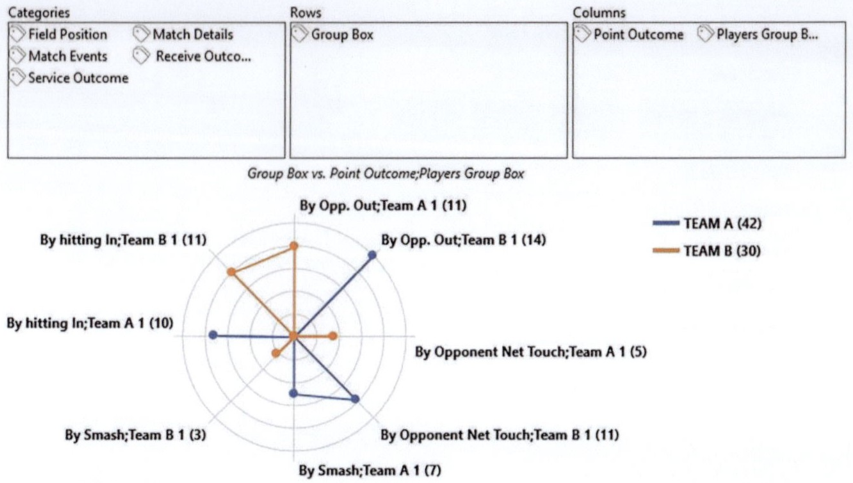

Fig. 3 Radial data visualization using the report window. In this, we can see the individual team's point outcome and the different ways in which points are scored for the team in two sets

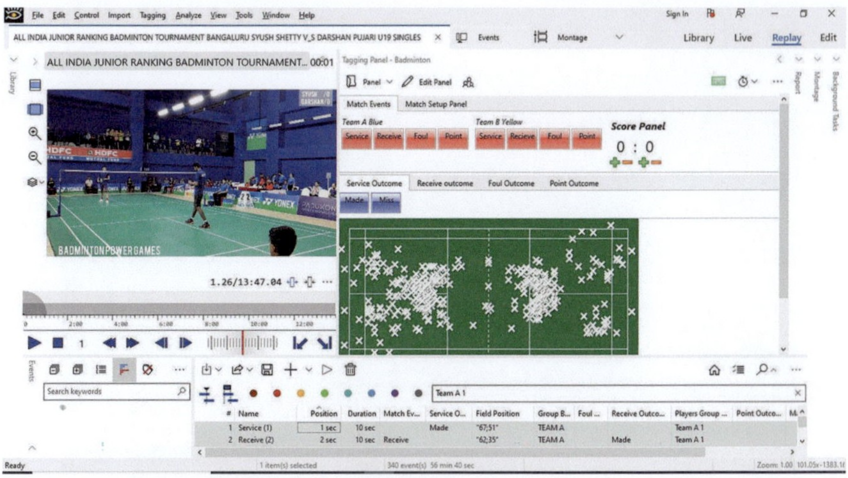

Fig. 4 Indicates the coordinates of the badminton court where Team A's players and the shuttle involved in service, point, receive, out, and hit the court

145, out of which 7 times he missed receiving the shuttle and 138 times he received it successfully, and Darshan Pujari's total received is 150, out of which 9 times he missed the shuttle and 141 times he received the shuttle successfully. Finally, the A team's player, Syush Shetty, won the match, leading the score in two games, which is one sided (Table 2, Fig. 9).

In the singles match that happened in the Chief Minister's Trophy for para-badminton in the category of Standing Upper SU 5 (which indicates impairment

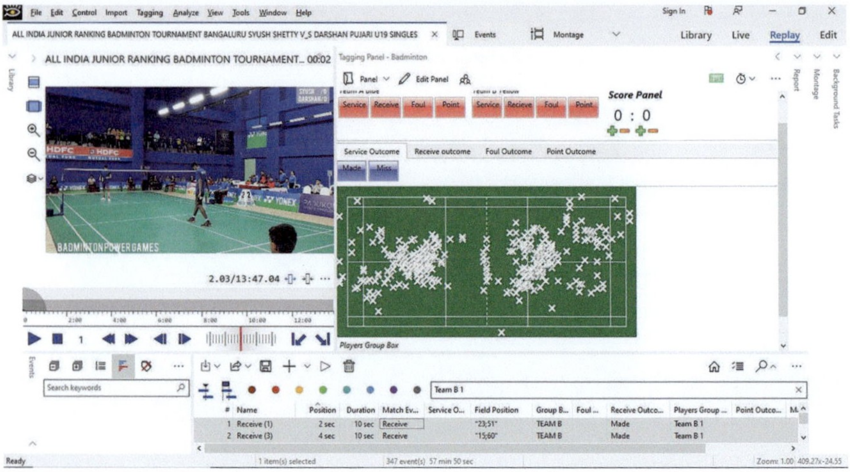

Fig. 5 Indicates the coordinates of the badminton court where Team B's players and the shuttle involved in service, point, receive, out, and hit the court

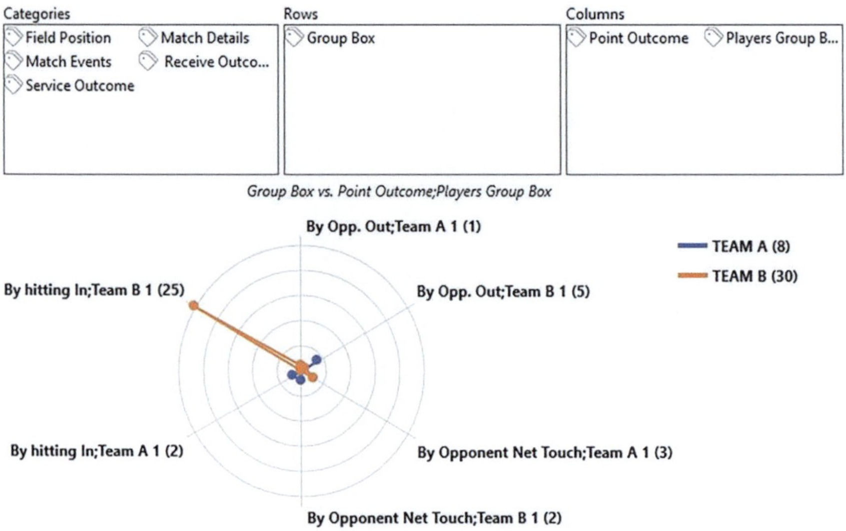

Fig. 6 Radial data visualization using the report window. In this, we can see the individual team point outcome and different ways in which points are scored for the team in all the two sets of para-badminton matches that took place in the Chief Minister's Trophy 2023. Other data is also listed below

Fig. 7 Indicates the coordinates of the badminton court where Para-Badminton Team A's players and the shuttle involved in service, point, receive, out, and hit in the court (Abian-Vicen et al. 2013)

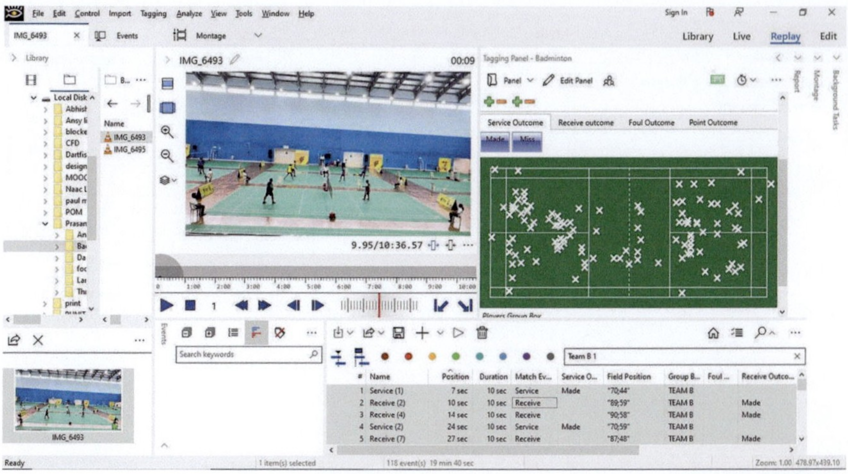

Fig. 8 Indicates the coordinates of the badminton court where Para-Badminton Team B's players and the shuttle involved in service, point, receive, out, and hit in the court (Abian-Vicen et al. 2013)

Table 1 First game data of normal match between Sayush Shetty and Dharshan Pujari

S.No	Total service	Total receive	Received	Missed receive
Sayush Shetty(21)	20	122	118	4
Dharshan Pujari(14)	14	127	120	7

Table 2 Second Game data of normal match between Sayush Shetty and Dharshan Pujari

S.No	Total service	Total receive	Received	Missed receive
Sayush Shetty(21)	20	145	138	7
Dharshan Pujari(16)	17	150	141	9

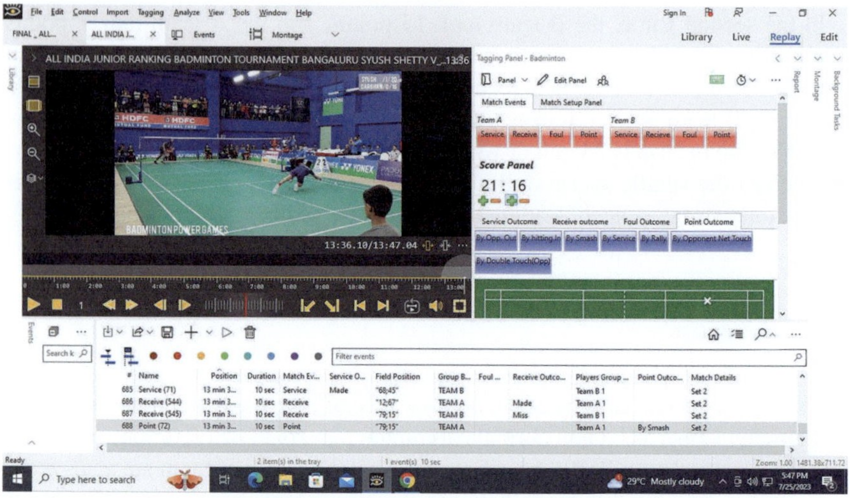

Fig. 9 Shows the window that is updated with the score at the second set of the game

in the upper limb either in playing or non-playing hand) (International Paralympic Committee 2023), which is state level. The player who reaches 15 points is declared to be the winner of the game, or else the player who leads by two points first is the winner of the best of three games (Table 3).

In this para-badminton match, the B team's player from Coimbatore took 15 points, and the A team's player on the other end took 3 points in the first game. Totally, 18 services happened in the first game; out of those, 14 were from the B team's player and 4 were from the A team's player who started from the left of the court in video. The total received is 54, and the total received by the B team is 25 out of which 1 time he missed receiving the shuttle and 24 times he received it successfully, and the A team's total received is 29 out of which 3 times he missed the shuttle and 26 times he received the shuttle successfully (Table 4).

Table 3 First game data of Para-badminton match

S.No	Total service	Total receive	Received	Missed receive
A Team(3)	4	29	26	3
B Team(15)	14	25	24	1

Table 4 Second game data of para-badminton match

S.No	Total service	Total receive	Received	Missed receive
A Team(5)	5	33	27	6
B Team(15)	15	32	32	0

In the second game, the B team took 15 points, and the A team took 5 points. Totally, 20 services happened in the first game, out of which 15 were from the B team and 5 from the A team. The total received in the second game is 65, out of which the receive by B team is 32, out of which all 32 times he received it successfully, and team A's total receive is 33, out of which 6 times he missed the shuttle and 27 times he received the shuttle successfully. Finally, the B team's player who represented Coimbatore won the match, leading the score in two games, which is one sided.

4 Discussion

This study shows the analysis of the performance of both para-badminton players in the Standing Upper SU 5 and normal badminton players in two different events that occurred (International Paralympic Committee 2023). Both of them are analyzed to compare and find the difference between the events that occur in two matches of different categories. We have provided the zone mapping of the shuttle badminton court, which will also enable the player and coach to figure out the opportunity to improve in landing the shuttle in a better position. In the results, we have clearly shown the number of events that occurred in both matches. In which it shows that the normal players are always having higher number of events in their match than the Standing Upper SU 5 category para-badminton match (International Paralympic Committee 2023). However, this is not for the purpose of comparing the events. Also, this will help the player identify the weak areas where they have missed opportunities to perform well in their game or match. The analyzed matches can be shared with the players who want to improve their performance with the help of cloud storage, and it will generate a link in Dartfish. Using that link, they can go through with their performance and improve their skills with the help of the coach to perform better in the upcoming matches.

5 Conclusion and Future Work

Dartfish Video Analysis is one of the best software programs that provides video analysis services to players and coaches to improve their game and coaching strategies, respectively. In this work, two different matches are selected randomly and imported into the Dartfish software, and they are analyzed to get data about the game and the player's performance. Here, the two matches were analyzed, and it was seen that the Standing Upper SU 5 category para-badminton team has performed very well (International Paralympic Committee 2023; Abián et al. 2014). Even though they have been trained in this, the performance of the Standing Upper SU 5 category para-badminton from Coimbatore is good (International Paralympic Committee 2023). The instinct and strategy were very good. When coming into the normal match analysis and Standing Upper SU 5 category para-badminton analysis comparison from a video analyst point of view, the total number of events count is 689 for a normal match, and the Standing Upper SU 5 category para-badminton player event is 196. According to this data, the normal match's events are approximately 351.6% higher. It might vary based on the performance increase and decrease of the player in other matches if the events are calculated like this. However, this study doesn't provide additional information that could provide physical and psychological data that will affect the player's performance. For future work, with the help of sensors, we can have a detailed analysis of the heart rate, fatigue level, and other issues that might affect the players to keep them informed about them. With this kind of analysis, we can aim for a world-class performer in terms of standing Upper SU 5 and other categories in para-badminton. Also, in future the sensors and other technologies should be incorporated to improve the player's performance.

References

Abdullahi Y, Coetzee B (2017) Notational singles match analysis of male badminton players who participated in the African Badminton Championships. Int J Perform Anal Sport 17:1–2, 1–16.https://doi.org/10.1080/24748668.2017.1303955

Abian-Vicen J, Castanedo A, Abian P, Sampedro J (2013) Temporal and notational comparison of badminton matches between men's singles and women's singles. Int J Perform Anal Sport 13(2):310–320. https://doi.org/10.1080/24748668.2013.11868650

Abián P, Castanedo A, Feng XQ, Sampedro J, Abian-Vicen J (2014) Notational comparison of men's singles badminton matches between olympic games in Beijing and London. Int J Perform Anal Sport 14(1):42–53. https://doi.org/10.1080/24748668.2014.11868701

Badminton World Federation (BWF). Rules of Badminton. http://www.bwfbadminton.org/, last accessed 2023/09/26.

Barreira J, Chiminazzo JGC, Fernandes PT (2016) Analysis of point difference established by winners and losers in games of badminton. Int J Perform Anal Sport 16(2):687–694. https://doi.org/10.1080/24748668.2016.11868916

Cabello Manrique D, González-Badillo JJ (2003) Analysis of the characteristics of competitive badminton. Br J Sports Med 37:62–66

Gawin W, Beyer C, Seidler M (2015) A competition analysis of the single and double disciplines in world-class badminton. Int J Perform Anal Sport 15(3):997–1006. https://doi.org/10.1080/24748668.2015.11868846

International Paralympic Committee. https://www.paralympic.org/badminton/classification, last accessed 2023/09/26.

Prasanth VV, Nallavan G (2022) Analysis of football players' performance using python and dartfish 9:2395–0056

Salim MS, Lim HN, Salim MSM, Baharuddin MY (2010) Motion analysis of arm movement during badminton smash. In: 2010 IEEE EMBS conference on biomedical engineering and sciences (IECBES), Kuala Lumpur, Malaysia, pp 111–114. https://doi.org/10.1109/IECBES.2010.5742210

Applying Convolutional Neural Networks (CNNs): A Machine Learning Method in Modern Sports

Muhamad Ridzuan Radin Muhamad Amin, Abdul Nasir, M. H. Muhammad Sidik, Ahmad Shahir Bin Jamaludin, and Ainur Munira Rosli

Abstract This review paper aims to examine the applications of Convolutional Neural Network (CNN) in the field of sports biomechanics. CNN, which is a specific category of deep learning models, has exhibited considerable promise in the analysis of intricate data structures. As a result, they have become indispensable instruments in various domains. Machine learning has been utilized in the field of sports biomechanics to gain insights and improve athletic performance. CNN has been effectively employed in sports analytic studies, specifically in movement algorithm and sports video analysis. CNN has demonstrated its efficacy in the domain of sports rehabilitation, in terms of application and sensor-based sports injury rehabilitation. Notwithstanding these advancements, there remain gaps and prospects for additional research, particularly in the realm of integrating the various applications of CNN. This review aims to identify gaps in the existing literature and suggest potential areas for future research, specifically focusing on the application of CNN in the field of sports biomechanics.

Keywords Convolutional neural networks · Sports biomechanics · Rehabilitation · CNN-based action recognition

M. R. R. M. Amin · A. Nasir · A. M. Rosli
Faculty of Electrical and Electronic Engineering Technology, Universiti Malaysia Pahang Al-Sultan Abdullah, 26600 Pekan, Pahang, Malaysia

M. H. Muhammad Sidik (✉)
Faculty of Mechanical and Automotive Engineering Technology, Universiti Malaysia Pahang Al-Sultan Abdullah, 26600 Pekan, Pahang, Malaysia
e-mail: hanafisidik@umpsa.edu.my

A. S. B. Jamaludin
Faculty of Manufacturing and Mechatronic Engineering Technology, Universiti Malaysia Pahang Al-Sultan Abdullah, 26600 Pekan, Pahang, Malaysia

Z. Mohamed et al. (eds.), *Proceedings of the 2nd International Conference on Innovation and Technology in Sports, ICITS 2023, 27–28 November, Malaysia*, Lecture Notes in Bioengineering, https://doi.org/10.1007/978-981-97-3741-3_36

1 Introduction

The emergence of machine learning, particularly convolutional neural networks (CNNs) has brought about significant transformations in diverse domains, such as to enhance the quality of medical imaging to identify diseases in the field of healthcare (Akinyelu et al. 2022; Xiang et al. 2022), mapping of landslide susceptibility and the classification of coastal wetlands in geospatial research (Jamali and Mahdianpari 2022), efficacy in the classification of malware in domain of cybersecurity (Lin and Yeh 2022) and successfully employed for the purpose of acoustic monitoring in robotic facilities in manufacturing sector (Bynum and Lattanzi 2021). The utilization of sophisticated computational models has been proven to possess the capability to analyze intricate data structures and produce profound predictions, rendering them indispensable instruments for both researchers and practitioners (Shetty et al. 2022; Fan and Sun 2022). In sports field, machine learning has been utilized in the field of sports biomechanics to gain insights and improve athletic performance (Meng and Wang 2022).

The success of CNN in the image domain has inspired research on CNN-based action video recognition (Ravanbakhsh et al. 2015; Muhammad et al. 2021). The methodologies and CNN architectures extended from image to video have undergone significant advancements. Therefore, CNN-based action recognition has become an active research field, with numerous CNN-based approaches, which have also achieved great success, recently emerging (Pajak et al. 2022; Akan and Varlı 2023). CNN's success in images domain has become the inspirations of research in CNN-based video recognition (Host and Ivašić-Kos 2022). In addition, a rapid advancement can be seen in methodologies and CNN architecture that extends from image to video based. Accordingly, there are numerous numbers of CNN-based action recognition research related to sport application been conducted which have achieved great success and the trend is increasing from year 1993 to 2023 (November) based on database from Scopus (www.Scopus.com) as shown in Fig. 1.

Despite the notable progress made in this field, there remain certain gaps and prospects for additional investigation, specifically concerning the incorporation of various applications of CNN. The purpose of this review is to examine the potential

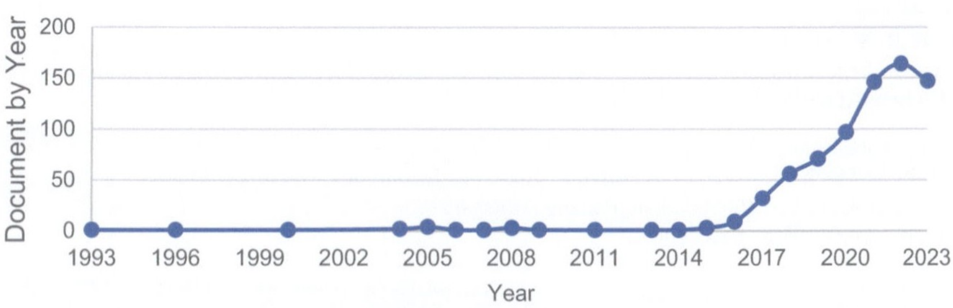

Fig. 1 Total publication in Scopus for application CNN in sports field

opportunities in the field of sports science, specifically in relation to the utilization of CNN.

The objective of this review is to offer a comprehensive examination of the existing body of research concerning the utilization of CNN across diverse domains, with specific emphasis on the domain of sports biomechanics. Our objective is to analyze the principal methodologies, findings, current applications, while also emphasizing potential avenues for future research. The review's focus is restricted to scholarly articles published from 2020 onwards, which have employed CNN in the domains of health care, sports biomechanics, sports analytics, and sports injury rehabilitation. Our attention will be directed toward research studies that have showcased novel applications of CNN and have made noteworthy advancements in their respective domains.

2 CNN-Based Action Recognition

CNN is a biologically inspired variant of multilayer perception, which feeds into artificial neural networks. It is made up of an output layer, input layer, and a series of hidden layers. Convolutional, pooling, or fully connected are the hidden layers. Convolutional layers apply a convolution procedure and bias added into input data, which then passes results from one layer to another using the activation function. The equation of convolution in y-axis (jth) feature map in x-axis (ith) layer as Eq. (1).

$$v_{il}^{xy} = \varphi\left(b_{i,j} + \sum_{m} \sum_{p=0}^{P_i-1} \sum_{q=0}^{Q_i-1} w_{i,j,m}^{p,q} v_{(i-1),m}^{(x+p),(y+q)}\right) \quad (1)$$

φ = non-linear activation function

w = weight matrix

P = height of the kernel

Q = width of the kernel.

One of the non-linear down-sampling methods is pooling layer. High-level reasoning in the CNN starts with convolutional, pooling layers and lastly fully connected layers where every neuron is connected to all previous layers' activation. History in advancement of CNN architecture when LeNet-5 was developed in order to numbers in document in 1998 (LeCun et al. 1998). CNN was developing at a slow pace, but until its huge breakthrough in 2012 with image classification. CNN's ability to classify images and detect objects has been significantly improved in recent years. ResNets, ZFNet, VGG, BN-Inception, GoogLeNet, AlexNet, and many more have been introduced. These CNN architectures provide pre-trained models that are obtained from pre-training large datasets. Further training must be conducted to increase accuracy of pre-trained model when there is modality or new small-scale of dataset. Due to the performance of CNN in image recognition, many researchers started to apply it into action recognition. Consequently, now there are many CNN-based action recognition research that have been published and showed a good performance with the result.

As shown in Eq. (1), initially CNN has been used in 2D space and is proven to be very powerful with the extraction of spatial information from still images. Then, CNN was adopted in extracting spatial information for action recognition, which is considered as the auxiliary cue and combined with features such as Improved Dense Trajectories (iDT) to enhance action recognition (Wang et al. 2014). CNN was mainly used for the extraction of spatial features, but video is viewed as 3D spatiotemporal signal and makes CNN low performance because of insufficient temporal information (Yao et al. 2019). The exploitation of temporal information is the main thing that must be addressed in order to extend the application of CNN from image to video (Suresha et al. 2020). 3D convolution, use motion-related information as input for CNN, and fusion are strategies that can be used to exploit the lack of temporal information.

3 CNN in Sports Biomechanics

The utilization of machine learning, specifically deep learning models such as CNN, has experienced a growing trend within the domain of sports biomechanics (Xiang et al. 2022; Yan et al. 2022; Hsu et al. 2019). Computational models present the capacity to analyze intricate biomechanical data, thereby offering valuable insights that have the potential to optimize athletic performance and mitigate the likelihood of sustaining injuries (Ramesh and Mahesh 2020).

Bynum et al. (Bynum and Lattanzi 2021) undertook a comprehensive analysis of the advancements made in the utilization of machine learning techniques in the domain of lower limb running biomechanics, specifically focusing on the integration of wearable technology (Xiang et al. 2022). The potential of utilizing machine learning techniques for the analysis of biomechanical data is to forecast running injuries and enhance running performance. The research also highlighted the importance of utilizing more extensive and varied datasets in order to enhance the applicability of machine learning models in the field of sports biomechanics (Xiang et al. 2022). Figure 2 shows that deep learning algorithms occupied 57% of total machine learning approaches, compared with 43% of traditional machine learning algorithms. CNN was the most frequently used algorithm (seven times), followed by ANN and long short-term memory (LSTM).

One of the challenges in computer vision is video classification (Islam et al. 2020; Bhatt et al. 2021). For sports video, common actions are shown between categories of sport such as jumping is a common action among diving and high jumps. To identify the type of action, a tool or a model with high accuracy rate is needed. The objective of video classification is to provide videos of sportsperson's interest to improve their training and at the same time increase their performance in the future (Feng et al. 2019).

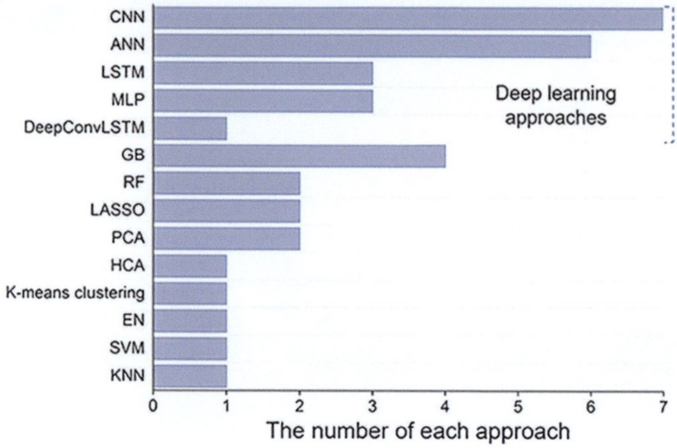

Fig. 2 Machine learning approaches (Xiang et al. 2022)

A research been conducted on video recognition to compare pre-trained neural networks which are GoogLeNet and AlexNet with CNN (Ramesh and Mahesh 2020). This study used UCF101 which is an action recognition data set of realistic action videos, collected from YouTube, having 101 action categories (Ramesh and Mahesh 2019). The result shows that the pre-trained neural networks are classifying with higher accuracy (GoogLeNet 91.67% and AlexNet 92.68%) when compared with CNN where it is just 88.75%. The classification accuracy can be increased by adding numbers of layer and changing the option of training with another dataset.

4 CNN in Sports Analytics

4.1 CNN-Based Human Movement Recognition Algorithm

CNNs have become a potent tool in the field of computer vision and artificial intelligence for understanding human movement. For example as shown in Fig. 3, CNN-based human movement recognition algorithm is transforming how we study and comprehend human behavior (Ullah et al. 2021).

This algorithm makes use of the complex CNN architecture to analyze visual data quickly and reliably. By examining patterns and characteristics in individual picture or video frames, it has the capacity to recognize and categorize a broad variety of human motions, from tiny gestures to challenging athletic actions. What sets this algorithm apart is its versatility. It finds applications across diverse domains, including sports analysis, health care, security, and even entertainment. In sports, it aids coaches and athletes by providing insights into performance and injury prevention. The CNN-based human movement recognition algorithm continues to evolve,

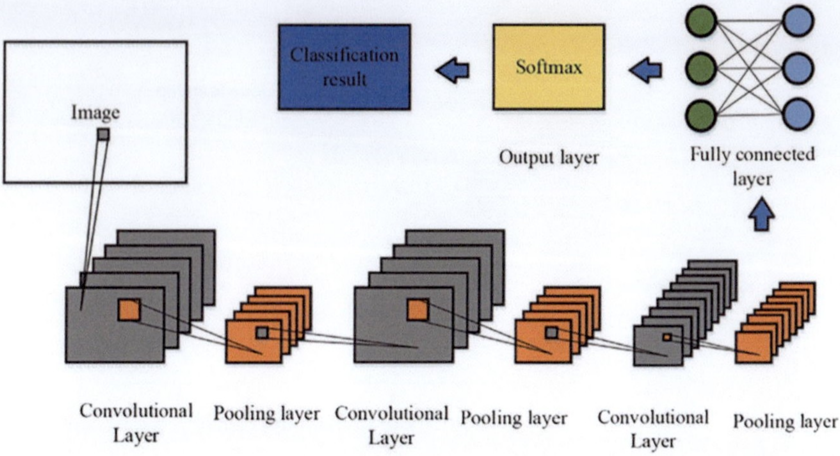

Fig. 3 Structure of the optimized C3D (Ullah et al. 2021)

thanks to ongoing advancements in deep learning and computer vision. Its potential to transform industries and enhance our understanding of human motion remains limitless. As the world increasingly relies on visual data, this algorithm serves as a cornerstone in unlocking the insights hidden within our movements.

4.2 Deep Learning in Sports Video Analysis

In the world of sports, every move, every play, and every gesture can make the difference between victory and defeat. Understanding these intricacies has long been a challenge, but with the advent of deep learning, the game is changing. Table 1 shows the comparison between 2D-CNN, 3D-CNN, Recurrent Neural Network (RNN) and Long Short-Term Memory (LTSM) deep learning models. At first, CNN developed to extract information from images and 2D-CNN is for videos. But due to 2D-CNN failure to extract temporal information in videos, 3D-CNN is created for spatial and temporal information extraction. Then RNN model was built in action recognition but, lack of short-term memory and can't use in real time application. LTSM was developed to overcome this issue and now widely used in computer vision applications. Deep learning is transforming sports video analysis (Rangasamy et al. 2020), offering a new level of insight into athlete performance and strategy. At the heart of this revolution is the application of CNN and Recurrent Neural Networks (RNNs) (Lin and Yeh 2022). These deep learning models can process vast amounts of video data, tracking player movements, ball trajectories, and game dynamics with remarkable precision.

In real-time, coaches and analysts can now access a wealth of information, from player positioning to passing accuracy, and even predict opponent strategies. This

Table 1 Comparison between deep learning models

Model	Advantage	Drawback
2D-CNN	Automatically capture the spatial information in the image patches	Could not be able to capture the temporal information in video data
3D-CNN	Automatically capture both spatial and temporal information	Expensive model due to 3D
RNN	Automatically capture the temporal information in sequential data	Has short memory ability, could not apply in real situation Gradient explosion
LSTM	Automatically capture the temporal information in sequential data	Nil

level of detail not only aids in performance optimization but also plays a crucial role in injury prevention and recovery. Furthermore, deep learning enhances the fan experience. Augmented reality overlays, player tracking, and advanced statistics are all made possible through these technologies. Fans can now immerse themselves in the game on a whole new level, thanks to the insights provided by deep learning in sports video analysis. As deep learning continues to advance, the possibilities are endless. With every game, athletes and fans alike are benefiting from the powerful synergy of sports and technology.

4.3 Sports Health Management Model Based on Deep Learning

The application of deep learning techniques has also been extended to the development of models for managing sports health such as introducing a sports health management model that utilizes deep learning techniques (Meng and Wang 2022; Xiao et al. 2022; Ma and Zhang 2022). The proposed model employs CNN to analyze data obtained from wearable devices. This analysis enables the generation of personalized exercise recommendations and health management strategies tailored specifically for athletes. The research shows the capacity of deep learning techniques in the field of sports health management, with a specific focus on personalized training and injury prevention. Nevertheless, it also acknowledged the necessity for additional research in order to authenticate and enhance the model, emphasizing the significance of practical experimentation and ongoing model enhancement (Meng and Wang 2022).

5 CNN in Sports Injury Rehabilitation

5.1 Advanced CNN Applications in Sports Injury Rehabilitation

Within the realm of sports injury rehabilitation, modern technology is fundamentally altering how athletes recover and return to their peak performance (Zhu and Sun 2019). CNN has arisen as a potent force, introducing a level of precision and effectiveness never seen in the rehabilitation process. CNN was originally designed for image recognition, are being harnessed to analyze intricate biomechanical movements. By processing visual data from motion capture systems and wearable sensors, CNN can monitor an athlete's movements with unparalleled accuracy. This technology enables therapists and sports scientists to tailor rehabilitation programs with a level of personalization that was once unimaginable.

One of the most significant advantages of CNN-based rehabilitation (Fan and Sun 2022) is real-time feedback. Athletes can receive immediate insights into their movements, helping them correct errors and prevent further injury (Liang et al. 2023; Chen and Yuan 2021). This proactive approach not only expedites recovery but also reduces the risk of reinjury. Moreover, CNN facilitates remote monitoring, allowing experts to oversee an athlete's progress from anywhere in the world. This level of accessibility ensures continuous care and guidance, even during off-site training or competitions. As we delve deeper into the realm of sports injury rehabilitation, the role of advanced CNN applications becomes increasingly crucial. These technologies are not merely tools, they are game changers. Offering athletes, a path to recovery that is more precise, personalized, and efficient than ever before.

5.2 CNN Approaches for Sensor-Based Sports Rehabilitation

Utilizing sensor-based data, deep learning algorithms analyze an athlete's movements with unparalleled accuracy (Meng and Wang 2022). These sensors, often integrated into wearable devices, capture every nuance of motion, from joint angles to muscle activation. Deep learning models process this wealth of data to provide personalized rehabilitation plans and real-time feedback.

One of the most significant advantages of deep learning in sports rehabilitation is adaptability. These models continuously adapt to an athlete's progress, modifying exercises and routines accordingly. This dynamic approach expedites recovery, reduces the risk of reinjury, and enhances overall performance. Moreover, remote monitoring capabilities allow experts to oversee an athlete's progress from anywhere, ensuring continuous care and guidance. With deep learning, sensor-based sports rehabilitation is not just a treatment but a personalized and data-driven journey to peak athletic potential.

6 Gaps and Future Directions

6.1 Identified Gaps in the Current Literature

The field of sports analytics now has a notable deficiency in effectively incorporating a wide range of data sources (Chmait and Westerbeek 2021). The existing body of literature predominantly centers around the utilization of CNN within the field of biomechanics. However, it does not extensively address the complex task of integrating physiological, psychological, and environmental data to achieve a comprehensive study of athlete performance. Furthermore, the existing discussion fails to adequately address the examination of real-time performance optimization in the context of live sporting events. The lack of available tools in the market and approaches for prompt insights hinders the effective use of CNN in enhancing athlete performance in real-time situations.

Even though in recent years, several machine learning models have been introduced and applied to recognize the movement of athletes, there are some parts of it where further study is needed by researchers and experts. It is where a machine learning classifier combines with other classifiers to improve the accuracy (Ali et al. 2022). A study conducted by combining 2 machine learning classifiers showed a better performance rate of accuracy (Li et al. 2023).

6.2 Proposed Solutions and Future Research Directions

Future research endeavors in the field of sports analytics should prioritize the investigation of solutions aimed at achieving a smooth and efficient integration of varied data sources. The comprehensive examination of athlete performance necessitates the integration of physiological measures, psychological variables, and environmental situations. It is imperative to direct research endeavors toward the development of real-time analytics systems that utilize CNN. These systems are expected to provide real-time analysis on performance optimization and injury prevention during live sporting events. Furthermore, it is imperative for academics to delve into innovative uses of artificial intelligence (AI) in the realm of sports that extend beyond the scope of performance analysis (Naser and Alavi 2021; Araújo et al. 2021). Exploring the potential of AI technologies, such as CNN to boost fan engagement through personalized experiences, augmented reality applications, and interactive content is a compelling direction for future academic inquiry. The developed system should only be limited to one machine learning classifier (Ullah et al. 2021; Li et al. 2023). Should there be a few options that users can choose which are producing the highest accuracy in recognizing the type of motions?

Acknowledgements The author wishes to express profound gratitude to the Universiti Malaysia Pahang Al-Sultan Abdullah for their invaluable support and financial assistance under the grant RDU Number RDU230328.

References

Akan S, Varlı S (2023) Use of deep learning in soccer videos analysis: survey. Multim Syst 29(3):897–915

Akinyelu AA et al (2022) Brain tumor diagnosis using machine learning, convolutional neural networks, capsule neural networks and vision transformers, applied to MRI: a survey. J Imag 8(8):205

Ali OMA, Kareem SW, Mohammed AS (2022) Evaluation of electrocardiogram signals classification using CNN, SVM, and LSTM algorithm: a review. In: 2022 8th international engineering conference on sustainable technology and development (IEC). IEEE

Araújo D et al (2021) Artificial intelligence in sport performance analysis. Routledge

Bhatt D et al (2021) CNN variants for computer vision: history, architecture, application, challenges and future scope. Electronics 10(20):2470

Bynum J, Lattanzi D (2021) Combining convolutional neural networks with unsupervised learning for acoustic monitoring of robotic manufacturing facilities. Adv Mech Eng 13(4):16878140211009016

Chen X, Yuan G (2021) Sports injury rehabilitation intervention algorithm based on visual analysis technology. Mob Inf Syst 2021. https://doi.org/10.1155/2021/9993677

Chmait N, Westerbeek H (2021) Artificial intelligence and machine learning in sport research: an introduction for non-data scientists. Front Sports Act Living 3:363

Fan S, Sun Y (2022) Early detection of Parkinson's disease using machine learning and convolutional neural networks from drawing movements. In: CS & IT conference proceedings. CS & IT conference proceedings

Feng X et al (2019) Computer vision algorithms and hardware implementations: a survey. Integration 69:309–320

Host K, Ivašić-Kos M (2022) An overview of human action recognition in sports based on computer vision. Heliyon

Hsu Y-L, Chang H-C, Chiu Y-J (2019) Wearable sport activity classification based on deep convolutional neural network. IEEE Access 7:170199–170212

Islam MS, Sultana S, Kumar Roy U, Al Mahmud J (2020) A review on video classification with methods, findings, performance, challenges, limitations and future work. J Ilm Tek Elektro Komput Dan Inform (JITEKI) 6:47–57

Jamali A, Mahdianpari M (2022) Swin transformer and deep convolutional neural networks for coastal wetland classification using sentinel-1, sentinel-2, and LiDAR data. Remote Sens 14(2):359

LeCun Y, Bottou L, Bengio Y, Haffner P (1998) Gradient-based learning applied to document recognition. Proc IEEE 86(11):2278–2324

Li H et al (2023) Subject-independent estimation of continuous movements using CNN-LSTM for a home-based upper limb rehabilitation system. IEEE Robot Autom Lett

Liang C et al (2023) Effective automatic detection of anterior cruciate ligament injury using convolutional neural network with two attention mechanism modules. BMC Med Imag 1–13. https://doi.org/10.1186/s12880-023-01091-6

Lin W-C, Yeh Y-R (2022) Efficient malware classification by binary sequences with one-dimensional convolutional neural networks. Mathematics 10(4):608

Ma X, Zhang Z (2022) Research on sports health care information system based on computer deep learning algorithm. Comput Intell Neurosci

Meng J, Wang S (2022) The construction of sports health management model based on deep learning. Appl Bionics Biomech
Muhammad K et al (2021) Human action recognition using attention based LSTM network with dilated CNN features. Futur Gener Comput Syst 125:820–830
Naser M, Alavi AH (2021) Error metrics and performance fitness indicators for artificial intelligence and machine learning in engineering and sciences. Architect Struct Constr 1–19
Pajak G et al (2022) An approach to sport activities recognition based on an inertial sensor and deep learning. Sens Actuators, A 345:113773
Ramesh M, Mahesh K (2019) Sports video classification with deep convolution neural network: a test on UCF101 dataset. Int J Eng Adv Technol 8(4S2):2249–8958
Ramesh M, Mahesh K (2020) A performance analysis of pre-trained neural network and design of CNN for sports video classification. In: 2020 international conference on communication and signal processing (ICCSP). IEEE
Rangasamy K, As'ari MA, Rahmad NA, Ghazali NF, Ismail S (2020) Deep learning in sport video analysis: a review. Telkomnika (Telecommunication Computing Electronics and Control) 18(4):1926–1933. https://doi.org/10.12928/TELKOMNIKA.V18I4.14730
Ravanbakhsh M et al (2015) Action recognition with image based CNN features. arXiv preprint. arXiv:1512.03980
Shetty B et al (2022) Skin lesion classification of dermoscopic images using machine learning and convolutional neural network. Sci Rep 12(1):18134
Suresha M, Kuppa S, Raghukumar D (2020) A study on deep learning spatiotemporal models and feature extraction techniques for video understanding. Int J Multim Inf Retr 9:81–101
Ullah M et al (2021) Attention-based LSTM network for action recognition in sports. Electron Imag 33:1–6
Wang L, Qiao Y, Tang X (2014) Action recognition and detection by combining motion and appearance features. THUMOS14 Action Recognition Challenge 1(2):2
Xiang L et al (2022) Recent machine learning progress in lower limb running biomechanics with wearable technology: a systematic review. Front Neurorobot 16:913052
Xiao L et al (2022) Management and analysis of sports health level of the elderly based on deep learning. Comput Intell Neurosci
Yan S, Chen J, Huang H (2022) Biomechanical analysis of martial arts movements based on improved pso optimized neural network. Mob Inf Syst
Yao G, Lei T, Zhong J (2019) A review of convolutional-neural-network-based action recognition. Pattern Recogn Lett 118:14–22
Zhu P, Sun F (2020) Sports athletes' performance prediction model based on machine learning algorithm. In: International conference on applications and techniques in cyber intelligence ATCI 2019: applications and techniques in cyber intelligence 7. Springer

Optimizing 3D Printing Manufacturing Process for Sports Instrumentation Production

Muhammad Nur Farhan Saniman, Norasilah Karumdin, Ahmad Shahir Jamaludin, Mohd Nizar Mhd Razali, and Abdul Nasir

Abstract The utilization of 3D printing technology has emerged as a significant catalyst for change across diverse manufacturing industries. This innovative technique is distinguished by its capacity to generate complex designs, facilitate personalized production, and expedite the creation of prototypes. This analysis examines the inherent obstacles and constraints associated with 3D printing techniques, with a specific focus on the complexities of material selection and optimization. The selection of materials is a critical factor in determining the overall quality of the product. However, this process is often challenging due to the wide range of available options and the necessity to consider several factors, including cost, mechanical qualities, and suitability. The process of optimization poses challenges in terms of computational requirements and comprehending the intricate interactions among process factors. It is worth noting that there exists a substantial research vacuum in the field, specifically with the utilization of 3D printing technology for the creation of sports instruments. Despite the distinct requirements of the sports business, such as the need for robustness, precision, and customization, there has been limited attention given to this area of study. The exploration of this discrepancy possesses the capacity to transform the manufacturing of sports equipment, hence guaranteeing improved safety measures and customized solutions for players. Future research areas emphasize the necessity of investigating novel material combinations and customizing optimization methods to suit the intricacies of sports equipment production.

M. N. F. Saniman
Mechanical Engineering Section, Universiti Kuala Lumpur Malaysia France Institute, 43650 Bandar Baru Bangi, Malaysia

N. Karumdin (✉) · A. S. Jamaludin · M. N. M. Razali
Faculty of Manufacturing and Mechatronic Engineering Technology, Universiti Malaysia Pahang Al-Sultan Abdullah, 26600 Pekan, Pahang, Malaysia
e-mail: norasilahkarumdin@gmail.com

A. Nasir
Faculty of Electrical and Electronic Engineering Technology, Universiti Malaysia Pahang Al-Sultan Abdullah, 26600 Pekan, Pahang, Malaysia

Z. Mohamed et al. (eds.), *Proceedings of the 2nd International Conference on Innovation and Technology in Sports, ICITS 2023, 27–28 November, Malaysia*, Lecture Notes in Bioengineering, https://doi.org/10.1007/978-981-97-3741-3_37

Keywords 3D printing · Sports instrumentation · Material selection · Optimization techniques · Additive manufacturing

1 Introduction to 3D Printing Manufacturing for Sports Instrumentation Production

Three-dimensional (3D) printing, commonly referred to as additive manufacturing, is a revolutionary method employed in industrial production that facilitates the fabrication of components and systems with enhanced strength and reduced weight (Ping 2021; Raam Kumar et al. 2021; Babu et al. 2023, 2022; Hasya et al. 2022; Gomes et al. 2022; Nasir et al. 2013; Vidakis et al. 2022; Ramli et al. 1060; Hasdiansah 2022; Atalla et al. 2022; Chohan et al. 2022; Raja et al. 2022). The procedure entails the creation of three-dimensional solid things from a digital file with additive manufacturing techniques (Raja et al. 2022; Md Kamil et al. 2022; Atalla et al. 2020; Jamaludin and Yassin 2013; Hameed et al. 2022; Pyteraf et al. 2022; Jamaludin et al. 2017; Kim et al. 2022; Alsibai et al. 2018; Shariff et al. 2021; Abidin et al. 2022; Abdullah et al. 2022; Baharuddin et al. 2022; Farag et al. 2019; Nasir et al. 2015; Babu et al. 2021; Engler et al. 2022). In the procedures, the fabrication of an item is achieved through the sequential deposition of material layers till its finalization (Ping 2021; Raam Kumar et al. 2021; Babu et al. 2023, 2022; Hasya et al. 2022; Gomes et al. 2022). One notable benefit of 3D printing is in its capacity to fabricate complex geometries that are otherwise unachievable using conventional manufacturing methods like casting, forging, and machining (Engler et al. 2022; Ali et al. 2022; Rosli et al. 2021; Li et al. 2022; Jamaludin et al. 2018; Norazman et al. 2020; Abas et al. 2022; Saniman et al. 2020, 2022; Rutzen et al. 2021). In addition, it provides manufacturers with the opportunity to engage in mass customization, hence facilitating the customization of products to meet the specific requirements of individual consumers, as shown in Fig. 1 (Raam Kumar et al. 2021; Babu et al. 2023, 2022; Hasya et al. 2022; Gomes et al. 2022; Nasir et al. 2013; Vidakis et al. 2022).

The utilization of 3D printing technology in the field of sports instrumentation production has significant promise. The utilization of this technology has the potential to streamline the development of customized equipment tailored specifically to athletes, resulting in improved performance and reduced susceptibility to injuries. However, to fully use the capabilities of 3D printing in this field, it is imperative to do a thorough optimization of the 3D printing process. The critical procedures encompass the careful choice of materials, precise adjustment of printing conditions, and utilization of appropriate post-processing techniques to achieve the intended mechanical qualities and surface finishes (Gomes et al. 2022; Nasir et al. 2013; Vidakis et al. 2022; Ramli et al. 1060; Hasdiansah 2022; Atalla et al. 2022; Chohan et al. 2022), as shown in Table 1 (Hasdiansah 2022).

The review aims to summarize the latest developments in optimization techniques for 3D printing, with a specific focus on the manufacture of sports equipment. In this

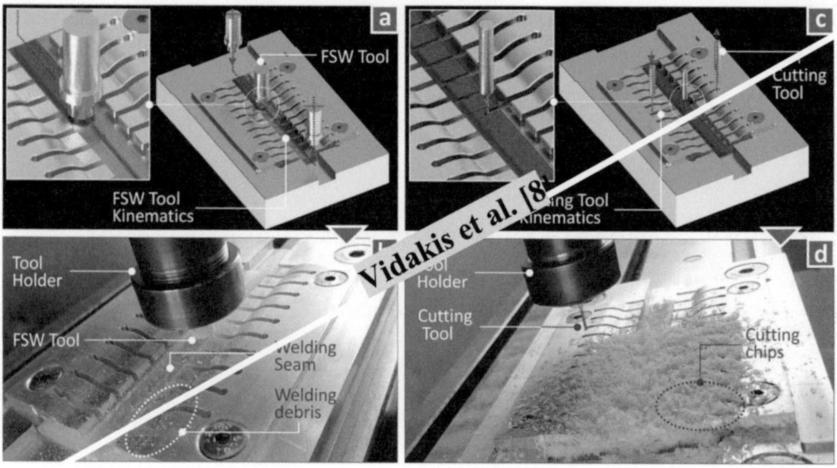

Fig. 1 Example of customization of products to meet the specific requirements of individual consumers from micromilling process of plastic material (Vidakis et al. 2022)

Table 1 Factors that affect mechanical qualities and surface finishes of 3D printed product (Hasdiansah 2022)

Code	Factor	Level		
		1	2	3
A	Nozzle temperature (°C)	256	258	260
B	Bed temperature (°C)	95	98	100
C	Flowrate (%)	90	95	100
D	Printing speed (mm/s)	25	30	35
E	Layer thickness (mm)	0.2	0.24	0.28
F	Wall thickness (mm)	0.8	1.0	1.2
G	Overlap (%)	5	10	15
H	Infill density (%)	20	25	30
I	Infill speed (mm/s)	20	25	30

study, we investigate the intricacies of material selection, examine the ideal printing parameters, and analyze the importance of post-processing processes (Vidakis et al. 2022; Raja et al. 2022; Md Kamil et al. 2022; Atalla et al. 2020; Jamaludin and Yassin 2013; Hameed et al. 2022). Moreover, the discourse expands to emphasize the numerous applications of 3D printing in the realm of sports equipment, identifying areas that show potential and should be further investigated (Vidakis et al. 2022; Chohan et al. 2022; Pyteraf et al. 2022). The review extracts valuable lessons from recent research that use optimization approaches, specifically the genetic algorithm, to enhance the efficiency of 3D printing procedures (Kim et al. 2022). Significantly, the materials that serve as the foundation for these advancements, such as polylactic acid (PLA) and acrylonitrile butadiene styrene (ABS), are given particular emphasis due to their crucial functions and their appropriateness for sports instruments (Chohan et al. 2022; Shariff et al. 2021).

2 Applications of 3D Printing and 3D Printing Materials

2.1 Application of 3D Printing onto Sports Instrumentation Production

The utilization of 3D printing technology has the capacity to bring about significant changes in the domain of sports equipment manufacturing. Customized equipment, designed to accommodate the unique anatomical characteristics of athletes, has the potential to not only improve athletic performance but also reduce the risk of injuries. With the utilization of 3D printing technology, it has been possible to manufacture customized safety equipment, such as helmets and pads that conform precisely to the anatomical features of individual athletes (Raam Kumar et al. 2021; Babu et al. 2023, 2022; Hasya et al. 2022; Gomes et al. 2022). This practice guarantees not only enhanced comfort but also maximum protection effectiveness. In addition, the utilization of 3D printing technology enables the production of custom-fit footwear, which has the potential to optimize an athlete's performance while simultaneously reducing the occurrence of foot-related injuries (Ping 2021; Raam Kumar et al. 2021; Babu et al. 2023, 2022; Hasya et al. 2022; Gomes et al. 2022; Nasir et al. 2013; Vidakis et al. 2022; Ramli et al. 1060; Hasdiansah 2022; Atalla et al. 2022, 2020; Chohan et al. 2022; Raja et al. 2022; Md Kamil et al. 2022; Jamaludin and Yassin 2013; Hameed et al. 2022; Pyteraf et al. 2022).

One of the most appealing benefits of incorporating 3D printing technology in the field of sports lies in its capacity to produce sports equipment with complex geometries that are typically unachievable using conventional manufacturing techniques. As an example, the utilization of technology enables the creation of golf clubs that exhibit intricate internal configurations, precisely engineered to optimize the performance attributes of the club (Ping 2021). In addition to its utilization in the field of sports, the utilization of 3D printing is increasingly being observed in a diverse range of applications. The field of personalized medicine is currently experiencing a transformative period, driven by advancements in 3D printing technology. A previous study had conducted a significant investigation that explored the complexities associated with the utilization of daytime polymer 3D printing in the production of personalized pharmaceutical items (Raam Kumar et al. 2021). The results of their study highlight the significant potential of 3D printing technology in the development of personalized medicine to meet the specific requirements of individual patients (Ping 2021; Raam Kumar et al. 2021).

Furthermore, the construction industry is increasingly adopting the revolutionary capabilities of 3D printing. According to a previous study, 3D printing has emerged as a highly effective manufacturing technique for carbon fiber-reinforced cementitious composites, demonstrating remarkable flexural strength (Rutzen et al. 2021). This observation implies the potential of 3D printing technology in the development of durable and resilient construction materials. Previous research has demonstrated the multifunctionality of materials used in 3D printing and their potential for optimization. An example of this may be seen in acrylonitrile styrene acrylate (ASA) specimens produced using Fused Deposition Modeling (FDM), where it has been seen that an augmentation in infill percentage results in an elevation of tensile strength along the direction of printing (Raam Kumar et al. 2021). In a similar vein, the management of the characteristics of three-dimensional printed components derived from recycled thermoplastics can effectively address the issue of material deterioration (Gomes et al. 2022). This example illustrates the wide range of research endeavors aimed at enhancing the manufacturing process of 3D printing for diverse applications, hence underscoring its significance in contemporary production methodologies.

2.2 3D Printing Materials and Its Properties

Polylactic Acid (PLA) is a thermoplastic aliphatic polyester that is both biodegradable and bioactive. It is created from renewable resources, including corn starch, cassava roots, and sugar cane. The substance is commonly employed in the field of 3D printing due to its user-friendly nature, environmentally sustainable properties, and favorable mechanical characteristics (Raja et al. 2022). In contrast, scholarly investigations pertaining to the enhancement of 3D printing process parameters for PLA in Fused Deposition Modeling (FDM) have revealed that the extruder temperature, bed temperature, layer height, and printing speed are crucial variables that significantly impact the mechanical properties and dimensional accuracy of PLA-based prints (Raja et al. 2022). Previous researchers had conducted a comprehensive study to examine the relationship between several parameters of 3D printing and properties such as tensile strength, modulus, and energy absorption rate for PLA, where it can serve as a guide to produce parts with enhanced strength and endurance (Raja et al. 2022; Engler et al. 2022).

Acrylonitrile Butadiene Styrene (ABS) is another widespread thermoplastic polymer in 3D printing, respected for its robustness, adaptability, and high-temperature resistance. Chohan et al. (2022) conducted a relevant investigation that examined the impact of distinct FDM printing parameters on the surface polish, thickness, and external dimensions of ABS samples. These observations can play a crucial role in enhancing the 3D printing techniques specifically designed for manufacturing sporting equipment made from ABS material (Chohan et al. 2022; Raja et al. 2022; Md Kamil et al. 2022; Atalla et al. 2020; Jamaludin and Yassin 2013; Hameed et al. 2022; Pyteraf et al. 2022).

Nylon, renowned for its remarkable tensile strength, flexibility, and long-lasting nature, also holds significance within the domain of 3D printing. The study had focused on the optimization of 3D printing parameters specifically for nylon filaments (Hasdiansah 2022). The study emphasized the importance of many parameters, such as nozzle temperature, bed temperature, layer thickness, and printing speed, in enhancing the quality of 3D printed sports equipment (Hasdiansah 2022).

Carbon fiber-reinforced materials are increasingly being utilized in the field of 3D printing, mostly due to their exceptional flexural strength, which makes them highly suitable for applications that require significant durability and lifespan. The study had shed light on the integration of tensile reinforcement inside 3D printing methodologies, where the possibility of refining 3D printing techniques had been demonstrated for the production of sports equipment by utilizing carbon fiber-reinforced materials (Rutzen et al. 2021).

In short, there exists a wide range of materials for 3D printing, each possessing unique characteristics and offering various possibilities for practical utilization. The enhancement of performance and dependability of sports equipment can be achieved by the optimization of printing parameters for these materials, as indicated by several research (Hasdiansah 2022; Atalla et al. 2022; Chohan et al. 2022; Raja et al. 2022; Shariff et al. 2021; Rutzen et al. 2021).

3 Optimization Techniques in 3D Printing

3.1 Taguchi Method and Genetic Algorithms

The Taguchi technique is a statistical methodology developed with the objective of improving the quality of manufactured products. The utilization of this methodology has been observed in the field of 3D printing, as it aids in comprehending the impact of many parameters on the overall quality of the result. The mechanical properties of Acrylonitrile Styrene Acrylate (ASA) in 3D printing were investigated by a previous study (Vidakis et al. 2022) through the utilization of the Taguchi method. The researchers' findings underscored the significant impact that the appropriate selection and adjustment of printing parameters can have on enhancing the mechanical qualities of 3D printed components. Additionally, it was determined that variables such as infill density and layer height significantly impacted the outcome, as seen by the attainment of maximum tensile strengths reaching 51.86 MPa (Vidakis et al. 2022).

Genetic algorithms (GAs) are widely employed in optimization issues across diverse areas, including the field of 3D printing. These algorithms draw inspiration from natural selection and genetics, making them a fundamental tool in this area. Search-based algorithms demonstrate a notable proficiency in identifying the most optimal solutions within a somewhat extensive solution space. Chohan et al. (2022) conducted a study utilizing genetic algorithms to optimize the process parameters of Fused Deposition Modeling (FDM) printing. The study specifically aimed to improve the surface smoothness, thickness, and exterior dimensions of ABS polymer specimens. The study not only emphasized the effectiveness of genetic algorithms (GAs) in improving the 3D printing process, but it also established a clear correlation between printing parameters and outcomes. This was further supported by conducting practical test tests, which ultimately validated their findings (Shariff et al. 2021).

In addition to the Taguchi method and genetic algorithms, the field of 3D printing has observed the implementation of several optimization techniques. The Design of Experiments (DoE) methodology, for example, offers a systematic approach to determining the relationship between multiple elements that impact a process and the resulting outcome. In their study on the optimization of individualized medical products by daylight polymer 3D printing, Pyteraf et al. (2022) employed the Design of Experiments (DoE) technique. In contrast to the reference provided, the actual suitability of the Design of Experiments (DoE) technique was demonstrated in a research endeavor focused on optimizing the parameters of 3D printing to attain enhanced tensile strength, modulus, and energy absorption rate. The study emphasized the considerable influence of many parameters, including layer height, infill percentage, and infill pattern, on the optimization strategies in the field of 3D printing (Pyteraf et al. 2022).

4 Challenges and Limitations in 3D Printing

4.1 Challenges in Material Selection

The careful selection of suitable materials for 3D printing is of utmost importance as it significantly influences the overall quality of the end result (Raam Kumar et al. 2021; Babu et al. 2023, 2022; Hasya et al. 2022; Gomes et al. 2022; Nasir et al. 2013; Vidakis et al. 2022; Ramli et al. 1060; Hasdiansah 2022; Atalla et al. 2022; Chohan et al. 2022; Raja et al. 2022). The complexity of this selection arises from the wide range of materials that are accessible, as well as the need to consider many factors such as mechanical qualities, cost, and suitability for the intended application. A comprehensive investigation was carried out to analyze the mechanical characteristics of a specimen made of Acrylonitrile Styrene Acrylate (ASA) using the Fused Deposition Modeling (FDM) technique (Raam Kumar et al. 2021). The study underscored the significance of taking into account the infill % and build orientation in order to achieve the appropriate levels of tensile and flexural strength. This sheds light on the complex correlation between the choice of materials and the subsequent parameters used in printing in order to achieve the best possible results (Raam Kumar et al. 2021).

Moreover, the viability of materials utilized in conjunction with particular processes assumes a crucial significance. A previous study examined the controlling characteristics of 3D printed components made from recycled thermoplastics, providing insights into the difficulties presented by material degradation (Gomes et al. 2022). The results of their study highlight the significance of addressing the deterioration of materials in distributed recycling, which emerges as a key factor in the choice of materials for sustainable 3D printing initiatives (Gomes et al. 2022).

Engler et al. (2022) conducted a separate investigation whereby they explored the difficulties associated with material selection. Specifically, they examined the process of 3D printing PLA and Halloysite Nanotube (HNT) composites utilizing APF. The researchers' findings underscored the need of taking into account factors such as material composition, machine settings, deposition angles, and axis orientations in order to attain the most favorable mechanical properties, particularly in the context of dealing with composite materials.

In his study on optimizing process parameters for achieving dimensional precision, Hasdiansah (2022) emphasized the complexities associated with the selection of appropriate materials for 3D printing, specifically focusing on nylon filaments. The study highlighted the difficulties posed by materials that may demonstrate dimensional instability while being printed and emphasized the importance of comprehending the ideal parameters, such as nozzle temperature and bed temperature, in order to address these concerns effectively. In brief, although 3D printing presents a wide range of potential applications, the process of selecting appropriate materials remains a complex one, necessitating a thorough understanding of the material's characteristics, its compatibility with the chosen 3D printing methodology, and the desired qualities of the end product (Hasdiansah 2022; Atalla et al. 2022, 2020; Chohan et al. 2022; Raja et al. 2022; Md Kamil et al. 2022; Jamaludin and Yassin 2013; Hameed et al. 2022; Pyteraf et al. 2022; Jamaludin et al. 2017; Kim et al. 2022; Alsibai et al. 2018; Shariff et al. 2021).

4.2 Limitations in the Optimization Process

The optimization of 3D printing is crucial for attaining desired outcomes, but it can pose constraints, particularly in terms of computational requirements and the complexities associated with specific methodologies. Li et al. (2022) conducted a study that investigated the examination of the BP Neural Network for optimizing 3D printing parameters in order to minimize material deformations. One of the primary observations revealed that the BP Neural Network Algorithm, despite its excellent precision indicated by a correlation coefficient of 0.99985, necessitates a substantial volume of data input. Furthermore, the computing demands of this task may provide challenges in various production environments or for different user groups.

Moreover, the study conducted by Abas et al. (2022) emphasizes the significance of parametric investigation in understanding the influence of various process factors on the final results, specifically with regard to dimensional variations in the fused deposition modeling of 3D printed components. This study has demonstrated that although advantageous, conducting parametric investigations can be a time-consuming endeavor. Furthermore, in order to enhance optimization strategies derived from these research, it is important to possess a comprehensive comprehension of the 3D printing procedure. The research revealed that the impacts of infill density and layer height on several characteristics of the print were diverse, highlighting the intricate nature of parameter interactions (Abas et al. 2022).

Furthermore, it is worth noting that the research did not delve into the topic of material choices and their interaction with optimization methodologies. As an illustration, prior studies (Raam Kumar et al. 2021) have examined materials such as acrylonitrile styrene acrylate (ASA), which have demonstrated alterations in mechanical properties, specifically tensile strength, in response to fluctuations in parameters like infill percentage (Engler et al. 2022; Ali et al. 2022; Rosli et al. 2021; Li et al. 2022). These interactions serve to complicate the optimization process and emphasize the necessity for thorough investigations and comprehension. In brief, the utilization of optimization approaches in the context of 3D printing is highly valuable. However, it is crucial to acknowledge and comprehend the inherent limits associated with these techniques, specifically with regard to temporal constraints, computational requirements, and the complex interdependencies among various process parameters (Li et al. 2022; Jamaludin et al. 2018; Norazman et al. 2020; Abas et al. 2022; Saniman et al. 2020, 2022; Rutzen et al. 2021).

5 Research Gaps and Future Directions

5.1 Identified Research Gaps

The utilization of 3D printing technology has attracted considerable interest across many industrial industries owing to its capacity to generate complex designs, enable product customization, and expedite the process of creating prototypes. Numerous research have been conducted to investigate the optimal strategies for enhancing 3D printing parameters, examining various materials, and assessing post-printing interventions. For example, researchers have undertaken endeavors to comprehend the mechanical characteristics of acrylonitrile styrene acrylate (ASA) specimens fabricated by Fused Deposition Modeling (FDM). Their investigations have indicated that augmenting the infill percentage can enhance the tensile strength of these specimens. In a similar vein, scholarly investigations pertaining to the enhancement of the 3D printing procedure utilizing nylon filaments have successfully identified the most favorable parameters, including a nozzle temperature of 256 °C and a bed temperature of 96 °C. Significant advancements have also been made in comprehending the prospective utilization of recycled thermoplastics within the realm of 3D printing, as well as in devising strategies to alleviate material deterioration.

Notwithstanding these developments, there is a conspicuous deficiency in research, namely, a scarcity of literature that particularly delves into the optimization of 3D printing methods for the fabrication of sports instrumentation. The reviewed research, although highly valuable, generally focus on the larger aspects of 3D printing, including parameters, materials, and applications, without delving into the specialized intricacies and unique demands of sports instruments. Given the distinct requirements inherent in the sports business, including the imperative for durability, lightweight design, precision, and customized fit, it becomes evident that there exists a noteworthy prospect for further investigation and exploration in this domain.

In light of the indisputable advantages offered by 3D printing in the realm of sports, such as the ability to produce personalized equipment, guarantee the suitability of designs for certain sports, and incorporate intricate geometries, it is imperative to address this existing research need. By optimizing 3D printing methodologies and variables specifically for the sports industry, there exists the possibility of transforming the production of sports equipment, hence enhancing safety, efficiency, and customization for athletes.

5.2 *Future Directions*

Dedicated study is necessary to address the existing knowledge gaps in the field of 3D printing technology for the manufacturing of sports instruments, given its significant advancements. The aforementioned studies have yielded significant contributions in the field of optimization techniques, including the utilization of genetic algorithms and the Design of Experiments (DoE) method. However, it is important to note that the application of these techniques to the production of sports instrumentation may necessitate adaptations or novel approaches.

Moreover, it is noteworthy to mention that a wide range of materials have been extensively investigated for their potential use in 3D printing applications. The research included the utilization of materials such as acrylonitrile styrene acrylate (ASA), which revealed that an augmentation in infill percentage leads to a corresponding enhancement in tensile strength along the print direction. Additionally, investigations were conducted on nylon filaments to determine the ideal parameters for achieving dimensional precision. Furthermore, a study conducted on polylactic acid (PLA) revealed that its suitability for filament manufacturing can be enhanced by optimizing parameters according to the results of tensile tests. Moreover, when PLA is mixed with halloysite nanotube (HNT) composites, it demonstrates exceptional mechanical capabilities under specific deposition angles and axis orientations. The ABS specimens yielded noteworthy observations on the impact of printing speed on surface roughness. Furthermore, there is intriguing research that suggests the potential of utilizing carbon fiber-reinforced techniques to enhance the flexural strength of items produced by 3D printing.

Although these materials have demonstrated considerable potential in diverse applications, the unique requirements and constraints of sports instrumentation, such as durability, adaptability, and weight considerations, may require the investigation of novel materials or the creative combination of existing ones. In addition, it is imperative to comprehend the mechanical characteristics of these materials, including tensile strength, flexural strength, and impact strength, as evidenced by research conducted on acrylonitrile styrene acrylate (ASA) specimens.

In summary, in order to enhance the efficiency of 3D printing in the production of sports instrumentation, it is imperative to consider not only the selection of appropriate materials but also to comprehend and improve the pertinent printing characteristics associated with each material. Further investigation is warranted to conduct a more comprehensive examination of novel materials and the customization of optimization methods in order to effectively meet the distinct demands of sporting equipment.

Acknowledgements The authors are very grateful to the Ministry of Higher Education (MoHE) Malaysia and the Universiti Malaysia Pahang Al-Sultan Abdullah for their great financial support, which made this study possible. This study was made possible by grants from the Fundamental Research Scheme (FGRS) FRGS/1/2022/TK10/UMP/02/57 and the Universiti Malaysia Pahang Al-Sultan Abdullah Fundamental Research Grant RDU220365.

References

Abas M, Habib T, Noor S, Salah B, Zimon D (2022) Parametric investigation and optimization to study the effect of process parameters on the dimensional deviation of fused deposition modeling of 3D printed parts. Polymers 14:3667

Abdullah NA, Abdullah FF, Sufian AH, Abidin AN, Jamaludin AS, Razali MN (2022) Effect of degradation by temperature onto nitrile rubber elastomer mechanical properties. Mater Today: Proc 48:1941–1946

Abidin AN, Azmi AH, Kassim KA, Jamaludin AS, Razali MN (2022) A review on automotive tires significant characteristic identification for general consumers. In: Proceedings of the 2nd energy security and chemical engineering congress, pp 375–385

Ali MA, Sivarao, Abdullah Z, Izamshah R, Kassim MS, Jamaludin AS (2022) Multi-response optimization of machining simulation approach using grey relational analysis (Lecture notes in mechanical engineering), pp 154–157

Alsibai MH, Hamran NNN, Nasir A, Abdullah AA (2018) A monitoring system for EPW safe use. J Telecommun, Electron Comput Eng (JTEC) 10(1–2):131–135

Atalla Y, Hashim Y, Ghafar ANA, Jabbar WA (2020) Temperature characteristics of FinFet based on channel fin width and working voltage. Int J Electr Comput Eng (IJECE) 10:5650

Atalla Y, Hashim Y, Ghafar ANA (2022) The impact of channel fin width on electrical characteristics of Si-FinFET. Int J Electr Comput Eng (IJECE) 12:201

Babu D, Nasir A, Jamaludin AS, Rosle MH (2021) Holding, grasping and sensing of prosthetic robot arm like a real human hand, a journey beyond limits: an extensive review. In: Human-centered technology for a better tomorrow, pp 485–504

Babu D, Nasir A, Farag M, Sidik MH, Rejab SB (2022) Development of prosthetic robotic arm with patient monitoring system for disabled children; preliminary results. In: 2022 9th international conference on electrical and electronics engineering (ICEEE)

Babu D, Nasir A, Ravindran, Farag M, Jabbar WA (2023) 3D printed prosthetic robot arm with grasping detection system for children. Int J Adv Sci, Eng Inf Technol 13:226

Baharuddin NJ, Abdul Manaf AR, Jamaludin AS (2022) Study on die shoulder patterning method (DSPM) to minimise springback of U-bending. Int J Automot Mech Eng 19:9509–9518

Chohan JS, Kumar R, Yadav A, Chauhan P, Singh S, Sharma S, Li C, Dwivedi SP, Rajkumar S (2022) Optimization of FDM printing process parameters on surface finish, thickness, and outer dimension with ABS polymer specimens using Taguchi orthogonal array and genetic algorithms. Math Probl Eng 2022:1–13

Engler LG, Crespo JS, Gately NM, Major I, Devine DM (2022) Process optimization for the 3D printing of PLA and HNT Composites with Arburg plastic freeforming. J Compos Sci 6:309

Farag M, Azlan NZ, Alsibai MH, Ghafar AN (2019) Slippage detection for grasping force control of robotic hand using force sensing resistors. In: Proceedings of the 2019 5th international conference on computer and technology applications

Gomes TE, Cadete MS, Dias-de-Oliveira J, Neto V (2022) Controlling the properties of parts 3D printed from recycled thermoplastics: a review of current practices. Polym Degrad Stab 196:109850

Hameed AZ, Aravind Raj S, Kandasamy J, Shahzad MA, Baghdadi MA (2022) 3D printing parameter optimization using Taguchi approach to examine acrylonitrile styrene acrylate (ASA) mechanical properties. Polymers 14:3256

Hasdiansah H (2022) Optimization of process variables in 3D printing on dimensional accuracy using nylon filaments. IJNMT (Int J New Media Technol) 1–5

Hasya N, Razali MNM, Jamaludin AS, Saffe SNBM, Hadi MA (2022) Micromilling performance analysis—cold air assisted aluminium machining behaviour observation. In: Innovative manufacturing, mechatronics and materials forum 2022 (iM3F 2022), Online conference, Pekan, Malaysia, pp 124–130

Jamaludin AS, Yassin A (2013) Analysis of laser sintered materials using finite element method. Sains Malaysiana 42(12):1727–1733

Jamaludin AS, Hosokawa A, Furumoto T, Koyano T, Hashimoto Y (2017) Evaluation of the minimum quantity lubrication in orthogonal cutting with the application of finite element method. Int J Mech Mechatron Eng IJMME-IJENS 17(01):104–109

Jamaludin AS, Hosokawa A, Furumoto T, Koyano T, Hashimoto Y (2018) Study on the effectiveness of extreme cold mist MQL system on turning process of stainless steel AISI 316. IOP Conf Ser: Mater Sci Eng 319 012054

Kim SE, Yoo SC, Kim H-G (2022) An optimization of manufacturing process of SS304L using direct energy deposition method. Korean J Metals Mater 60:819–826

Li Y, Ding F, Tian W (2022) Optimization of 3D printing parameters on deformation by BP neural network algorithm. Metals 12:1559

Md Kamil NH, Jamaludin AS, Mhd Razali MN, Ghaffar ANA (2022) Temperature and heat flow analysis in a drying chamber through finite element method (Lecture notes in mechanical engineering), pp 309–316

Nasir A, Akagi T, Dohta S, Ono A, Masago Y (2013) Development of low-cost wearable servo valve using buckled tube driven by servo motor. Appl Mech Mater 393:532–537

Nasir A, Akagi T, Dohta S, Ono A (2015) Analysis of low-cost wearable servo valve using buckled tubes for optimal arrangement of tubes. In: 2015 IEEE international conference on advanced intelligent mechatronics (AIM)

Norazman SHB, Nakamura T, Yamaguchi M (2020) Digital holography-assisted 3-D bright-field image reconstruction and refocusing. Opt Rev 27:455–464

Ping G (2021) Detection of power data tampering attack based on gradient boosting decision tree. J Phys: Conf Ser 1846:012057

Pyteraf J, Pacławski A, Jamróz W, Mendyk A, Paluch M, Jachowicz R (2022) Application and multi-stage optimization of daylight polymer 3D printing of personalized medicine products. Pharmaceutics 14:843

Raam Kumar S, Sridhar S, Venkatraman R, Venkatesan M (2021) Polymer additive manufacturing of ASA structure: influence of printing parameters on mechanical properties. Mater Today: Proc 39:1316–1319

Raja S, Agrawal AP, Patil PP, Thimothy P, Capangpangan RY, Singhal P, Wotango MT (2022) Optimization of 3D printing process parameters of polylactic acid filament based on the mechanical test. Int J Chem Eng 2022:1–7

Ramli MR, Razak NA, Ismail I, Jamaludin AS, Manaf AR (2022) Effect of dimple size onto wear rate of mild steel AISI 1060 surface (Lecture notes in mechanical engineering), pp 99–102

Rosli AM, Jamaludin AS, Razali MNM (2021) Recent study on hard to machine material—micromilling process. Evergreen 8:445–453

Rutzen M, Schulz M, Moosburger-Will J, Lauff P, Fischer O, Volkmer D (2021) 3D printing as an automated manufacturing method for a carbon fiber-reinforced cementitious composite with outstanding flexural strength (105 N/mm^2). Mater Struct 54

Saniman MNF, Bidin MF, Nasir RM, Shariff JM, Harimon MA (2020) Flexural properties evaluation of additively manufactured components with various infill patterns. Int J Adv Sci Technol 29(8s):4646–4657. http://sersc.org/journals/index.php/IJAST/article/view/25532

Saniman MN, Dzulkifli NA, Wahid KA, Muhamad WM, Mohamad KA, Alias EA, Shariff JM (2022) Water retention properties of a fused deposition modeling based 3D printed polylactic acid vessel. In: Advanced maritime technologies and applications, pp 311–323

Shariff ZA, Lokesh M, Mayandi K, Saravanan AK, Ramalingam PS, Kanna SK (2021) A hybrid multi-objective optimization of 3D printing process parameters using genetic algorithm. AIP Conf Proc 2395(1):050008

Vidakis N, Petousis M, Mountakis N, Kechagias JD (2022) Optimization of friction stir welding parameters in hybrid additive manufacturing: weldability of 3D-printed poly(methyl methacrylate) plates. J Manuf Mater Process 6:77

Influence of Injection Molding Manufacturing Process on Sports Equipment Durability

Norasilah Karumdin, Mohd Syamim Mazlan, Muhammad Nur Farhan Saniman, and Muhamad Ridzuan Radin Muhamad Amin

Abstract This review explores the substantial role that injection molding plays in the sports equipment industry. The application of injection molding, a manufacturing method characterized by the introduction of molten materials into a mold to produce intricate shapes, is of paramount importance in the production process of various athletic equipment, encompassing items such as protective gear and gym apparatus. The durability and safety of athletic equipment are heavily influenced by various aspects, including the composition of materials, mold design, and specific parameters utilized during the injection molding process. These considerations play a significant role in determining the longevity of these goods. This study highlights the significance of multiple characteristics, such as injection speed, pressure, temperature, and cooling time, in ascertaining the quality and longevity of the product. Recent research has examined many approaches to enhance process optimization, including the integration of conformal cooling channels and the exploitation of innovative materials, in order to improve durability. However, there are still knowledge gaps that need to be addressed, particularly in relation to knowing the impact of different manufacturing procedures on product durability and the use of recycled materials. This study underscores the need for additional research in these areas and emphasizes the promise of sustainable approaches in the field of injection molding. The case studies of injection molding demonstrate its capacity to effectively address diverse production requirements, as evidenced by its application in the fabrication of face shields among the COVID-19 pandemic and its utilization in recycling polymer waste.

N. Karumdin (✉) · M. S. Mazlan
Faculty of Manufacturing and Mechatronic Engineering Technology, Universiti Malaysia Pahang Al-Sultan Abdullah, 26600 Pekan, Pahang, Malaysia
e-mail: norasilahkarumdin@gmail.com

M. N. F. Saniman
Mechanical Engineering Section, Universiti Kuala Lumpur Malaysia France Institute, 43650 Bandar Baru Bangi, Malaysia

M. R. R. M. Amin
Faculty of Electrical and Electronic Engineering Technology, Universiti Malaysia Pahang Al-Sultan Abdullah, 26600 Pekan, Pahang, Malaysia

Z. Mohamed et al. (eds.), *Proceedings of the 2nd International Conference on Innovation and Technology in Sports, ICITS 2023, 27–28 November, Malaysia*, Lecture Notes in Bioengineering, https://doi.org/10.1007/978-981-97-3741-3_38

Keywords Injection molding · Sports equipment durability · Industrial Internet of Things · Sustainability · Recycled materials

1 Introduction to the Importance of Injection Molding in Sports Equipment Manufacturing

The utilization of injection molding technology is of great significance within the sports equipment industry owing to its capacity to manufacture intricate forms with remarkable precision and effectiveness (Kang et al. 2022; Dämmer et al. 2022; Alsibai et al. 2018; Chaabene et al. 2022; Nasir et al. 2015; Abidin et al. 2022; Ridloi and Jakaria 2021; Babu et al. 2021, 2022; Jamaludin et al. 2018a; Md Kamil et al. 2022; Hasya et al. 2022; Rosli et al. 2021; Atalla et al. 2022; Thompson et al. 2022). The procedure entails the introduction of a liquid substance into a specifically designed cavity, which is then followed by the subsequent cooling and solidification of the material in order to achieve the desired form (Thompson et al. 2022; Rosli et al. 2021; Kuo et al. 2022; Jamaludin et al. 2017, 2018b; Jamaludin and Yassin 2013; Kuo and Chen 2021; Rosli et al. 2019; Atalla et al. 2020; Taddese et al. 2021; Ali et al. 2022; Ramli et al. 1060; Vieten et al. 2021; Babu et al. 2023; Nasir et al. 2013; Sztorch et al. 2021). The materials employed in this procedure consist of a diverse range of polymers and metals, which are selected according to the particular specifications of the sporting equipment under consideration (Kang et al. 2022; Dämmer et al. 2022; Chaabene et al. 2022; Sztorch et al. 2021; Lee et al. 2021; Madhuka Priyashan 2020; Baharuddin et al. 2022, 2023; Binti Kamarudin et al. 2021; Farag et al. 2019; Kunkel et al. 2020; Sufian et al. 2021; Abdullah et al. 2022; Synyuk et al. 2020; Kanbur et al. 2020; Kuo et al. 2021). Injection molding is a pivotal process in the manufacturing of various sporting equipment, encompassing vital components of larger equipment such as gym machines, as well as protective clothing like helmets and pads. The utilization of injection molding technology provides manufacturers with the opportunity to tailor equipment according to the unique requirements of various sports and athletes, hence leading to enhanced safety and performance metrics (Kang et al. 2022; Dämmer et al. 2022; Chaabene et al. 2022; Sztorch et al. 2021; Kunkel et al. 2020). The fundamental concern in the design and production of sports equipment is the emphasis placed on durability (Kang et al. 2022; Dämmer et al. 2022; Alsibai et al. 2018; Chaabene et al. 2022; Nasir et al. 2015; Abidin et al. 2022; Ridloi and Jakaria 2021; Babu et al. 2021, 2022; Jamaludin et al. 2018a; Md Kamil et al. 2022; Hasya et al. 2022; Rosli et al. 2021; Atalla et al. 2022; Thompson et al. 2022). The use of durable equipment not only decreases the necessity for frequent replacements but also improves the safety and efficacy of athletes (Kang et al. 2022; Dämmer et al. 2022; Chaabene et al. 2022). An illustration of this concept may be seen in the use of a helmet that has been engineered to endure substantial impacts without experiencing fractures, so serving as a pivotal measure in protecting players from grave injuries (Thompson et al. 2022; Rosli et al. 2021; Kuo et al. 2022; Jamaludin et al. 2017, 2018b; Jamaludin and Yassin 2013; Kuo and Chen 2021; Rosli et al. 2019; Atalla et al. 2020; Taddese et al. 2021; Ali et al. 2022; Ramli et al. 1060; Vieten et al. 2021; Babu et al. 2023; Nasir et al. 2013; Sztorch et al. 2021). The durability of the end product in the context of injection molding is subject to influence from multiple elements, such as the material composition, mold design, and specific parameters of the injection molding process (Sztorch et al. 2021; Lee et al. 2021; Madhuka Priyashan 2020; Baharuddin et al. 2022, 2023; Binti Kamarudin et al. 2021; Farag et al. 2019; Kunkel et al. 2020; Sufian et al. 2021; Abdullah et al. 2022; Synyuk et al. 2020; Kanbur et al. 2020; Kuo et al. 2021). By engaging in the process of deciphering and enhancing these attributes, manufacturers possess the capacity to develop sports equipment that incorporates elevated levels of durability and dependability.

Several elements have been examined in current research lines within the subject of injection molding, particularly in regard to durability. The aspects described above involve the optimization of process parameters, as evidenced by the improvement of cooling consistency in thermoplastic injection molds by the utilization of conformal cooling channels (Chaabene et al. 2022; Kanbur et al. 2020; Kuo et al. 2021). In addition, the examination of novel materials, such as carbon fiber and polyphenylene sulfide thermoplastic composites, has exhibited the potential to expedite the thermal cycle, diminish processing duration, and augment longevity (Kang et al. 2022). Notwithstanding these perceptive findings, there remain deficiencies in the existing corpus of literature, specifically on the application of state-of-the-art technologies for augmenting the durability of sports equipment manufactured by injection molding. This scenario provides a potential avenue for further scholarly investigations within this particular topic. The purpose of the review paper is to explore and elucidate the critical role of the injection molding process in the sports equipment industry, particularly in relation to the durability of the products produced. The paper aims to the following:

- Highlight the significance of injection molding in producing a wide range of sports equipment, emphasizing its ability to produce complex shapes with precision and efficiency.
- Discuss the various factors and parameters of the injection molding process that influence the durability and quality of sports equipment.
- Present current research directions in the field, showcasing advancements and innovations that have been made to enhance equipment durability.
- Identify existing gaps in the current body of research, particularly in areas related to specific manufacturing methods and the use of recycled materials.
- Propose areas for future research, emphasizing the need to delve deeper into the direct effects of diverse methods on sports equipment durability and the potential of sustainable practices in injection molding.
- Provide real-world case studies, such as the production of face shields during the COVID-19 pandemic and the use of recycled polymer waste, to demonstrate the practical applications and adaptability of injection molding in various scenarios.

The primary objective of this review paper is to present a thorough and inclusive examination of the injection molding process within the realm of sports equipment production. This analysis places particular emphasis on highlighting the significance of this method, current research directions, and potential avenues for future investigation.

2 Injection Molding Process and Parameters

2.1 Understanding the Injection Molding Process

Injection molding is a fundamental manufacturing method that holds significant importance, notably in the realm of sports equipment manufacturing. The complex procedure involves the introduction of liquefied substances, primarily polymers or particular metals, into a properly crafted mold. After undergoing the process of cooling and solidification, the mold is then opened, revealing the intricately fabricated component (Kang et al. 2022; Chaabene et al. 2022; Sztorch et al. 2021; Kunkel et al. 2020). The process of injection molding can be divided into many key steps, namely, clamping, injection, dwelling, cooling, and ejection. The clamping step involves the convergence and secure clamping of the two parts of the mold. During the injection phase, the molten material is introduced into the cavity of the mold through infusion. The dwelling phase is responsible for the thorough distribution of the material throughout the mold, with the assistance of carefully applied pressure. The cooling stage is responsible for facilitating the transformation of the material from a liquid to a solid state, while the ejection stage marks the culmination of the process by which the final product is released from its mold (Kang et al. 2022; Chaabene et al. 2022; Sztorch et al. 2021; Kunkel et al. 2020). The quality and longevity of the final product depend on numerous parameters inherent to the injection molding process. These include the following:

- **Injection Speed**: This parameter delineates the velocity at which the molten material is channeled into the mold. An optimal speed is quintessential to ensure uniform and complete filling of the mold, thereby influencing the structural integrity and durability of the final product (Kang et al. 2022; Kuo et al. 2022).
- **Pressure**: Exerted during the dwelling phase, the pressure ensures the thorough filling of the mold cavity. The magnitude of this pressure can modulate the density and robustness of the resultant product, which is crucial for sports equipment that demands high durability (Dämmer et al. 2022; Kuo et al. 2022).
- **Temperature**: A pivotal parameter, the temperature, dictates the molten state of the material and the cooling rate of the mold. For instance, the manufacturing of carbon fiber/polyphenylene sulfide thermoplastic composites using induction-heating molding has underscored the significance of a quicker thermal cycle, which not only trims down processing time but also augments durability (Kang et al. 2022).

- **Cooling Time**: This phase, pivotal for the material's solidification within the mold, has a profound impact on the dimensional precision and inherent physical properties of the product. Innovations like conformal cooling channels have been spotlighted for their prowess in enhancing cooling uniformity, thereby influencing the product's durability (Chaabene et al. 2022; Kanbur et al. 2020; Kuo et al. 2021).

Moreover, researchers have investigated the practicality and possible durability advantages of integrating soft tooling with automated slides in the field of injection molding (Jamaludin et al. 2018a). The versatility and adaptability of injection molding in meeting various production requirements are highlighted by the combination of fused deposition modeling (FDM) and injection molding processes. This is particularly evident in the mass production of items such as face shields during urgent situations like the COVID-19 pandemic (Sztorch et al. 2021; Kunkel et al. 2020).

2.2 *Impact of Process Parameters on Product Quality and Durability*

The modification of injection molding process parameters significantly influences the longevity and general excellence of sports equipment. The utilization of induction-heating molding in the production of carbon fiber/polyphenylene sulfide thermoplastic composites has highlighted the importance of a shorter thermal cycle. This not only reduces processing time but also enhances the overall durability of the composites (Kang et al. 2022). The speed of injection is a crucial parameter. Excessive velocity has the potential to result in incomplete mold filling or the introduction of flaws in the final product. Conversely, an excessively low speed might lead to the premature cooling of the material, resulting in a final product that lacks dimensional accuracy (Kang et al. 2022; Kuo et al. 2022).

Pressure, particularly in the phase of habitation, is an additional crucial element. Sufficient pressure is necessary to enable the entire filling of the mold cavity, whereas inadequate pressure may result in the formation of voids, hence compromising the structural integrity and durability of the product (Dämmer et al. 2022; Kuo et al. 2022). The adjustment of temperature, both in the material and the mold, is critical. Insufficient heating of a material might result in a lack of smooth flow into the mold, hence causing possible flaws. Conversely, inadequate cooling of the mold can have negative consequences on the solidification process of the product, leading to a decrease in its dimensional precision and inherent physical characteristics (Chaabene et al. 2022; Kuo and Chen 2021).

The significance of recent developments, such as the implementation and refinement of conformal cooling channels, has garnered attention due to their capacity to improve the uniformity of cooling, consequently impacting the longevity of the product (Kanbur et al. 2020). Furthermore, there has been a proposal for the advancement of quick injection molds that provide superior cooling capabilities. This proposal highlights the significance of designing optimal conformal cooling channels to achieve reductions in cooling time, thus extending the longevity of sporting equipment (Kuo et al. 2021). The thorough calibration and adjustment of injection molding process parameters are essential for the production of durable and high-quality sports equipment.

3 Case Studies and Applications

3.1 Case Study: Production of Face Shields During COVID-19 Pandemic

The COVID-19 epidemic has presented unique obstacles that emphasize the need for prompt and effective manufacturing of personal protective equipment (PPE), particularly face shields. A comparison investigation was undertaken to determine the effectiveness of two primary manufacturing techniques: fused deposition modeling (FDM) or 3D printing, and injection molding (IM), as shown in Figs. 1 and 2, respectively (Kunkel et al. 2020). During the initial phase of the epidemic, FDM (Fused Deposition Modeling) appeared as a feasible solution, particularly because of its adaptable nature and efficient setup capabilities. The mentioned factor enabled efficient and frequent production, effectively meeting the sudden increase in consumer demand (Kunkel et al. 2020). Nevertheless, with the prolonged duration of the pandemic and the substantial increase in the need for face shields, there was a notable shift in the production approach towards injection molding. The process of Injection Molding (IM), known for its notable speed, effectiveness, and uniformity, emerged as a crucial component in the manufacturing of face shields on a wide scale, guaranteeing the provision of face shields of superior quality (Kunkel et al. 2020). This case study examines the significant significance of injection molding, in comparison to other manufacturing processes, in meeting emerging industrial needs amid global crises.

3.2 Application: Use of Recycled Polymer Waste in Injection Molding

The utilization of recycled polymer waste in injection molding is a significant application within the field of sustainable manufacturing. In regard to this particular issue, an extensive investigation carried out by a previous study has provided insight into the cutting-edge technological procedures and specialized equipment designed for the injection molding of polymer goods that use recycled polymer waste (Synyuk et al. 2020). In contrast to prevailing concerns, the research emphasized that the integration of recycled polymer waste did not damage the quality of the final product. The aforementioned disclosure highlights the capacity of recycled materials to serve

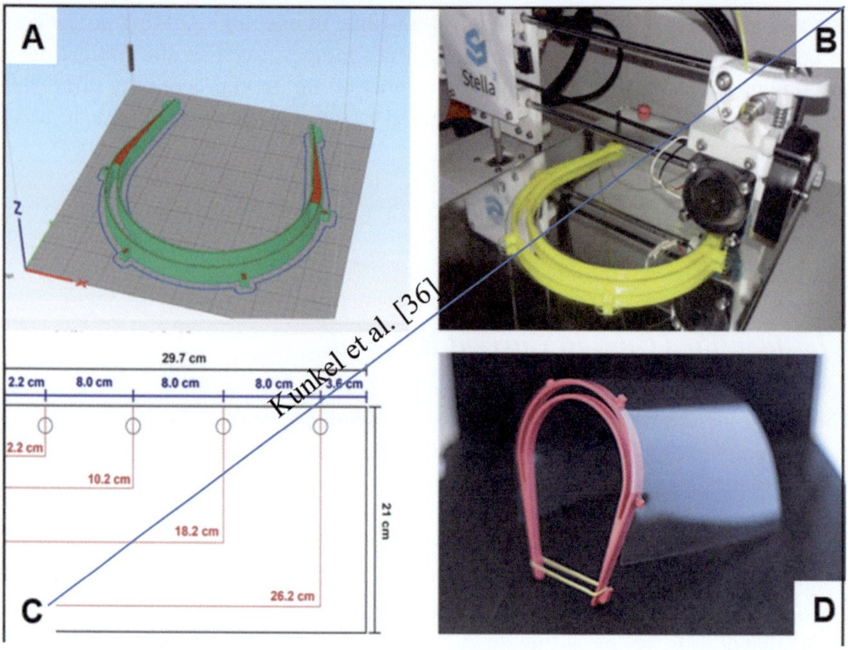

Fig. 1 FDM 3D printed face shield (Kunkel et al. 2020)

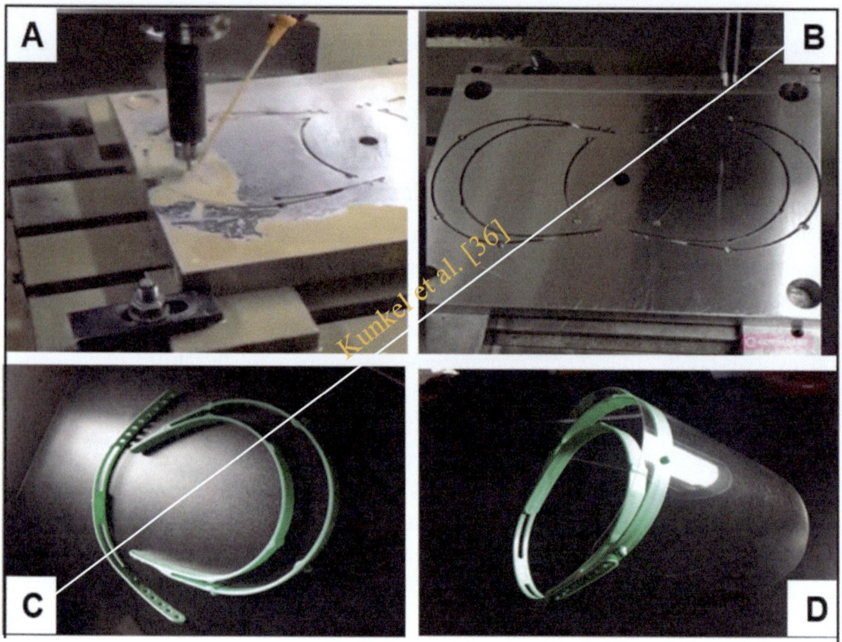

Fig. 2 Injection molding fabricated face shield (Kunkel et al. 2020)

as a viable and environmentally friendly substitute in injection molding procedures. In addition to the apparent environmental advantages, such as the mitigation of carbon emissions in production and the promotion of circular economy principles, this strategy also offers a practical resolution to the growing issue of plastic waste management (Synyuk et al. 2020).

4 Gaps and Future Research

4.1 Identified Gaps in Current Research

The extant literature provides a wealth of knowledge regarding the complexities of the injection molding process, particularly in relation to the durability of sports equipment. Nevertheless, there are still certain loopholes that have not been addressed. One area that has received insufficient attention is the lack of comprehensive investigation into the direct impact of particular manufacturing techniques, such as induction-heating molding of carbon fiber/polyphenylene sulfide thermoplastic composites, on the long-term resilience of sports equipment (Kang et al. 2022). Moreover, it is worth noting that despite the increasing interest in using recycled materials into the process of injection molding, there is a lack of extensive research examining the extended performance and durability of sports equipment manufactured using such materials (Synyuk et al. 2020). Moreover, there is a noticeable lack of comprehensive case studies that shed light on sustainable practices within the injection molding industry, particularly in the context of manufacturing sports equipment (Taddese et al. 2021).

4.2 Proposed Areas for Future Research

There is a need to expand the scope of future research in order to further investigate the direct impacts of various techniques on the long-term durability of sports equipment after the injection molding process. One potential avenue for investigation involves conducting experimental studies to compare the durability of equipment created through traditional manufacturing methods with those manufactured utilizing advanced techniques, such as metal fused filament fabrication for high-performance alloys (Thompson et al. 2022). Further investigation is warranted to explore the possibilities of recycled and bio-based materials in the context of injection molding. This inquiry should specifically prioritize the evaluation of long-term performance parameters of sports gear manufactured using these resources (Synyuk et al. 2020). With a focus on sustainability, future research endeavors may aim to document further case studies that provide insights into the adoption of sustainable strategies within the field of injection molding. These studies would examine the resulting effects on environmental, economic, and societal results (Taddese et al. 2021).

4.3 Potential Solutions to Address Identified Gaps

The process of addressing the identified gaps requires a collaborative effort between academia and industry in order to achieve synergy. Manufacturers have the potential to adopt sophisticated methodologies, such as the implementation of design and optimization techniques for conformal cooling channels, in their injection molding processes. Additionally, they can diligently monitor the durability trajectory of the resulting sports equipment (Kanbur et al. 2020). Additionally, they have the potential to be at the forefront of utilizing a wide range of recycled and bio-based materials in the process of injection molding. This would involve doing thorough assessments of the performance and durability of the resulting products (Synyuk et al. 2020). In order to advance the discourse on sustainability, manufacturers have the opportunity to advocate for and implement sustainable practices within the injection molding sector. By recording their experiences and outcomes, manufacturers may provide valuable insights to the existing body of information on sustainable paradigms in this industry (Taddese et al. 2021).

Acknowledgements The authors are very grateful to the Ministry of Higher Education (MoHE) Malaysia and the Universiti Malaysia Pahang Al-Sultan Abdullah for their great financial support, which made this study possible. This study was made possible by grants from the Fundamental Research Scheme (FGRS) FRGS/1/2022/TK10/UMP/02/57 and the Universiti Malaysia Pahang Al-Sultan Abdullah Fundamental Research Grant RDU220365.

References

Abdullah NA, Abdullah FF, Sufian AH, Abidin AN, Jamaludin AS, Razali MN (2022) Effect of degradation by temperature onto nitrile rubber elastomer mechanical properties. Mater Today: Proceedings 48:1941–1946

Abidin AN, Azmi AH, Kassim KA, Jamaludin AS, Razali MN (2022) A review on automotive tires significant characteristic identification for general consumers. In: Proceedings of the 2nd energy security and chemical engineering congress, pp 375–385

Rosli AM, Jamaludin AS, Razali MN (2021) Recent study on hard to machine material—micromilling process. Evergreen 8:445–453

Ali MA, Sivarao, Abdullah Z, Izamshah R, Kassim MS, Jamaludin AS (2022) Multi-response optimization of machining simulation approach using grey relational analysis. Lecture notes in mechanical engineering, pp 154–157

Alsibai MH, Hamran NNN, Nasir A, Abdullah AA (2018) A monitoring system for EPW safe use. J Telecommun, Electron Comput Eng (JTEC) 10(1–2):131–135

Atalla Y, Hashim Y, Abd. Ghafar AN, Jabbar WA (2020) Temperature characteristics of finfet based on channel fin width and working voltage. Int J Electr Comput Eng (IJECE) 10:5650

Atalla Y, Hashim Y, Abd. Ghafar AN (2022) The impact of channel fin width on electrical characteristics of Si-FinFET. Int J Electr Comput Eng (IJECE) 12:201

Babu D, Nasir A, Jamaludin AS, Rosle MH (2021) Holding, grasping and sensing of prosthetic robot arm like a real human hand, a journey beyond limits: An extensive review. Hum-Cent Technol Better Tomorrow, pp 485–504

Babu D, Nasir A, Farag M, Sidik MH, Rejab SB (2022) Development of prosthetic robotic arm with patient monitoring system for disabled children; preliminary results. In: 2022 9th International Conference on Electrical and Electronics Engineering (ICEEE)

Babu D, Nasir A, Ravindran, Farag M, Jabbar WA (2023) 3D printed prosthetic robot arm with grasping detection system for children. Int J Adv Sci, Eng Inf Technol 13:226

Baharuddin NJ, Abdul Manaf AR, Jamaludin AS (2022) Study on die shoulder patterning method (DSPM) to minimise springback of U-bending. Int J Automot Mech Engineering 19:9509–9518

Baharuddin NJ, Manaf ARA, Jamaludin AS (2023) Study of springback behavior on U-bending part using die shoulder patterning method (DSPM). AIP Conf Proc 2544(1):040020

Binti Kamarudin SN, Lee H, Ishak I, Manaf AR, Jamaludin AS, Shaharudin MA, Zawawi MZ (2021) Rapid direct continuous method for hot embossing of Glass microlens array combined with CO_2 laser irradiation and external preheating/cooling. Lecture notes in mechanical engineering, pp 669–675

Chaabene A, Chatti S, Ben Slama M (2022) Optimization of the cooling of a thermoplastic injection mold. Ann Dunarea Jos Univ Galati Fasci-Cle XII Weld Equip Technology 32:61–70

Dämmer G, Bauer H, Neumann R, Major Z (2022) Design, additive manufacturing and component testing of pneumatic rotary vane actuators for lightweight robots. Rapid Prototyp J 28:20–32

Farag M, Azlan NZ, Alsibai MH, Ghafar AN (2019) Slippage detection for grasping force control of robotic hand using force sensing resistors. In: Proceedings of the 2019 5th international conference on computer and technology applications

Hasya N, Razali MNM, Jamaludin AS, Saffe SNBM, Hadi MA (2022) Micromilling performance analysis—cold air assisted aluminium machining behaviour observation. In: Innovative manufacturing, mechatronics & materials forum 2022 (iM3F 2022), Online Conference, Pekan, Malaysia, pp 124–130

Jamaludin AS, Yassin A (2013) Analysis of laser sintered materials using finite element method. Sains Malaysiana 42(12):1727–1733

Jamaludin AS, Hosokawa A, Furumoto T, Koyano T, Hashimoto Y (2017) Evaluation of the minimum quantity lubrication in orthogonal cutting with the application of finite element method. Int J Mech Mechatron Eng IJMME-IJENS 17(01):104–109

Jamaludin AS, Hosokawa A, Furumoto T, Koyano T, Hashimoto Y (2018a) Study on the effectiveness of extreme cold mist MQL system on turning process of stainless steel AISI 316. In: IOP conference series: Materials science and engineering, pp 319:012054

Jamaludin AS, Akira H, Furumoto T, Koyano T, Hashimoto Y (2018b) High precision estimation on physical behavior for cutting with various tool rake angle by finite element method. Lecture notes in mechanical engineering, pp 715–723

Kanbur BB, Suping S, Duan F (2020) Design and optimization of conformal cooling channels for injection molding: A review. Int J Adv Manuf Technology 106:3253–3271

Kang C-S, Shin H-K, Chung Y-S, Seo M-K, Choi B-K (2022) Manufacturing of carbon fibers/ polyphenylene sulfide composites via induction-heating molding: Morphology, mechanical properties, and flammability. Polymers 14:4587

Kunkel ME, Vasques MT, Perfeito JA, Zambrana NR, Bina T dos, Passoni LH, Ribeiro TV, Rodrigues SM, Castro RO, Ota NH (2020) Additive manufacturing and injection molding process for mass-production of face shields during COVID-19 pandemic: A comparative study. NBSP

Kuo C-C, Chen W-H (2021) Improving cooling performance of injection molding tool with conformal cooling channel by adding hybrid fillers. Polymers 13:1224

Kuo C-C, Nguyen T-D, Zhu Y-J, Lin S-X (2021) Rapid development of an injection mold with high cooling performance using molding simulation and rapid tooling technology. Micromachines 12:311

Kuo C-C, Tasi Q-Z, Xie B-X, Huang J-M, Qiu S-X (2022) Improving the quality of a circular cooling channel fabrication by fused filament fabrication using Taguchi methods. Int J Adv Manuf Technology 120:3353–3364

Lee H, Binti Kamarudin SN, Ishak I, Manaf AR, Jamaludin AS, Shaharudin MA, Zawawi MZ (2021) Feasibility study of wafer scale laser assisted thermal imprinting of glass nanostructures. Lecture notes in mechanical engineering, pp 917–923

Priyashan WDM, Thilakarathne NN (2020) IIOT framework for SME level injection molding industry in the context of industry 4.0. Int J Eng Manag Research 10:61–68

Md Kamil NH, Jamaludin AS, Mhd Razali MN, Abd. Ghaffar AN (2022) Temperature and heat flow analysis in a drying chamber through finite element method. Lecture notes in mechanical engineering, pp 309–316

Nasir A, Akagi T, Dohta S, Ono A, Masago Y (2013) Development of low-cost wearable servo valve using buckled tube driven by servo motor. Appl Mech Mater 393:532–537

Nasir A, Akagi T, Dohta S, Ono A (2015) Analysis of low-cost wearable servo valve using buckled tubes for optimal arrangement of tubes. In: 2015 IEEE International Conference on Advanced Intelligent Mechatronics (AIM)

Ramli MR, Razak NA, Ismail I, Jamaludin AS, Manaf AR (2022) Effect of dimple size onto wear rate of mild steel AISI 1060 surface. Lecture notes in mechanical engineering, pp 99–102

Ridloi M, Jakaria RB (2021) Totaliproductive maintenance (TPM) analysis using the overall equipment effectiveness (OEE) method and six big losses on an injection molding machine. Procedia Eng Life Sci 1

Rosli AM, Jamaludin AS, Razali MN, Sani AS, Hamzah SB, Osman MS (2019) Modelling of fuzzy inference system for micro milling—a preliminary study through FEM. Lecture notes in mechanical engineering, pp 445–456

Rosli AM, Jamil N, Jamaludin AS, Razali MN, Yusoff AR (2021) Tool wear observation during unconventional low speed machining using low cost micromilling. Lecture notes in mechanical engineering, pp 589–597

Sufian AH, Xun TZ, Abidin AN, Jamaludin AS, Razali MN (2021) Study on tire tread design effect onto tire-road contact behavior through FEM. Lecture notes in mechanical engineering, pp 893–902

Synyuk O, Musiał J, Zlotenko B, Kulik T (2020) Development of equipment for injection molding of polymer products filled with recycled polymer waste. Polymers 12:2725

Sztorch B, Brząkalski D, Jałbrzykowski M, Przekop RE (2021) Processing Technologies for crisis response on the example of COVID-19 pandemic—injection molding and FFF case study. Processes 9:791

Taddese G, Durieux S, Duc E (2021) Sustainability performance evaluation of faceshield bracket manufacturing by using the analytic hierarchy process. Sustainability 13:13883

Thompson Y, Zissel K, Förner A, Gonzalez-Gutierrez J, Kukla C, Neumeier S, Felfer P (2022) Metal fused filament fabrication of the nickel-base superalloy in 718. J Mater Sci 57:9541–9555

Vieten T, Stahl D, Schilling P, Civelek F, Zimmermann A (2021) Feasibility study of soft tooling inserts for injection molding with integrated automated slides. Micromachines 12:730

The Development of Mini Timing Gate for Sprint Performance Assessment

Jorrye Jakiwa, Muhammad Faiz Md Din, Shahrulfadly Rustam, Mohd Syrinaz Azli, Siti Azilah Atan, Ahmad Bisyri Husin Musawi Maliki, Nur Diyanah Binti Mohamad Zahid, Tan Jia Min, Tharani A. P. Elango, and Intan Nurdiyana Binti Mohd Zanom

Abstract In the discipline of speed measurement, methods such as camera systems and timing gates have proven effective but are often quite expensive. Therefore, the objective of this study was to address the need for a cost-effective solution by developing a sprint test device referred to as 'mini timing gates' for assessing speed performance among Cadet Officer. The device's development incorporates an input sensor, utilizing an Infrared (IR) sensor and a microcontroller known as Arduino Uno. The microcontroller efficiently processes sensor signals, resulting in multiple output indicators, including Light Emitting Diodes (LEDs), a Liquid-Crystal Display (LCD) screen and a buzzer. Remarkably, this study successfully achieves the creation of mini timing gates with a budget-friendly cost, totaling less than 500.00 Malaysian Ringgit. The design prioritizes accessibility and affordability, catering to a broad user base encompassing individuals from various backgrounds and categories. Beyond cost-efficiency, this device represents a significant leap forward in terms of energy conservation while consistently delivering reliable and precise results. The mini timing gates offer a promising solution for assessing speed performance without the burden of extremely high costs, making them a valuable tool not only in military but also in sports and athletic settings.

Keywords Sport innovation · Physical test · Speed measurement · Physical fitness

J. Jakiwa (✉) · S. Rustam · M. S. Azli · S. A. Atan · A. B. H. M. Maliki
Defence Fitness Academy, National Defence University of Malaysia, 57000 Kuala Lumpur, Malaysia
e-mail: jorrye@upnm.edu.my

M. F. M. Din · N. D. B. M. Zahid · T. J. Min · T. A. P. Elango · I. N. B. M. Zanom
Department of Electrical and Electronic Engineering, Faculty of Engineering, National Defence University of Malaysia, 57000 Kuala Lumpur, Malaysia

Z. Mohamed et al. (eds.), *Proceedings of the 2nd International Conference on Innovation and Technology in Sports, ICITS 2023, 27–28 November, Malaysia*, Lecture Notes in Bioengineering, https://doi.org/10.1007/978-981-97-3741-3_39

1 Introduction

The requirement to be physically fit among cadet officers is very crucial (Jakiwa et al. 2020). Physical fitness is important for cadet officers as they can maintain good health and engage in daily tasks without difficulties, boost their productivity in both academic and training, help ease stress and build teamwork (Kyröläinen et al. 2018; Nadzmi et al. 2023; Jakiwa et al. 2022). One of the very important components of physical fitness that need to be possessed by cadet students is the speed of running. Speed can be defined as an ability to move body parts as quickly as possible from one point to another point. In military, speed of running is important because they may face a tough situation such as going into battle, fighting with the enemy and also be able to execute the mission such as saving lives (Crawley et al. 2016).

Therefore, improving and measuring the speed ability is clearly crucial in military. Cadet students must possess a minimum speed required to prove that they are healthy and fit enough to stay in the service. Basically, Physical Trainer (PT) conducted a sprint test to measure the speed ability among cadet students. Nevertheless, PT still uses a traditional method to measure speed which is via stopwatch. The limitations of stopwatches have been discussed extensively in previous research including the inconsistency in time recorded and different testers producing different results (Chen et al. 2021; Dewanti and Hermawan 2022). Accuracy and reliability of speed testing are paramount to provide cadet students with valid feedback. Failure to achieve high standards of reliability by using stopwatches compromises the assessment of performance in students and may in fact provide invalid feedback which may prove detrimental to the training plan (Hetzler et al. 2008).

The use of sophisticated technology, such as electronic devices, to measure sprint ability is highly recommended (Colino et al. 2019). However, the current sprint test devices, most popularly called timing gates, are too expensive (Tillaar et al. 2022). For example, the price of timing gates by smartSPEED ranges from RM 15,000.00 to RM 45,000.00. Hence, the innovation of a new electronic sprint test device that is more affordable is required, as it provides cost-effective timing gates accessible to all, including cadet officers and athletes. Therefore, the main objective of this study is to develop a low-cost mini timing gate to assess sprint ability among cadet officers at the National Defence University of Malaysia (NDUM).

2 Methodology

2.1 *Steps for Developing the Design of Device*

The initial stages involve identifying the problem, establishing objectives and devising solutions for these issues. The primary task is to determine the main problem as outlined in the problem statement through collaboration among team members, followed by the execution of the objectives to be accomplished by the conclusion of the study. Additionally, the team will engage in brainstorming to develop a comprehensive design solution for the study, which includes identifying suitable components necessary for measuring running speed.

The second step is to gather information. Information of the customer requirements as well as the engineering characteristic for the product are identified by reading some related articles related to the study. In addition, the team gathered information on the existing product and made comparison through benchmarking. The next step is generation of prototype using Fritzing. Three possible solutions that have been discussed among team members are designed in Tinkercad to get a clear overview of the system. Different types of sensors have been used to create three different design solutions and the best design workflow among the three designs is being chosen.

Next is Apps development for the study. MIT App Inventor is used to create apps for iPhone and Android. It is an intuitive, visual programming environment that allows everyone to build fully functional apps for smartphones and tablets. This application will be connected to the Arduino so the result can be executed. The fifth step is installation of Arduino IDE software. Arduino IDE needs to be installed in order to upload the coding from Tinkercad to the Arduino UNO board. In this software, the coding can be checked and verified as it can detect if there is any error in the coding.

The final step is product development and debugging. The actual product is being set up using the components listed in Table 1. The arrangement of the product will be based on the prototype made in Tinkercad and Fritzing. After the execution of actual product, all team members took part in testing the whole system as well as locating and fixing the error or bug made either in the product or software.

2.2 *List of Hardware Components and Tools Used*

2.3 *Simulation Tools Used*

See Table 2.

2.4 *Embodiment Design*

2.4.1 Final Design Concept

From Fig. 1, it was expected that the design would consist of three main components, forming the core of the device. These components played a crucial role in the study as they made the design efficient, accurate and user-friendly. The Arduino Uno controlled all the components and their operations. The code written in the Arduino

Table 1 List of hardware components

Components	Image	Function
Arduino Uno		A programmable board that can be used to build interactive circuits. It consists of a piece of software that runs on computer, used to write and upload computer code to the physical board
IR Proximity sensor		A sensor is used to detect object movements and sense human presence. It also measured the distance between two points and execute speed as the result
HC-05 Bluetooth module		Used for many applications mainly for wireless connection. It is also to connect the apps with Arduino in order to obtain the result
HC-12 wireless module		Used to transmit and receive signal from two different network. It is also for producing communication and connecting between two different Arduino
Buzzer		Makes sounds at different frequencies. It is also will turn on when IR sensor detects any motion
LCD 16 × 2		A liquid crystal display capable of displaying two line of 16 characters. Used to display the speed and time taken by the user
9V battery clip connector		To connect battery power supplies to prototype circuits. It also connects the 9V battery with the Arduino board
9V battery		Voltage supply for the system. Used as an alternative for USB port as it needs to be connected to a PC
Jumper wires		To connect the port in Arduino board to the respective components. Male to female jumper is used for IR sensor, buzzer, LCD and potentiometer. Female to female jumper is used for LED
Resistor 1 kΩ		Limit the current flow to avoid circuit from damage. Connected to LED and LCD for controlling the current flow through them

(continued)

Table 1 (continued)

Components	Image	Function
LED (Green and Red)		Light-emitting diode that lights up when electricity passes through it in the correct direction. Green LED is turned on when no movement is detected while Red LED is turned on when there is movement detected from any sensor
Potentiometer		A type of resistor whose resistance changes at the turn of a knob. Used to adjust the light intensity of LCD
Enclosure plastic box (20 × 11.5 × 8.5 cm)		Used to protect the insides from impact and lightweight compared to metal enclosure. To hold the components to be connected to the tripod
Tripod		A three-legged stand designed to support a camera. It is used as it provides better stability compared to monopod. Used to hold the enclosure plastic box so the sensor can detect movement more accurately
Breadboard		Used to build the temporary circuit which makes it easier to remove and replace components. To build the main circuit of the device
Capacitor		Temporarily stores electrical energy by distributing charged particles on (generally two) plates to create a potential difference. To charge up faster than battery and release energy quickly

IDE software was uploaded to the physical board through this unit. Additionally, the HC-05 Bluetooth Module was utilized to facilitate communication from the board to computers or phones. This module assisted in transmitting the data collected from the Arduino to an application developed using MIT App Inventor. Finally, for the first and second Arduinos to communicate with each other, the HC-12 Wireless module served as the connector. With this module, data from the first Arduino, positioned at the starting point, was sent to the second Arduino located at the end of the track field. Consequently, the time taken was recorded, enabling the calculation of speed.

Table 2 List of simulation tools used in the study

Simulation tools	Function
Tinkercad	Tinkercad is used to develop the 3D design for the product as seen in Sect. 3.3. The design specification in this section is made according to the real product measurement. In addition to that, it is also being used to design system prototype before being executed as a real product. This step is crucial as it can help us detect any mistake in the wire connection and coding beforehand. Therefore, it will make the process of product development more efficient as it is less time-consuming
Arduino IDE	For this study, Arduino IDE is used to upload the coding that has been written in Tinkercad software to the Arduino Uno board. Other than that, it also helps in compiling and debugging the code before being uploaded onto the board. It will check if there is any error occurring in the code and notify us. This software helps the real product to function according to the correct workflow of the program
MIT app inventor	In order to create an app for the system, this cloud-based service is used. It can be accessed by using a web browser such as Chrome, Firefox or Safari. The app inventor is a free intuitive platform that helps in building a fully functional app that will connect the Arduino systems to the application created. The data of the results will be executed inside the app and can be downloaded as an Excel file
Fritzing	Fritzing is primarily a software tool for designing and documenting electronic circuits, and it does not provide a built-in programming environment. However, Fritzing does support the use of microcontrollers. Therefore, it uses third-party software tools such as Arduino integrated development environment (Arduino IDE) to program and control the microcontrollers

2.4.2 Key Subsystems

HC-05 Bluetooth Module with Arduino

Figure 2 shows the connectivity of HC-05 Bluetooth module with the Arduino Uno. There are four legs for the HC-05 Bluetooth module in which one round leg is connected to the pin of Arduino Uno and the other leg will be connected to the power 5 V at the Arduino Uno. All the legs are connected to the pins of Arduino Uno using jumpers where the 2 green jumpers are connected to the input pin of Arduino Uno and the black colour jumper is connected to the ground while the red colour jumper is connected to the power source +5 V of the Arduino Uno.

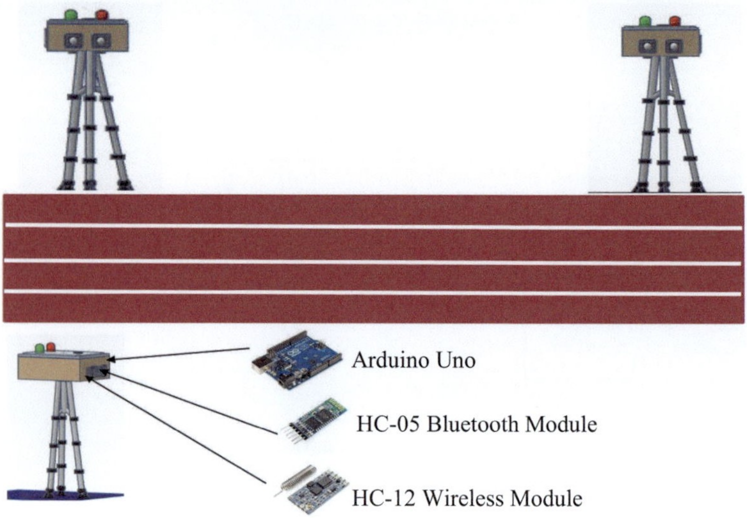

Fig. 1 Final design concept

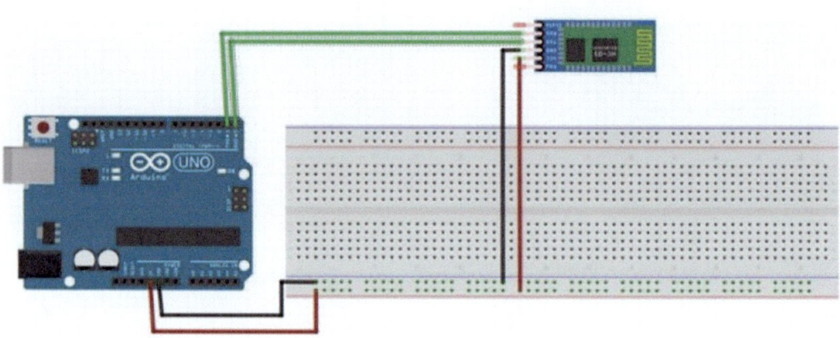

Fig. 2 Arduino Uno connected to HC-05 bluetooth module

Crystal Liquid Screen (LCD) and Potentiometer with Arduino

Based on the diagram above, 6 pins are connected at the input pins of the Arduino Uno from the LCD using light blue jumpers. the 10 K potentiometer is connected at the first terminal to the Arduino's 5 V pin by using a red jumper wire and the second terminal (middle pin) to the LCD's pin 3 and the third terminal to the Arduino's GND pin with black jumper wire.

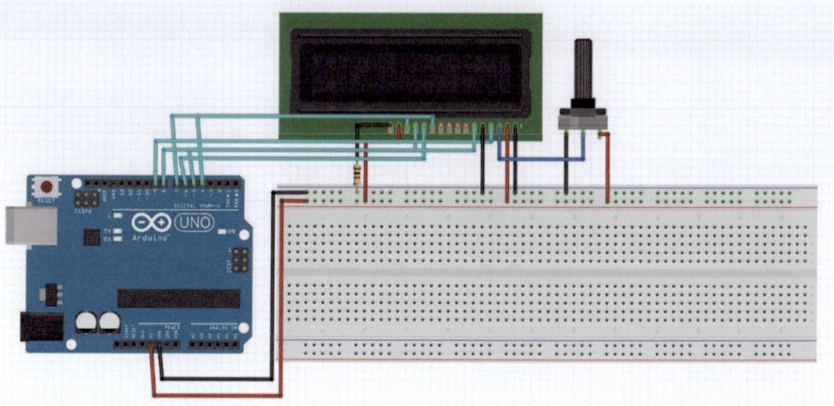

Fig. 3 Arduino Uno connected to crystal liquid screen (LCD) and potentiometer

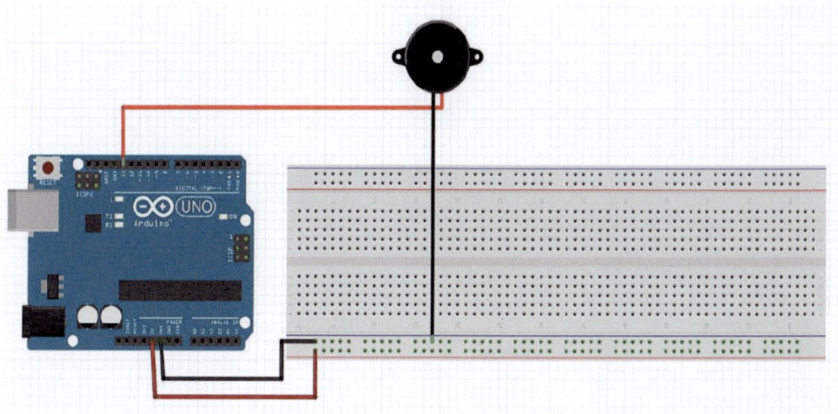

Fig. 4 Arduino Uno connected with buzzer

Buzzer with Arduino

Figure 4 shows the buzzer connected with the Arduino Uno. The buzzer has two legs where one is connected to the input pin of Arduino, and another is connected to the ground. As shown in figure above, the black colour jumper is represented as ground while the other leg with red jumper is connected to the power source of 5 V. The red colour jumper wire is connected to the VCC of the Arduino.

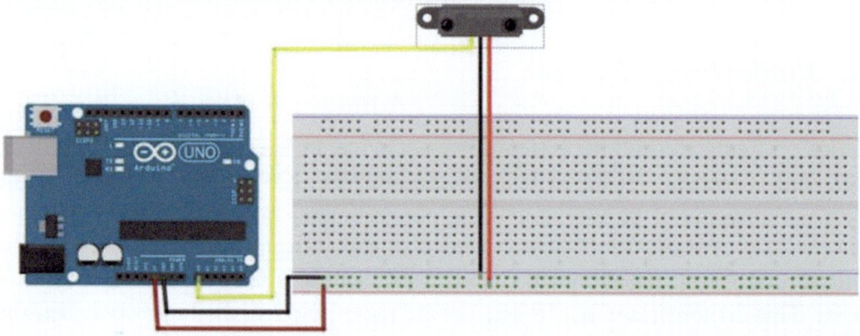

Fig. 5 Shows Arduino Uno connected with IR sensor

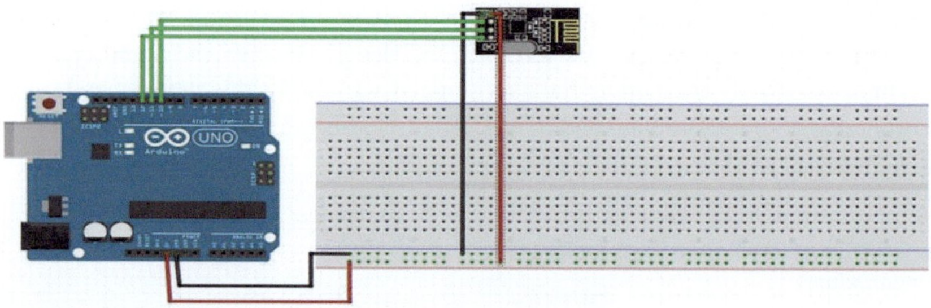

Fig. 6 Arduino Uno connected with NRF24L01 wireless module

Infrared Sensor with Arduino

Based on Fig. 5 the IR sensor has three pins which GND, V and OUT. This works under power supply of 5 V. GND pin of IR sensor is connected to GND of Arduino.

NRF24L01 Wireless Module with Arduino

Based on the figure above, the NRF wireless module has 5 legs where 3 legs are connected to the input pins of the Arduino Uno with green colour jumpers. Meanwhile, the fourth leg is connected to power supply pin which is 5 V of Arduino from the NRF wireless module with a red jumper. Finally, a black colour jumper wire connects the NRF wireless module to the GND pin of Arduino Uno. NRF24L01 was used as replacement for HC-12 wireless module as HC-12 wireless module was not available in the Fritzing app as all the wiring connections of both components are similar.

3 Results

3.1 Product Architecture

The above figure showed the different views of the architecture. The LCD, buzzer, LED, and IR sensor, HC-05 Bluetooth module, HC-12 Wireless Module are the major components of prototype and the potentiometer and resistor are the minor ones. Instead of using an ultrasonic sensor in this design, it's required to employ an IR sensor. An electronic device called an IR sensor is used to find any object or barrier. This demonstrates that IR sensors are more practical than ultrasonic sensors since they can gauge speed from any angle. The red LED illuminates when the runner goes by the IR sensor, and the green LED illuminates when the IR sensor detects no movement. Each LED also has a specific function. The LCD will also display the runner's pace and time when they reach the end points.

Additionally, the apparatus's height can be structurally adjusted because the tripod leg used in this design is adjustable (see Fig. 8). A portable tripod, which has three legs, can support any weight. Its three-legged design ensures the stability of the box connected to it. To compute the user's speed accurately, this device needs to be positioned at both the start and end points. Accurate timing for both the start and end is also necessary.

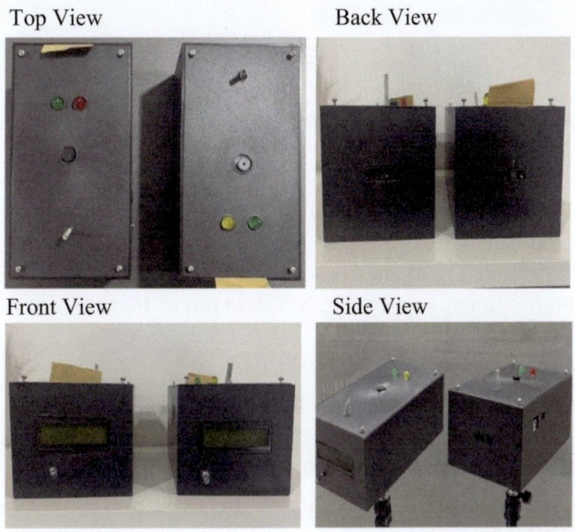

Fig. 7 Different views of the architecture

Fig. 8 Final product with tripod

3.2 Device Prototype

Figure 9 illustrates two different devices placed at a distance from each other. The gap between the two devices is approximately 20 m, and this setup is intended to demonstrate how the timing gate device will be utilized. As each box is equipped with a wireless module, data can be transmitted from the starting point to the ending point. However, there are some drawbacks to the final design, such as delays in sensor detection and data transmission. These delays may be mitigated by implementing non-blocking code structures, which allow the Arduino to perform multiple tasks simultaneously. Rather than using delay functions, millis or micros should be employed to create timed intervals for various tasks, enabling the code to run continuously while it waits for certain conditions or timing events to occur.

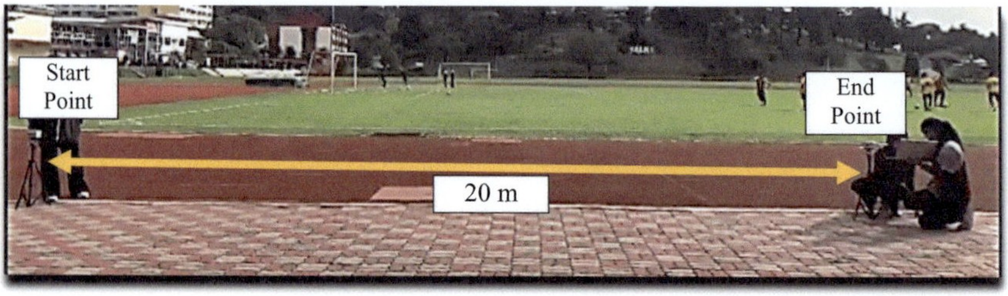

Fig. 9 Placement of the device at start and end point

Table 3 Running speed test using a prototype mini timing gate

Trail	Actual speed (s)	Software speed (s)	Differences (s)
1st	3.56	4.51	−0.95
2nd	4.10	4.77	−0.67
3rd	4.30	4.90	−0.60
4th	4.25	5.10	−0.85
5th	5.10	5.25	−0.15

3.3 *Sprint Test Using a Prototype Mini Timing Gate*

Table 3 presents results from a running speed test using a prototype mini timing gate, comparing actual speeds with those recorded by the software across five trials. The actual speed was captured with a video camera, while the software speed was recorded by the prototype of the mini timing gate. The actual speeds ranged from 3.56 to 5.10 s. The software speeds were consistently higher, ranging from 4.51 to 5.25 s. The differences between the actual speeds and the software speeds indicate that the software recorded slower times, with discrepancies ranging from −0.15 to −0.95 s. The smallest difference occurred in the 5th trial (−0.15 s), and the largest difference was in the 1st trial (−0.95 s).

3.4 *Cost Analysis*

Cost estimation plays a crucial role in manufacturing. The material, dimensions, tolerances and finish of each item are essential pieces of information needed to establish manufacturing costs, and detailed drawings enable the determination of final cost estimates. Table 4 displays the cost estimation for the development of the mini timing gate. To complete this mini timing gate, approximately less than 500.00 Malaysian Ringgit is needed. Therefore, this amount is considered cheaper compared to the currently available timing gate technology.

4 Discussion

The primary aim of this study was to develop an affordable mini timing gate to assess sprint abilities among cadet officers at the National Defence University of Malaysia. The design and creation of this mini timing gate are especially significant given the financial constraints often faced not only in military training environments but also in sports. A thorough cost analysis of the necessary components was conducted, revealing that the total cost to develop the prototype would be RM 326.90. This figure is notably lower than that of existing timing gate technologies, which are typically

Table 4 The cost analysis of all the items

No	Component	Quantity	Price per unit	Component
1	Arduino Uno	2	45.90	91.80
2	LCD 16 × 2	2	8.50	17.00
3	Buzzer	2	1.50	3.00
4	IR Proximity	2	35.50	71.00
5	Sensor	2	1.00	2.00
6	9V battery clip	2	3.90	7.80
7	Connector	2	2.00	4.00
8	Breadboard	1	3.70	3.70
9	9V battery	4	1.00	4.00
10	Jumper wires	2	4.90	9.80
11	Resistor 1 kΩ	2	8.50	17.00
12	Shipping fee	2	13.90	27.80
13	Tripod	4	1.00	4.00
14	HC-05 Bluetooth	2	1.00	2.00
15	Module	2	1.00	2.00
16	LED	2	15.00	30.00
17	Capacitor	2	15.00	30.00
TOTAL				RM 326.90

more expensive (Dyer 2015). By offering a more economical alternative, this study stands to provide significant benefits to the cadet officers' training regimen, allowing for a wider and more accessible implementation of advanced training technologies.

Furthermore, the cost-effectiveness of this mini timing gate aligns with broader observations in the field of sports and health interventions (Gc et al. 2019). As noted by Cobiac et al. (2009), affordability plays a crucial role in the adoption and success of physical activity interventions, directly contributing to improved health outcomes. In the context of sports technology, cost-effectiveness not only ensures the maintenance of performance quality but also enhances engagement and participation in sports activities (Kos et al. 2018). This aspect is particularly relevant in the training of cadet officers, where the integration of advanced, yet affordable, technology can lead to more effective training methods (Siekańska et al. 2021; Thomas and Gilbert 2016). The development of this mini timing gate thus not only meets a practical need at the National Defence University of Malaysia but also contributes to the ongoing discourse on the importance of cost-effectiveness in promoting physical health and improving sports performance.

Throughout this study, each phase of the mini timing gates' development was accompanied by thorough testing to ensure their effective functioning. This rigorous testing process was vital in assessing the reliability and accuracy of the gates in real-world conditions (Earp and Newton 2012). Despite the overall satisfactory performance of the gates, it identified areas for improvement through these tests. Consequently, it's important to implement several minor but impactful modifications to both the design and the components. These adjustments were targeted at enhancing the gates' responsiveness and precision, ensuring that they meet the high standards required for accurate sprint timing (Balsalobre-Fernández et al. 2014).

In the prototype's sprint trials, it observed a notable discrepancy between the device's timing and the times recorded by a video recorder. Specifically, the device demonstrated a slower reaction compared to the video-recorded times. This finding aligns with research by Tillar et al. (2022), which suggests that different types of sensors in timing gates can yield varying results. In our current design, it's used an IR sensor; however, the insights gained from these trials have led us to plan a shift to laser technology for the next stage of development. Anticipation of the change will address the issues of delayed reaction time and further refine the accuracy of our timing gates, thereby better serving the needs of our users in accurately measuring sprint times (Colino et al. 2019; Earp and Newton 2012).

This process of making changes and improvements isn't just about fixing problems, it's about understanding exactly what the cadet officers need and making sure the tool works perfectly for them (Bond et al. 2017). By always looking to make the study better and by paying close attention to feedback, the team shows a strong commitment to creating something that is not just good enough, but really great for the cadet officers (Siekańska et al. 2021; Balsalobre-Fernández et al. 2014). It's all about making sure the tool helps in the best way possible, making it a valuable part of the cadet officers' training and development.

5 Conclusion

In conclusion, while there is still work to be done in testing the prototype, this study has successfully developed a cost-effective mini timing gate. It should be recognized that further testing and refinements are necessary to perfect the prototype, ensuring it meets all the required standards and functions as intended. However, the progress made thus far is noteworthy. Having managed to create a timing gate that is not only functional but also affordable, making it a valuable tool for measuring sprint times, especially in settings where budget constraints are a significant consideration. This achievement marks an important step in the development of accessible sports technology. The ability to offer an affordable solution without compromising on quality or effectiveness is a key outcome of this study. Looking ahead, the continued testing and refinement of the prototype will further enhance its reliability and usability. The ultimate goal is to provide a tool that can be widely used to accurately assess sprint performance, particularly in environments like training programmes at educational institutions where resources may be limited. This study lays a solid foundation for achieving that goal, bringing us closer to making efficient and cost-effective sports technology available to a broader audience.

Acknowledgements This research is fully supported by GPJP grant, UPNM/2022/GPJP/SS/2. The authors fully acknowledge Ministry of Higher Education (MOHE) and National Defence University of Malaysia (NDUM) for the approved fund which makes this important research viable and effective.

References

Balsalobre-Fernández C, Tejero-González CM, Del Campo-Vecino J, Bavaresco N (2014) The concurrent validity and reliability of a low-cost, high-speed camera-based method for measuring the flight time of vertical jumps. J Strength Cond Res 28(2):528–533. https://doi.org/10.1519/JSC.0B013E318299A52E

Bond CW, Willaert EM, Noonan BC (2017) Comparison of three timing systems: reliability and best practice recommendations in timing short-duration sprints. J Strength Cond Res 31(4):1062–1071. https://doi.org/10.1519/JSC.0000000000001566

Chen Z, Bian C, Liao K, Bishop C, Li Y (2021) Validity and reliability of a phone app and stopwatch for the measurement of 505 change of direction performance: a test-retest study design. Front Physiol 12. https://doi.org/10.3389/FPHYS.2021.743800

Cobiac LJ, Vos T, Barendregt JJ (2009) Cost-effectiveness of interventions to promote physical activity: a modelling study. PLoS Med 6(7). https://doi.org/10.1371/JOURNAL.PMED.1000110

Colino E et al (2019) Validity and reliability of a commercially available indoor tracking system to assess distance and time in court-based sports. Front Psychol 10(SEP). https://doi.org/10.3389/FPSYG.2019.02076

Crawley AA, Sherman RA, Crawley WR, Cosio-Lima LM (2016) Physical fitness of police academy cadets: baseline characteristics and changes during a 16-week academy. J Strength Cond Res 30(5):1416. https://doi.org/10.1519/JSC.0000000000001229

Dewanti RA, Hermawan I (2022) A comparative study on measuring running speed using timing gates and stopwatches to reduce human error. J Phys Educ Sport 22(12):3080–3085

Dyer B (2015) The controversy of sports technology: a systematic review. Springerplus 4(1). https://doi.org/10.1186/S40064-015-1331-X

Earp JE, Newton RU (2012) Advances in electronic timing systems: considerations for selecting an appropriate timing system. J Strength Cond Res 26(5):1245–1248. https://doi.org/10.1519/JSC.0B013E3182474436

Gc VS, Suhrcke M, Atkin AJ, Van Sluijs E, Turner D (2019) Cost-effectiveness of physical activity interventions in adolescents: model development and illustration using two exemplar interventions. BMJ Open 9(8):e027566. https://doi.org/10.1136/BMJOPEN-2018-027566

Hetzler RK, Stickley CD, Lundquist KM, Kimura IF (2008) Reliability and accuracy of handheld stopwatches compared with electronic timing in measuring sprint performance. J Strength Cond Res 22(6):1969–1976. https://doi.org/10.1519/JSC.0B013E318185F36C

Jakiwa J, Azli MS, Zainuddin AA, Atan SA (2020) Time spent in physical activity and abdominal muscle endurance level between cadets and civilian students of national defense university of Malaysia. J Mil Med 22(8):873–882. https://doi.org/10.30491/JMM.22.8.873

Jakiwa J, Atan SA, Azli MS, Rustam S, Hamzah N, Zainuddin AA (2022) The level of sports participation and academic success among Malaysian student-athletes. Int J Learn Teach Educ Res 21(6):122–137. https://doi.org/10.26803/ijlter.21.6.8

Kos A, Wei Y, Tomažič S, Umek A (2018) The role of science and technology in sport. Procedia Comput Sci 129:489–495. https://doi.org/10.1016/J.PROCS.2018.03.029

Kyröläinen H, Pihlainen K, Vaara JP, Ojanen T, Santtila M (2018) Optimising training adaptations and performance in military environment. J Sci Med Sport 21(11):1131–1138. https://doi.org/10.1016/J.JSAMS.2017.11.019

Nadzmi A et al (2023) Development of motivation model towards anthropometric, soccer skills, maturity and physical fitness using machine learning. In: Lumpur 129–141. https://doi.org/10.1007/978-981-99-0297-2_10.

Siekańska M, Bondár RZ, di Fronso S, Blecharz J, Bertollo M (2021) Integrating technology in psychological skills training for performance optimization in elite athletes: a systematic review. Psychol Sport Exerc 57:102008. https://doi.org/10.1016/J.PSYCHSPORT.2021.102008

Thomas SV, Gilbert JE (2016) Integrating technology to enhance athlete development: a literature review. J High Educ Athl Innov 1(1):73–84. https://doi.org/10.15763/ISSN.2376-5267.2016.1.1.73-84

van den Tillaar R, Haugen ME, Falch HN (2022) A comparison of sprint mechanical parameters measured with timing gates and a laser gun. Front Sport Act Living 4:877482. https://doi.org/10.3389/FSPOR.2022.877482

Exploring the Use of Polyurethane as an Engineering Material in Sports Surfaces

Ainur Munira Rosli, Aina Najwa Azmi, Mohamad Rusydi Mohamad Yasin, Abdul Nasir, and Norain Binti Abdullah

Abstract Polyurethane (PU) has emerged as a highly prevalent polymer within the sports business because to its exceptional characteristics, such as its remarkable durability, flexibility, and resistance to external factors. This article provides a thorough analysis of the synthesis, characteristics, and advancements of polyurethane (PU), with a specific emphasis on its application in sports surfaces and equipment. The integration of 4D printing technology has resulted in a notable paradigm shift in the application of polyurethane (PU), particularly in the manufacturing of customized sports gear like mouthguards. Moreover, this study investigates the prospective applications of polyurethane (PU) in the domains of sound absorption, thermal insulation in athletic equipment, and its role in contemporary sports surfaces. While polyurethane (PU) offers certain advantages, it is crucial to recognize the presence of specific issues such as wear, degradation, and the need for enhanced durability. The conclusion of the article underscores the imperative of continued research in order to address these challenges effectively and optimize the utilization of polyurethane in sports-related contexts.

Keywords Polyurethane · Sports surfaces · Equipment innovation · 4D printing · Anti-fingerprint

---

A. M. Rosli (✉) · A. N. Azmi · M. R. M. Yasin
Faculty of Manufacturing and Mechatronic Engineering Technology, Universiti Malaysia Pahang Al-Sultan Abdullah, 26600 Pekan, Pahang, Malaysia
e-mail: ainurmunira@gmail.com

A. Nasir
Faculty of Electrical and Electronic Engineering Technology, Universiti Malaysia Pahang Al-Sultan Abdullah, 26600 Pekan, Pahang, Malaysia

N. B. Abdullah
Bio-Coke Research Institute, Kindai University, Osaka, Japan

Z. Mohamed et al. (eds.), *Proceedings of the 2nd International Conference on Innovation and Technology in Sports, ICITS 2023, 27–28 November, Malaysia*, Lecture Notes in Bioengineering, https://doi.org/10.1007/978-981-97-3741-3_40

1 Introduction to the Importance of Polyurethane in Sports Surfaces

Polyurethane (PU) is a highly adaptable polymer that finds extensive use in a wide range of industries, including sports and biomedical sectors (Hada et al. 2023; Guo 2023; Feng and Wang 2022; Gopal and Sudarshan 2022; Li and Wang 2022; Xu et al. 2022; Caldiroli et al. 2023; Sheng and Wen 2022; Verma et al. 2021; Nasir et al. 2013; Baharuddin et al. 2023; Yoon et al. 2023; Hasya et al. 2022; Khalid et al. 2021; Rosli et al. 2019, 2021; Atalla et al. 2022; Wang 2022; Sufian et al. 2021; Yang et al. 2023; Abidin et al. 2022; Jamaludin et al. 2017, 2018a; Al Rashid et al. 2020; Liu 2022; Ramli et al. 1060; Akram et al. 2020). The word "polyurethane" is employed to refer to polymers that consist of a significant quantity of urethane (ethyl-carbamate) groups. The urethane bond is formed through the reaction between the hydroxyl group of the macrodiol/polyol and the isocyanate group, as illustrated in Fig. 1 (Akram et al. 2020). The sports surfaces are favored due to their distinctive characteristics, which encompass durability, suppleness, and resistance to external forces (Akram et al. 2020; Chatre et al. 2022; Kupčák and Zouhar 2020; Md Kamil et al. 2022; Yin et al. 2022; Jin et al. 2022a; Alsibai et al. 2018; Sidorov and Iljin 1239; Ecevit et al. 2022; Jothibasu et al. 2018; Atalla et al. 2020; Abdullah et al. 2022; Wu et al. 2022; Tomin and Kmetty 2021). For example, the resilience and adaptability of polyurethane (PU) have resulted in its extensive application in the construction of running tracks, playgrounds, and artificial turfs, thereby ensuring the safety and comfort of athletes (Tomin and Kmetty 2021).

Nevertheless, similar to other substances, polyurethane (PU) is not exempt from encountering obstacles (Sheng and Wen 2022; Tomin and Kmetty 2021; Yeh and Chen 2020; Gao et al. 2022; Jin et al. 2022b; Ali et al. 2022; Muna and Mieloszyk 2021; Das and Mahanwar 2020; Azman et al. 1917; Ainur Munira Rosli 2021; Zhang et al. 2022; Fathi-Karkan et al. 2022; Babu et al. 2021, 2023; Baharuddin et al. 2022; Binti Kamarudin et al. 2021; Stergiou et al. 2022; Lee et al. 2021; Franta and Zając 2022; Jamaludin et al. 2018b; Nasir et al. 2015; Čubrić et al. 2021). The performance of the object can be compromised due to wear and degradation that occurs over time, which highlights the need for ongoing study to improve its durability (Čubrić et al. 2021; Farag et al. 2019; Scholle et al. 2021; Yan et al. 2021; Babu et al. 2022; Jamaludin and Yassin 2013; Kurapati and Natarajan 2022; Zhang et al. 2020; Eryilmaz and Sancak 2021; Ahirwar et al. 2022). The purpose of this paper is to

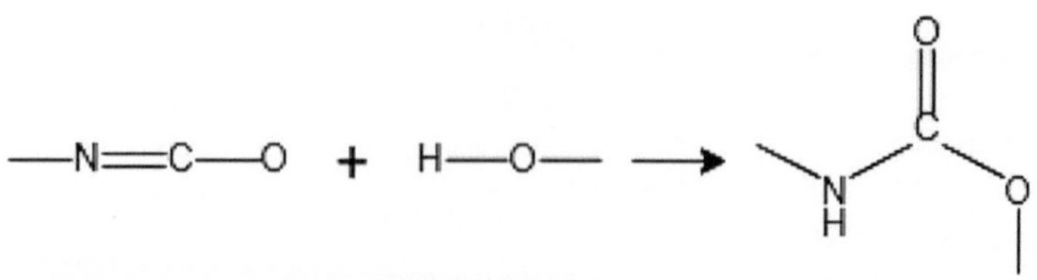

Fig. 1 A single urethane linkage (Akram et al. 2020)

provide a comprehensive examination of the current applications of polyurethane (PU), including its production, characteristics, and recent breakthroughs. In addition, we will elucidate the difficulties linked to PU and emphasize promising paths for future investigation.

2 Advancement of Polyurethane as an Engineering Material in Sports Surfaces

2.1 *Properties of Polyurethane (PU)*

Polyurethane (PU) is a highly adaptable polymer that is synthesized through the reaction of polyols, which are molecules containing several hydroxyl groups, with diisocyanates. Various categories of polyols and isocyanates can be employed to precisely adjust this chemical reaction, hence enabling the production of polyurethane materials with a diverse spectrum of characteristics, spanning from pliable and supple to firm and inflexible (Akram et al. 2020; Das and Mahanwar 2020).

PU possesses intrinsic characteristics that render it highly desirable in numerous industries. These traits include exceptional resistance to abrasion, good mechanical capabilities, a high capacity to bear loads, and resistance to a range of solvents, including oil and grease (Ahirwar et al. 2022). One of the noteworthy developments in polyurethane (PU) technology pertains to its use in sports surfaces. For example, the adaptability and durability of PU have been utilized in the development of running tracks and playgrounds, offering players with durable surfaces capable of enduring significant stress and strain (Tomin and Kmetty 2021). Table 1 shows advantages of polyurethane when compared to conventional materials.

Furthermore, polyurethane (PU) has emerged as a pivotal material in the formulation of composites tailored for sound absorption. These composites are adept at intercepting sound waves, thereby mitigating noise. Such attributes render them invaluable in diverse settings, notably in sports facilities where noise attenuation is paramount (Yin et al. 2022). The inherent sound-absorbing prowess of these composites is rooted in the distinct structure of PU. Through meticulous engineering, PU can be endowed with bespoke acoustic characteristics, paving the way for innovative sound-absorbing materials (Yin et al. 2022).

2.2 *4D-Printing and Laser Removing Technique in Synthesis of Polyurethane Product*

The manufacturing of sporting equipment has undergone a radical change as a result of the incorporation of 4D printing technology. Compared to conventional fabrication techniques, this technology delivers a protection level that is noticeably higher (Hada

Table 1 Advantages of polyurethane when compared to conventional materials (Sheng and Wen 2022; Akram et al. 2020; Chatre et al. 2022; Kupčák and Zouhar 2020; Md Kamil et al. 2022; Yin et al. 2022; Jin et al. 2022a, 2022b; Alsibai et al. 2018; Sidorov and Iljin 1239; Ecevit et al. 2022; Jothibasu et al. 2018; Atalla et al. 2020; Abdullah et al. 2022; Wu et al. 2022; Tomin and Kmetty 2021; Yeh and Chen, 2020; Gao et al. 2022; Ali et al. 2022; Muna and Mieloszyk 2021; Das and Mahanwar 2020; Azman et al. 1917; Ainur Munira Rosli 2021; Zhang et al. 2022; Fathi-Karkan et al. 2022; Babu et al. 2021, 2023; Baharuddin et al. 2022; Binti Kamarudin et al. 2021; Stergiou et al. 2022; Lee et al. 2021; Franta and Zając 2022; Jamaludin et al. 2018b; Nasir et al. 2015; Čubrić et al. 2021)

Advantages	Versus rubber	Versus metal	Versus plastic
High abrasion resistance	✓		✓
High cut & tear resistance	✓		
Superior load bearing	✓		
Thick section molding	✓		✓
Colorability	✓		
Oil resistance	✓		
Ozone resistance	✓		
Radiation resistance	✓		✓
Broader hardness range	✓		
Castable nature	✓		
Low pressure tooling	✓		
Lightweight		✓	
Noise reduction		✓	✓
Abrasion resistance		✓	✓
Less expensive fabrication		✓	
Corrosion resistance		✓	
Resilience		✓	✓
Impact resistance		✓	
Flexibility		✓	
Easily moldable		✓	
Non-conductive		✓	
Non-sparking		✓	
High impact resistance			✓
Elastic memory			✓
Variable coefficient of friction			✓
Low temperature resistance			✓
Cold flow resistance			✓
Lower cost tooling			✓

et al. 2023). Fundamentally, 4D printing is improvement of the widely used 3D printing technology. However, in the context of 4D printing, the resultant three-dimensional structure possesses the capability to undergo shape transformation in reaction to external stimuli (Hada et al. 2023). The fourth dimension in this case refers to the temporal aspect of the shape alteration subsequent to the printing process. Its capacity to account for time-induced transformations—which permits materials to alter in form or quality over time after printing—sets it apart. This adaptable capacity has created new opportunities in the sports equipment industry (Hada et al. 2023). The flexible polymer polyurethane (PU) has become a key component in this creative process.

Sports mouthguards are one of the most significant uses of this integration between PU and 4D printing. Mouthguards that are manufactured using 3D technology can be precisely crafted to fit the unique features of each athlete's mouth, as the traditional mouthguards frequently have a one-size-fits-all design, as shown in Fig. 2. This provides unmatched protection against potential injuries in addition to guaranteeing a snug fit (Hada et al. 2023). Meanwhile, 4D printing offers advantages for sporting gear that go beyond improved protection, where 4D-printed equipment is customizable, athletes can benefit from increased comfort levels. The mutually beneficial method is a glaring example of the progressive path that sports equipment design and production are taking. It is expected that in the future, sports equipment will be extensively customized to meet the needs of each athlete and will be both more effective and efficient (Hada et al. 2023).

Additionally, in terms of material removing technique, other than machining process with tools, the non-invasive feature of the laser-removal technology is causing a stir on another technical front. Conventional techniques for eliminating coatings or layers from substrates frequently entail physical contact, which may result in mechanical harm. This risk is eliminated by the laser-removal process, which makes it particularly useful in situations where the substrate is fragile or prone to harm. The sports industry is one of the main areas that could benefit greatly from this non-contact approach to surface engineering (Akram et al. 2020). One example would be the laser removal techniques created especially for micro-thickness polyurethane coatings. These techniques are capable of carefully removing PU coatings from a range of substrates, such as aluminum alloy. The accuracy guarantees that the

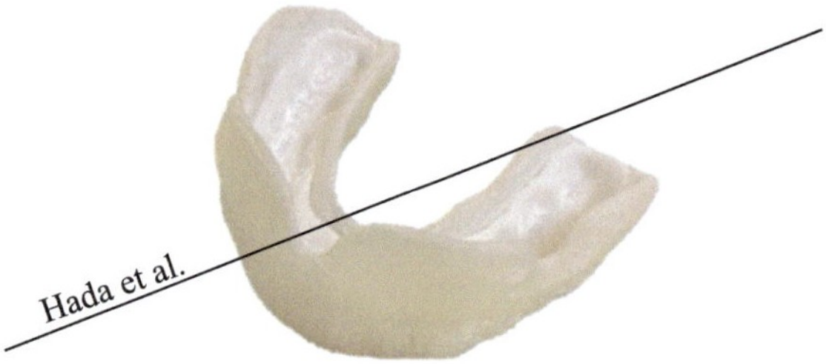

Fig. 2 A completely 4D printed mouthguard (Hada et al. 2023)

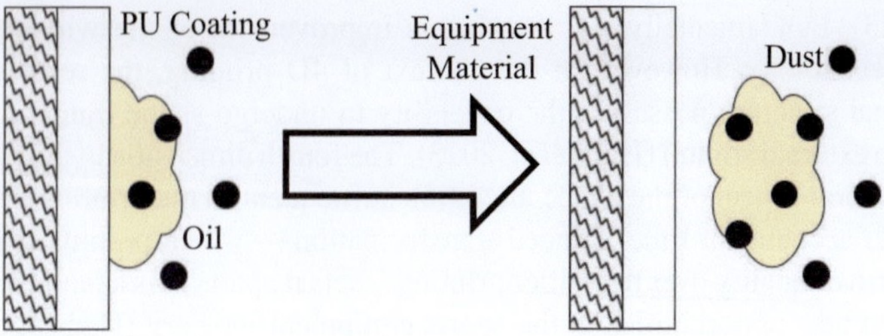

Fig. 3 Polyurethane surfaces as anti-fingerprint and smudge preventer (Guo 2023)

underlying substrate is preserved, underscoring the effectiveness and security of this state-of-the-art method (Akram et al. 2020).

3 Innovation, Opportunity, and Challenges of Polyurethane as Sports Material

3.1 Polyurethane Surfaces in Sports Equipment: Addressing the Smudge Challenge

In the world of sports, equipment aesthetics and functionality go hand in hand. Equipment with PU surfaces, especially those frequently handled, are susceptible to fingerprints and smudges. This becomes a significant concern in high-contact sports where a smudged surface, like a visor, can hinder an athlete's performance by obstructing vision. To tackle this, the industry has witnessed groundbreaking advancements in surface engineering. Anti-fingerprint treatments tailored for PU surfaces have been developed, ensuring equipment remains smudge-free, as shown in Fig. 3. A prime example is the advent of smudge-proof and self-cleaning glasses technology, which can be integrated into visors, goggles, and even helmets. This not only preserves the equipment's aesthetic appeal but also guarantees optimal functionality (Gopal and Sudarshan 2022). Furthermore, the potential of PU-based polymeric foams in energy absorption has been a focal point of research. Their ability to dissipate energy makes them invaluable in high-impact sports, offering enhanced protection against injuries (Tomin and Kmetty 2021).

3.2 Innovations in Polyurethane-Based Sports Apparel: Beyond Comfort

The field of sports apparel is ever-evolving, with innovations aimed at amplifying athlete performance and ensuring unparalleled comfort. PU, with its versatile

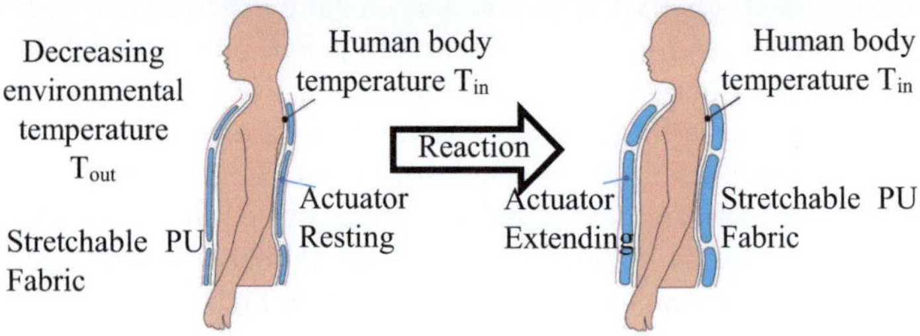

Fig. 4 PIC reaction to decreasing external temperature (Čubrić et al. 2021)

mechanical properties, has been a game-changer in this sector. A standout innovation is the introduction of polyurethane inflated chambers (PICs) for advanced thermal insulation as show in Fig. 4 (Čubrić et al. 2021). These PICs, when embedded in sports attire, offer dynamic temperature regulation. Whether it's the biting cold of winter sports or the scorching heat of summer games, PICs ensure athletes remain in their comfort zone. The design intricacies, such as the direction and number of ultrasonic joints in these PICs, are pivotal. By fine-tuning these design elements, sportswear manufacturers can customize PICs to cater to specific sport's needs, ensuring athletes experience optimal comfort and peak performance (Čubrić et al. 2021).

3.3 Enhancing Durability and Longevity of Polyurethane in Sports Equipment

One of the promising directions in polyurethane (PU) research for sports applications is improving its fatigue resistance. In sports, materials are often subjected to repeated stress, especially in protective gear. The introduction of 4D-printing technology has revolutionized the fabrication of sports equipment, simplifying complex processes and employing shape memory materials (Hada et al. 2023). However, to ensure the longevity and durability of such equipment, enhancing PU's fatigue resistance is paramount. This can be achieved through innovative PU formulations or by integrating reinforcement techniques. For instance, the functionalization of PU with chitosan-fatty acid (CS-FA) derivatives has shown potential in enhancing its properties, making it more resilient to wear and tear (Ecevit et al. 2022).

3.4 Polyurethane in Advanced Sports Surfaces

While PU is widely recognized for its potential in sports applications, there remains a significant research gap concerning its role as an advanced engineering material for sports surfaces. Sports courts, tracks, and fields require materials that offer optimal grip, shock absorption, and durability. PU, with its unique properties, can be an ideal candidate for such applications. However, comprehensive studies focusing on the design, fabrication, and performance of PU in sports surfaces are limited. Addressing this gap can lead to the development of state-of-the-art PU sports surfaces, ensuring enhanced safety and performance for athletes. The advanced energy-absorbing capabilities of polymeric foams, including polyurethane, further highlight their potential in safeguarding athletes from potential injuries during high-impact sports (Tomin and Kmetty 2021).

4 Conclusion and Future Prospects

The extensive examination of polyurethane (PU) in the context of sports has shed light on its considerable prospects and inherent difficulties. The adaptability of PU is apparent through its novel applications in the field of 4D-printed sports equipment, as well as its potential role in the development of new sports surfaces. The unique characteristics of this material, such as its ability to withstand fatigue, compatibility with biological systems, and strong mechanical properties, make it highly suitable for a wide range of sports-related uses. However, it is important to note that each material presents its own unique set of obstacles, and polyurethane (PU) is certainly no exception in this regard. Issues such as the production of blood clots in polyurethane-based vascular replacements and the pursuit of improved longevity in athletic gear underscore the imperative for continuous scholarly investigation and advancement.

The utilization of emerging surface engineering techniques, particularly the functionalization of polyurethane (PU) with chitosan-fatty acid derivatives, shows great potential for enhancing the characteristics of PU and expanding its range of applications. Nevertheless, there is a significant research deficit, specifically in comprehending the potential of polyurethane (PU) in the context of sports surfaces. By addressing this gap in knowledge, it is possible to facilitate the development of novel ideas and advancements that have the potential to improve athlete safety, optimize performance, and enhance the overall sports experience.

Acknowledgements The authors are very grateful to the Ministry of Higher Education (MoHE) Malaysia and the Universiti Malaysia Pahang Al-Sultan Abdullah for their great financial support, which made this study possible. This study was made possible by grants from the Fundamental Research Scheme (FGRS) FRGS/1/2022/TK10/UMP/02/57 and the Universiti Malaysia Pahang Al-Sultan Abdullah Fundamental Research Grant RDU220312.

References

Abdullah NA, Abdullah FF, Sufian AH, Abidin AN, Jamaludin AS, Razali MN (2022) Effect of degradation by temperature onto nitrile rubber elastomer mechanical properties. Mater Today: Proc 48:1941–1946

Abidin AN, Azmi AH, Kassim KA, Jamaludin AS, Razali MN (2022) A review on automotive tires significant characteristic identification for general consumers. In: Proceedings of the 2nd energy security and chemical engineering congress, pp 375–385

Ahirwar D, Telang A, Purohit R, Namdev A (2022) A short review on polyurethane polymer composite. Mater Today: Proc 62:3804–3810

Rosli AM, Jamaludin AS, Razali MN (2021) Recent study on hard to machine material—Micromilling process. Evergreen 8:445–453

Akram N, Zia KM, Mumtaz N, Saeed M, Usman M, Rehman S (2020) Polyurethane coatings. Polym Coat 135–157

Al Rashid A, Khalid MY, Imran R, Ali U, Koc M (2020) Utilization of banana fiber-reinforced hybrid composites in the sports industry. Materials 13:3167

Ali MA, Sivarao, Abdullah Z, Izamshah R, Kassim MS, Jamaludin AS (2022) Multi-response optimization of machining simulation approach using grey relational analysis. In: Lecture notes in mechanical engineering, pp 154–157

Alsibai MH, Hamran NNN, Nasir A, Abdullah AA (2018) A monitoring system for EPW safe use. J Telecommun Electron Comput Eng (JTEC) 10(1–2):131–135

Atalla Y, Hashim Y, Abd. Ghafar AN, Jabbar WA (2020) Temperature characteristics of FinFET based on channel fin width and working voltage. Int J Electr Comput Eng (IJECE) 10:5650

Atalla Y, Hashim Y, Abd. Ghafar AN (2022) The impact of channel fin width on electrical characteristics of Si-FinFET. Int J Electr Comput Eng (IJECE) 12:201

Azman MA, Asyraf MR, Khalina A, Petr? M, Ruzaidi CM, Sapuan SM, Wan Nik WB, Ishak MR, Ilyas RA, Suriani MJ (2021) Natural fiber reinforced composite material for product design: a short review. Polymers 13:1917

Babu D, Nasir A, Jamaludin AS, Rosle MH (2021) Holding, grasping and sensing of prosthetic robot arm like a real human hand, a journey beyond limits: an extensive review. In: Human-centered technology for a better tomorrow, pp 485–504

Babu D, Nasir A, Farag M, Sidik MH, Rejab SB (2022) Development of prosthetic robotic arm with patient monitoring system for disabled children; preliminary results. In: 2022 9th international conference on electrical and electronics engineering (ICEEE)

Babu D, Nasir A, Ravindran, Farag M, Jabbar WA (2023) 3D printed prosthetic robot arm with grasping detection system for children. Int J Adv Sci Eng Inf Technol 13:226

Baharuddin NJ, Abdul Manaf AR, Jamaludin AS (2022) Study on die shoulder patterning method (DSPM) to minimise springback of U-bending. Int J Autom Mech Eng 19:9509–9518

Baharuddin NJ, Manaf ARA, Jamaludin AS (2023) Study of springback behavior on U-bending part using die shoulder patterning method (DSPM). AIP Conf Proc 2544(1):040020

Binti Kamarudin SN, Lee H, Ishak I, Manaf AR, Jamaludin AS, Shaharudin MA, Zawawi MZ (2021) Rapid direct continuous method for hot embossing of glass microlens array combined with CO2 laser irradiation and external preheating/cooling. In: Lecture notes in mechanical engineering, pp 669–675

Caldiroli A, Cappelletti S, Birarda G, Redaelli A, Adele Riboldi S, Stani C, Vaccari L, Piccirilli F (2023) Infrared nanospectroscopy depth-dependent study of modern materials: morpho-chemical analysis of polyurethane/fibroin binary meshes. Analyst 148:3584–3593

Chatre A, Bagewadi S, Ramchandani D, Chitnis A, Kulkarni S (2022) Application of epoxy resin and glass fiber composites in cricket bats. J Mech Eng 19:221–233

Čubrić G, Salopek Čubrić I, Rogale D, Firšt Rogale S (2021) Mechanical and thermal properties of polyurethane materials and inflated insulation chambers. Materials 14:1541

Das A, Mahanwar P (2020) A brief discussion on advances in polyurethane applications. Adv Ind Eng Polym Res 3:93–101

Ecevit K, Silva E, Rodrigues LC, Aroso I, Barros AA, Silva JM, Reis RL (2022) Surface functionalization of ureteral stents-based polyurethane: engineering antibacterial coatings. Materials 15:1676

Eryilmaz O, Sancak E (2021) Effect of silane coupling treatments on mechanical properties of epoxy based high-strength carbon fiber regular (2 × 2) braided fabric composites. Polym Compos 42:6455–6466

Farag M, Azlan NZ, Alsibai MH, Ghafar AN (2019) Slippage detection for grasping force control of robotic hand using force sensing resistors. In: Proceedings of the 2019 5th international conference on computer and technology applications

Fathi-Karkan S, Banimohamad-Shotorbani B, Saghati S, Rahbarghazi R, Davaran S (2022) A critical review of fibrous polyurethane-based vascular tissue engineering scaffolds. J Biol Eng 16

Feng Q, Wang L (2022) The effect of polymer composite materials on the comfort of sports and fitness facilities. J Nanomater 2022:1–10

Franta A, Zając D (2022) Carbon fiber reinforced polymer and tensegrity structures in search of model architectural and engineering solutions. Tech Trans 2022:1–18

Gao X, Wen M, Liu Y, Hou T, Niu B, An M (2022) Synthesis and characterization of Pu/PLCL/cmcs electrospun scaffolds for skin tissue engineering. Polymers 14:5029

Gopal L, Sudarshan T (2022) Surface engineering for anti-fingerprint applications. Surf Eng 38:571–575

Guo J (2023) Structural stiffness and mechanical analysis of fiber wound composite sports equipment reinforced with carbon fiber materials. Sci Adv Mater 15:695–702

Hada T, Komagamine Y, Kanazawa M, Minakuchi S (2023) Fabrication of sports mouthguards using a semi-digital workflow with 4D-printing technology. J Prosthodont Res

Hasya N, Razali MNM, Jamaludin AS, Saffe SNBM, Hadi MA (2022) Micromilling performance analysis—Cold air assisted aluminium machining behaviour observation. In: Innovative manufacturing, mechatronics & materials forum 2022 (iM3F 2022), online conference, Pekan, Malaysia, pp 124–130

Jamaludin AS, Yassin A (2013) Analysis of laser sintered materials using finite element method. Sains Malaysiana 42(12):1727–1733

Jamaludin AS, Hosokawa A, Furumoto T, Koyano T, Hashimoto Y (2017) Evaluation of the minimum quantity lubrication in orthogonal cutting with the application of finite element method. Int J Mech Mechatron Eng IJMME-IJENS 17(01):104–109

Jamaludin AS, Hosokawa A, Furumoto T, Koyano T, Hashimoto Y (2018) Study on the effectiveness of extreme cold mist MQL system on turning process of stainless steel AISI 316. IOP Conf Ser: Mater Sci Eng 319:012054

Jamaludin AS, Akira H, Furumoto T, Koyano T, Hashimoto Y (2018) High precision estimation on physical behavior for cutting with various tool rake angle by finite element method. In: Lecture notes in mechanical engineering, pp 715–723

Jin L, Zhou Z, Li K, Zhang G, Liu Q, Yao B, Fang Y (2022b) Applying evolutionary multitasking for process parameter optimization in polymerization process of carbon fiber production. Appl Sci 12:9378

Jin Y, Zheng T, Yan C (2022) Preparation of composite multi-iron material 0.2NiFe2O4/0.8BaTiO3 and its application in sports practice. Integr Ferroelectr 229:313–330

Jothibasu S, Mohanamurugan S, Vijay R, Lenin Singaravelu D, Vinod A, Sanjay M (2018) Investigation on the mechanical behavior of areca sheath fibers/jute fibers/glass fabrics reinforced hybrid composite for light weight applications. J Ind Text 49:1036–1060

Khalid MY, Al Rashid A, Arif ZU, Sheikh MF, Arshad H, Nasir MA (2021) Tensile strength evaluation of glass/jute fibers reinforced composites: an experimental and numerical approach. Results Eng 10:100232

Kupčák R, Zouhar J (2020) Application of composite materials in sports optics. Manuf Technol 20:200–209

Kurapati R, Natarajan U (2022) Role of chemical linkage in solvation of polyurethanes in organic solvents studied by explicit molecular dynamics simulations. Ind Eng Chem Res 61:16883–16894

Lee H, Binti Kamarudin SN, Ishak I, Manaf AR, Jamaludin AS, Shaharudin MA, Zawawi MZ (2021) Feasibility study of wafer scale laser assisted thermal imprinting of glass nanostructures. In: Lecture notes in mechanical engineering, pp 917–923

Li Z, Wang F (2022) Influence of polymer properties on sports equipment performance from the perspective of composite materials. Sci Adv Mater 14:1613–1621

Liu E (2022) Application and performance discussion of fiber-reinforced composite materials in sports equipment based on image processing technology. Sci Program 2022:1–12

Md Kamil NH, Jamaludin AS, Mhd Razali MN, Abd. Ghaffar AN (2022) Temperature and heat flow analysis in a drying chamber through finite element method. In: Lecture notes in mechanical engineering, pp 309–316

Muna II, Mieloszyk M (2021) Temperature influence on additive manufactured carbon fiber reinforced polymer composites. Materials 14:6413

Nasir A, Akagi T, Dohta S, Ono A, Masago Y (2013) Development of low-cost wearable servo valve using buckled tube driven by Servo Motor. Appl Mech Mater 393:532–537

Nasir A, Akagi T, Dohta S, Ono A (2015) Analysis of low-cost wearable servo valve using buckled tubes for optimal arrangement of tubes. In: 2015 IEEE international conference on advanced intelligent mechatronics (AIM)

Ramli MR, Razak NA, Ismail I, Jamaludin AS, Manaf AR (2022) Effect of dimple size onto wear rate of mild steel AISI 1060 surface. In: Lecture notes in mechanical engineering, pp 99–102

Rosli AM, Jamaludin AS, Razali MN, Sani AS, Hamzah SB, Osman MS (2019) Modelling of Fuzzy Inference system for micro milling—A preliminary study through FEM. In: Lecture notes in mechanical engineering, pp 445–456

Rosli AM, Jamil N, Jamaludin AS, Razali MN, Yusoff AR (2021) Tool wear observation during unconventional low speed machining using low cost micromilling. In: Lecture notes in mechanical engineering, pp 589–597

Scholle P, Rüther S, Sinapius M (2021) Comparison of electrical contacting techniques to carbon fiber reinforced plastics for self-strain-sensing applications. C. 7:81

Sheng Z, Wen K (2022) Preparation and performance study of carbon fiber composite electroplated carbon badminton racket. J Nanomater 2022:1–11

Sidorov A, Iljin I (2022) Investigation of the resistance, durability and abrasion of printed elements applied to sports equipment. IOP Conf Ser: Mater Sci Eng 1239:012003

Stergiou V, Konstantopoulos G, Charitidis CA (2022) Carbon fiber reinforced plastics in space: life cycle assessment towards improved sustainability of space vehicles. J Compos Sci 6:144

Sufian AH, Xun TZ, Abidin AN, Jamaludin AS, Razali MN (2021) Study on tire tread design effect onto tire-road contact behavior through FEM. In: Lecture notes in mechanical engineering, pp 893–902

Tomin M, Kmetty Á (2021) Polymer foams as advanced energy absorbing materials for sports applications—A review. J Appl Polym Sci 139:51714

Verma M, Verma A, Kumhar V (2021) Evaluation of ultimate strength and fundamental frequency of fiber/epoxy and fiber/AL6061 composite plate. Biointerface Res Appl Chem 12:3390–3406

Wang Y (2022) Fiber nanocomposite material used in college tennis training and preparation method thereof. Integr Ferroelectr 225:266–281

Wu Y, Gao X, Nguyen TT, Wu J, Guo M, Liu W, Du C (2022) Green and low-cost natural lignocellulosic biomass-based carbon fibers—Processing, properties, and applications in sports equipment: a review. Polymers 14:2591

Xu X, Wang R, Zhang D, Feng Y (2022) Clinical application of graphene composite in internal fixation of ankle fracture in sports. Adv Mater Sci Eng 2022:1–9

Yan J, Huang Y, Zhang X, Gong X, Chen C, Nie G, Liu X, Liu P (2021) Mos2-decorated/integrated carbon fiber: phase engineering well-regulated microwave absorber. Nano-Micro Lett 13

Yang L, Ou Z, Jiang G (2023) Research progress of elastomer materials and application of elastomers in drilling fluid. Polymers 15:918

Yeh J-T, Chen W-J (2020) Characteristics of carbon fiber for badminton gears. In: 2020 IEEE Eurasia conference on IOT, communication and engineering (ECICE)

Yin D, Liu Y, Wang Y, Gao Y, Hu S, Liu L, Zhao X (2022) Finite element solution for dynamic mechanical parameter influence on underwater sound absorption of polyurethane-based composite. Int J Mol Sci 23:14760

Yoon B, Choo H, Lee C (2023) Synthesis and evaluation of engineering properties of polymer-coated glass beads. Materials 16:4476

Zhang X, Li X, Wang H, Zhang T (2020) Multi-objective optimization of machining parameters during milling of carbon-fiber-reinforced polyetheretherketone composites using grey relational analysis. Adv Mech Eng 12:168781402096623

Zhang S, Song H, Xu L, Cai K (2022) Application research on the lightweight design and optimization of carbon fiber reinforced polymers (CFRP) floor for automobile. Polymers 14:4768

Enhancing Piezoelectric Sensors in Sports Instrumentation: Addressing Durability with Nano-Coating Techniques

Syukran Hakim, Abdul Nasir, Ahmad Noor Syukri Zainal Abidin, Mohamad Rusydi Mohamad Yasin, and Waheb A. Jabbar

Abstract Piezoelectric sensors have significantly transformed the field of sports instrumentation by providing an unprecedented level of accuracy in the measurement and analysis of dynamic loads, motion, and vibrations. In the past, the utilization of flexible sensors in sports has been limited due to their inherent rigidity. However, significant progress in the field of material science has paved the way for a transformative era, enabling the development of flexible sensors with enhanced flexibility. This breakthrough has substantially broadened the range of applications for these sensors within the realm of sports. This review examines the impact of nano-coatings on improving the longevity of sensors, particularly in the challenging context of sports. The examination of specific instances, such as the composite of CNT/PVA/ZnO and the incorporation of ZnO nanostructures into sensors based on PVDF-TrFE, serves to underscore the collaborative relationship between novel material compositions and the effectiveness of sensors. Nevertheless, there are still existing gaps in our understanding, namely, about the enduring impacts of nano-coatings and their use in practical sporting settings. As the relationship between sports science and material engineering becomes more robust, there is a growing potential for customized sensor

S. Hakim · A. Nasir
Faculty of Electrical and Electronic Engineering Technology, Universiti Malaysia Pahang Al-Sultan Abdullah, 26600 Pekan, Pahang, Malaysia

A. N. S. Z. Abidin (✉) · M. R. M. Yasin
Faculty of Manufacturing and Mechatronic Engineering Technology, Universiti Malaysia Pahang Al-Sultan Abdullah, 26600 Pekan, Pahang, Malaysia
e-mail: ansyukri@miros.gov.my

A. N. S. Z. Abidin
Malaysian Institute of Road Safety Research (MIROS), Institut Penyelidikan Keselamatan Jalan Raya Malaysia (MIROS), Taman Kajang Sentral, Lot 125-135, Jalan TKS 1, 43000 Kajang, Selangor, Malaysia

W. A. Jabbar
School of Engineering and Built Environment, Birmingham City University, Birmingham, West Midlands, England, UK

Z. Mohamed et al. (eds.), *Proceedings of the 2nd International Conference on Innovation and Technology in Sports, ICITS 2023, 27–28 November, Malaysia*, Lecture Notes in Bioengineering, https://doi.org/10.1007/978-981-97-3741-3_41

systems that will enhance the integration of laboratory research with on-field applications. This paper offers a comprehensive analysis of the present status, obstacles, and prospective pathways of piezoelectric sensors that have been improved through the utilization of nano-coating techniques in the field of sports instrumentation.

Keywords Piezoelectric sensors · Nano-coating techniques · Sports instrumentation · Material durability · Flexible sensor design

1 Introduction to Piezoelectric Sensors Application in Sports

Piezoelectric sensors have become a crucial technology in the field of sports instrumentation due to their capability to accurately measure dynamic and static loads, motion, and vibrations (Chen et al. 2020; Kim et al. 2020; Babu et al. 2023; Sufian et al. 2021; Rosli et al. 2021a; Ma et al. 2019; Baharuddin et al. 2022; Patel et al. 2020; Abidin et al. 2022; Abdullah et al. 2022; Nasir et al. 2013; Su et al. 2020). These sensors, which possess the ability to transform mechanical energy into electrical signals, have been utilized in several domains, encompassing the monitoring of athletes' performance as well as the assurance of the structural soundness of sporting equipment (Su et al. 2020; Taborda and López 2020; Baharuddin et al. 2023; Atalla et al. 2022; Hasya et al. 2022; Ryu et al. 2021; Babu et al. 2022, 2021; Rosli et al. 2019; Liu et al. 2021; Zhao et al. 2022; Jamaludin et al. 2017; Ma et al. 2022). The CNT/PVA/ZnO composite has demonstrated promise in the monitoring of static and dynamic loads on concrete structures, suggesting its suitability for use in sports infrastructure (Chen et al. 2020). Furthermore, the emergence of wearable piezoelectric yarns has created opportunities for the analysis of gait data, hence offering valuable insights into the motions and biomechanics of athletes, as depicted in Fig. 1 (Ryu et al. 2021).

In the past, the utilization of piezoelectric sensors in the field of sports was constrained by their inflexibility and restricted flexibility (Ma et al. 2022; Rosli et al. 2021b; Jamaludin et al. 2018a, b; Zhang and Shi 2022; Nasir et al. 2015; Zhang et al. 2022; Elvirsson et al. 2022; Jamaludin and Yassin 2013; Chen et al. 2022; Md Kamil et al. 2022; Alsibai et al. 2018; Ramli et al. 2022; Homayounfar et al. 2022). Nevertheless, the field of material science and engineering has made significant progress in the development of sensors that possess flexibility and adaptability (Elvirsson et al. 2022; Jamaludin and Yassin 2013; Chen et al. 2022; Md Kamil et al. 2022; Alsibai et al. 2018; Ramli et al. 2022; Homayounfar et al. 2022; Binti Kamarudin et al. 2021; Ali et al. 2022; Liu et al. 2023; Farag et al. 2019; Atalla et al. 2020; Lee et al. 2021; Li et al. 2023; Nguyen et al. 2023; Wu et al. 2023; Saniman and Ihara 2016, 2014; Wang et al. 2023). The utilization of flexible sensors in sports has been enhanced by the transition from rigid sensors to more adaptable alternatives, such as binary spiky/spherical nanoparticle films (Kim et al. 2020) and cotton fabric (Ma et al. 2019).

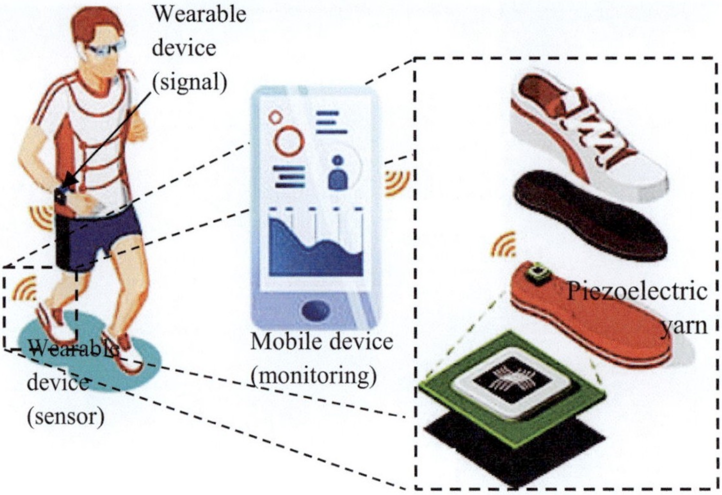

Fig. 1 Wearable piezoelectric yarn technology has made it possible to analyze gait data and gain insights into the biomechanics and motions of athletes (Ryu et al. 2021)

The aforementioned advancements have facilitated the ability to monitor players in real-time, hence improving training protocols, mitigating the risk of injuries, and conducting comprehensive performance analysis (Kim et al. 2020; Babu et al. 2023; Sufian et al. 2021; Rosli et al. 2021a; Ma et al. 2019).

The field of sports instrumentation has experienced notable progress in the sphere of piezoelectric sensor technologies. The developments in question are not solely confined to the design of the sensor, but also involve the selection of materials and the composition of their structures. An example of an improvement in the piezoelectric characteristics of PVDF-TrFE-based sensors is observed with the incorporation of ZnO nanostructures (Liu et al. 2023). In a similar vein, the employment of magnesium-doped zinc oxide nanorods has led to the creation of flexible piezoelectric pressure sensors with the ability to detect human mobility and vital signs in real-time (Li et al. 2023). Moreover, the advancement of triboelectric nanogenerators integrated into garments represents a shift toward the incorporation of these sensors directly into athletic apparel, hence facilitating uninterrupted data gathering, as illustrated in Fig. 2 (Wang et al. 2023).

Although piezoelectric sensors possess several advantages, their longevity, particularly in the challenging context of sports, continues to be a subject of concern (Liu et al. 2023; Farag et al. 2019; Atalla et al. 2020; Lee et al. 2021; Li et al. 2023; Nguyen et al. 2023; Wu et al. 2023; Saniman and Ihara 2016, 2014; Wang et al. 2023). The performance of these entities might deteriorate with time due to several factors, including repetitive mechanical stress, exposure to climatic conditions, and the effects of wear and tear. In order to tackle this issue, scholars have resorted to employing nano-coating methodologies. An example of the assessment of cementitious materials utilizing PZT sensors through the EMI approach revealed that

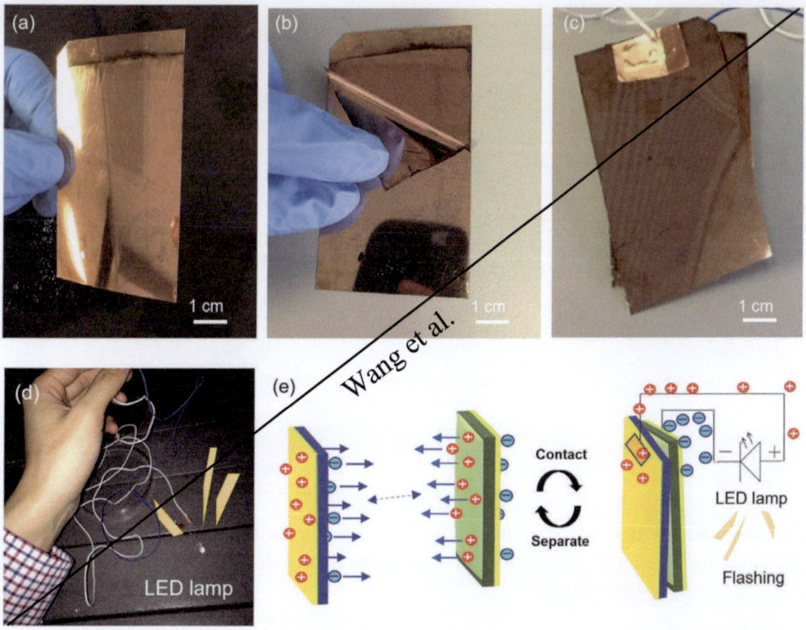

Fig. 2 **a**, **b**, and **c** Depict the two triboelectric layers that were initially bound together, respectively, before, during, and after separation. The two triboelectric layers—the PET and PE layers—had their exterior surfaces covered with a homogeneous, extremely sticky thin copper film that was created using a chemical procedure. **d** Picture displaying an LED bulb that is lit and powered by a sturdy, flexible TENG. **e** Diagram illustrating the workings of the flexible TENG, which has the ability to run an LED light (Saniman and Ihara 2016)

polymer-coated sensors, particularly those with polyester coatings, exhibit enhanced longevity compared to their uncoated counterparts (Su et al. 2020). In a similar vein, the application of nitride nano-coatings has been found to improve the thermal stability of sheathed platinum resistance sensors, thereby ensuring their compliance with international requirements (Zhao et al. 2022). The utilization of nano-coating techniques serves to augment the robustness of the sensors, hence broadening their scope of prospective applications within the realm of sports instrumentation (Zhao et al. 2022; Jamaludin et al. 2017, 2018a, b; Ma et al. 2022; Rosli et al. 2021b; Zhang and Shi 2022; Nasir et al. 2015; Zhang et al. 2022; Elvirsson et al. 2022).

The objective of this review article is to present a thorough overview of the use, evolution, and advancements of piezoelectric sensors in the field of sports. Special attention will be given to the issues associated with durability and the potential solutions provided by nano-coating techniques. The present study aims to provide a comprehensive overview of the fundamental principles behind piezoelectric sensors and their various applications.

2 Fundamentals of Piezoelectric Sensors and Their Applications

2.1 *Basic Principles of Piezoelectricity*

Piezoelectricity refers to the phenomena in which specific materials produce an electric charge when subjected to mechanical stress (Chen et al. 2020; Kim et al. 2020; Babu et al. 2023; Sufian et al. 2021; Rosli et al. 2021a; Ma et al. 2019; Baharuddin et al. 2022; Patel et al. 2020; Abidin et al. 2022; Abdullah et al. 2022; Nasir et al. 2013; Su et al. 2020). The distinctive characteristic of this property has been utilized in diverse applications, particularly within the field of sensors. The CNT/PVA/ZnO composite has been extensively studied due to its noteworthy piezoresistive and piezoelectric properties. The study revealed that the piezoresistive sensitivity of CNT/PVA films is subject to variation based on the degree of PVA doping. Moreover, the combination of carbon nanotubes (CNT), polyvinyl alcohol (PVA), and zinc oxide (ZnO) demonstrated considerable promise in the detection and measurement of both stationary and moving loads, particularly in relation to concrete structures, as depicted in Fig. 3 (Chen et al. 2020).

Moreover, there has been a significant interest in organic ferroelectric materials owing to their distinct characteristics and prospective utilization in piezoelectric sensors. One example of such a material is poly(vinylidene fluoride-trifluoroethylene) (P(VDF-TrFE)), which has been extensively investigated for its method of nanostructure fabrication. The findings of the study indicate that the nanostructured organic ferroelectric film composed of P(VDF-TrFE) exhibits notable piezoelectric and ferroelectric properties. Furthermore, the efficacy of this substance was improved by a substantial increase of 24.3% (Zhang and Shi 2022).

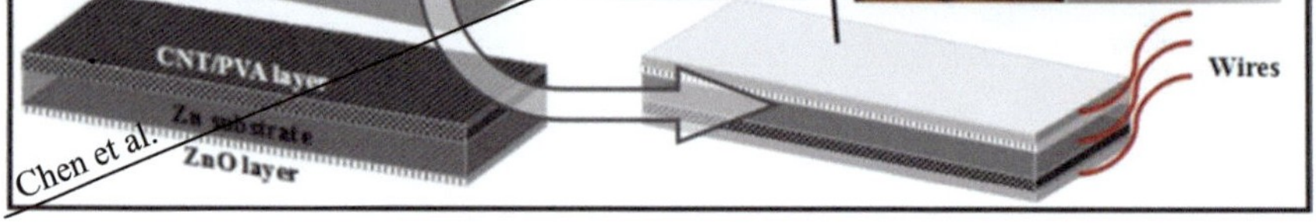

Fig. 3 The CNT/PVA/ZnO composite construction process is schematically diagrammed in (**a**), the CNT/PVA/ZnO configuration is shown in (**b**), and the CNT/PVA film and nano-ZnO array are shown in (**c**) from the front (① silver electrode, ② ③ adhesive tape, ④ wires, ⑤ CNT/PVA layer, and ⑥ nano-ZnO layer) (Chen et al. 2020)

2.2 Applications in Sports and Health Care

There has been a notable increase in the demand for motion monitoring systems within the domain of sports and health care. Flexible sensors have become increasingly popular due to their adaptability and comfort, making them a preferred option (Ma et al. 2019). One significant advancement in this field is the creation of a sensor that utilizes cotton cloth, as seen in Fig. 4 (Ma et al. 2019). The sensor exhibits not only a high level of sensitivity but also provides a wide range of sensing capabilities and a quick reaction time. The versatility of this technology is evident in its prospective applications, which encompass motion monitoring and the identification of vocal cord vibrations. This makes it a valuable instrument in the fields of sports and medical diagnostics (Ma et al. 2019). In addition, the utilization of gait analysis holds significant importance within the realm of sports, as it serves to optimize performance and mitigate the risk of injuries (Ma et al. 2019; Baharuddin et al. 2022, 2023; Patel et al. 2020; Abidin et al. 2022; Abdullah et al. 2022; Nasir et al. 2013; Su et al. 2020; Taborda and López 2020; Atalla et al. 2022; Hasya et al. 2022; Ryu et al. 2021). The subject has been significantly transformed with the advent of wearable piezoelectric yarn, commonly referred to as IEPY. IEPY is distinguished by its remarkable tensile strength and longevity, rendering it well-suited for the purpose of continuous gait data collection and processing. The prospective applications of this technology encompass various domains, including sports training, rehabilitation, and clinical diagnostics (Ryu et al. 2021).

Moreover, within the healthcare industry, it is of utmost importance to maintain a constant surveillance of human movement and critical indicators (Su et al. 2020; Taborda and López 2020; Baharuddin et al. 2023; Atalla et al. 2022; Hasya et al.

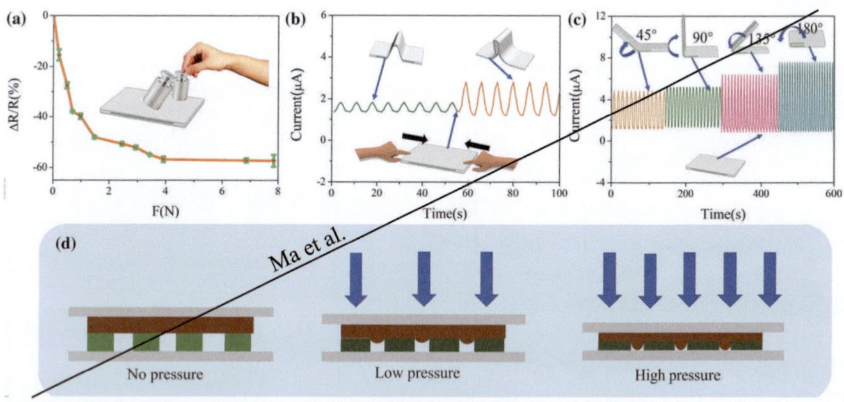

Fig. 4 Dynamic response tests include the following: **a** the sensor's relative resistance variation–pressure curve when applied at varying pressures; **b** a schematic diagram showing how the sensor's current changes as the angle of extrusion varies; **c** a schematic diagram showing how the sensor's current changes as the angle of folding varies; and **d** a schematic representation showing how the sensor deforms as pressure is applied (Ma et al. 2019)

2022; Ryu et al. 2021; Babu et al. 2022, 2021; Rosli et al. 2019; Liu et al. 2021; Zhao et al. 2022). The development of flexible piezoelectric pressure sensors, particularly those produced using Mg-doped ZnO nanorods, has facilitated the achievement of this outcome. The sensors demonstrate a remarkably high sensitivity of 3.1 V $N - 1$, which allows for the real-time monitoring of human mobility status and vital signs. These technological improvements show potential in the areas of remote patient monitoring, post-operative care, and sports performance analysis (Li et al. 2023).

3 Advancements in Sensor Design and Material Compositions

3.1 Innovative Sensor Designs

There has been a notable trend in sensor technology toward the development of designs that prioritize flexibility while maintaining high sensitivity. One significant progress in this field pertains to the creation of pressure sensors with binary spiky/spherical nanoparticle films, as depicted in Fig. 5. The sensors have exhibited notable sensitivity and a linear response within a broad spectrum of pressure levels. The architecture of these devices renders them highly suitable for use in healthcare monitoring and airflow detection, hence addressing the growing need for non-invasive and effective monitoring solutions (Kim et al. 2020).

The advent of wearable technology has significantly transformed the manner in which we observe and analyze biopotentials and bodily mobility. One notable advancement in this particular domain involves the innovation of electromechanical sensors that may be worn, which make use of the vapor deposition technique of PEDOT-Cl. The sensors possess the ability to detect a broad spectrum of pressures and have exhibited exceptional stability, maintaining 99.21% of their operational capacity following 70,000 cycles of bending. The aforementioned developments provide the ongoing monitoring of health and the study of sports performance, as seen in Fig. 6 (Homayounfar et al. 2022).

The incorporation of energy-harvesting devices into conventional garments has presented novel opportunities for the development of self-sustaining wearable electronics. Garment-integrated triboelectric nanogenerators are currently leading the way in this field of innovation (Wang et al. 2023). The devices, which have been created through the utilization of polymer-assisted metal deposition, possess qualities of both durability and flexibility. Additionally, they have the capability to be smoothly integrated into various types of apparel. The durability of their design enables them to endure regular usage and potential damage, such as exposure to water during wash testing, so presenting a bright prospect for the advancement of wearable technology (Wang et al. 2023).

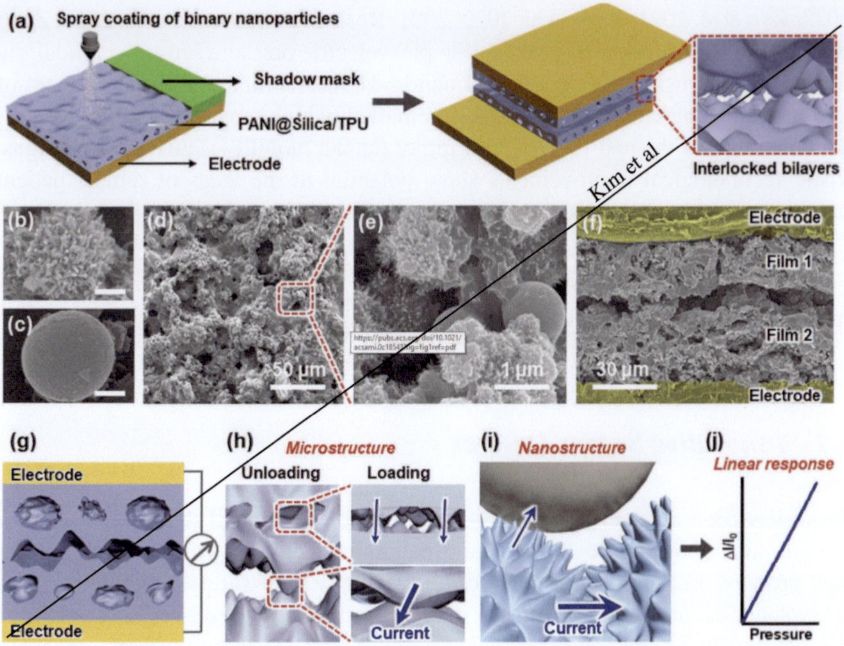

Fig. 5 The PANI@silica-based pressure sensor's, **a** diagram shows how the device is made. SEM pictures of the film surface, particles exposed on the film surface, a sea urchin-shaped PANI@silica particle, and a spherical PANI@silica particle (scale bar indicating 300 nm). The pressure sensor's cross-sectional SEM picture is shown in (**f**). **g** Interlocked porous bilayer film circuit diagram; **h** interlocked rough and porous microstructure current flow behavior under loading pressure; and **i** binary nanoparticle current flow. **j** Linear piezoresistive response induced by change in external pressure (Kim et al. 2020)

3.2 *Material Compositions and Their Impact*

The composition of materials significantly influences the functionality and efficiency of sensors (Homayounfar et al. 2022). The composite material consisting of CNT/PVA/ZnO demonstrates distinctive piezoresistive and piezoelectric properties. The sensitivity of the composite material, particularly in films composed of carbon nanotubes (CNT) and polyvinyl alcohol (PVA), has been found to exhibit variation depending on the level of PVA doping. Current research is investigating the possibilities of utilizing composites to monitor both static and dynamic stresses on structures, such as concrete. This exploration aims to provide real-time insights into the structural health of these materials (Chen et al. 2020).

The integration of nanostructures into sensor materials has the potential to greatly enhance their properties. For example, the incorporation of ZnO nanostructures into sensors based on PVDF-TrFE has led to a significant enhancement in their piezoelectric characteristics. In particular, it was observed that sensors containing 0.75 wt% ZnO NRs nanofiller demonstrated a piezoelectric response that was 2.4 times

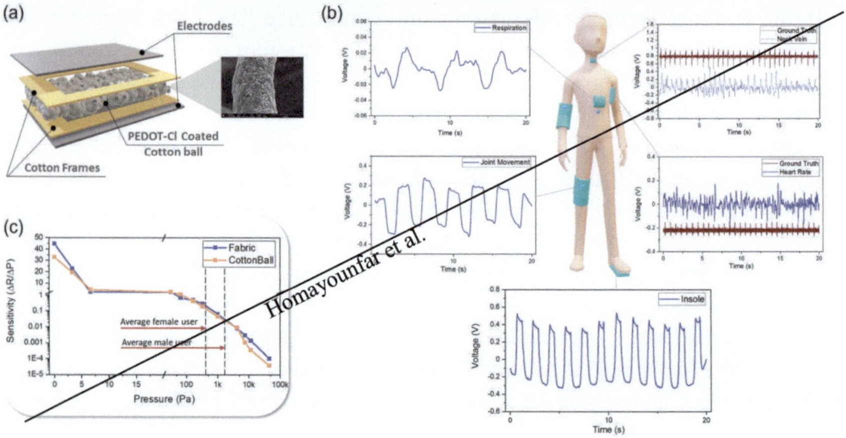

Fig. 6 **a** The sensor's schematic illustration, which includes a SEM image of the PEDOT-Cl coated layer; **b** signals obtained by the sensor that represent bodily movements and physiological processes; and **c** the sensor's sensitivity throughout a broad pressure range (Homayounfar et al. 2022)

higher than that of sensors made only of pure PVDF-TrFE. This finding highlights the significant influence of material composition on the performance of sensors (Liu et al. 2023).

The advancements in materials play a crucial role in enhancing the functionalities of piezoelectric sensors. An exemplary illustration can be observed in the utilization of magnesium-doped zinc oxide nanorods for the production of pliable piezoelectric pressure sensors. The sensors have successfully attained a remarkably high sensitivity of 3.1 V $N - 1$, hence facilitating their capability to continuously monitor the real-time motion status and vital signs of individuals. The progress made in the development of material compositions is establishing new standards in the field of health monitoring and diagnostics (Li et al. 2023).

4 Addressing Durability with Nano-Coating Techniques

4.1 Role of Nano-Coatings in Enhancing Durability

Nano-coatings have recently emerged as a crucial approach for augmenting the thermal stability of sensors. An exemplary illustration can be found in the sheathed platinum resistance sensor, which, upon application of a nitride nano-coating, satisfies global standards pertaining to platinum resistance. The aforementioned improvement guarantees the sensor's stability and accuracy, even when subjected to fluctuating thermal conditions. As a result, it becomes a dependable option for applications that require a high level of precision (Zhao et al. 2022).

In the realm of medicine, the paramount significance lies in the durability and user-friendliness of surgical instruments. The application of composite coatings on surgical needles has demonstrated a notable decrease in both the pressures required for insertion and retraction (Su et al. 2020; Taborda and López 2020; Baharuddin et al. 2023; Atalla et al. 2022; Hasya et al. 2022; Ryu et al. 2021; Babu et al. 2022, 2021; Rosli et al. 2019; Liu et al. 2021; Zhao et al. 2022; Jamaludin et al. 2017; Ma et al. 2022). Nano-coatings have been found to have a significant impact on the efficiency of surgical procedures, as well as the comfort and safety of patients. This highlights the potential of nano-coatings to bring about dramatic changes in medical applications (Patel et al. 2020). The adaptability of nano-coatings is demonstrated by the advancement of superhydrophobic, self-healing photothermal deicing, and photodetectors featuring a distinctive "papillary structure." The coatings, which are produced using a layer-by-layer assembly spray method utilizing nanomaterials of different dimensions (0D, 1D, and 2D), exhibit remarkable characteristics such as self-cleaning abilities, prolonged durability, and exceptional photothermal deicing capabilities. These advancements serve to emphasize the versatile possibilities of nano-coatings in many applications (Zhang et al. 2022).

4.2 Case Studies: Successful Implementations of Nano-Coatings

In the domain of construction and infrastructure, it is of utmost importance to ensure the quality of cementitious materials. PZT sensors, when coated with polymers and utilized in conjunction with the electromagnetic interference (EMI) approach, have demonstrated efficacy in detecting alterations in the mechanical characteristics of these materials. It is worth noting that polyester coatings have demonstrated superior durability compared to PDMS. This highlights the significance of carefully choosing the appropriate nano-coating for certain applications, as depicted in Fig. 7 (Su et al. 2020).

The investigation of incorporating graphene into low-emissive paint has been undertaken in order to improve its mechanical resilience. The study observed that graphene and graphene oxide (GO) of inferior quality exhibited enhanced mechanical durability and infrared (IR) emissivity. However, certain samples of greater quality have shown the potential to reduce emissivity. This particular case study highlights the intricate effects of nano-coatings and emphasizes the significance of material quality in attaining intended results (Elvirsson et al. 2022).

Bearings are of utmost importance in a wide range of mechanical systems, and it is crucial to monitor their bidirectional load in order to maintain optimal performance. A novel smart sensor coating has been created to effectively satisfy this requirement, wherein the performance of the coating is significantly controlled by the specific material properties and structural design. This particular case study serves as an

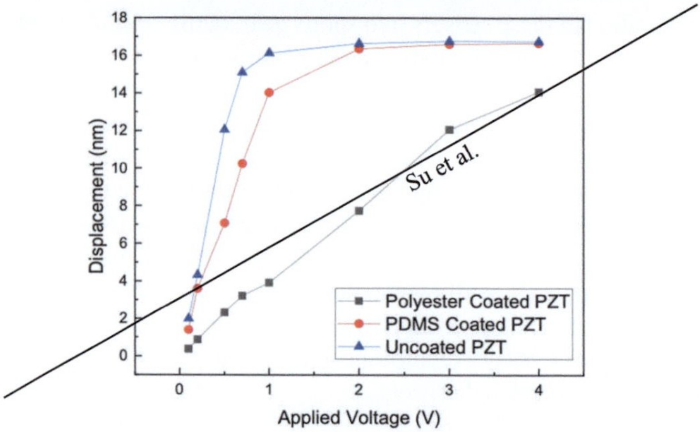

Fig. 7 Results of three different types coating of sensors' displacement measured by a laser Doppler vibrometer (LDV) at various applied voltages (Su et al. 2020)

illustration of how nano-coatings can be customized to fulfill precise monitoring needs within engineering applications (Nguyen et al. 2023).

5 Summary, Gaps, and Future Directions

The advancement of sensor technology, specifically in the domain of piezoelectric sensors, has had a significant impact. The incorporation of new materials and inventive designs has been a significant driving force behind advancements. The CNT/PVA/ZnO composite has demonstrated considerable potential in the monitoring of static and dynamic stresses on concrete structures, owing to its piezoresistive and piezoelectric properties. In a similar vein, the advancement of pressure sensors that possess flexibility and sensitivity, namely, those utilizing binary spiky/spherical nanoparticle films, has significantly broadened the scope of healthcare monitoring and airflow detection. The utilization of nano-coatings plays a crucial role in augmenting the longevity and efficacy of these sensors. Nano-coatings have demonstrated significant transformative effects, ranging from enhancing the thermal stability of sheathed platinum resistance sensors to enhancing the mechanical endurance of low-emissive paints by the use of graphene.

Although nano-coatings have demonstrated immediate advantages in terms of durability and performance, there is a lack of study about their long-term impacts. There are still unresolved inquiries pertaining to the probable deterioration of nano-coated sensors over time, the impact of environmental factors on their performance, and the overall duration of their functionality. The utilization of nano-coated sensors in sports applications holds significant promise, including a wide range of benefits such as the monitoring of athletes' performance and the assurance of equipment safety. Nevertheless, there exists a scarcity of scholarly investigations about the seamless integration of these sensors into sports gear, outfit, or equipment, as well as their performance under the demanding circumstances seen in professional sports.

The field of nanotechnology is constantly advancing and developing. Thus, it is imperative to investigate novel materials and inventive methodologies for the application of nano-coatings. In order to establish a connection between laboratory research and practical applications in the field of sports, it is imperative to adopt a collaborative approach. By facilitating collaboration between sports scientists, who possess comprehensive knowledge of the intricacies of athletic performance, and material engineers, who specialize in sensor technology, customized solutions can be devised. These collaborations have the potential to result in the development of sensors that not only assess performance but also contribute to the avoidance of injuries, optimization of equipment, and other related aspects.

Acknowledgements The authors are very grateful to the Ministry of Higher Education (MoHE) Malaysia and the Universiti Malaysia Pahang Al-Sultan Abdullah for their great financial support, which made this study possible. This study was made possible by grants from the Fundamental Research Scheme (FGRS) FRGS/1/2022/TK10/UMP/02/57 and the Universiti Malaysia Pahang Al-Sultan Abdullah Fundamental Research Grant RDU220365.

References

Abdullah NA, Abdullah FF, Sufian AH, Abidin AN, Jamaludin AS, Razali MN (2022) Effect of degradation by temperature onto nitrile rubber elastomer mechanical properties. Mater Today: Proc 48:1941–1946

Abidin AN, Azmi AH, Kassim KA, Jamaludin AS, Razali MN (2022) A review on automotive tires significant characteristic identification for general consumers. In: Proceedings of the 2nd energy security and chemical engineering congress, pp 375–385

Ali MA, Sivarao, Abdullah Z, Izamshah R, Kassim MS, Jamaludin AS (2022) Multi-response optimization of machining simulation approach using grey relational analysis. In: Lecture notes in mechanical engineering, pp 154–157

Alsibai MH, Hamran NNN, Nasir A, Abdullah AA (2018) A monitoring system for EPW safe use. J Telecommun Electron Comput Eng (JTEC) 10(1–2):131–135

Atalla Y, Hashim Y, Abd. Ghafar AN, Jabbar WA (2020) Temperature characteristics of FinFET based on channel fin width and working voltage. Int J Electr Comput Eng (IJECE) 10:5650

Atalla Y, Hashim Y, Abd. Ghafar AN (2022) The impact of channel fin width on electrical characteristics of Si-FinFET. Int J Electr Comput Eng (IJECE) 12:201

Babu D, Nasir A, Jamaludin AS, Rosle MH (2021) Holding, grasping and sensing of prosthetic robot arm like a real human hand, a journey beyond limits: an extensive review. In: Human-centered technology for a better tomorrow, pp 485–504

Babu D, Nasir A, Farag M, Sidik MH, Rejab SB (2022) Development of prosthetic robotic arm with patient monitoring system for disabled children; preliminary results. In: 2022 9th international conference on electrical and electronics engineering (ICEEE)

Babu D, Nasir A, Ravindran, Farag M, Jabbar WA (2023) 3D printed prosthetic robot arm with grasping detection system for children. Int J Adv Sci Eng Inf Technol 13:226

Baharuddin NJ, Abdul Manaf AR, Jamaludin AS (2022) Study on die shoulder patterning method (DSPM) to minimise springback of U-bending. Int J Autom Mech Eng 19:9509–9518

Baharuddin NJ, Manaf ARA, Jamaludin AS (2023) Study of springback behavior on U-bending part using die shoulder patterning method (DSPM). AIP Conf Proc 2544(1):040020

Binti Kamarudin SN, Lee H, Ishak I, Manaf AR, Jamaludin AS, Shaharudin MA, Zawawi MZ (2021) Rapid direct continuous method for hot embossing of glass microlens array combined with CO2 laser irradiation and external preheating/cooling. In: Lecture notes in mechanical engineering, pp 669–675

Chen S, Luo J, Wang X, Li Q, Zhou L, Liu C, Feng C (2020) Fabrication and piezoresistive/ piezoelectric sensing characteristics of carbon nanotube/PVA/nano-zno flexible composite. Sci Rep 10

Chen F, Han L, Dong Y, Wang X (2022) Biomimetic self-adhesive structures for wearable sensors. Biosensors 12:431

Elvirsson J, Hallberg T, Karlsson LH (2022) Adding graphene to low-emissive paint. In: Target and background signatures VIII

Farag M, Azlan NZ, Alsibai MH, Ghafar AN (2019) Slippage detection for grasping force control of robotic hand using force sensing resistors. In: Proceedings of the 2019 5th international conference on computer and technology applications

Hasya N, Razali MNM, Jamaludin AS, Saffe SNBM, Hadi MA (2022) Micromilling performance analysis—Cold air assisted aluminium machining behaviour observation. In: Innovative manufacturing, mechatronics & materials forum 2022 (iM3F 2022), online conference, Pekan, Malaysia, pp 124–130

Homayounfar SZ, Kiaghadi A, Ganesan D, Andrew TL (2022) (Digital Presentation) harnessing wide-range, highly stable pressure sensitivity via PEDOT-Cl vapor printed textiles for health monitoring. In: ECS meeting abstracts. MA2022-02, pp 2286–2286

Jamaludin AS, Yassin A (2013) Analysis of laser sintered materials using finite element method. Sains Malaysiana 42(12):1727–1733

Jamaludin AS, Hosokawa A, Furumoto T, Koyano T, Hashimoto Y (2017) Evaluation of the minimum quantity lubrication in orthogonal cutting with the application of finite element method. Int J Mech Mechatron Eng IJMME-IJENS 17(01):104–109

Jamaludin AS, Akira H, Furumoto T, Koyano T, Hashimoto Y (2018a) High precision estimation on physical behavior for cutting with various tool rake angle by finite element method. In: Lecture notes in mechanical engineering, pp 715–723

Jamaludin AS, Hosokawa A, Furumoto T, Koyano T, Hashimoto Y (2018b) Study on the effectiveness of extreme cold mist MQL system on turning process of stainless steel AISI 316. IOP Conf Ser: Mater Sci Eng 319:012054

Kim Y-R, Kim MP, Park J, Lee Y, Ghosh SK, Kim J, Kang D, Ko H (2020) Binary spiky/spherical nanoparticle films with hierarchical micro/nanostructures for high-performance flexible pressure sensors. ACS Appl Mater Interfaces 12:58403–58411

Lee H, Binti Kamarudin SN, Ishak I, Manaf AR, Jamaludin AS, Shaharudin MA, Zawawi MZ (2021) Feasibility Study of wafer scale laser assisted thermal imprinting of glass nanostructures. In: Lecture notes in mechanical engineering, pp 917–923

Li M, Yao R, Liu Y (2023) A flexible and ultra-highly sensitive tactile sensor based on MG-doped zno nanorods for human vital signs and activity monitoring. Semicond Sci Technol 38:095012

Liu M, Liu Y, Zhou L (2021) Novel flexible PVDF-TrFE and PVDF-TrFE/zno pressure sensor: fabrication, characterization and investigation. Micromachines 12:602

Liu Y, Zeng X, Zhu L, Wang C, Geng K, Yao R (2023) Effect of nanostructure morphology and concentration on the piezoelectric performance of flexible pressure sensor based on PVDF-TRFE/nano-ZNO composite thin film. Curr Nanosci 19

Ma Z, Wang W, Yu D (2019) Assembled wearable mechanical sensor prepared based on cotton fabric. J Mater Sci 55:796–805

Ma H, Liu Z, Lou J, Ding Q, Jiang Y, Li X, Han W (2022) Bacterial cellulose/MWCNT coatings for highly sensitive and flexible paper-based humidity sensors. Cellulose 30:1193–1204

Md Kamil NH, Jamaludin AS, Mhd Razali MN, Abd. Ghaffar AN (2022) Temperature and heat flow analysis in a drying chamber through finite element method. In: Lecture notes in mechanical engineering, pp 309–316

Nasir A, Akagi T, Dohta S, Ono A, Masago Y (2013) Development of low-cost wearable servo valve using buckled tube driven by Servo Motor. Appl Mech Mater 393:532–537

Nasir A, Akagi T, Dohta S, Ono A (2015) Analysis of low-cost wearable servo valve using buckled tubes for optimal arrangement of tubes. In: 2015 IEEE international conference on advanced intelligent mechatronics (AIM)

Nguyen V-C, Le M-Q, Bernadet S, Hebrard Y, Mogniotte J-F, Capsal J-F, Cottinet P-J (2023) Design rules of bidirectional smart sensor coating for condition monitoring of bearings. Polymers 15:826

Patel KI, Zhu L, Reddy Gidde ST, Ren F, Hutapea P (2020) Influence of composite polymer coatings on the insertion force of needle-like structure in soft materials. Biomed Biotechnol 5

Ramli MR, Razak NA, Ismail I, Jamaludin AS, Manaf AR (2022) Effect of dimple size onto wear rate of mild steel AISI 1060 surface. In: Lecture notes in mechanical engineering, pp 99–102

Rosli AM, Jamaludin AS, Razali MN, Sani AS, Hamzah SB, Osman MS (2019) Modelling of fuzzy inference system for micro milling—A preliminary study through FEM. In: Lecture notes in mechanical engineering, pp 445–456

Rosli AM, Jamaludin AS, Razali MNM (2021a) Recent study on hard to machine material—Micromilling process. Evergreen 8:445–453

Rosli AM, Jamil N, Jamaludin AS, Razali MN, Yusoff AR (2021b) Tool wear observation during unconventional low speed machining using low cost micromilling. In: Lecture notes in mechanical engineering, pp 589–597

Ryu CH, Cho JY, Jeong SY, Eom W, Shin H, Hwang W, Jhun JP, Hong SD, Kim T, Jeong IW, Sung TH (2021) Wearable piezoelectric yarns with inner electrodes for energy harvesting and signal sensing. Adv Mater Technol 7

Saniman MN, Ihara I (2014) Application of air-coupled ultrasound to noncontact evaluation of paper surface roughness. J Phys: Conf Ser 520:012016

Saniman MN, Ihara I (2016) Feasibility Study on characterization of non-Gaussian rough surface by ultrasonic reflection method with the Kirchhoff theory. Mech Eng J 3

Su Y-F, Han G, Kong Z, Nantung T, Lu N (2020) Embeddable piezoelectric sensors for strength gain monitoring of cementitious materials: the influence of coating materials. Eng Sci

Sufian AH, Xun TZ, Abidin AN, Jamaludin AS, Razali MN (2021) Study on tire tread design effect onto tire-road contact behavior through FEM. In: Lecture notes in mechanical engineering, pp 893–902

Taborda JA, López EO (2020) Research perspectives on functional micro and nano scale coatings. Data Anal Med 1076–1109

Wang S-C, Zhang B, Kang L, Liang C, Chen D, Liu G, Guo X (2023) Flexible and robust triboelectric nanogenerators with chemically prepared metal electrodes and a plastic contact interface based on low-cost pressure-sensitive adhesive. Sensors 23:2021

Wu W, Wang Y, Sun J (2023) A bioinspired multilayer film with angle dependent and humidity responsive color-changing performances based on its crosslink networks. J Mater Sci 58:5808–5821

Zhang J, Shi F (2022) Preparation of organic ferroelectric material P (VDF-Trfe) nanostructure and its application in process simplification. Integr Ferroelectr 229:138–157

Zhang L, Zhang H, Yu X, Xu L, Wang D, Lu X, Zhang A (2022) Superhydrophobic MXene coating with biomimetic structure for self-healing photothermal deicing and photoelectric detector. ACS Appl Mater Interfaces 14:53298–53313

Zhao R, Luo W, Zhou A, Jiang Z (2022) Thermal stability evaluation of sheathed platinum resistance sensor based on nitride nano-coating. In: 2022 international conference on wireless communications, electrical engineering and automation (WCEEA)

Anxiety Monitoring Parameters Assessment for Archery Athletes Among Universities in Malaysia

Nur Khalijah Kamarudin, Wan Nurlisa Wan Ahmad, and Mohd Azrul Hisham Mohd Adib

Abstract The development of archery sport in Malaysia is very encouraging and many universities are actively promoting this sport. However, the difficulty of these archery athletes in concentrating during a competition or match is very limited and they tend to get anxiety. This study aims to investigate the anxiety assessment among archer athletes based on skill level and assess the potential benefits of utilizing an anxiety monitoring detector device in managing anxiety. The study focuses specifically on archery athletes, collecting data through a quantitative survey including the Competitive State Anxiety Inventory-2 Revised (CSAI-2R) questionnaire to evaluate anxiety levels. The analysis was performed using Statistical Package for the Social Sciences (SPSS) aims to determine the impact of anxiety on athletes and explore the effectiveness of an anxiety monitoring detector device in helping athletes manage their anxiety. The findings of this research contribute to a better understanding of anxiety in sports and provide insights into the potential use of monitoring devices for anxiety management among athletes.

Keywords Anxiety · Monitoring · Athletes · Parameters · Detector

N. K. Kamarudin · W. N. W. Ahmad · M. A. H. M. Adib (✉)
Human Engineering Group (HEG), Faculty of Mechanical and Automotive Engineering Technology, Universiti Malaysia Pahang Al-Sultan Abdullah, 26600 Pekan, Pahang, Malaysia
e-mail: azrul@umpsa.edu.my

W. N. W. Ahmad
e-mail: nurlisa@ftv.upsi.edu.my

M. A. H. M. Adib
Medical Engineering and Health Intervention Team (MedEHiT), Centre for Advanced Industrial Technology, Universiti Malaysia Pahang Al-Sultan Abdullah, 26600 Pekan, Pahang, Malaysia

W. N. W. Ahmad
Faculty of Technical and Vocational, Department of Engineering Technology, Universiti Pendidikan Sultan Idris, Perak, Malaysia

Z. Mohamed et al. (eds.), *Proceedings of the 2nd International Conference on Innovation and Technology in Sports, ICITS 2023, 27–28 November, Malaysia*, Lecture Notes in Bioengineering, https://doi.org/10.1007/978-981-97-3741-3_42

1 Introduction

Anxiety is a prevalent psychological phenomenon that affects individuals across various domains, including sports performance. In the context of sports, anxiety can have a significant impact on athletes' psychological well-being and athletic performance. Archery, as a precision-focused sport, requires athletes to maintain steady composure, concentration, and precise execution (Oudejans and Pijpers 2009). Understanding the specific challenges and implications of anxiety in archery athletes is vital for developing effective strategies to manage and mitigate its negative effects (Marwat et al. 2021). The existing body of research on anxiety among archery athletes focuses on understanding the nature of anxiety, its potential causes, and the impact it may have on various aspects of athletes' performance. Furthermore, the authors review to explore potential risk factors and triggers of anxiety in archery athletes, including factors related to the competitive nature of the sport, performance pressure, and personal characteristics (Wan Ahmad et al. 2022a, b, 2023a, b; Wan Ahmad and Adib 2022). Understanding these factors will provide insights into the unique challenges faced by archery athletes and how these challenges contribute to anxiety. The impact of anxiety on different dimensions of archery performance will also be explored. This includes assessing the relationship between anxiety and shooting accuracy, consistency, focus, and overall athletic achievement (Horvath et al. 2022). The author (Bandelow et al. 2017), examines empirical studies that have investigated the link between anxiety and performance outcomes in archery, highlighting both direct and indirect effects. Additionally, the existing interventions and strategies employed to manage anxiety among archery athletes. This may include psychological interventions, cognitive-behavioral techniques, relaxation techniques, and the potential role of technology, such as anxiety monitoring detector devices, in assisting athletes with anxiety management.

Therefore, this study attentions on investigating the anxiety assessment among archer athletes based on skill level and evaluates the potential benefits focusing on the usage of monitoring devices to track anxiety levels.

2 Methodology

2.1 Quantitative Survey Using Competitive State Anxiety Inventory (CSAI)

In this study, a descriptive explanatory research approach was employed to assess anxiety levels among archery athletes from four universities: Universiti Tun Hussein Onn Malaysia (UTHM), Universiti Malaysia Pahang Al-Sultan Abdullah (UMPSA), Universiti Pendidikan Sultan Idris (UPSI), and Universiti Kebangsaan Malaysia (UKM). The data collection process involved the administration of a structured online survey using Google Forms in the first phase. The survey was carefully designed to capture relevant information on athletes' anxiety levels, their experiences, and the utilization of monitoring tools. The survey included the Competitive State Anxiety Inventory-2 Revised (CSAI-2R), which measures cognitive anxiety, somatic anxiety, and self-confidence. It also assessed the anxiety effect during training and competition, the physical condition of the athletes, and whether they utilized monitoring devices to track their physical and psychological states. Participant recruitment utilized convenience sampling methods to ensure wide participation, and ethical considerations, such as participant confidentiality, anonymity, and informed consent, were diligently addressed. The survey was conducted from April to June 2023. The comprehensive methodology aims to provide valuable insights into anxiety levels among archery athletes across different universities, skill levels, and genders. The collected data will contribute to a better understanding of anxiety in this specific population, leading to the development of targeted interventions and support strategies. Ultimately, the findings will enhance athletes' well-being and optimize their performance outcomes.

2.2 Data Analysis Using Statistical Package for the Social Sciences (SPSS)

The methodology employed in this study involved the use of SPSS (Statistical Package for the Social Sciences) for data analysis, focusing on the assessment of anxiety among archery athletes. Data were collected from four universities: UMPSA (Universiti Malaysia Pahang Al-Sultan Abdullah), UTHM (Universiti Tun Hussein Onn Malaysia), UPSI (Universiti Pendidikan Sultan Idris), and UKM (Universiti Kebangsaan Malaysia). To begin with, a comprehensive questionnaire was developed based on established anxiety scales and tailored specifically to the context of archery athletes. The questionnaire consisted of a series of validated items that aimed to capture various dimensions of anxiety experienced by the athletes during their participation in archery. Once the data collection phase was completed, the collected responses were inputted into the SPSS software for analysis. Descriptive statistics were utilized to examine the distribution and characteristics of the collected data, providing insights into the overall anxiety levels among archery athletes from the four universities. Furthermore, inferential statistics, such as t-tests or analysis of variance (ANOVA), were employed to determine potential differences in anxiety levels between universities. Post-hoc tests were conducted when significant differences were observed, allowing for a more detailed understanding of the variations among the universities. Additionally, correlation analyses were performed to explore potential relationships between anxiety levels and other relevant factors, such as years of experience in archery or training frequency. This analysis aimed to identify potential predictors or factors that may influence anxiety levels among archery athletes. Overall, the use of SPSS for data analysis enabled a systematic examination of the collected data, providing valuable insights into the anxiety levels among archery

Table 1 Distribution of archer athletes based on gender

University	Female	Male	Total number of archer athletes
UMPSA	4	4	8
UTHM	1	7	8
UKM	4	5	9
UPSI	3	3	6
Total	**12**	**19**	**31**

athletes from UMPSA, UTHM, UPSI, and UKM. The findings derived from this analysis will contribute to a better understanding of anxiety in this specific athlete population and can potentially inform interventions or support programs aimed at managing anxiety in archery athletes.

3 Results and Discussion

3.1 Data Distribution of Archer Athletes Across the University

Table 1 shows a total of 31 archery athletes were selected as participants in this study. The distribution of participants across the universities and their respective genders is as follows.

3.2 Skills Level of Archer Athletes Based on the University

The participants were further categorized based on their skill levels shown in Table 2.

In the archery anxiety assessment survey, we calculated descriptive statistics for the variable "Anxiety" across different universities (UKM, UMPSA, UPSI, and UTHM) and athletic levels (Beginner, Intermediate, and Expert). These statistics give us an understanding of the average anxiety scores and how they vary within each group. For Beginner-level athletes, the mean anxiety score was 39.25, and the standard deviation (SD) was 5.66, based on a total of 16 participants from all universities. At UKM, the mean score was 38.25 with an SD of 7.89 (4 participants). UMPSA had a mean score of 42.00 with an SD of 2.45 (6 participants). At UPSI, there was only one competitor with a score of 32.00. Finally, at UTHM, the mean score was 38.20 with an SD of 6.26 (5 participants). For Intermediate-level athletes, we had limited data, with only 2 participants. At UKM, the mean anxiety score was 42.00, and at UPSI, it was 35.00. The overall mean anxiety score for the Intermediate group, considering all universities, was 38.50, with an SD of 4.95. Expert level, the mean anxiety score at UKM was 43.75 with an SD of 2.87 (4 participants). At UPSI,

Table 2 Skills level of archer athletes based on different universities

Sports level	University	Total number of archer athletes
Beginners	UMPSA	6
	UTHM	5
	UKM	4
	UPSI	1
Intermediates	UMPSA	2
	UTHM	3
	UKM	4
	UPSI	4
Expert	UMPSA	–
	UTHM	–
	UKM	1
	UPSI	1
Total		**31**

the mean score was 44.00 with an SD of 1.41 (2 participants). At UTHM, the mean score was 40.75 with an SD of 4.35 (4 participants). Considering all universities and athletic levels, the mean anxiety score for the Expert group was 42.46, with an SD of 3.67, based on a total of 13 participants. These descriptive statistics provide an overview of the average anxiety levels and how they vary across different athletic levels and universities. Table 3 summarizes the data and highlights the differences in anxiety scores within each group.

3.3 *Descriptive Statistics Anxiety Based on Sport Level Among University*

See Table 3.

3.4 *Means of Anxiety Against University Based on Sport Level*

Figure 1 shows the results of an analysis that looked at different factors and their impact on anxiety levels. The overall analysis did not find any significant effects of the combined factors on anxiety levels ($p = 0.411$). However, there was a significant variation in anxiety scores across all participants ($p < 0.001$). Specifically, the factors of sports level and university did not show significant differences in anxiety scores. Overall, the analysis explained a small portion of the variability in anxiety scores

Table 3 Descriptive anxiety based on sports level among university

Sport level	University	Mean	Std. deviation	N
Beginner	UKM	38.2500	7.88987	4
	UMPSA	42.0000	2.44949	6
	UPSI	32.0000		1
	UTHM	38.2000	6.26099	5
	Total	39.2500	5.66274	16
Intermediate	UKM	42.0000		1
	UPSI	35.0000		1
	Total	38.5000	4.94975	2
Expert	UKM	43.7500	2.87228	4
	UMPSA	44.0000	1.41421	2
	UPSI	40.7500	4.34933	4
	UTHM	42.0000	5.19615	3
	Total	42.4615	3.66550	13
Total	UKM	41.1111	5.84047	9
	UMPSA	42.5000	2.32993	8
	UPSI	38.3333	5.12510	6
	UTHM	39.6250	5.82942	8
	Total	40.5484	4.99892	31

($R^2 = 0.318$), but when adjusted for the number of factors, it had limited explanatory value (adjusted $R^2 = 0.026$). In simpler terms, the study did not find any significant relationship between the variables and anxiety scores among participants.

3.5 Analysis of the Utilization of Monitoring Device for Anxiety

The analysis of the data presented in Table 4 focuses on the usage of monitoring devices to track anxiety levels. The table provides group statistics for two distinct groups: those who have utilized a monitoring device (Group: Yes) and those who have not (Group: No).

An independent samples t-test was conducted to compare the means of anxiety scores between the two groups. Before proceeding with the t-test, the assumption of equal variances was assessed using Levene's test for variance equality. The results of Levene's test indicated that the assumption of equal variances had been violated ($F = 2.282$, $p = 0.142$), suggesting that the variances of anxiety scores in the two groups were not equal. Consequently, the t-test was performed without assuming equal variances. The findings of the t-test revealed a statistically significant difference in

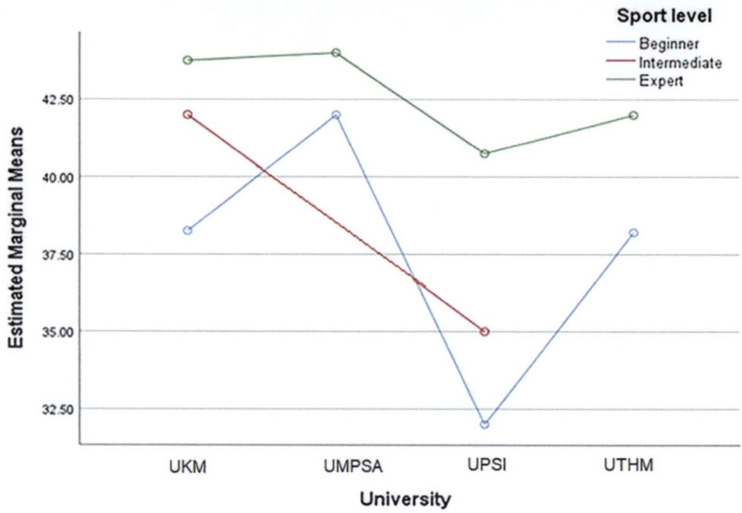

Fig. 1 Estimated marginal means against the university

anxiety levels between the two groups (t (28.443) = 2.123, p = 0.043). This indicates that the mean anxiety scores in the two groups were significantly different from each other. The standard error difference in anxiety ratings was 1.68993, suggesting the average amount of variability in the difference between the two groups' means. The mean difference in anxiety scores was 3.58750, indicating the average amount by which one group's anxiety scores were higher or lower than the other groups. The 95% confidence interval for the difference in anxiety scores ranged from 0.12826 to 7.04674. This means that we can be 95% confident that the true difference in anxiety levels between the two groups falls within this interval. Notably, the lower bound of the confidence interval is considerably greater than zero, further supporting the conclusion that there is a significant difference in anxiety levels between the two groups. Based on these results, the study found a significant difference in anxiety scores between the two groups, suggesting a noteworthy variation in anxiety levels.

This implies that utilizing monitoring devices to track anxiety levels in athletes may be beneficial in understanding and managing anxiety in the context of sports. Monitoring devices can provide valuable insights and support in assessing and addressing anxiety-related challenges, ultimately helping athletes optimize their performance.

Table 4 Analysis data on the usage of monitoring devices to track anxiety levels

Independent samples test

		Levene's test for equality of variances		t-test for equality of means						
		F	Sig	t	df	Sig (2-tailed)	Mean difference	Std. error difference	95% confidence interval of the difference	
									Lower	Upper
Anxiety	Equal variances assumed	3.2282	4.142	5.2108	6.29	7.044	8.358,750	9.170,155	10.10,743	11.70,675
	Equal variances not assumed	13.0	14.0	15.2123	16.28,443	17.043	18.358,750	19.168,993	20.12,826	21.70,467

	Have you ever used a monitoring device to track your anxiety levels?	N	Mean	Std. deviation	Std. error mean
Anxiety	Yes	15	42.4000	4.20544	1.08584
	No	16	38.8125	5.17969	1.29492

4 Conclusions

The findings reveal that anxiety is prevalent among archery athletes, indicating that it is a common experience among university archers. The anxiety assessment among archer athletes is based on skill level and evaluates the potential benefits focusing on the usage of monitoring devices to track anxiety levels. It is important to recognize that anxiety levels may vary among archery athletes due to individual differences such as prior experience, skill level, and personal disposition. High levels of anxiety can potentially impact performance outcomes by affecting focus, concentration, and coordination, leading to decreased shooting accuracy. Therefore, incorporating mental skills training programs that equip athletes with coping strategies, relaxation techniques, and other tools to effectively manage anxiety becomes crucial. Such interventions can enhance athletes' psychological resilience and overall well-being, ultimately improving their performance outcomes. Further research should focus on exploring the specific triggers of anxiety in archery, evaluating the effectiveness of different coping strategies, and developing tailored interventions to address anxiety-related challenges in this specific athletic context. This will contribute to a deeper understanding of anxiety in archery and aid in the development of targeted interventions to enhance athletes' psychological well-being and performance.

Acknowledgements This research work is strongly encouraged by the Internal Grant (RDU 190399) from University Internal Grant (RDU) Universiti Malaysia Pahang Al-Sultan Abdullah (UMPSA), which provided the research materials and equipment.

References

Bandelow B, Michaelis S, Wedekind (2017) Treatment of anxiety disorders. www.dialogues-cns.org

Damovska L, Anastasovski I. Anxiety among athletes basketball players and non-athletes during the medium adolescence, impact of national team identification on national identity in Montenegro and Macedonia. View project evolutionary morphology and cultural pressures view projects. https://doi.org/10.13140/2.1.1634.1129

Horvath E, Kovacs MT, Toth D, Toth L (2022) A study of the relationship between anxiety, cognitive emotion regulation and heart rate variability in athletes. J Phys Educ Sport 22(2):528–534. https://doi.org/10.7752/jpes.2022.02066

Marwat NM, Islam SZ, Luqman MS, Manzoor M, Irfanullah (2021) Effect of competition anxiety on athletes' sports performance: implication for coach. Hum Soc Sci Rev 9(3):1460–1464. https://doi.org/10.18510/hssr.2021.93146

Oudejans R, Pijpers J (2009) Training with anxiety has a positive effect on expert perceptual-motor performance under pressure. J Exp Psychol 62:1631–1647. https://doi.org/10.1080/17470210802557702

Wan Ahmad WN, Adib MAHM (2022) Integration of the health monitoring system with IoT application in sports technology: a review. Jurnal Kejuruteraan 5:101–105. https://doi.org/10.17576/jkukm-2022-si5(2)-11

Wan Ahmad WN, Ghazalli Z, Adib MAHM (2022) Anxiety performance among athletes in response to theories and standard instruments: a systematic review, pp 379–394. https://doi.org/10.1007/978-981-16-4115-2_30

Wan Ahmad WN, Adib MAHM, Txi MRS (2022) Enhanced the anxiety monitoring system among athletes with IoT for sports performance: a review. J Phys Educ Sport 22:2700–2707. https://doi.org/10.7752/jpes.2022.11344

Wan Ahmad WN, Adib MAHM, Txi MRS, Jaafar MHM (2023) Formulation of anxiety concentration index on athletes' performances using the application of IoT device. Int J Emerg Technol Adv Eng 13:44–47. https://doi.org/10.46338/ijetae0123_05

Wan Ahmad WN, Adib MAHM, Nashir IM, Txi MRS, Salleh FNM, Tang JR (2023) Anxiety status of junior archers in Covid-19 during training isolation period towards the shooting performance. Int J Public Health 12:361–370. https://doi.org/10.11591/ijphs.v12i1.21982

Philosophical Reflections on Artificial Intelligence Reshaping the "Field" of Physical Education Teaching

Xiaoxiao Liu, Tengku Fadilah Tengku Kamalden, and Mengyao Chao

Abstract Artificial Intelligence (AI) is reshaping physical education by introducing immersive virtual environments, significantly enhancing the teaching experience. This paradigm shift involves digitizing every aspect of physical training, presenting data-driven insights, and transforming traditional teacher-student interactions into interactive user relationships. The challenges lie in addressing the altered teacher-student dynamic and preserving the essential experiential and contextual aspects of physical education. Research methods, including literature review and logical reasoning, highlight the need for educators to discern technology's essence and return to a focus on nurturing. Future efforts should prioritize students' experiences, insights, and the quality of physical education, fostering a holistic approach beyond knowledge and skills transfer. This adaptation is crucial in navigating the changing landscape of AI in education.

Keywords Artificial intelligence · Physical education · Philosophical reflections · Field

1 Introduction

The real driver behind the integration of artificial intelligence (AI) into sports teaching is not merely technological breakthroughs but the continuous expansion and amplification of human reasoning. The amplification of human rationality, deeply rooted in human subjectivity, has reached unprecedented heights. Innovations such as AlphaGo's victory, the dominance of the "meta-universe," and ChatGPT's transformative impact on human-AI interaction and output result from the ongoing elevation of human intelligence, improved logical thinking abilities, and enhanced mastery over nature. This continuous increase in human intelligence and the improvement

X. Liu · T. F. T. Kamalden (✉) · M. Chao
Department of Sports Study, Faculty of Educational Studies, Universiti Putra Malaysia, 43400 Serdang, Selangor, Kuala Lumpur, Malaysia
e-mail: tengku@upm.edu.my

Z. Mohamed et al. (eds.), *Proceedings of the 2nd International Conference on Innovation and Technology in Sports, ICITS 2023, 27–28 November, Malaysia*, Lecture Notes in Bioengineering, https://doi.org/10.1007/978-981-97-3741-3_43

of logical thinking and managerial abilities contribute to the creation of new technologies and inventions. The utilization of scientific and technological achievements, from Descartes' "I think" and Leibniz's "perceptual monad" to Kant's "subject of unity" and Fichte's "ego," has not only enhanced human wisdom but also led to the invention of groundbreaking technologies. However, as AI propels human subjectivity to new heights, there is a risk of moving to the other extreme. In the context of AI in physical education transitioning from deep integration to heavy intervention, the potential problem of the alienation of "subjectivity" in sports teaching becomes increasingly apparent.

ChatGPT (chat generative pre-trained transformer) launch in November 2022, people from all walks of life have been drawn to it due to its vast knowledge base, logical reasoning, information gathering, and content creation capabilities. Additionally, there is currently a heated debate about the use of artificial intelligence in education. Artificial intelligence is becoming more and more "about the existence of the revolution" and is not just a straightforward technological revolution (Zhao 2018). The impact of machine mass production replacing the handicraft industry and workshop craftsmen was covered by Marx in "Capital—Critique of Political Economy" early. According to Marx, the technological revolution created a surplus labor force that was created by workers being displaced by machines, which urgently prompted a shift in the form of labor. Marx explained that the technological revolution consists of replacing manual labor with machines. As technology advances, this trend of replacing humans with machines in manual labor areas deepens while also gradually extending to mental labor. Artificial intelligence (AI) has an impact on all facets of life and has even altered the "field" of physical education and sports training. AI's creation of an intelligent, personalized, and customized learning environment will lead to structural changes in the education sector. Artificial intelligence has had a significant impact on the administration of physical education, sports training curriculum, instructional strategies, and other aspects of physical education. It has also gradually affected the physical attributes of sports venues, facilities, equipment, and personnel, including coaches, teachers, and other internal staff members. All aspects of the physical education ecosystem are gradually being affected by artificial intelligence, so it's critical to take the initiative to investigate how artificial intelligence and the "field" of physical education can work together, as well as to support the growth of physical education in light of the recent developments.

"The replacement of humans by machines is impossible; we have higher levels of thinking, emotional functioning, emotional skills, leadership, creativity, and interpersonal interactions that can only be realized by humans themselves." Since artificial intelligence is an object, it cannot truly possess this advanced ability that is uniquely human. "The subjective initiative of human beings, as well as the ability to think and autonomy that exists based on the biological body of human beings, cannot be accomplished by artificial intelligence or the non-human world in the current field space or era context." Artificial intelligence, by definition, cannot truly possess this advanced ability that is uniquely human.

2 Research Methodology

2.1 A Phenomenological Approach

The adoption of a qualitative phenomenological approach is justified by the primary aim of gathering data on biographical experiences, aiming to immerse the researchers in the participants' perspectives, personal theories, and the local, temporary, and situational context (Flick 2009). The objective is to facilitate a more profound analysis of the structures influencing participants' work and identities, thereby assisting them in reconstructing their life cycles through the narration of their narratives (Court et al. 2009). Phenomenological analysis is not limited to uncovering universal, invariant formal structures of the life-world but extends to investigating specific sociocultural life-worlds (Eberle 2012). It seeks to comprehend and analyze the essence of an experience by recognizing that wisdom and knowledge are rooted in educators' lived experiences (Smith 2015). In this context, the phenomenological approach enables a focus on academics' perspectives on the assessment process in which they are mandated to participate and, in some cases, collaborate, particularly in terms of management and design. Therefore, the phenomenon under investigation is the meaning that academics ascribe to the TPA process. The primary characteristic of this study lies in concentrating on the meaning constructed by the participants and their experiences, differentiating the phenomenological approach from other qualitative methodologies (Chan et al. 2013).

2.2 Logical Reasoning

In the context of Latour's Actor-Network Theory, the process of reflecting on one's own thoughts and actions based on the reactions of machines involves the emergence of artificial intelligence as an "actor." This AI, stimulated by its reactive capabilities, becomes an active participant in the consciousness of human activities. The interaction space is formed by both human and non-human entities, where machines play the role of "actors" and also act as "translators," interpreting human questions into symbolic answers that represent the existing knowledge level, serving as the basis for human reflective actions.

Human creativity, considered irreplaceable by artificial intelligence, is defined as the "ability to generate new knowledge through fission." The essence of creativity lies in disruption, disorder, and reconstruction, wherein existing knowledge structures are dismantled to construct new rule systems. The most challenging aspect of creativity is the construction of knowledge systems. From this perspective, the knowledge output capacity of artificial intelligence, achieved through repetitive imitation of human creative results, ultimately depends on the level of human creativity. Exploring human advantages in the interaction with machines, especially in the realm of creativity, may provide more confidence to sports educators. Creativity, as a comprehensive ability composed of complex elements, involves the destruction of existing knowledge systems and the reconstruction of new ones. This process originates from the practical experience of existing rules, discovering their limitations in explaining or handling real-world problems, and subsequently initiating a targeted reconfiguration from scratch. The practice of existing rules and the contemplation of newly formed rules are intricately linked to bodily experiences.

2.3 Code of Theoretical Framework

Based on logical judgment, the phenomenon's textual analysis is carried out by code, which can swiftly and more thoroughly comprehend the phenomena of physical education in the setting of AI. To increase the analysis's effectiveness, several phenomena' descriptive keywords can be automatically analyzed at the same time. It is feasible to achieve deep mining of data and philosophical concepts on the foundation of generic research methods, and to flexibly investigate data to uncover new insights and unidentified linkages. The theoretical framework code for this study is shown in Fig. 1.

3 Artificial Intelligence in Education—Impact and Evidence

The widespread application of artificial intelligence (AI) in the current educational sphere is providing innovative ways to enhance the effectiveness and efficiency of learning. AI analyzes students' learning habits, knowledge levels, and subject interests to offer personalized learning paths and resources, catering to individual needs and improving learning outcomes. Chatbots and virtual assistants, utilizing natural language processing and speech recognition, answer questions, provide explanations, and offer guidance, delivering more intuitive and personalized support. In addition, AI-driven educational games dynamically adjust difficulty levels based on student performance, making the learning process more engaging and interactive. In science and engineering, AI enables the creation of virtual labs and simulated environments, assisting students in conducting experiments and practical applications.

Fig. 1 Code of theoretical framework

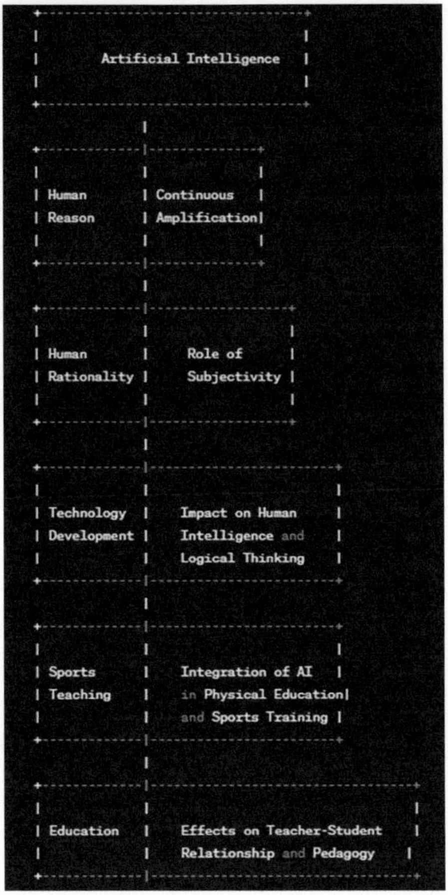

3.1 Artificial Intelligence in Physical Education, Training—Impact and Evidence

The adaptive nature of AI extends to assessments, automatically adjusting difficulty levels and formats based on student performance and providing instant feedback. Furthermore, AI's data analysis capabilities help extract valuable insights from large datasets, assisting educational institutions in making informed decisions about management and teaching strategies. In summary, AI's integration into education offers personalized learning experiences, interactive tools, adaptive assessments, and data-driven insights, transforming traditional teaching methods for more effective educational practices.

The application of artificial intelligence (AI) in the field of sports education and training is continually evolving, leveraging computer vision and motion tracking technologies. AI facilitates the analysis of students' movement skills in various sports activities. For instance, Prozone Sports offers a soccer skill analysis system that employs multiple cameras and algorithms to assess players' passing, shooting, and running abilities.

Virtual reality (VR) and augmented reality (AR) technologies enable AI to create virtual sports arenas and training grounds, providing real-time feedback to enhance students' understanding of tactics and skills. Holodia, a VR fitness application, offers virtual experiences of rowing, running, and cycling while recording users' movement data. AI processes extensive sports data, offering deeper insights for coaches and athletes. Catapult Sports' motion analysis platform utilizes sensors and machine learning to analyze athletes' movements, heart rate, and fatigue levels, supporting the optimization of training and performance. Machine learning algorithms enable AI to devise personalized training plans based on individual fitness and performance. AthleteMonitoring tailors training recommendations using athletes' physiological data and training history.

Furthermore, AI, through data analytics and machine learning, aids in predicting potential sports injury risks. Kinduct Technologies' platform utilizes big data and machine learning to monitor athletes' physical states, proactively identifying potential health issues. As the application of AI in sports education advances, it is crucial to address challenges such as privacy concerns, data security, and the technological training of educators to ensure sustainable and beneficial outcomes.

3.2 The Impact of Artificial Intelligence on the Physical Education Teaching "Field"

Artificial intelligence (AI) enriches the field of physical education by incorporating visualization, human–computer interaction, and intelligent displays, allowing students to engage in sports courses more scientifically and enjoyably. The richness of student physical education learning is manifested in several aspects.

Firstly, there is immersive sports teaching context. In traditional physical education, students primarily learn movement skills and acquire sports knowledge through methods such as live demonstrations by physical education teachers and textual images in textbooks. Over time, students may lose enthusiasm for sports due to the monotonous learning environment. Today, with the development of virtual technology and the application of intelligent sports, physical education teachers can bring new experiences to the classroom by integrating visualization techniques with virtual technologies (VR, AR, MR). Building a teaching context that is "virtualized, interactive, imaginative, and immersive" not only stimulates students' curiosity and thirst for knowledge but also addresses issues such as insufficient equipment and the singularity of teaching scenarios. It eliminates the constraint that the teaching environment is always limited to the real field of view, thus aiding physical education in transcending temporal and spatial constraints.

Secondly, there is personalized sports teaching content. The application of artificial intelligence in physical education is not only an innovation in the teaching environment but also a revolution in the output of teaching content. Traditional online education platforms, while providing students with a vast amount of learning resources, often struggle to accommodate individual differences. On this basis, the comprehensive introduction of artificial intelligence technology is steering online education towards intelligence. AI, with its robust learning capabilities and interdisciplinary integration, realizes the process of "adaptive learning," providing learning services tailored to the individual characteristics of different learners. It selects and dynamically adjusts teaching content based on students' learning goals and personal learning characteristics, thereby achieving personalized teaching objectives and alleviating issues related to differentiated student learning needs.

Thirdly, there is the scientific evaluation of physical education teaching. Evaluation in physical education has always been a widely discussed issue in teaching practice. The application of artificial intelligence technology in teaching evaluation involves collecting monitoring data and inputting evaluation standards to create a quantitative assessment model. This makes the evaluation more scientific and precise. For instance, physical health testing kits with visual analysis technology can monitor students' completion quality in real-time, providing more objective and accurate assessment of students' test scores. In terms of evaluating movement skills, the intelligent-assisted project developed, employs AI-based systems that analyze key frames and estimate body postures from students' uploaded sports videos. It offers professional guidance to improve movement quality and sports performance. Through the use of intelligent evaluation systems, comprehensive assessments, including formative assessment, process assessment, and summative assessment, are achieved, providing students with a more holistic understanding of their physical education learning assessments. As artificial intelligence continues to deepen its application in the field of teaching evaluation, physical education assessments are moving towards greater precision, scientific rigor, and fairness, truly realizing the guiding role that teaching evaluation should have in the teaching process.

3.3 *The Impact of Artificial Intelligence (AI) on the "Learning Domain"*

The impact of artificial intelligence (AI) on the "learning domain" in physical education is evident in various aspects, including personalized learning, real-time feedback, and interdisciplinary integration. AI technology analyzes students' learning styles, physical fitness levels, and interests, tailoring individualized training plans and academic courses for each student. This enhances student engagement, caters to individual differences, and encourages students to actively participate in both sports activities and academic learning. Holodia's VR fitness application, offering virtual rowing, running, and cycling experiences, has received positive user feedback

(Finkelstein et al. 2018). Sensor technology combined with AI enables real-time monitoring and analysis of students' athletic performance, providing instant feedback. Students can quickly assess their movement skill levels, allowing coaches to develop more targeted training plans based on data, thereby promoting rapid skill improvement. AI integrates multiple disciplines in physical education, such as physiology, psychology, and data science. Students not only cultivate athletic skills but also gain in-depth understanding of interdisciplinary knowledge related to sports, promoting overall competence.

Intelligent coaching assistants, utilizing algorithms to analyze student performance, offer personalized advice and training plans. For instance, applications like Coach's Eye provide detailed technical analysis and improvement suggestions based on student-recorded movements. The AthleteMonitoring platform successfully utilizes machine learning to provide personalized training recommendations, enhancing training effectiveness (Buchheit et al. 2014). This fosters deep interaction between coaches and students. Real-time motion analysis and feedback systems (e.g., PlaySight) (Palao et al. 2015) enable coaches to interact in real-time with students, analyze athletes' skills and tactics, and provide personalized advice and guidance.

Through intelligent sports applications, students can receive customized training plans tailored to individual differences, such as the Nike Training Club app (Darla 2019) Using sensors and smart rackets, like Babolat's Play Pure Drive, coaches can monitor players' swing actions in real-time, providing targeted technical feedback (Jordan et al. 2017). In training, students receive guidance closer to their individual needs, allowing coaches to more accurately monitor student progress and strengthen the coach-student collaboration. The use of virtual competition platforms (eSports) enables students to participate in global competitions and collaborative learning. Engaging in virtual competitions promotes cross-cultural exchange and the sharing of disciplinary knowledge (Yafei et al. 2021). This facilitates cross-cultural communication, allowing students to learn collaboratively with peers from different regions, broadening their perspectives, and fostering global awareness. In summary, the application of artificial intelligence in physical education enriches the learning domain, elevating the quality of education and making teaching more personalized, scientific, and interactive.

Education is a social practice activity based on the interaction between teachers and students with the goal of cultivating individuals. With the application of artificial intelligence in physical education, the convenience and effectiveness brought by intelligence allow teachers and students to enjoy enhanced sports teaching. However, concurrently, artificial intelligence has a certain impact on the emotional interaction between teachers and students in the physical education classroom. The widespread application of artificial intelligence in education makes knowledge dissemination easier, and it tends to diminish the authoritative status of teachers, turning teaching into a unidimensional knowledge transfer process. However, regarding knowledge acquisition, the educator Dewey believes that "the purpose of learning knowledge is not the knowledge itself but the ability to create knowledge to satisfy one's own needs" (Dewey 1922). Clearly, the process of acquiring knowledge is more important than the outcome of knowledge acquisition. Teaching is a generative activity of subject interaction, and the charm of education lies in the teaching behaviors and teaching art that teachers exhibit during the teaching process (Liu 2021).

Under the influence of digital technology, the absence of teacher-student behavioral interaction diminishes the emotional collisions and emotional fusion that were once inherent in the sports classroom. For example, as mentioned earlier, AR and VR virtual systems, sports teaching data collection systems, and mentor systems represented by ChatGPT in football virtual teaching scenarios show that although human–machine interaction in sports teaching applications can simulate natural responses to human emotions through simulation technology, it still belongs to the results of data induction statistics and intelligent algorithm operations, lacking self-awareness and motivation. As a result, it cannot achieve empathy and emotional feedback with individuals. The characteristics of dataization and virtualization created by artificial intelligence have to some extent eroded the essence of interpersonal communication in sports teaching activities. The original emotional experiences and interactions between teachers and students, once filled with wisdom and warmth, are gradually simplified into programmed and mechanical information symbols. This process inevitably leads to a decrease in individual empathy capabilities.

With the widespread application of sports data in physical education, data related to student physical conditions and athletic performance may raise concerns about data privacy and ethical issues. While Catapult Sports' motion analysis platform is extensively used in sports education, the associated data privacy issues have garnered attention (Mackenzie and Cushion 2013).

4 Discussion and Results

Indeed, it can be observed that sports emphasize the natural order. Sports activities, as a social representation of the biological body (Yu and Wang 2023), involve the interaction of natural biological bodies. When machines become the main actors in sports actions, it actually contradicts the inherent natural order of sports activities. From the perspective of technology replacing human labor, artificial intelligence actually provides a broad space for the expression of the inherent natural order of sports. The instrumental nature of technology determines its purpose, which is to replace human labor. Despite causing issues such as technological unemployment, looking at the replacement process itself, the human body occupied by labor is freed, allowing individuals to have more freedom in using their own bodies. In this regard, sports, as a social representation of human body activities, maintain social order by demonstrating the natural order.

In other words, in the era of artificial intelligence, sports becomes a crucial space for humans to derive meaning from life, connecting the social order based on the biological body, natural order, and interpersonal interactions. Specifically, the connection of sports to these two orders is realized through the following two aspects. First, the intrinsic nature of games in sports determines that it is a representation of human life itself, making sports activities a place that directly brings meaning to human life. Wittgenstein, in his "Philosophical Investigations," considers all domains of human life as games, emphasizing that "the term 'game' is meant to bring into prominence the following similarities: games are activities or forms of life" (Wittgenstein 2005). This implies a close connection between games and forms of life. Games are spaces where people can feel life and attain freedom, and only by linking games to people's life activities can the true meaning of games be understood. In the sense that life and games are mutually foundational, the humanistic significance of sports movement is highlighted. The term "humanistic spirit" refers to the core level of human cultural structure, including cultural factors such as human values, ways of thinking, beliefs, and religious sentiments, which profoundly influence the formation of human psychology and behavior Wittgenstein (1913). The humanistic spirit of sports is the modern pursuit of humanistic spirit in sports activities, aiming at freedom and comprehensive development of individuals, focusing on the existence and value of human beings, and caring about the fate of humanity. In simple terms, the humanistic spirit of sports lies in its intrinsic nature as a game, which means that sports itself is one of the expressions of human life forms. Therefore, the practice of sports behavior implies the direct acquisition of the meaning of human life. In this way, in the digital age where individuals gradually become isolated islands, sports reserve a place for modern individuals to settle their wandering bodies and minds.

Second, the interaction among individuals based on the biological body in sports is fundamental for sports to generate meaning in life. Merleau-Ponty, in his contemplation on the problem of the body, points out the holistic schema of the body and the intentionality of the body (Merleau-Ponty 1945). The former is the situationalization of the body discussed in "Phenomenology of Perception," which states, "My body forms an open space through movement," where human perception, body, and the world interact, and the intentionality of the body as a subject is the ability of the body as a subject to orient itself toward objects through movement, namely the fusion of the body and objects.

"Situationality" and "experientiality" are key elements that have been overlooked by past sports studies, and as a subjective behavior, the experientiality of the subject in sports action is precisely what sports studies truly need to focus on. The body is situational, and the body in sports, relying on this situationalization and coexistence with the external world and others, allows for an inseparable coexistence. Therefore, the neglect of "situationality" implies the separation of sports behavior from its social context, hovering only in the discussion of pure bodily actions. Today, the reduction of life meaning in the digital age is calling for sports studies and related research to pay attention to the situationality and experientiality of sports, that is, to return to the exploration of the value significance of the sports body, seeking the power to resist technology within the body that machines cannot reach.

The interaction among individuals based on the biological body in sports is fundamental for sports to generate meaning in life. Merleau-Ponty, in his contemplation on the problem of the body, points out the holistic schema of the body and the intentionality of the body (Merleau-Ponty 1945). The former is the situationalization of the body discussed in "Phenomenology of Perception," which states, "My body forms an open space through movement," where human perception, body, and the world interact, and the intentionality of the body as a subject is the ability of the body as a subject to orient itself toward objects through movement, namely the fusion of the body and objects. The body, as the holistic schema and sensory body of the body and the essence of the flesh, provides people with the ability to perceive the body, the perceived existence, and the empathy toward.

Issues for Further Consideration

The construction of a philosophical framework for physical education models emphasizes the subjectivity of students, advocating for their active participation and autonomous learning. This framework reflects the philosophical notions of individualism and student-centeredness in education. Guided by educational philosophy, the ultimate goal of physical education is clarified as the holistic development of students, encompassing physical, mental, moral, and social dimensions.

5 Conclusions

In the process of physical education teaching, the unique features of artificial intelligence lead to a gradual fading of the interactive communication between teachers and students. Traditional teaching elements such as explanations, demonstrations, corrective guidance, and quality monitoring are increasingly replaced, or even surpassed, by artificial intelligence technologies. However, in reality, the interactive relationship between teachers and students in physical education is a dynamic structure developed on the basis of unique emotional experiences.

What can be replaced by artificial intelligence is, in fact, only the specific content of teaching, such as explanations, demonstrations, corrective guidance, and quality monitoring. The shaping of students' emotions, emotions, and worldview, which is a process that teachers undertake, is irreplaceable by machines. Therefore, highlighting emotional interaction between teachers and students becomes crucial in the context of artificial intelligence.

Firstly, physical education teachers need to discern the essence of technology, returning to the core of nurturing individuals. Traditionally, education has often been equated with knowledge dissemination, a task that artificial intelligence can perform better than teachers. In the future, physical education teachers must fully realize that the purpose of physical education is not limited to the transmission of knowledge and skills but extends to the shaping of students' personalities, emotions, perspectives on life, and worldview. This aims to promote the holistic development of students' physical and mental health and cultivate lifelong habits of physical activity. Only by shifting the focus from a technological perspective to an educational perspective can teachers nurture individuals for free and comprehensive development within the realm of intelligent education.

Secondly, physical education teachers should build upon the physical learning environment and resources provided by artificial intelligence. Instead of solely emphasizing the completion of student tasks, teachers should shift their focus to caring about students' learning experiences, insights, and the formation of qualities unique to physical education that artificial intelligence cannot accomplish. Additionally, during the teaching process, physical education teachers should prioritize interactive communication among students. For example, leveraging virtual teaching scenarios and abundant resources provided by artificial intelligence systems to organize virtual sports competitions and online discussions on sports-related topics. This enables students to experience emotional interaction within the fusion of machines and individuals, fostering unique emotional experiences such as unity and cooperation inherent to sports. Students cannot sense meticulous care and love from cold machines; hence, teachers must consistently implement the fundamental task of cultivating morality and educating individuals. With a compassionate heart, teachers should actively enhance emotional communication between teachers and students, as well as among students, fostering a dialogue of life. This helps students maintain correct values throughout their lives.

References

Buchheit M (2014) Monitoring training status with HR measures: do all roads lead to Rome? Front Physiol

Chan ZCY, Fung YL, Chien WT (2013) Bracketing in phenomenology: only undertaken in the data collection and analysis process? Qual Rep 18(30):1–9

Court D, Merav L, Ornan E (2009) Preschool teachers' narratives: a window on personal-professional history, values, and beliefs. Int J Early Years Educ 17(3):207–217

Darla M (2019) Motor performance or opportunities to move? What do children need the most? J Sport Health Sci 004

Dewey J (1922) How we think. D.C. Heath & Co., Boston

Eberle TS (2012) Phenomenological life-world analysis and ethnomethodology's program. Hum Stud 35(2):279–304

Flick U (2009) An introduction to qualitative research, 4th edn. Sage Publications

Finkelstein S et al (2018) Exploring the impact of VR fitness with HOLOFIT: the MOTIVE study

Jordan LF et al (2017) A review of player monitoring approaches in basketball: current trends and future directions. J Strength Cond Res 31(7)

Liu Y (2021) The trap of excellence. J High Educ Res (11):17

Mackenzie R, Cushion C (2013) Performance analysis in football: a critical review and implications for future research. J Sports Sci 31(6):639–676

Merleau-Ponty M (1945) Phenomenology of perception. Routledge, London

Palao JM, Hastie PA, Ortega E (2015) Designing training programs for developing athletes. J Hum Kinet 49:1–11

Smith JA (2015) Qualitative psychology: a practical guide to research methods, 3rd edn. Sage Publications, London

Yafei S et al (2021) Investigating relationships among blended synchronous learning environments, students' motivation, and cognitive engagement: a mixed methods study. Comput Educ 168

Yu H, Wang Z (2023) Cultural overlap consensus: a sports science explanation for strengthening the awareness of the Chinese national community – Based on the analysis of the theme narrative of the "Sports and Science Academic Workshop". J Sports Sci 44(01):1–10

Wittgenstein L (1913) Notes on logic
Wittgenstein L (2005) Philosophical investigations (Chen, J., Trans.). Shanghai People's Publishing House, Shanghai
Zhao T (2018) Will artificial intelligence be a deadly problem? Open Times 2018(06)

Analysis of Knee Injuries in Tai Ji Quan Athletes Based on Selective Functional Movement Assessment Perspective

Xiaoxiao Liu, Tengku Fadilah Tengku Kamalden, and Mengyao Chao

Abstract This study investigates the musculoskeletal ramifications of knee pain in Tai Ji Quan practitioners, delving into the intricate relationship between knee pain, adjacent joint dysfunction, and gender-specific joint characteristics. Utilizing the SFMA, 87 participants from Chengdu Sport University underwent comprehensive assessments. The findings unveiled that 88.68% of individuals with knee pain exhibited dysfunction in adjacent joints, emphasizing a strong correlation. Specific movements, such as "swinging the foot and dropping the fork" and "turning around and swinging the lotus," were identified as challenging, prompting knee compensatory behaviors in the presence of hip and ankle function disorders. Gender disparities highlighted superior hip and ankle mobility in females but weaker ankle stability. Moreover, male participants demonstrated a lower likelihood of dysfunctions in specific movements. These insights underscore the need for tailored interventions to optimize musculoskeletal well-being in Tai Ji Quan practitioners, considering individualized characteristics and movement patterns.

Keywords SFMA · Injuries · Joint · Knee · Tai Ji Quan

1 Introduction

Tai Ji Quan, a traditional Chinese form of fitness, has gained immense popularity for its holistic benefits on the respiratory, cardiovascular, and skeletal systems, in addition to its metabolism-boosting effects, drawing millions of enthusiasts worldwide (Hu and Zheng 2018). 'Available data' suggests that more than 300 million people, spanning over 150 countries and regions, practice Tai Ji Quan, demonstrating its widespread appeal. However, recent reports from Tai Ji Quan athletes have highlighted a concerning issue—knee pain during practice (Li et al. 2004)

X. Liu · T. F. T. Kamalden (✉) · M. Chao
Department of Sports Study, Faculty of Educational Studies, Universiti Putra Malaysia, 43400, Selangor Kuala Lumpur, Malaysia
e-mail: tengku@upm.edu.my

Z. Mohamed et al. (eds.), *Proceedings of the 2nd International Conference on Innovation and Technology in Sports, ICITS 2023, 27–28 November, Malaysia*, Lecture Notes in Bioengineering, https://doi.org/10.1007/978-981-97-3741-3_44

While some scholars have tentatively linked this pain to irregular movements, a low center of gravity (Li et al. 2018), and excessive exercise, there is a notable dearth of studies assessing the physical function of Tai Ji Quan athletes (Fransen et al. 2007). In response, this study employs the Selective Functional Movement Assessment (SFMA) (Cook 2001) to systematically test and assess Tai Ji Quan athletes experiencing knee pain(Belo et al. 2009). The ensuing analysis aims to provide a profound understanding of the root causes of knee pain (Su et al. 2020) in Tai Ji Quan athletes and offer guidance on targeted destability exercises (Irnich and Bäumler 2023).

This research concentrates on the physical functional movement aspects of the athletes themselves (Kang et al. 2011). It underscores the necessity for further exploration on how effective exercises can be employed to ameliorate knee injuries in Tai Ji Quan athletes, elevating the overall quality of practice (Lee et al. 2009). Knee injuries, pervasive across various sports and physical activities (Altman et al. 2002), have emerged as a significant concern affecting athletes of diverse disciplines (Schumacher 2010). Despite Tai Ji Quan's reputation for its low-impact nature, practitioners, too, grapple with the prevalence of knee injuries (Wang 2010). This recognition underscores the urgency for a thorough examination of the issue. The Selective Functional Movement Assessment (SFMA) has gained prominence as an invaluable tool for deciphering movement patterns and uncovering potential dysfunctions within the kinetic chain (Gribble et al. 2013) Applying the SFMA perspective to scrutinize knee injuries in Tai Ji Quan athletes introduces an innovative methodology for identifying underlying movement imbalances contributing to injury occurrences.

This study embarks on a mission to unravel the intricacies of knee injuries in Tai Ji Quan practitioners. By employing the SFMA framework to systematically assess movement patterns, the aim is to deepen the understanding of specific biomechanical factors fueling knee injuries in Tai Ji Quan. Moreover, the research endeavors to inform injury prevention and rehabilitation strategies tailored to the distinct movement characteristics of this traditional martial art. Through a profound exploration of the relations articulatio coxae between Tai Ji Quan, knee injuries, and the SFMA perspective, this study aspires to contribute valuable insights to the fields of sports medicine, physical therapy, and the broader domain of martial arts research. The overarching goal is to enhance the well-being of Tai Ji Quan athletes and fortify the foundation of knowledge in this vital area of study.

2 Methods

2.1 Participants

This study focuses on conducting a preliminary survey at Chengdu Sport University, Sichuan, China, targeting students enrolled in the compulsory Tai Ji Quan program. The survey aims to investigate knee discomfort in participants who do not take public physical education courses. A random sampling method was employed to select participants. The study targeted students enrolled in the compulsory Tai Ji Quan program at Chengdu Sport University. The selection criteria included: Non-public physical education course participants; Random sampling approach. After the initial survey, a total of 87 (Male 39, Female 48) individuals met the sample criteria, reporting various degrees of knee discomfort and pain. The selected participants exhibit the following characteristics: Practicing Tai Ji Quan for over two years; Engaging in Tai Ji Quan practice four times a week. Each session lasting more than 90 min. This preliminary survey successfully identified 87 participants from Chengdu Sport University's compulsory Tai Ji Quan program who met the criteria of experiencing knee discomfort and pain. The sampled individuals showcase a significant level of commitment to Tai Ji Quan practice, with an average of two years of experience, a frequency of four sessions per week, and each session lasting over 90 min. The study focused on Tai Ji Quan practitioners experiencing knee pain. A selective sampling method was (Roddy et al. 2005) employed to recruit volunteers meeting the inclusion criteria. Informed consent was obtained from all participants.

2.2 *Selective Functional Movement Assessment (SFMA) Testing*

Participants underwent SFMA testing, a comprehensive assessment system designed to categorize movement patterns and identify corresponding dysfunctions and pain (Gribble et al. 2013). SFMA primarily revolves around pain and injury assessment, aiding operators in systematically identifying movement patterns causing pain and guiding appropriate treatment directions.

SFMA employs a hierarchical testing approach, yielding four results: Functional and Non-Painful (FN), Functional Painful (FP), Dysfunctional Non-Painful (DP), and Dysfunctional Painful (DN). For results of FP, DP, and DN, further decomposition of the movements was conducted to pinpoint the root issues preventing attainment of the FN result.

SFMA Hierarchical Testing Components

Neck Movement Pattern

Upper Limb Movement Pattern

Multi-Segmental Flexion

Multi-Segmental Extension

Multi-Segmental Rotation

Single Leg Stance

Overhead Deep Squat

SFMA Testing Procedures and Scoring Criteria

1. Multi-Segmental Flexion

Objective: Evaluate normal flexion of the articulatio coxae and spine.

Procedure: Participant stands with feet together, toes forward. Hinging at the articulatio coxaes, the participant reaches towards the toes without bending the knees, maintaining the gaze on the toes for 2 s.

Functional Criteria: Touching toes, weight shift backward, sacral angle at least 70°, no excessive force or movement control.

2. Multi-Segmental Extension

Objective: Assess normal extension ability of the articulatio coxae and spine during full shoulder flexion.

Procedure: Participant stands with feet together, toes forward. Arms raised overhead, the participant extends backward, articulatio coxaes moving forward. Maintain for 2 s.

Functional Criteria: Shoulder flexion reaching and sustaining 170°, Anterior Superior Iliac Spine (ASIS) beyond toes, scapular spine beyond heels, smooth spinal curvature, no excessive force or movement control.

3. Multi-Segmental Rotation

Objective: Evaluate flexibility in trunk, pelvis, articulatio coxaes, knees, and ankles.

Procedure: Participant stands with feet together, toes forward. Arms relaxed at sides, palms facing forward. Rotate trunk and arms simultaneously without moving the feet. Repeat to the other side. Hold for 2 s.

Functional Criteria: Total rotation of 100° (50° from pelvis, 50° from trunk), maintaining posture, sustaining foot position, no excessive force or movement control.

4. Single Leg Stance

Objective: Assess stability on one leg under static conditions.

Procedure: Participant stands with feet together, toes forward. Lift right knee off the ground and maintain balance on the left leg for at least 10 s. Repeat with eyes closed for an additional 10 s. Perform the same test with the left leg raised.

Functional Criteria: Maintain balance with eyes open for 10 s, maintain balance with eyes closed for 10 s, sustain posture, sustain foot position, no excessive force or movement control.

5. Overhead Deep Squat

Objective: Evaluate articulatio coxae flexion and rotation, as well as knee and ankle symmetry.

Procedure: Participant begins with knees together, feet together, arms in front of the body. Squat as deeply as possible while maintaining alignment. When thighs are below parallel (articulatio coxaes below knees), lower arms to touch the ground. The movement should occur only in the sagittal plane.

Functional Criteria: Thighs below parallel, fists touch the ground, stable and controlled movement in the sagittal plane, no excessive force or movement control.

6. Single Leg Stance

Objective: Assess stability on one leg under static conditions.

Procedure: Participant stands with feet together, toes forward. Lift right knee off the ground and maintain balance on the left leg for at least 10 s. Repeat with eyes closed for an additional 10 s. Perform the same test with the left leg raised.

Functional Criteria: Maintain balance with eyes open for 10 s, maintain balance with eyes closed for 10 s, sustain posture, sustain foot position, no excessive force or movement control.

7. Overhead Deep Squat

Objective: Evaluate articulatio coxae flexion and rotation, as well as knee and ankle symmetry.

Procedure: Participant begins with knees together, feet together, arms in front of the body. Squat as deeply as possible while maintaining alignment. When thighs are below parallel (articulatio coxaes below knees), lower arms to touch the ground. The movement should occur only in the sagittal plane.

Functional Criteria: Thighs below parallel, fists touch the ground, stable and controlled movement in the sagittal plane, no excessive force or movement control.

2.3 *SFMA Scoring*

Each movement was scored using the following abbreviations: Functional (F), Dysfunction (D), Pain (P), No Pain (N). SFMA assessment records one of the four options: Functional and Non-Painful (FN), Functional Painful (FP), Dysfunctional Non-Painful (DP), and Dysfunctional Painful (DN). Testing was conducted without warm-up, barefoot, and with meticulous attention to detail, allowing for quick demonstrations but not instructional guidance. For movements resulting in DN, FP, and DP, further decomposition was performed until the fundamental cause was identified (Table 1).

Table 1 Characterisation of the SFMA scoring

Scoring	Notes
FN	Functional and non-painful
FP	Functional painful
DP	Dysfunctional non-painful
DN	Dysfunctional painful

3 Results

3.1 Basic Information Regarding the Exam Subjects

The 39 males and 48 females who took the exam had an age difference of almost 3 years, although it was not statistically significant. The level of physical activity remained largely unchanged, as did the duration of participation in Tai Ji Quan practice (three semesters of courses), the instructors, and the training material. The focus of the exercise training was on Chen's and 42-style Tai Ji Quan, and each exercise included professional warm-up and stretch exercises prior to practice as well as relaxation and stretching exercises following practice.

According to Fig. 1, the examination of participants revealed that 5 individuals, comprising 5.74% of the sample, reported knee joint pain prior to engaging in Tai Ji Quan practice. The assessment based on the test results categorized the knee joint pain as sports-related injuries, characterized by heightened dysfunctions and persistent discomfort. This may be attributed to routine activities causing damage to the knee joints, resulting in wear and tear of the joints and surrounding ligaments, consequently instigating inflammation in the knee joints.

Throughout the Tai Ji Quan practice sessions, 59 instances of pain were reported, representing 67.8% of the participants, and 49 cases of post-practice pain were recorded, accounting for 56.3%. Notably, some participants experienced pain both during and after Tai Ji Quan practice. Certain subjects reported that during specific "rolling" movements in Tai Ji Quan, substantial force was applied to the knee joints to complete the actions, and the knees felt warm after the footwork changes. It is important to highlight that Tai Ji Quan practice emphasizes initiating movements with hip rotation, ensuring full hip abduction before guiding the movement of the hips and legs. However, practitioners may exert compensatory force to achieve the correct angle range, resulting in injuries.

Percentage of left and right legs knee pain

According to Fig. 2, the analysis reveals that a higher percentage of the individuals tested experienced pain exclusively in the right knee compared to those with pain solely in the left knee. Specifically, 34 individuals, constituting 39.08% of the sample, reported pain exclusively in the right knee, while 19 individuals, or 21.84%,

Fig. 1 Percentage of practitioners of Tai Ji Quan who have knee pain both before and after practice

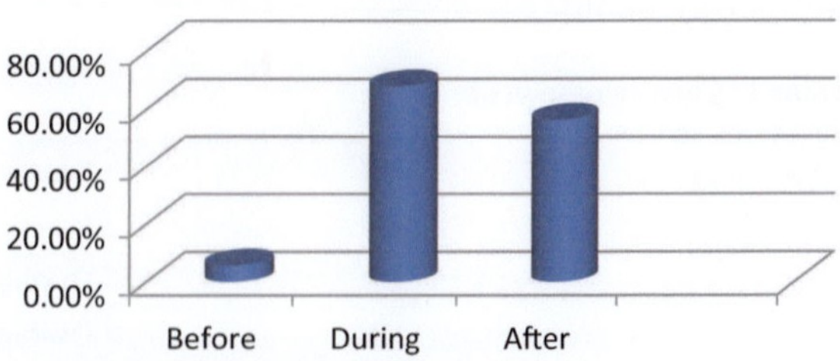

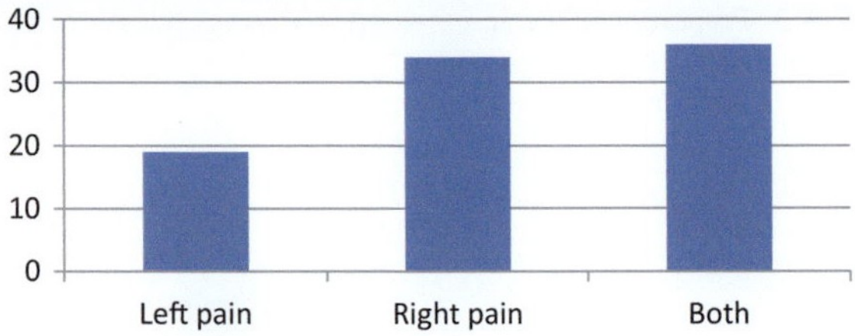

Fig. 2 Percentage of left and right legs knee pain

experienced pain solely in the left knee. Furthermore, 36 individuals, accounting for 41.38%, reported pain in both the right and left knee joints. The observed disparity suggests that individuals practicing Tai Ji Quan with knee joint pain are more prevalent in the right knee than the left knee. This discrepancy can be attributed to the fact that, during Tai Ji Quan practice, the right leg undergoes more active movements compared to the left leg. Additionally, the majority of individuals exhibit greater muscle strength in the right leg than in the left leg, leading them to actively choose the right leg to stabilize their posture during Tai Ji Quan practice.

Location of pain

The knee, being a complex joint, exhibits pain in specific locations indicative of corresponding issues. Considering that test subjects might not possess the capability to accurately diagnose their knee pain, this study focuses on the location of the patella, categorizing it as follows: anterior for the patella; lateral collateral ligaments for the knee; posterior for the posterior side of the knee and popliteal fossa; medial for the inner part of the knee, i.e., cruciate ligaments; inferior for the meniscus position of the knee and patellar ligament below the patella; and superior for the quadriceps muscle. As depicted in Fig. 3, it is evident that the largest number of individuals experienced pain on both sides of the knee, constituting the highest proportion. The knee, being a stable joint, primarily facilitates forward and backward flexion and extension. During Tai Ji Quan practice, adjustments to the center of gravity, opening and closing in virtual reality, hip rotations, and foot movements may lead to compensatory actions by the knee in the presence of dysfunctions in the hips and ankles. This, in turn, forces the knee joint into left and right rotations, posing a risk for knee injuries. The fewest individuals reported pain in the back of the knee, and there is also a minimal percentage of individuals experiencing pain in multiple parts of the knee.

Type of pain

Analysis of SFMA test findings in knee-painting Tai Ji Quan practitioners

Premier Tier of the SFMA Test

According to Fig. 4, it is evident that 93.1% of the tested individuals experienced indirect pain, while 6.9% reported persistent pain. With the duration of practice increasing, knee pain is exacerbated by the superimposition of incorrectly generated forces during practice. Key points in Tai Ji Quan practice include ensuring that "each

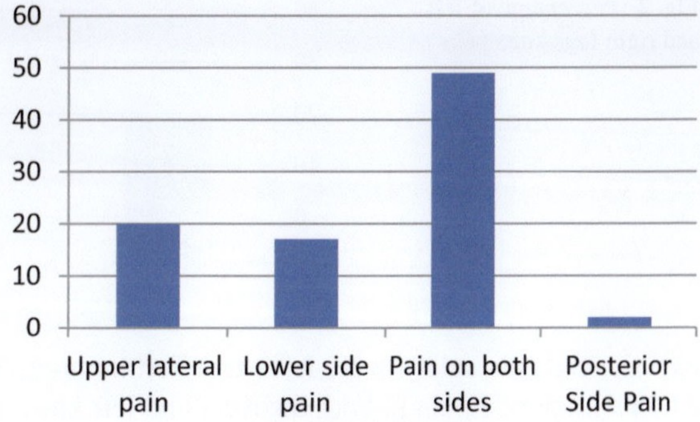

Fig. 3 Proportion of Tai Ji Quan practitioners with pain in the knee joints

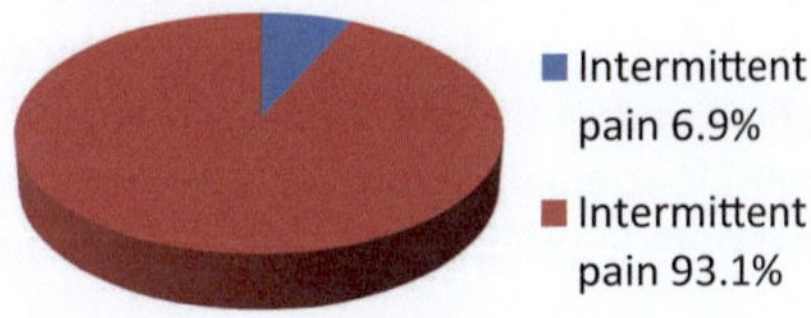

Fig. 4 Percentage of test subjects experiencing a particular type of knee pain graph

punch is executed under the drive of waist strength to achieve simultaneous upward and downward movement," "loosening the waist," "convergence of the buttocks," and maintaining full abduction of the hip joints when the center of gravity is in motion. During this movement, the hip joints control the motion of the thighs while maintaining full adduction. Due to the flexibility of the hip joint, insufficient abduction can result in inadequate leg control, easily leading to knee joint compensation, increased pressure on the knee joint, and the potential for intermittent knee joint pain (Table 2).

Analysis of SFMA Prime Tier Testing in Tai Ji Quan Practitioners with Knee Pain by Gender

According to Table 3, it is observed that the proportion of Functional and Non-Painful (FN) outcomes among male participants was higher than that among females in the single-leg standing test maneuver. Conversely, the FN proportion was lower than that

Table 2 SFMA primary tier testing status table for Tai Ji Quan practitioners with knee pain (%)

Test movements	Flexure	Stretching	Multi-part rotation		One-legged stand		Squat (exercise)
Test results			Left	Right	Left	Right	
FN	61.4	77.9	19.01	14.24	8.5	7.71	39.7
FP	0	2.76	0	0	0	0	8.76
DP	0	0	0	0	0	0	17.21
DN	36.92	13.44	84.45	88.96	92.31	90.62	45.51

among females in the other four test maneuvers: multi-site flexion, multi-site extension, multi-site rotation, and arm-dropping deep squat. In these four maneuvers, the proportion of Dysfunctional Non-Painful (DN) outcomes among male participants was higher than that among females.

Multi-site flexion decomposition results

When evaluating the movement pattern of multi-site flexion, 68 individuals did not meet the established standard. According to Fig. 5, it is evident that 78.2% of the tested individuals exhibited spinal flexion Dysfunction (MD). The proportion of active hip flexion Stability/Motor Control Disorder (SMCD) and core (pelvic direction) SMCD was comparatively lower. During the examination, it was observed that the physiological curvature of the lumbar spine was diminished, resulting in a flat and straight lumbar spine. The lumbar muscles exhibited reduced elasticity and had become rigid. Additionally, some individuals displayed herniated lumbar discs, which compromised the flexibility of the spine. These factors were identified as the primary reasons for not meeting the criteria for multiple flexion. (Note: MD—Flexibility Dysfunction, SMCD—Stability/Motor Control Disorder.)

Table 3 Analysis of SFMA prime tier testing in Tai Ji Quan practitioners with knee pain by gender (%)

Test movements	Flexure	Stretching	Multi-part rotation		One-legged stand		Squat (exercise)
Test results			Left	Right	Left	Right	
FN	40.1/ 62.66	73.01/ 84.21	7.42/ 19.22	7.32/ 15.55	17.27/ 7.01	17.27/ 4.67	16.22/ 42.51
FP	0/5.12	7.91/0	0/0	0/0	0/0	0/0	0/8.76
DP	0/0	0/1.43	0/0	0/0	0/0	0/0	6.53/ 18.07
DN	57.31/ 37.32	17.67/ 14.03	92.23/ 83.06	92.23/ 87.17	84.21/ 93.08	93.33/ 92.12	11.84/ 40.55

Notes(Male/Female)

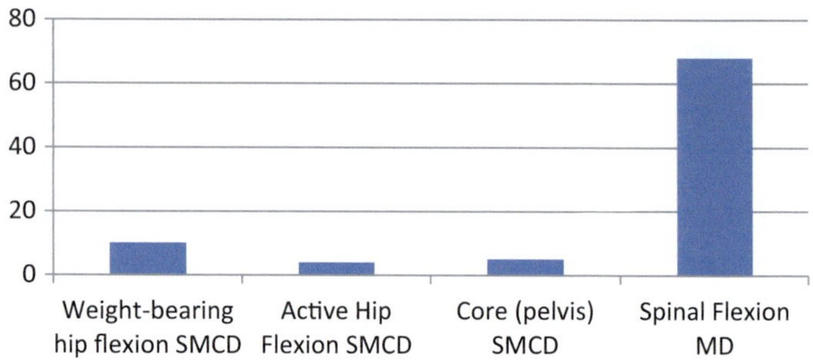

Fig. 5 Multi-site flexion decomposition results

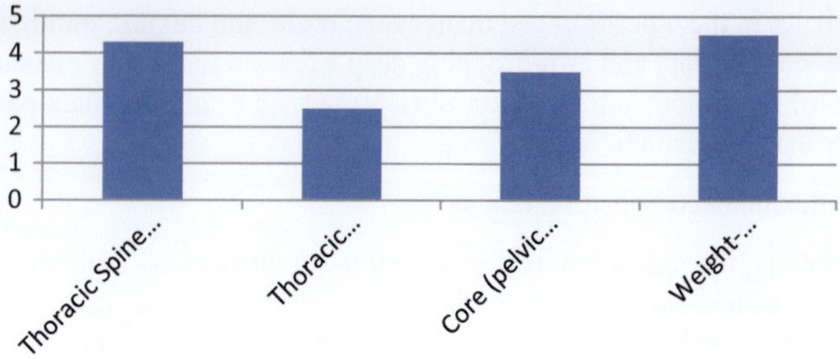

Fig. 6 Scale diagram of the results of multi-site stretching decomposition

Multi-site Stretch Breakdown Results

Analyzing knee discomfort in Tai Ji Quan practitioners from an FMS standpoint the degree of multi-site extension on the column extension SMCD is limited by hip extension dysfunction; the degree of extension is also influenced by weak core stability and an inability to effectively control one's body stability (Fig. 6).

Multi-site rotational decomposition results

When assessing the movement pattern of multi-site rotation, 79 individuals did not meet the established standard. According to Fig. 7, among those who did not meet the standard, excluding the thoracic spine, the percentage of individuals with Stability/Motor Control Disorder (SMCD) in weight-bearing or active left hip external rotation, weight-bearing or active right hip external rotation, and core (pelvic direction) SMCD was higher, accounting for 13.88%, 10.02%, and 17.40%, respectively. Additionally, the proportion of Stability/Motor Control Disorder (SMCD) with internal rotation was lower than that of weight-bearing or active left hip external rotation SMCD. The proportion of Flexibility Dysfunction (MD) with internal rotation of the left tibia was higher than that of MD with external rotation of the right tibia. In Tai Ji Quan practice, the hip joint often requires active external rotation. If individuals have Stability/Motor Control Disorder (SMCD) or Flexibility Dysfunction (MD) in hip external rotation, insufficient function may lead to compensation by the two stable joints adjacent to the hip joint, increasing the risk of sports injuries.

One-Leg Stand Breakdown Results

When evaluating the movement pattern of standing on one leg, a total of 24 individuals failed to meet the established standard. According to Fig. 8, it can be observed that the inability of many test subjects to meet the standard of standing on one leg is not attributed to ankle joint function. Instead, it may be associated with vestibular dysfunction, reduced balance ability, and weak core stability in individuals. A higher percentage of participants exhibited ankle valgus dysfunction compared to ankle varus dysfunction, potentially linked to the anatomical structure of the ankle joints, where the outer ankles are longer and lower than the inner ankles. The percentage of individuals with right ankle dorsiflexion dysfunction was slightly higher than those

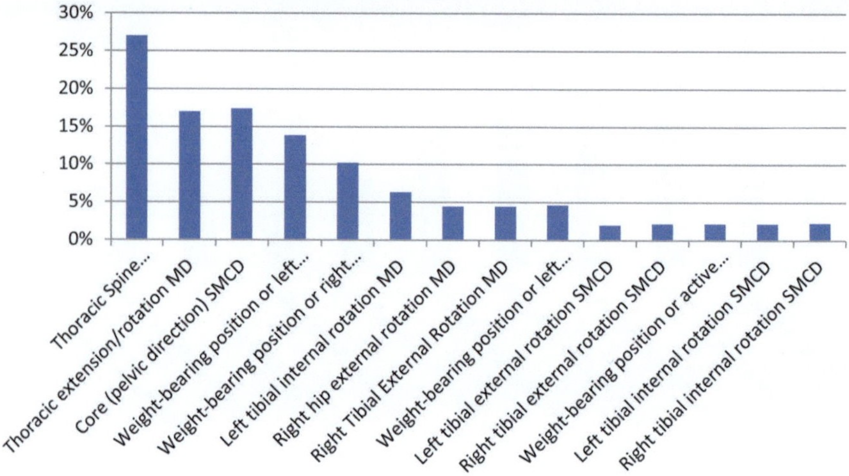

Fig. 7 Scale diagram of multi-part rotational decomposition results

with left ankle dorsiflexion dysfunction, possibly related to the extensibility of the left and right calf muscles. During tai chi practice, numerous movements require maintaining balance on a single leg. If stability is compromised, leading to body instability, increased pressure on the knee joint may occur, increasing the risk of injury.

Dropped Arm Squat Breakdown Results

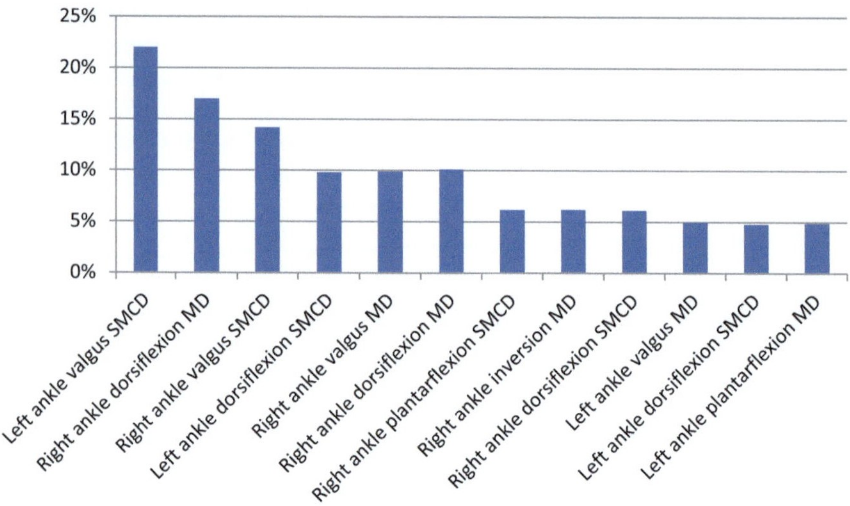

Fig. 8 Scale diagram of the results of the decomposition of the one-legged stand

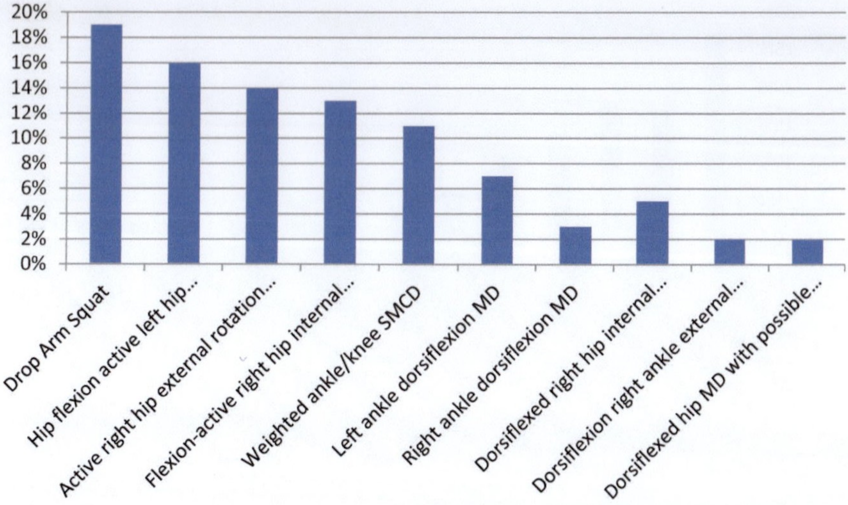

Fig. 9 Dropped arm squat breakdown results scale chart

A total of 16 individuals did not meet the standard when testing the movement pattern of the drop arm squat. According to Fig. 9, among the individuals who failed to meet the standard in the movement pattern of the drop arm squat, the percentage of those with hip flexibility disorder was not high. The highest percentage was observed in hip external rotation control stability dysfunction in the flexed hip position. Inadequate flexibility in the hip and ankle joints can hinder squatting, while poor control stability in the trunk, hip-knee-ankle joint complex can lead to posture instability and overexertion. During Tai Ji Quan practice, adhering to the principles of "sinking the waist and sitting on the hips," and initiating movements with hip rotations places significant demand on hip external rotation ability. In cases of hip external rotation dysfunction, compensatory movements in other kinetic chains may become inevitable to complete the desired actions. "Empty" in the table signifies that at the primary level, the arm drop squat movement did not meet the standard, and no dysfunction was identified during the decomposition.

4 Discussions

In addition to the primary findings, a noteworthy observation from the SFMA test on Tai Ji Quan practitioners with knee pain is the high prevalence of dysfunction in adjacent joints. A substantial 88.68% of Tai Ji Quan practitioners experiencing knee pain demonstrated dysfunction in neighboring joints. This statistic underscores the intricate interplay between various joints and their contribution to knee pain in this specific population.

The strong correlation observed between knee pain and dysfunction in adjacent joints further emphasizes the interconnected nature of musculoskeletal health in Tai Ji Quan practitioners. The intricate movements involved in Tai Ji Quan, such as the challenging "swinging foot and dropping the fork" and "turning around and swinging the lotus," place considerable demands on the body's functional capacities. Notably, these movements require coordinated actions of the hips, knees, and ankles, making practitioners susceptible to knee compensatory behaviors when faced with hip and ankle function disorders.

For practitioners experiencing left knee pain, the dysfunction of the hip and ankle predominantly manifests in the left leg. Conversely, those with right knee pain exhibit dysfunction concentrated in the right leg. This tailored insight into the lateralization of dysfunction contributes to a more targeted approach in addressing specific joint impairments based on the location of knee pain.

Moreover, gender differences in joint mobility and stability were observed among testers. Female testers demonstrated superior hip and ankle mobility but exhibited weaker ankle stability compared to their male counterparts. The SFMA's premier tier test revealed that male testers, in comparison to female testers, were less likely to exhibit Dysfunctional (DN) outcomes in the single-leg stance maneuver. Additionally, men showed lower rates of DN in multiple flexions, extensions, rotations, and arm dips in the deep squat, highlighting distinct patterns of joint function and dysfunction based on gender.

The findings presented in this discussion broaden our understanding of the nuanced relationships between knee pain, adjacent joint dysfunction, and gender-specific joint characteristics among Tai Ji Quan practitioners. These insights contribute to the development of targeted interventions, personalized training approaches, and preventive strategies aimed at enhancing the overall musculoskeletal well-being of individuals engaged in Tai Ji Quan practice.

5 Conclusions

In conclusion, this study utilized SFMA to comprehensively assess movement patterns and identify dysfunctions contributing to knee injuries in Tai Ji Quan practitioners. The examination of pain locations, prevalence, and impact of prolonged practice provided a nuanced understanding of the factors influencing knee health in this population.

The study's findings underscore the importance of tailored interventions, including targeted exercises and modifications to practice regimens, to address specific movement dysfunctions and mitigate the risk of knee injuries. Instructors and practitioners alike should prioritize maintaining balance between the lower limbs, emphasizing proper technique, and incorporating adequate rest periods to prevent overuse injuries.

Ultimately, this research contributes valuable insights to the field of sports medicine, providing a foundation for future studies on injury prevention and rehabilitation specific to Tai Ji Quan practitioners. By addressing the identified dysfunctions through informed strategies, the broader goal is to enhance the overall well-being and longevity of individuals engaged in this traditional martial art practice.

Acknowledgements I would like to express my gratitude to athletes who helped me during the test of this study.

References

Altman R, Asch E, Bloch D, et al (2002) Development of criteria for the classification and reporting of osteoarthritis. Classification of osteoarthritis of the knee. Diagnostic and Therapeutic Criteria Committee of the American Rheumatism Association. Arthritis Rheum 29:1039–1049

Belo JN, Berger MY, Koes BW et al (2009) The prognostic value of the clinical ACR classification criteria of knee osteoarthritis for persisting knee complaints and increase of disability in general practice. Osteoarthr Cartil 17:1288–1292

Cook G (2001) Selective Functional Movement Assessment (SFMA). Functional Movement Systems

Fransen M, Nairn L, Winstanley J et al (2007) Physical activity for osteoarthritis management: a randomized controlled clinical trial evaluating hydrotherapy or Tai Chi classes. Arthritis Rheum 57:407–414

Gribble PA, Kelly SE, Refshauge KM, Hiller CE (2013) Interrater reliability of the functional movement screen. J Strength Cond Res 27(4):978–981

Hu Z, Zheng X (2018) Application of SFMA in elite male volleyball players in Hebei Province. Contemp Sports Technol 8(31):225–226

Kang JW, Lee MS, Posadzki P et al (2011) T'ai Chi for the treatment of osteoarthritis: a systematic review and meta-analysis. BMJ Open 1:e000035

Irnich D, Bäumler P (2023) Effect of Tai Chi on knee pain and muscle strength in middle-aged and older adults with knee osteoarthritis: a randomized controlled trial protocol. BMC Complement Med Ther 37:413–425

Lee HJ, Park HJ, Chae Y et al (2009) Tai Chi Qigong for the quality of life of patients with knee osteoarthritis: a pilot, randomized, waiting list controlled trial. Clin Rehabil 23:504–511

Li F, Harmer P, Fischer KJ et al (2004) Tai Chi and fall reductions in older adults: a randomized controlled trial. J Gerontol 60A:187–194

Li F, Harmer P, Fitzgerald K et al (2018) Transforming traditional Tai Ji Quan techniques into integrative movement therapy—Tai Ji Quan: moving for better balance. J Sport Health Sci 7(1):1–8

Roddy E, Zhang W, Doherty M et al (2005) Evidence-based recommendations for the role of exercise in the management of osteoarthritis of the hip or knee—the MOVE consensus. Rheumatology (Oxford) 44:67–73

Schumacher HR (2010) Tai Chi improves physical function in older Chinese women with knee osteoarthritis: retraction. J Clin Rheumatol 16:357

Su W, Liu G, Liu X, et al. (2020) Angiogenesis stimulated by elevated PDGF-BB in subchondral bone contributes to osteoarthritis development. JCI Insight 5(8)

Wang C (2010) Serious concerns related to the article entitled "Tai Chi improves physical function in older Chinese women with knee osteoarthritis." J Clin Rheumatol 16:356

Automated Volunteer Scheduler for Sporting Events Using First Come First Serve Approach

Norfadila Mahrom, Ahmad Faizal Annuar, Suhizaz Sudin, and Choong Ting Quan

Abstract Sporting events heavily rely on volunteers for effective operations. As the demand for volunteers' surges, the necessity for structured management of their information, task preferences, and proficiencies becomes a necessity to avoid potential operational challenges. Current manual scheduling methods, although traditional, often fall short, often resulting in inefficiencies and, at times, an under-utilization of skilled volunteers across varying sports events. In light of these challenges, this research embarked on an in-depth evaluation of the First Come First Serve (FCFS) algorithm, specifically concerning volunteer scheduling. Key performance indicators assessed included accuracy, response duration, waiting periods, and turnaround times. Results were promising where the FCFS algorithm demonstrated 100% accuracy rate in the volunteer allocation process. Furthermore, the algorithm showed proficiency in systematically pairing volunteers with appropriate sports and venues, primarily based on their individual skill sets and task affinities. An interesting observation was the proportional increase in the performance metrics of the FCFS algorithm with an increase in volunteer count. To cap it off, the newly introduced web-centric Sports Volunteer Scheduling System, integrated with the FCFS algorithm, offers a robust solution, ensuring volunteers with matching skills and preferences are strategically aligned to suitable sporting events and venue assignments.

N. Mahrom (✉) · S. Sudin · C. T. Quan
Faculty of Electronic Engineering and Technology, University Malaysia Perlis (UniMAP), Pauh Putra, 02600 Arau, Malaysia
e-mail: norfadila@unimap.edu.my

N. Mahrom · S. Sudin
Sports Engineering Research Centre (SERC), Centre of Excellence, University Malaysia Perlis, Pauh Putra, 02600 Arau, Malaysia

Advanced Computing, Centre of Excellence University Malaysia Perlis, Pauh Putra, 02600 Arau, Malaysia

A. F. Annuar
Faculty of Mechanical Engineering and Technology, University Malaysia Perlis, Pauh Putra, 02600 Arau, Malaysia

Advanced Manufacturing System Research Group, Faculty of Mechanical Engineering and Technology, University Malaysia Perlis, Pauh Putra, 02600 Arau, Malaysia

Z. Mohamed et al. (eds.), *Proceedings of the 2nd International Conference on Innovation and Technology in Sports, ICITS 2023, 27–28 November, Malaysia*, Lecture Notes in Bioengineering, https://doi.org/10.1007/978-981-97-3741-3_45

Keywords First come first serve scheduling · Sports event · Sports scheduling

1 Introduction

A volunteer willingly undertakes tasks or services. Schedulers are software tools designed for task organization. In the past, sporting event volunteer management was often executed manually, employing methods like Excel spreadsheets or Google Forms for registration, alongside traditional pen and paper for data recording and retrieval. This approach, involving countless spreadsheets, form submissions, and physical records missing volunteer information, made data organization challenging, laborious, and error-prone (Irwin et al. 2019), leading to considerable time and resource wastage.

At present, the University Malaysia Perlis (UniMAP) Sports Centre assigns volunteers for its sporting events using data from Google Forms, spreadsheets, and physical document archives. There isn't a dedicated online platform where volunteers can register, review their assigned schedules, or see the timetable for sports events. Currently, only student athletes are privileged with volunteering opportunities for these events, given their familiarity with such environments in contrast to their non-athlete counterparts.

Hence, the main objective of this project was to establish a web-based sports volunteer scheduling platform utilizing an efficient scheduling algorithm. The system is designed to expediently manage and log all activities, significantly reducing time and effort. Importantly, it offers every student, athlete or otherwise, a system to volunteer in sporting events (Taks et al. 2015). It's essential to note that being an athlete does not guarantee expertise in all sports domains. For instance, a student proficient in football might volunteer for a basketball event, potentially introducing them to unfamiliar challenges in this new capacity.

Students have an opportunity to broaden their horizons by embracing roles they're unaccustomed to. For instance, by engaging in new tasks hands-on, students acquire fresh insights and skills. This encourages them to step outside their comfort zones, paving the way for potential networking opportunities that might be pivotal for their future leadership aspirations (Dickson and Darcy 2022). A centralized database will store all volunteer data, including event details, venues, and more. While volunteers have online-only access to the system, the sports volunteer coordinator can access both the database and the online platform at their convenience from any location.

The sports volunteer coordinator, often referred to as the volunteer coordinator, possesses the capability to operate the scheduler, thereby assigning volunteers to respective sports and venues in alignment with their skill sets and task predilections. Conversely, volunteers can consult the system to glean essential information about their assignments, including sport, venue, date, and time. The focal point of this system is devising solutions to scheduling challenges, particularly when a multitude of volunteers have identical skills and preferences for a certain sport but there's a dearth of proficient volunteers in other domains (Waters 2021).

In the realm of scheduling algorithms, two primary categories exist: preemptive and non-preemptive. Non-preemptive algorithms allow a process to run uninterrupted once initiated until its completion. In contrast, preemptive algorithms prioritize higher-ranked processes, causing low-priority ones to execute tasks with a higher priority (Thakur 2019). Round-robin algorithm is categorized as the pre-emptive algorithm and has been implemented in sports related projects such as in Raof et al. (2018). For this project, we adopted the First Come First Serve (FCFS) scheduling algorithm, with the objectives to assessing its efficacy through metrics like accuracy, response time, and both actual and average waiting periods, among others.

2 Methodology

2.1 System Requirements

The scope of this project is to develop a system that can schedule volunteers for different sports at different venues based on their skill and task preference based on FCFS principle. A web-based development platform integrated with the database is used for this project. The target users are the sports volunteer coordinator and volunteers in UniMAP Sports Centre. The planned sports are badminton, basketball, and football. On the other hand, the planned venues are Kubang Gajah Multipurpose Hall, UniMAP Sports Centre and Perlis Youth and Sports Complex. Each sport will have two sessions which are conducted on different days and times.

Next, the target users including the sports volunteer coordinator and volunteers are determined. The volunteers are made up of students from UniMAP. Hence, data sets used in this project were obtained from the UniMAP Sports Centre. The data covers volunteers from different sport events that have been organized by UniMAP. Five job scopes available for the volunteers to choose in the system are listed in Table 1. Each of the sports events such as badminton, basketball and football will have volunteers for these five-job scope.

The sports volunteer coordinator can access the system to run the scheduler for scheduling algorithms and assigning schedules to the volunteers. There are five categories of volunteer skills: computing, athletic, medical, event planner, and accountancy. The scheduler process is implemented by using FCFS scheduling algorithms. It is based on the allocated volunteers as the input to generate the performance metrics and later stored in the database. The sports volunteer coordinator can register the volunteer or update their details in case the volunteer is unable to perform the mentioned function. The sports volunteer coordinator can view the volunteer details as well as the schedule of the sports event assigned to the volunteers.

Volunteers must register their information in the database via a registration form in the system before being given the access. The information to be stored in the database will contain the name of the student, student matric number, faculty, gender, age, email, phone number, hostel name, skill, medical history, task preference, password

Table 1 Job scopes for the sports event

Job scopes	Descriptions
Crowd management	The role of crowd management is to control the crowd and traffic in sports events. This is to ensure that safety and security measures are implemented to guarantee the protection of the volunteers, staff and participants. This also ensures the sports event can operate smoothly
Technical	The role of technical is to keep records of meetings, update the result of the sports event, schedule the sports event and organise as well as set up venue preparation
Sports medicine	The role of sports medicine is to assist athletes or staff who sustain injuries by giving first aid, sports massage or physiotherapy while participating in any kind of sports event
Event planning	The role of event planning is to offer suggestions for improving the effectiveness of the sports event and specify the locations of the staff on duty. This role also includes planning, designing and delivering the opening, ongoing, medal and closing ceremonies
Sports event promoter	The role of a sports event promoter is to manage the sports event's budget planning, execution and follow-up. Additionally, they will be in charge of tasks like predicting participation and profits, visiting and choosing venues, and budget creation

and retype password. Additional information is added to the database when the sports volunteer coordinator runs the schedulers such as the performance metrics of the scheduling algorithms, sports and assigned schedules. The sports table in the database is defined by the sports volunteer coordinator and will contain the sport name, venue name, date and time, skill capacities and task preference.

2.2 System Design

The project development of this project is based on the Software Development Life Cycle (SDLC) Agile model. The SDLC Agile model combines incremental and iterative process models with a focus on user satisfaction and adaptability of processes through the quick delivery of functional software. The Agile software development model assisted developers in creating projects that quickly alter to changing requests over time (SDLC 2023). In the planning stage of SDLC, the objectives and problem statements of the volunteer scheduling system for the sporting event are determined. Meanwhile, the specification of the system should be determined in the requirement stage (Yi et al. 2579).

In the system, the volunteers are allowed to update their information. The volunteer will have the Accept/Reject the Assigned Schedule feature to respond to the assigned schedule. They are also able to view the sports schedule and decide whether to accept or reject the assigned schedule. The system is developed by using Visual Studio Code with Python programming language. For front-end web development frameworks such as HTML, CSS and JavaScript are used. For the back-end development, Python programming language and Django framework are used. PostgreSQL as the database software application is used for integrating with web development.

The web-based sports volunteer scheduling system allows both users to log in and log out of the system. However, only the sports volunteer coordinator has features to add, update, and delete volunteer details, along with running the scheduler and allocating volunteers. At the other end, the volunteer has the authority to accept or reject the assigned schedule. Both users can register volunteers, view volunteer personal information, update their details, and view sports schedules.

There are two important actions that can be performed only by the sports volunteer coordinator in the system. The first one is to allocate or re-allocate the volunteers to a specific sport and venue, and the second is to run the scheduler. If the coordinator chooses to allocate the volunteers, then the system will validate whether the coordinator has allocated the volunteers before. If the coordinator has allocated the volunteers once, then the system will display an error message. Otherwise, the system allocates the volunteers and displays the result of allocated and unallocated volunteers as well as the time generated for the assigned schedule for the coordinator. If the coordinator clicks to re-allocate the volunteers, then the system will validate whether the coordinator has allocated the volunteers before. If the coordinator has not allocated the volunteers then the system will display an error message together with the previously allocated and unallocated volunteers as well as the time generated to the coordinator. Otherwise, the system re-allocates the volunteers and displays the latest result of allocated and unallocated volunteers as well as the time generated to the coordinator.

Once the coordinator decides to run the scheduler, the system will use the allocated volunteers and their registration time to generate the schedule. Each job scope requires a specific number of volunteers, hence the scheduler will fulfill this number accordingly. In the system, different skills could be considered for a specific task and this process is done before FCFS generates the schedule. During scheduling, the system will also generate the performance metrics to analyze the performance of the FCFS algorithm. The generated performance metrics are displayed to the coordinator and added to the database. Other information such as allocated volunteers, unallocated volunteers, and time generated will also be displayed to the coordinator.

2.3 *Performance Metrics*

There are several performance metrics equations involved for scheduling the volunteers, which are accuracy, response time, actual turnaround time, and average turnaround time (Singhal 2018; Alhaidari et al. 2019). The accuracy of the FCFS algorithm in scheduling the volunteers is computed by dividing the number of scheduled volunteers by the total number of allocated volunteers. Equation (1) is used to calculate the accuracy.

$$\text{Accuracy} = \frac{\text{number of scheduled volunteers}}{\text{total allocated volunteers}} \tag{1}$$

The response time of the scheduler is calculated by using Eq. (2). The start time of the scheduler is subtracted from the end time. The start time is the initialized time when the scheduler is selected to run, while the end time is the time for the scheduler to complete running.

$$\text{Response time} = \text{end time} - \text{start time} \tag{2}$$

The total waiting time is the time used by the scheduler when allocating each volunteer who waited before their execution began in the scheduling queue. To obtain the total waiting time, the sum of the waiting time is computed. The waiting time is computed by getting the maximum value between 0 and the difference between the current time and arrival time as shown in Eqs. (3) and (4). Note that the arrival time is the time at which the allocated volunteer arrives in the ready scheduling queue.

$$\text{Waiting time} = \max(0, \text{current time} - \text{arrival time}) \tag{3}$$

$$\text{Total waiting time}+ = \text{waiting time} \tag{4}$$

The actual turnaround time of the FCFS is calculated as shown in Eq. (8). The actual turnaround time is the total turnaround time for all the scheduled volunteers. To obtain the total turnaround time, the sum of the turnaround time is computed. The turnaround time is computed by summing the computed waiting time and the execution time as shown in Eqs. (5) and (6). The execution time is defined for each skill in which the execution time for computing is 1 s, athletic is 2 s, medical is 3 s, event planner is 4 s, and accountancy is 5 s.

$$\text{Turnaround time} = \text{waiting time} + \text{execution time} \tag{5}$$

$$\text{Total turnaround time}+ = \text{turnaround time} \tag{6}$$

$$\text{Actual turnaround time} = \text{total turnaround time} \tag{7}$$

The average turnaround time of the FCFS is calculated as shown in Eq. (8). The average turnaround time is computed by dividing the total turnaround time by the number of scheduled volunteers.

$$\text{Average turnaround time} = \frac{\text{total turnaround time}}{\text{number of scheduled volunteers}} \tag{8}$$

3 Results and Discussion

The home page of the automated Sports Volunteer Scheduling System is shown in Fig. 1. The sidebar menu on the left offers four options to the sports coordinator, which are: Register Volunteer, View Volunteer Details, Run Scheduler and Allocate Volunteers, or View Sports Schedule. Initially, no volunteers were allocated in the system. Hence, no responses need to be submitted by any volunteers. As a result, no volunteer response is displayed in the *Notification Board* and the system displays message of '*No volunteer responses were found*'.

After the system has allocated the volunteers and the sports coordinator chooses to run the FCFS scheduling algorithm, the schedule and the performance metrics will be generated. Note that the input to run the scheduler is the allocated volunteers. The generated volunteers' schedule is shown in Fig. 2. The volunteers are assigned to each session of a specific game. For a specific game session, the skill and task preference of each volunteer are also displayed to ease the sports coordinator in managing the volunteers during the running session.

Meanwhile, Fig. 3 shows the performance metrics measured to assess the efficiency of the FCFS algorithm. The list of allocated and unallocated volunteers are also displayed along with the time generated.

The scheduler has been tested to allocate and schedule different numbers of volunteers as shown in Tables 2 and 3. It is observed that the scheduler has achieved an accuracy value of 1.00 (100%) which indicates it managed to assign the sports and venues to all volunteers in the database based on their skill and task preference. The waiting time indicates that the volunteers must wait for a specific amount of time

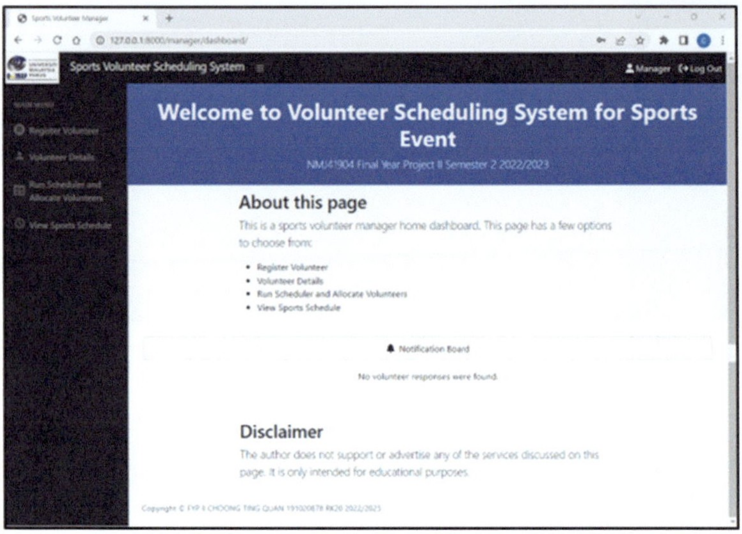

Fig. 1 Home page of the sports volunteer scheduling system

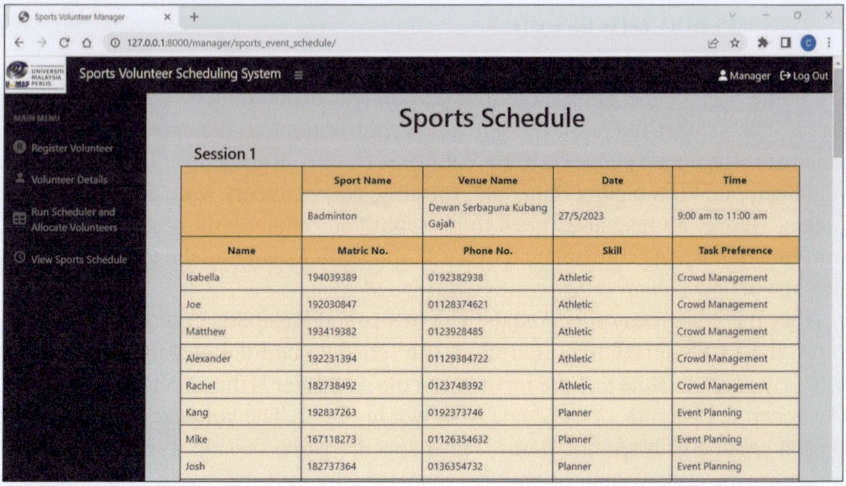

Fig. 2 Generated volunteers schedule

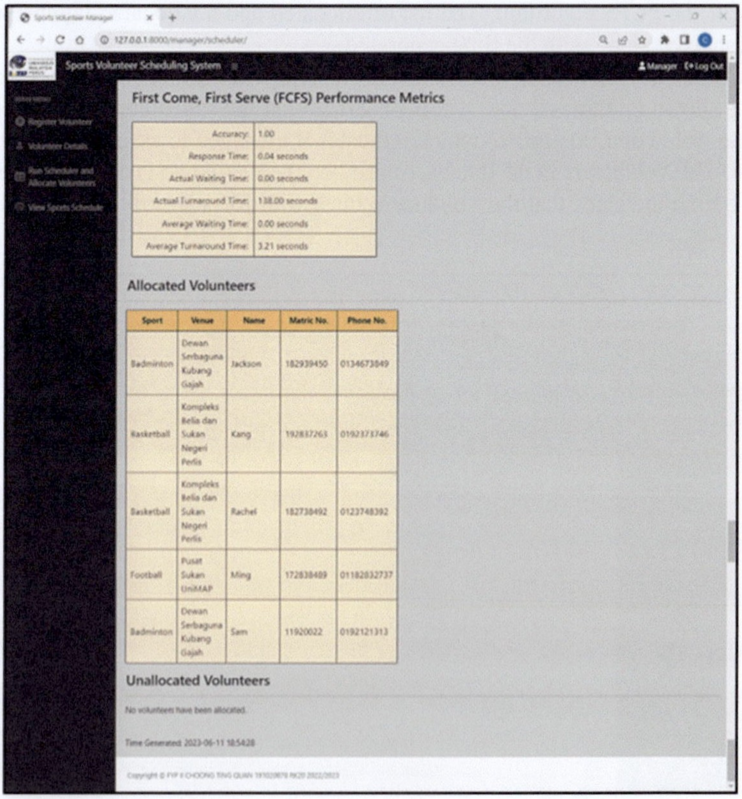

Fig. 3 FCFS performance metrics for volunteer allocation

in the queue before being executed. However, it was found that FCFS scheduling algorithm obtained 0.00 s of waiting time for different sets of volunteers to schedule, indicating that all volunteers were assigned to the sports and venues as soon as they arrived for execution.

The difference in the actual turnaround time obtained from different sets of volunteers to schedule is due to the sum of waiting time and execution time for each volunteer. This is equivalent to the total turnaround time for all scheduled volunteers. Same goes with the average turnaround time where the FCFS algorithm needs to consider both the execution time and arrival time. Overall, the results show an increasing trend in the response time and turnaround time as the number of volunteers increased.

Our findings suggest that the FCFS scheduling algorithm is one of the best scheduling algorithms for the system implementation due to a few factors. First, the FCFS scheduling algorithm has deterministic behavior. Theoretically, the FCFS scheduling algorithm is carried out in the order they are received in the scheduling queue. When implementing the FCFS scheduling algorithm, fewer errors or inconsistencies in volunteer allocation might occur which will increase accuracy.

Next, the FCFS scheduling algorithm does not have overhead in context switching. The volunteer at the head of the queue will begin the allocation and keeps going until it has been assigned to a sport and venue. By eliminating the overhead associated with context switching, volunteer allocation can be performed well.

Lastly, even though the allocation and scheduling duration proportionately increases as the volunteer count increases, the waiting time for different sets of volunteers is 0.00 s. This contributed to the shorter average turnaround time.

Table 2 Comparison between 250, 500 and 1000 volunteers

Parameters	250	500	1000	
Accuracy	1.00	1.00	1.00	
Response time (s)	0.07	0.07	0.08	
Actual turnaround time (s)	318.00	806.00	1592.00	
Average turnaround time (s)	3.18	3.22	3.18	

Table 3 Comparison between 1500, 2500 and 3500 volunteers

Parameters	1500	2500	3500	
Accuracy	1.00	1.00	1.00	
Response time (s)	0.18	0.19	0.20	
Actual turnaround time (s)	5011.00	8400.00	11,949.00	
Average turnaround time (s)	3.34	3.36	3.41	

4 Conclusion

The primary focus of this research is the creation of a web-oriented Sports Volunteer Scheduling System, harnessing the capabilities of the FCFS scheduling mechanism. The chosen development tools were the Python programming language in tandem with the Django framework. Integration was accomplished with the PostgreSQL database, facilitating the tasks of data storage, manipulation, and retrieval. The system has been effectively launched on a local server.

The research's secondary objective analyzes the FCFS algorithm's accuracy and response latency in the context of volunteer scheduling for sporting events. Utilizing the designated volunteer data as input, the scheduler was engaged, subsequently deriving performance metrics. The outcomes indicate a consistent accuracy rate of 100% for both initial volunteer assignment and subsequent reallocations.

Subsequently, the third objective of this study is to analyze the response time, actual and average turnaround time of the FCFS scheduling algorithms in scheduling volunteers for the sports event. After re-allocation of the volunteers, the obtained actual and average turnaround time were the same as the volunteer allocation. Finally, during observation, as the number of volunteers increases, the performance of the FCFS scheduling algorithm simultaneously increased proportionally.

References

Alhaidari F, Balharith T, AL-Yahyan E (2019) Comparative analysis for task scheduling algorithms on cloud computing. In: 2019 International conference on computer and information sciences (ICCIS), pp 1–6

Dickson TJ, Darcy S (2022, November) Surviving and thriving from the volunteer involvement in mega-sport events. Org.au website. https://volunteeringstrategy.org.au/wp-content/uploads/2022/11/VRP_Surviving-and-Thriving-from-the-Volunteer-Involvement-in-Mega-Sport-Events.pdf. Accessed 07 Jan 2023

Irwin NJ, Bennett A, Carlos K, Kiswani J, Harris CR, Dascalu S, Harris FC (2019) ARIA 3.0: a modern approach to web-based music festival registration systems. In: Proceedings of the ISCA 28th international conference on software engineering and data engineering (SEDE 2019). EPIC Series in Computing, vol 64, September 30–October 2, San Diego, CA, pp 215–224

Raof RAA, Sudin S, Mahrom N, Rosli ANC (2018) Sport tournament automated scheduling system. MATEC Web Conf 150:05027. https://doi.org/10.1051/matecconf/201815005027

SDLC—Agile Model (n.d.) Tutorialspoint.com website. https://www.tutorialspoint.com/sdlc/sdlc_agile_model.htm. Accessed 07 Jan 2023

Singhal A (2018, October 19) Round robin. Gatevidyalay.com; Gate Vidyalay. https://www.gatevidyalay.com/round-robin-round-robin-scheduling-examples/

Taks M, Chalip L, Green BC (2015) Impacts and strategic outcomes from non-mega sport events for local communities. Eur Sport Manag Q 15(1):1–6. https://doi.org/10.1080/16184742.2014.995116

Thakur D (2019, January 8) Operating system scheduling algorithms. Computer notes website. https://ecomputernotes.com/fundamental/disk-operating-system/scheduling-algorithms. Accessed 19 Dec 2022

Waters A (2021, April 28) Volunteer scheduling: how to boost efficiency using tech. Galaxydigital.com. https://www.galaxydigital.com/blog/volunteer-scheduling
Yi LY, Mahrom N, Calvin L (2023) Android-based timetable manager for University students using rule-based algorithm. AIP Conf Proc 2579(1):020011. https://doi.org/10.1063/5.0114159

Computer Vision-Based Automated Archery Performance Logging System

Jonathan Then Sien Phang, King Hann Lim, Basil Andy Lease, and Dar Hung Chiam

Abstract In the sport of archery, a stable shooting stance is crucial to obtain high-precision shooting. However, wearable motion sensors might interfere with the shooting performance of an archer. In this paper, a markerless-based motion capture system using a deep learning approach is proposed to retrieve the keypoint of archer shooting in a continuous shooting phase. The correlation of shooting posture is studied with the scoring precision and grouping. Studies are conducted on two types of archery shooting sessions using recurve and compound bow. The joint angle profile can be decomposed into six phases, i.e. stance, pre-draw, draw & anchoring, transfer, aiming, and follow through. Important parameters and joints, particularly the angles and their consistencies, are analyzed to study archers shooting performances. An algorithm is proposed in this study to automate the determination of the archer's anchoring and release sequence. Several qualitative results of joint kinematics and parameters and visualization are presented to further understand the bio-mechanical aspect of archery. In addition, this study reveals a direct positive correlation in posture consistency between arrow precision and grouping. Hence, it indicates a consistent shooting posture is a significant attribute of an archer's performance.

Keywords Archery stance · Computer vision · Posture correction · Automated scoreboard detection · Deep learning pose estimation

1 Introduction

The stance in archery (Vendrame et al. 2022; Ogasawara et al. 2023) is a critical element in achieving a high score during competition. A good shot is always associated with posture consistency on how a stable and repetitive stance can be maintained in a continuous shot. Through human visual inspection, the postural consistency of archers can be spotted at a glimpse, but the effectiveness of monitoring is subjective

J. T. S. Phang (✉) · K. H. Lim · B. A. Lease · D. H. Chiam
Curtin University, CDT 250, 98009 Miri Sarawak, Malaysia
e-mail: jonathan.phang@curtin.edu.my

Z. Mohamed et al. (eds.), *Proceedings of the 2nd International Conference on Innovation and Technology in Sports, ICITS 2023, 27–28 November, Malaysia*, Lecture Notes in Bioengineering, https://doi.org/10.1007/978-981-97-3741-3_46

to the attention of a coach. Therefore, the use of camera solutions in archery field is needed to provide a precise observation and objective analysis of the archery posture. To improve the consistency of an archer, repetitive training is highly required to train the muscle memory of archers on their posture and drawing pattern. Hence, using sports science knowledge (Hung et al. 2009) to analyze the archer's posture and drawing pattern may result in a better understanding of the archery shooting cycle and scoring.

The detection of scoring system (Ogasawara et al. 2023; Ligutan et al. 1391; Peng et al. 2021; Issa et al. 2021) is one of the measurement metrics to co-relate with the posture correctness. Numerous studies to detect scoreboards using image processing are commonly performed in five basic steps, i.e. image pre-processing, scoreboard detection, circle detection, arrow detection, and scoring counting. Rudzinski et al. (2012) proposed an automatic scoring algorithm using Prewitt edge detection and Hough transformation. Nguyen et al. (2011) proposed a line and gap information to locate and orient the arrow and pinpoint the piercing point on the target. Zin et al. (2013) proposed an image subtraction technique to detect arrows in scoreboard from an initial state with the additional operations such as dilation, erosion and dynamic thresholding. Peng et al. (2021) proposed the use of YOLOv4 model to detect arrows on an archery target using smartphone. However, the development is currently incomplete and it requires more training samples at the field to keep the accuracy of detection high.

In general, the detection of posture is crucial in archery as an incorrect alignment of the bow arm can negatively impact the exerted force (Debnath and Debnath 2016). In (Debnath and Debnath 2016), an image processing technique is used to detect the angle and position of the drawarm elbow as well as the draw force line. However, full-body stance monitoring is not covered to study the correlation between the consistencies with the score. On the other hand, Phang et al. (2022) proposed the use of a deep learning model to detect archer's posture for kinematics measurement. In this paper, an automated posture detection and scoreboard detection system is proposed to study the archer shooting phases with its scoring. First, the proposed system captures human posture from consistent shooting. Ten sets of keypoints are detected to understand the consistency of an archer in performing continuous shots. An automated scoreboard detection algorithm is performed to log the score of each shot in the database. Subsequently, several measurement metrics are performed to study the correlation between posture and score in the field. This process simplifies the monitoring of posture and keeps the objective observation from one candidate to another (Fig. 1).

2 Automated Archery Performance Logging System

To perform a postural study in the archery field, Fig. 1 shows the proposed automated archery performance logging system to automatically log the arrow score and archer's posture data for performance analysis. In order to log the data synchronously, the

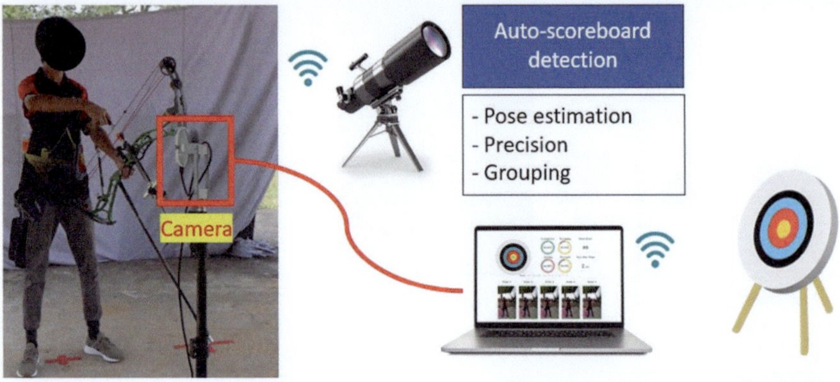

Fig. 1 Proposed computer vision-based automated archery performance logging system

system comprises two modules: (a) Automated scoreboard detection and (b) Deep learning pose estimation.

2.1 Automated Scoreboard Detection

In existing works (Peng et al. 2021; Rudzinski and Luckner 2012; Nguyen and Lin 2011) to automate arrow scoring system, they commonly require ideal operating conditions, such as unused target face and clear image view. Moreover, most existing works rely on statistical methods such as Hough transform for target circle detection. These implementations can pose issues when the target face is skewed, hence the target circle is distorted. As a result, incorrect arrow scoring can occur. This paper proposes a two-stage processing technique to avoid the use of circle detection for more robust arrow scoring.

The two-stage system shown in Fig. 2 comprises: (a) Initiation stage takes the first image of the target face as a template as frame $t-1$ and extracts each true circle of the target, (b) Image processing stage detects and locate consecutive arrow in frame t. The advantage of two-stage approach allow more robust target circle detection, regardless of image distortion, hence allowing better localization of the arrow for automatic score counting for precision score. Subsequently, the grouping score is computed using the collected scores. The grouping of arrows is associated with the spread of arrows relative to the precision of the arrows shot. Hence, the grouping score can be obtained by computing the normalized standard deviation of scores. First, each score is viewed as a complex number vector z as follows,

$$z = a + bi, \tag{1}$$

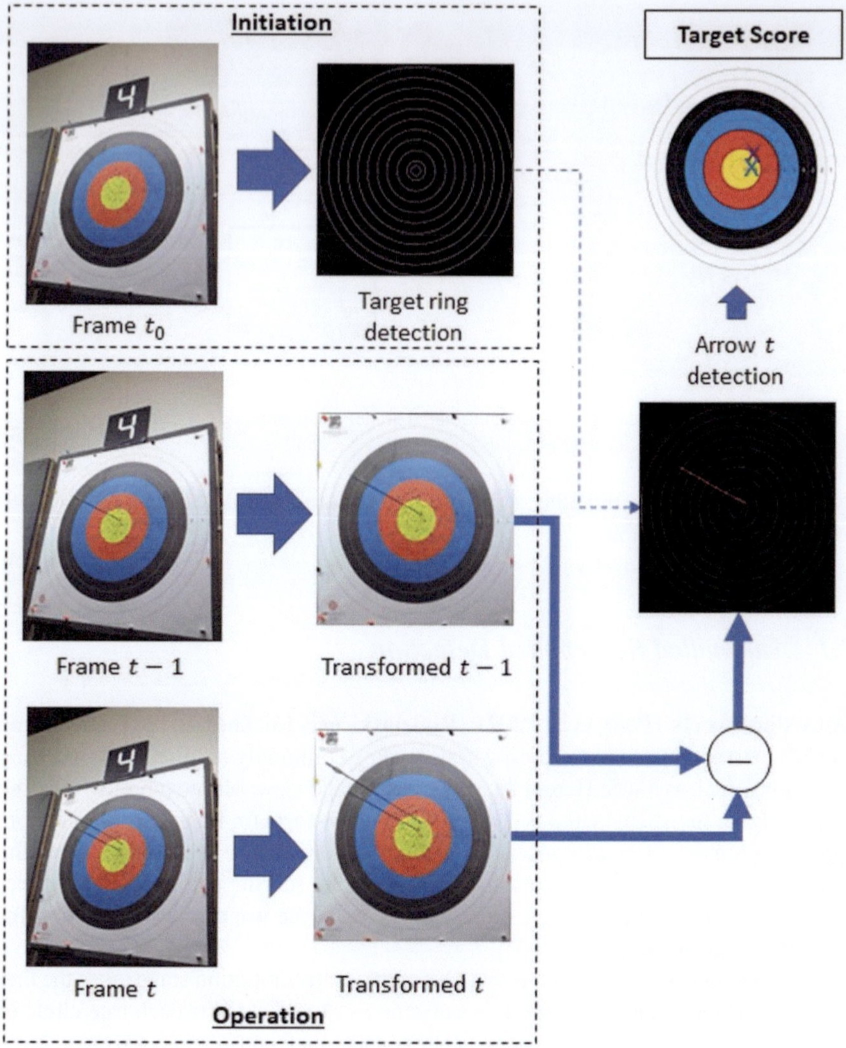

Fig. 2 Automated scoreboard detection using a proposed two-stage computer vision process

where $a = r \cdot \cos(\theta)$ and $b = r \cdot \sin(\theta)$, which represents real component and imaginary component of z respectively. The symbol θ is an angle in radian defined by the position of an arrow in the Cartesian plane. Subsequently, the grouping score is computed as follows,

$$\mathrm{G} = \frac{\sum_{n=1}^{N} (10 - \mathrm{std}(z_n))}{N} \times 100 \tag{2}$$

where N is the total number of shots performed by an archer and std(·) is the standard deviation. In our study, $N = 10$ is the set for the performance evaluation.

2.2 Deep Learning Pose Estimation

To perform posture consistency analysis, the body posture data of an archer is obtained through deep learning pose estimation. Deep learning models (Liu et al. 2018; Dang et al. 2019; Chen et al. 2020; He et al. 2020) are devised by implementing neural networks, and datasets, to automatically estimate human keypoints from input images. Hence, by implementing pose estimation using a deep learning model in image processing, sports evaluation can be applied, particularly in archery. In existing work, Saptadeep and Subir (Debnath and Debnath 2016) implemented image processing in different shooting stages to estimate the drawarm elbow and the draw force line from an archer. Meanwhile, Phang et al. (2022) implemented Keypoint RCNN to obtain human keypoint for posture consistency analysis in archery. Therefore, deep learning pose estimation allows posture data acquisition in a non-intrusive manner and maintains a consistent posture estimation across all subjects. In this study, Keypoint RCNN (He et al. 2020) is implemented to estimate the posture of an archer with the setting as shown in Table 1.

To compute the consistency score of archers, the aiming period is utilized for evaluation. Figure 3 shows the archery phases detected from the joint angle profile between right shoulder, right elbow and left wrist. By observing through the joint angle profile, an archery shooting cycle can be decomposed into six phases, i.e. stance, pre-draw, draw & anchoring, transfer, aiming, and follow through. The lines denoted as l_0, l_1, l_2, l_3, and l_4 are the transition from stance to pre-draw, pre-draw to draw to anchoring, anchoring to holding, holding to aiming, and aiming to follow through. The recorded movement sequences from the deep learning pose estimation are clipped from the anchoring phase ($l3$) to the release phase ($l4$). From the clipped

Table 1 The keypoint joints selected for the consistency study

Joints	Starting joint	Reference joint	Ending joint
J1	Right shoulder	Right elbow	Left wrist
J2	Left shoulder	Left elbow	Left wrist
J3	Head	Mid shoulder	Left wrist
J4	Mid shoulder	Mid hip	Right hip
J5	Right elbow	Right wrist	Left wrist
J6	Right wrist	Mid shoulder	Left wrist
J7	Right elbow	Mid shoulder	Left wrist
J8	Left wrist	Mid shoulder	Mid hip
J9	Right ankle	Mid hip	Left ankle
J10	Right shoulder	Left shoulder	Left hip

sequences, the angle of each joint set θ_{Ji} is computed as follows,

$$\theta_{ji} = \arccos\left(\frac{\overrightarrow{AB} \cdot \overrightarrow{BC}}{\left\|\overrightarrow{AB}\right\|\left\|\overrightarrow{BC}\right\|}\right) \tag{3}$$

To compute the consistency score for J_i, the scoring threshold $\tau = 0.05$ is estimated by observing the average performance of elite archers. Subsequently, the consistency score C_{Ji} is computed using Eq. (4).

$$C_{ji} = \begin{cases} \frac{(\tau \times 180) - \text{std}(\theta_{j_i}) \times 100\%,}{\tau \times 180} & \text{if } \tau > \text{std}(\theta_{j_i}) \\ 0, & \text{otherwise} \end{cases} \tag{4}$$

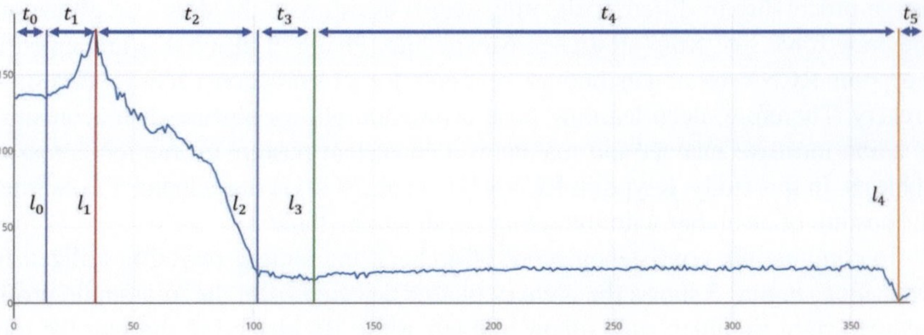

(a) Compound bow phases.

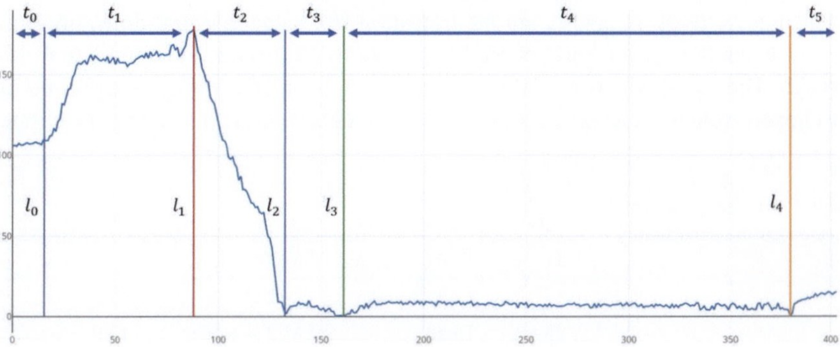

(b) Recurve bow phases.

Fig. 3 Archery shooting phases detected from the right shoulder, right elbow, and left wrist derived from KSL shot cycle (KSL n.d)

where $\overline{\theta}_{j_i}$ is the mean value of joint set θ_{j_i} shown in Eqs. (5) and (6) respectively.

$$\mathrm{std}\left(\theta_{j_i}\right) = \frac{\sum_f \left(\theta_{j_i} - \overline{\theta}_{j_i}\right)^2}{F} \tag{5}$$

$$\overline{\theta}_{j_i} = \frac{1}{F}\sum_{f}^{F} \theta_{f,j_i}, \tag{6}$$

where F is the total number of frames in a sequence.

3 Experiment Setups

A testing ground is set up with the archer's ages ranging from 10 to 20 years old in the field to analyze an archer's performance by correlating posture consistency relative to arrow precision and grouping. The compound bow archers are given a 50-m shooting range, while the recurve bow archers are given a 30-m shooting range. Our data acquisition setup comprises an automated scoreboard detection system that logs arrow scores and synchronously logs the posture data of the archer. For the automated scoreboard system setup, a wireless camera enabled telescope is pointed to the target board in which the archer is using. The telescope is connected to a computer in parallel to an external camera to synchronously collect posture data of the archer. The camera is placed tangent to the performing archer in order to get the full view of the body. Each archer is required to perform 10 arrow shots while the camera is recording the performance. In post-processing, the footage of the archer is clipped from anchoring phase to release phase to obtain the posture consistency information. Subsequently, the correlation R of consistency between precision and grouping is computed using Eq. (7) as follows,

$$R(X, Y) = \frac{\sum(x - \overline{x})(y - \overline{y})}{\sqrt{\sum(x - \overline{x})^2 \sum(y - \overline{y})^2}}, \tag{7}$$

where $\overline{x}$ and $\overline{y}$ are the mean value of array X and Y respectively.

4 Results and Discussions

This study aims to analyze the performance of an archer by correlating posture consistency relative to arrow precision and grouping. Table 2 shows the bow type used by each archer, arrow precision, grouping, and posture consistency. In Table 2, it is observed archer #7 is potentially the best-performing archer in the compound bow category by achieving arrow precision of 95. In addition, the archer also has the best grouping score out of all archers in that category. On the other hand, archers #4 and #5 are performing less ideally with precision of 54 and 69. Upon examining their grouping score, they are relatively poor with scores of 48.62% and 67.81%. These poor performances can be attributed to factors such as preparedness of the mental and body, and bow tuning. Furthermore, examining the posture consistency, archer #6 has the most consistent body posture during the shooting session. Overall, all archers showed above-average posture consistency in the compound bow category.

In the recurve bow category, the archers posed a higher degree of difficulty in shooting as one must maintain a more stringent body posture and stance compared to compound bow users. This is due to the recurve bow's lack of mechanical advantage that eases tension during anchoring. Additionally, a recurve bow archer demands higher focus while required to hold the tension of their bow.

From the acquired posture and score data, it is observed archer #7 performed the best among all archers, with a precision of 97 and a grouping of 95%. Some below-average archers are identified as archers #9 and # 19 with precisions of 78 and 74.

Table 2 The Bow types, i.e. Compound (**C**) and Recurve (**R**) used by archers, score precision, score grouping, body posture consistency, and correlation of precision relative to score precision and grouping

Archer #	Bow type	Precision	G (%)	C_{Ji} (%)
1	C	92	90.85	93.17
2	C	81	80.43	91.14
3	C	88	86.55	89.47
4	C	54	48.62	88.06
5	C	69	67.81	86.18
6	C	89	85.13	95.68
7	C	95	91.87	92.78
8	R	89	90.48	93.44
9	R	78	78.70	92.19
10	R	94	91.81	87.26
11	R	97	95.00	91.07
12	R	90	87.54	90.50
13	R	93	90.10	90.23
14	R	89	86.11	88.27
15	R	86	87.46	88.58
16	R	84	76.50	82.60
17	R	90	92.83	89.13
18	R	84	82.11	88.16
19	R	74	74.27	87.20
Correlation	–	–	0.36	0.41

It is also observed their groupings proportionally fell below average with scores of 78.7 and 74.27%. In posture consistency, archer #8 has the most consistent posture throughout the shooting session.

In the correlation analysis between posture consistency and arrow score, it is observed there is a positive correlation. This indicates that by maintaining a consistent posture throughout a shooting session, an archer has a better chance of scoring a higher score. On the other hand, a positive correlation between posture consistency and grouping. This suggests if archers are able to maintain their posture during shooting sessions, a high grouping of arrows can be achieved. Notably, posture consistency is directly attributed to high precision and high grouping in archery.

5 Conclusion

Postural consistency is crucial in achieving high precision in archery. The study of archer's stance can be performed with the proposed computer vision-based automated archery performance logging system, comprising an automated scoreboard detection, and a deep learning posture estimation. The automated scoreboard detection system is a camera-enabled telescope that can automatically log the arrow score. In addition, the deep learning posture estimation system logs the posture data in synchronous with the auto-scoreboard. Several metrics, such as precision, grouping, and posture consistency are devised to analyze the performance of archers. The study consists of compound bow and recurve bow categories of archers. The analysis revealed best-performing archers in each category by determining their precision, grouping, and posture consistently. Moreover, the study also revealed a positive correlation of posture consistency between precision and grouping. This suggests the system can be implemented to periodically monitor archer's posture performance during training sessions to progressively improve their muscle memory.

References

Chen Y, Tian Y, He M (2020) Monocular human pose estimation: A survey of deeplearning-based methods. Comput vis Image Underst 192:102897

Dang Q, Yin J, Wang B, Zheng W (2019) Deep learning based 2d human poseestimation: A survey. Tsinghua Sci Technol 24(6):663–676

Debnath S, Debnath S (2016) Performance evaluation by image processing techniques in archery—a case study. Int J New Technol Sci Eng 3(1)

HeK, Gkioxari G, Dolla'r R, Girshick R (2020) Mask r-cnn. IEEE Trans Pattern Anal Mach Intell 42(2):386–397

Hung TM, Tang WT, Shiang TY (2009) A case study of integrated sportsciences for an olympic archer. J Med Biol Eng 29(4):164–171

Issa AH, Hasan SD, Ali WH (2021) Automation of real-time target scoringsystem based on image processing technique. J Mech Eng Res Dev 44(2):316–323

KSL The ksl shot cycle. https://www.kslinternationalarchery.com/Technique/KSLShotCycle/KSLShotCycle-USA.html. Accessed 24 May 2023

Ligutan DD, Abad AC, Cabatuan M, Llorente C, Dadios EP (2019) Fpgaimplementation of archery target detection using color sequence recognition algorithm. Int J Recent Technol Eng 8(2) (Special Issue 8):1391

Liu Y, Xu Y, Li S-B (2018) 2-d human pose estimation from images based ondeep learning: a review. In: 2018 2nd IEEE Advanced Information Management, Communicates, Electronic and Automation Control Conference (IMCEC), pp. 462–465, IEEE

Nguyen C, Lin I (2011) Arrowsmith: Automatic archery scorer. Department of Computer Science, Stanford University, Tech. Rep.

Ogasawara T, Fukamachi H, Aoyagi K, Kumano S, Togo H, Oka K, Yamaguchi M (2023) Automatic shooting detection in archery from acceleration data for score prediction. Sport Eng 26(1):9

Peng J-S, Chen Y-J, Lin W-Y, Chen H-C, Liao C-N (2021) The development and implementation of a smartphone based archery analysis system. In 2021 IEEE International Conference on Consumer Electronics-Taiwan (ICCE-TW). IEEE, pp 1–2

Phang JTS, Lim KH, Lease BA, Chiam DH (2022) Deep learning pose estimation for kinematics measurement in archery. In: 2022 International Conference on Green Energy, Computing and Sustainable Technology (GECOST). IEEE, pp 298–302

Rudzinski J, Luckner M (2013) Low–cost computer vision based automatic scoring of shooting targets. In: Knowledge engineering, machine learning and lattice computing with applications: 16th International Conference, KES 2012, San Sebastian, Spain, September 10–12, 2012, Revised Selected Papers 16, pp. 185–195. Springer

Vendrame E, Belluscio V, Truppa L, Rum L, Lazich A, Bergamini E, Mannini A (2022) Performance assessment in archery: a systematic review. Sport Biomech:1–23

Zin TT, Oka I, Sasayama T, Ata S, Watanabe H, Sasano H (2013) Imageprocessing approach to automatic scoring system for archery targets. In: 2013 Ninth international conference on intelligent information hiding and multimedia signal processing. IEEE, pp 259–262

Local In-Lab-Made IoT Sports Device Acceptance Survey Instrument Using a Modified UTAUT Model: A Pilot Study

Nurul Syuhadah Khusaini, Abdul Rahman Hemdi, Raja Muhammad Firhad Raja Azidin, Mohd Hanif Mohd Ramli, and Zulkifli Mohamed

Abstract Internet of Things (IoT) devices are smart, interconnected devices that utilize sensors, actuators, and internet connectivity to collect and exchange data. These devices enable real-time monitoring, control, and automation. Muscle performance IoT devices offer opportunities for monitoring, analyzing, and improving muscle performance and physical therapy, contributing to advancements in healthcare, fitness, and rehabilitation. However, the level of acceptance of local in-lab-made IoT sports devices to monitor health and muscle performance among undergraduate students in public university is still scarce. Hence, this paper focuses on conducting a survey using a modified well-known technology acceptance model called as the "Unified Theory of Acceptance and Use of Technology" (UTAUT) to investigate the acceptance level of local in-lab-made IoT sports device. Three new independent variables are introduced in this study namely perceived risk, perceived aesthetics, and perceived comfort. A total of 36 valid responses received from the College of Engineering undergraduate students and the data is analyzed using Rasch Measurement Model (RMM). From the summary statistics, the item reliability stands at a strong value of 0.74. The person reliability yields a convincing value of 0.87, which separates the respondent into close to 3 groups. Averagely, the response shows that local in-lab-made sports device is accepted, and the new variables introduced in

N. S. Khusaini · M. H. M. Ramli · Z. Mohamed (✉)
College of Engineering, School of Mechanical Engineering, Universiti Teknologi MARA 40450 Shah Alam, Selangor, Malaysia
e-mail: zulkifli127@uitm.edu.my

N. S. Khusaini · A. R. Hemdi · R. M. F. R. Azidin · M. H. M. Ramli · Z. Mohamed
Sports Engineering and Artificial Intelligence Centre, College of Engineering, Universiti Teknologi MARA 40450 Shah Alam, Selangor, Malaysia

A. R. Hemdi
School of Mechanical Engineering, Universiti Teknologi MARA Cawangan Pulau Pinang Kampus Permatang Pauh, 13500 Permatang Pauh, Pulau Pinang, Malaysia

R. M. F. R. Azidin
Faculty of Sports Science and Recreation, Universiti Teknologi MARA 40450 Shah Alam, Selangor, Malaysia

Z. Mohamed et al. (eds.), *Proceedings of the 2nd International Conference on Innovation and Technology in Sports, ICITS 2023, 27–28 November, Malaysia*, Lecture Notes in Bioengineering, https://doi.org/10.1007/978-981-97-3741-3_47

the modified UTAUT Model is indeed a deciding factor for users to consider. As a conclusion, all local in-lab-made sports device for muscle performance should focus on the aspect of security, affordability, aesthetic and comfort for the device to be marketable.

Keywords Survey · Acceptance · UTAUT · Rasch measurement model · Internet of Things · Sports device

1 Introduction

The Internet of Things (IoT) quickly progressed causing major changes in various sectors. It added smart interconnected devices to our daily lives. Within this new era, lab-made sports tools have become a significant area of interest. They provide adaptable, affordable, and custom solutions for targeted research and practical needs.

Yet, lab-made sports tools bring some unique hurdles like reliability and acceptance. The quality of these tools can fluctuate based on expertise and resources in the lab. Unlike factory-made products that go through fierce testing and meet quality standards, lab-made sports tools could be lacking standard production systems, affecting their performance dependability.

In addition, winning recognition for lab-made sports tools in the wider scientific and industrial circles can be hard. Usually, professionals and researchers turn to trusted, available commercial devices. It might require solid proof of their accuracy, exactness, and durability to persuade them to use lab-made sports devices.

When it comes to technology adoption among Generation Z in Malaysia, Tjiptono's study suggests that this group can be described as Electronically engaged, Educated, Entrepreneurial, and Empowered (Tjiptono et al. 2020). However, it is still unknown whether Generation Z in Malaysia is opened to accept and to purchase local in-lab-made sports devices for muscle performance monitoring. Hence, this paper intends to explore the acceptance of local in-lab-made sports device among Generation Z Bumiputera in Malaysia public university.

2 Materials and Methods

2.1 *Targeted Respondents*

The survey study aims to gauge College of Engineering (COE) undergraduates. With 6439 COE undergrads total, we must establish a suitable sample size for the preliminary study. Experts in education and behavioural research commonly advise including no less than 30 participants. Adhering to this guideline will help ascertain an appropriate sample for our pilot survey (Whitehead et al. 2016; Suter 2014). Therefore, to acquire a reply amount of no less than 30 and to guarantee this questionnaire reaches the intended population, a technological approach is implemented; whereby the survey is constructed in Google Form and distributed through online platforms and applications such as email and WhatsApp. The response form (Google Form) connection is circulated with the assistance of College of Engineering lecturers and personnel.

This pilot study utilizes convenient sampling, a distinct non-probability sampling method where in the selected sample consists of population members who are readily accessible and willing to actively participate in the research. Convenient sampling is a frequently used approach in research domains. It has been chosen for this study due to participants' high level of accessibility, which facilitates recruitment efforts. Academia often gravitates toward this sampling technique given its reputation as a straightforward, cost-effective, and efficient option.

This preliminary survey gathered thirty-eight responses, exceeding the minimum requirement suggested. However, two responses were discarded since respondents selected "Strongly Agree" for every question, representing extreme responses. The remaining thirty-six responses will provide insight into whether this instrument is functioning as intended. Furthermore, the main objective of a pilot study is not hypothesis testing but improving the instrument and individual items.

2.2 *Model Construction*

The model used in this pilot study is named as Unified Theory of Acceptance and Use of Technology (UTAUT); and it consists of two direct determinants to Use Behaviour (UB) namely Behavioural Intention (BI) and Facilitating Condition (FC); as shown in Fig. 1. Also, Venkatesh includes three determinant factors that may affect BI which are Performance Expectancy (PE), Effort Expectancy (EE) and Social Influence (SI); which is shown in Fig. 1. However, there are studies pointing out that cost, risk, aesthetics, and comfort have impact on behavioural intention (Park et al. 2016; Adebesin and Mwalugha 2020; Piculo dos Santos et al. 2020; Pandey et al. 2022; Bhandari et al. 2019; Shimura et al. 2022; Esquivel et al. 2023; Chen et al. 2023; Jeong et al. 2021; Raman and Aashish 2022). Hence, four (4) additional determinants are included in this study.

As for moderating variables, the original UTAUT Model introduced four (4) variables namely gender, age, experience, and voluntariness of use. However, for this study, only gender in taken as the moderating variables since this study has put focus to Generation Z age range; and it has been emphasized in previous studies where gender is an essential moderating variable in Technology Acceptance.

The UTAUT Model from its inception sought to account for key factors involved in acceptance of new technologies. However, the model in its original form accounted for approximately 70% of variance observed, indicating additional elements likely

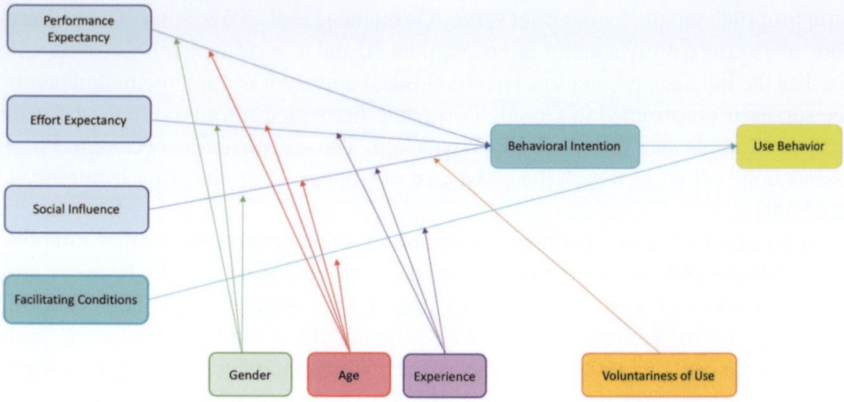

Fig. 1 The original (UTAUT) model

influence whether individuals choose to adopt technological innovations. Consequently, researchers should consider supplementing the core UTAUT Model with context-specific variables tailored to the unique aims and circumstances of each study.

Table 1 demonstrates the differences between the original UTAUT Model and two modified versions, including the one used in this study. The reason Table 1 is presented is to highlight the necessary changes to link the UTAUT Model with acceptance of locally developed sports devices within the lab. While other researchers have examined IoT acceptance among university students, none have focused specifically on acceptance of locally developed sports devices designed to measure muscle performance. Therefore, this study constructed a new acceptance model (Fig. 2) to investigate the relationships between all determinants in the modified UTAUT Model.

The following provides clarification on the variables used in this study, as depicted in Fig. 2, to enhance comprehension. Our adapted UTAUT Model incorporates

Table 1 Difference between the original UTAUT model with two types of modified version

UTAUT model		
Original independent variable	Modified independent variable I (Lin 2019)	Modified independent variable II (Used in this study)
Performance expectancy	Performance expectancy	Performance expectancy
Effort expectancy	Effort expectancy	Effort expectancy
Social influence	Social influence	Social influence
Facilitating conditions	Facilitating conditions	Facilitating conditions
	Perceived cost	Perceived cost
	Individual innovation	Perceived risk
		Perceived aesthetics
		Perceived comfort

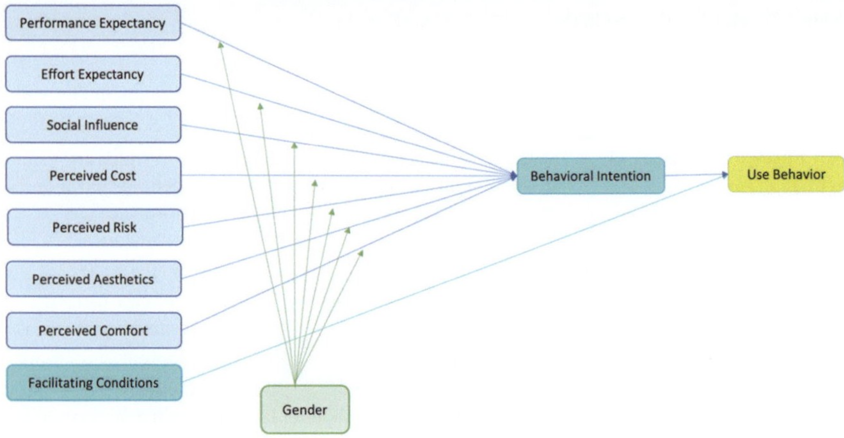

Fig. 2 The modified (UTAUT) model used in this study

four additional factors relative to the original UTAUT Model. The supplementary four factors are perceived cost, perceived risk, perceived aesthetics, and perceived comfort. The perceptions of cost, risk, aesthetics, and comfort considerably influence the adoption and application of wearable devices. Individuals are more inclined to adopt and use wearable technologies deemed convenient and offering optimal value (Park et al. 2016). Costly wearables that seem risky or inelegant to use are less likely to be taken up. Likewise, unwieldy forms less pleasing to employ will meet resistance. Only through understanding what drives acceptance can improvements be made to expand the appropriate and appealing scope of such emerging technologies. By proactively acknowledging and managing foreseeable risks such as privacy and security considerations, it is possible to enhance users' confidence and willingness to embrace the given solution (Park et al. 2016; Adebesin and Mwalugha 2020).

The way wearable devices look and are designed greatly impact the feelings and general satisfaction of people who use them. Wearable technology that has pleasing design features and matches what users want are more likely to be embraced. Products seen as attractive gain approval while those viewed as unattractive face rejection. Getting the appearance and design right encourages acceptance. A wearable's visual appeal and construction affect whether people enthusiastically adopt and continue using it (Piculo dos Santos et al. 2020; Pandey et al. 2022; Bhandari et al. 2019). The design elements that impact comfort are crucial for wearable devices to be useful tools, especially during prolonged collaborative work. How at ease users feel wearing these devices may strongly influence their willingness to keep using them. Comfort level and adoption potential are interconnected, together defining how readily wearables are accepted in the workplace (Shimura et al. 2022; Esquivel et al. 2023). Designers and developers should consider these factors to optimize user acceptance and satisfaction. Recent studies have highlighted the importance of these four variables (Chen et al. 2023; Jeong et al. 2021; Raman and Aashish 2022) (Table 2).

Table 2 Each variable used in the modified UTAUT model

UTAUT model		
Variables	Definition from original UTAUT model	Specific definition in this study
Performance expectancy (3 items)	The degree to which an individual thinks he or she can benefit from using a system and meet expectation	The degree to which an individual believes his or her muscle performance can improve by using local in-lab-made sports device
Effort expectancy (2 items)	The degree of ease of using the system	The degree of ease using the local in-lab-made sports device
Social influence (3 items)	The degree to which a person perceives the importance of others thinking he or she should use the system	The degree of influence from the surrounding social landscape to the thinking of an individual
Facilitating conditions (3 items)	The degree to which a person believes the infrastructure related to organization and technique supports the use of the system	The degree to which users perceive that the economic and time costs affect their use of local in-lab-made sports device
Perceived cost (3 items)	N/A	The degree to which users perceive the cost of the local in-lab-made sports device affects their use of that device
Perceived risk (3 items)	N/A	The degree to which users perceive the privacy and security affect their use of local in-lab-made sports device
Perceived aesthetics (2 items)	N/A	The degree to which users perceive qualities that appear in the physical properties of the in-lab-made sports devices
Perceived comfort (2 items)	N/A	The degree to which users perceive comfort wear of the in-lab-made sports devices

This study was designed to evaluate a small number of factors related to acceptance, with two to three items per variable and a total of 19 items. Each item was assigned a specific abbreviation followed by its number. All items in this instrument focused on measuring levels of acceptance through subjective responses regarding preferences and feelings. A common approach for collecting subjective responses is through Likert Rating scales. This particular study used a 5-point Likert Rating where responses could be classified into five categories: 1–Strongly Disagree, 2–Disagree, 3–Somewhat Agree, 4–Agree, and 5–Strongly Agree.

Also, the structure of the survey instrument is arranged as follows:

1. Section A—Acceptance Level

2. Section B—Demographic Details

Section A will be the section where respondents will answer all items related to the variables shown above. This section is where the 5-Likert Rating applies. Section B deals with the demographic data. The demographic data asked will only be on gender.

2.3 Data Analysis

The results from the preliminary research were analyzed using an Item Response Theory method referred to as the Rasch Measurement Model (RMM). RMM is a probabilistic approach and was created by Georg Rasch, serving as a well-known way to evaluate how effectively a tested instrument measures what it aims to. In this paper, the data collected during the pilot study was organized, prepared, and cleaned before analysis using Winstep software version 4.7.1. Winstep is one program applied to examine data through an RMM approach.

Summary Statistics. In RMM, the first analysis that will be observed is from the summary statistics. The summary statistic is the data summary for both Item (survey question) and Person (respondent). The data summary presented in Summary Statistics are Item and Person mean, Item and Person minimum and maximum measure, Cronbach Alpha (α).

When analyzing assessment data through the Rasch measurement model, or RMM, a key measurement unit is used, known as the "Logit." Logit is a statistical technique that allows researchers to examine response patterns and compute logit values. Essentially, taking the logarithm of the odds ratio for selecting a particular answer category provides insight into where an item or person falls on the latent trait continuum being evaluated. By applying a logarithmic function to the probability ratio, the RMM is able to scale respondents and items on the same scale, thus facilitating comparisons and helping gauge relative positions. In short, logit values distil complex response data into a unified measurement that charts positions along the dimension assessed (Wong et al. 2018). In the context of the Rasch model, the logit unit is employed to denote the level of difficulty exhibited by test items, as well as the abilities of individuals, both measured along a continuous scale (Zhang et al. 2015). This facilitates the assessment of latent traits at regular intervals, a prerequisite for conducting various statistical analyses (Zhang et al. 2015).

As shown in Table 3, the item and person reliability are +0.74 and +0.87 respectively. A value equal and more than 0.7 indicates good reliability index (Lee et al. 2023). The reliability of each item is an indication of how well that item measures acceptance across all respondent levels. On average, respondents accept using locally made in-lab sports devices to monitor muscle performance, as shown by the mean person value of 1.61 logits. The mean item value serves as a threshold set to zero on the logit scale. The maximum item measure is 1.35 logits, while the maximum person measure is higher at 4.79 logits. There are sufficient items as the minimum

Table 3 Summary statistics

	Item (Logit)	Person (Logit)
Reliability	+0.74	+0.87
Maximum measure	+1.35	+4.79
Minimum measure	–0.72	–0.25
Mean	0.00	+1.61
Separation	1.70	2.56
Cronbach alpha (α)	0.92	

item measure is –0.72 logits, compared to the minimum person measure of –0.25 logits. As Fig. 2 demonstrates, the instrument can distinguish people well with a person separation of 2.56. Both the person and item mean squares and standardized values are near 1 and 0, respectively, indicating a good fit and a valid instrument that measures what was intended. Furthermore, the Cronbach's alpha of 0.92 allows for additional analysis.

Item Fit and Misfit. This portion demonstrates how well a particular element assesses the attribute it intends to gauge. If an individual has difficulty comprehending a specific element, it is probable that the element necessitates improvement or substitution since it does not correspond with their degree of comprehension (Wilmskoetter et al. 2019). The value of PTMEA CORR can be utilized to assess the congruence of items (Boone et al. 2014). The PTMEA CORR index shows an excellent performance when the values are over 0.40. With values between 0.30 and 0.39, its performance is adequate, and with values between 0.20 and 0.29, it's moderate. If the values fall within the range of 0–0.19, the respondent struggles to distinguish between different items. If the PTMEA CORR value is 0 or less, researchers need to review their survey items (Alagumalai et al. 2005).

Table 4 presents the PTMEA CORR index for this study. All items demonstrate high PTMEA CORR; which ranges from 0.43 to 0.73. Based on these PTMEA CORR indices, all of the items can be used to measure the acceptance of in-lab-made sports device.

Certain steps must be taken with regards to items that do not fit well. The relevance of items in gauging the concept can be seen through outfit and infit Mean-Square (MNSQ) values. MNSQ values ought to fall between 0.6 and 1.4 to confirm the item's aptness in gauging the concept (Bond and Fox 2015). If the MNSQ index value exceeds 1.4; therefore, the item is a misfit, while an index value less than 0.6 indicates students easily accept the item. Z scores or Z Standardized (ZSTD) are also essential for finding misfit items, and the value should be from −2 to +2 (Bond and Fox 2015). When the MNSQ value is accepted, the ZSTD can be ignored (Linacre 2016). If the MNSQ condition is not fulfilled, the items should either be refined or removed. Table 5 shows the misfit items in the study.

The examination of Table 5 reveals that the infit MNSQ ranges from 0.32 to 1.88, whereas the outfit MNSQ spans between 0.33 and 1.94. There are only five items that fall beyond the acceptable range of MNSQ. There are two items with values

Table 4 Point measure correlation for all items

PTMEA CORR	Item	Indexing category
0.43	PC1_WSD price	Excellent
0.54	PC2_WSD affordable	Excellent
0.49	SI2_WSD listened to	Excellent
0.50	PC3_WSD cheaper	Excellent
0.62	FC2_WSD study env	Excellent
0.49	SI3_WSD older	Excellent
0.59	FC1_WSD living env	Excellent
0.67	FC3_WSD lifestyle	Excellent
0.57	PR1_WSD risky	Excellent
0.60	PA1_WSD feel good	Excellent
0.59	PA2_WSD urge buy	Excellent
0.64	PC2_WSD feel	Excellent
0.58	PR2_WSD no trouble	Excellent
0.65	SI1_WSD influence behaviour	Excellent
0.63	PC1_WSD comfort imp	Excellent
0.59	PE1_WSD useful	Excellent
0.66	PR3_WSD privacy no risk	Excellent
0.65	PE2_WSD improve health	Excellent
0.70	EE1_WSD ease of use	Excellent
0.70	EE2_WSD quick adapt	Excellent
0.73	PE3_WSD improve QoL	Excellent

exceeding 1.4, namely PC2_WSD Affordable (1.48) and PC1_WSD Price (1.88). Three additional items have values lower than 0.6, specifically PE3_WSD Improve Quality of Life (0.32), EE2_WSD Quick Adaptation (0.45), and EE1_WSD Ease of Use (0.51). Due to the fact that the infit and outfit MNSQ scores for each item are above the acceptable range, the researchers have decided to improve all five items. The enhancements were made to the language and sentence structure of the items.

Unidimensionality. A unidimensional measurement evaluates a single underlying attribute without being impacted by outside influences or additional aspects. Such an assessment concentrates solely on appraising a single inherent trait and remains untouched by extraneous variables or collateral dimensions (She et al. 2021). To provide clarity, an exclusive measure stresses a single underlying trait and neglects other factors that could impact how someone replies to the given items (She et al. 2021). Principal Component Analysis of Residual (PCAR) is a statistical method used to evaluate how well a measurement instrument measures a single underlying trait or dimension. To pass the unidimensionality test using PCAR, the instrument must demonstrate that it is primarily assessing one key concept and not multiple

Table 5 Misfit items analysis

Measure	Infit		Outfit		PTMEA CORR	Item
	MnSq	ZStd	MnSq	ZStd		
0.34	1.88	3.02	1.94	3.23	0.43	PC1_WSD price
1.35	1.48	1.82	1.55	2.06	0.54	PC2_WSD affordable
0.6	1.31	1.25	1.49	1.88	0.49	SI2_WSD listened to
0	1.35	1.41	1.47	1.82	0.50	PC3_WSD cheaper
0.2	1.36	1.45	1.36	1.44	0.62	FC2_WSD study env
0.86	1.22	0.94	1.25	1.05	0.49	SI3_WSD older
0.2	1.22	0.93	1.24	1.01	0.59	FC1_WSD living env
0.54	1.12	0.55	1.13	0.59	0.67	FC3_WSD lifestyle
–0.64	1.13	0.63	1.11	0.54	0.57	PR1_WSD risky
–0.64	0.96	–0.12	0.93	–0.24	0.60	PA1_WSD feel good
–0.35	0.95	–0.15	0.93	–0.22	0.59	PA2_WSD urge buy
0.27	0.89	–0.4	0.93	–0.21	0.64	PC2_WSD feel
–0.21	0.91	–0.32	0.89	–0.4	0.58	PR2_WSD no trouble
0	0.87	–0.49	0.85	–0.6	0.65	SI1_WSD influence behaviour
–0.72	0.83	–0.71	0.8	–0.82	0.63	PC1_WSD comfort imp
–0.5	0.79	–0.91	0.77	–0.97	0.59	PE1_WSD useful
0.54	0.75	–1.08	0.78	–0.92	0.66	PR3_WSD privacy no risk
–0.57	0.62	–1.84	0.61	–1.89	0.65	PE2_WSD improve health
–0.28	0.51	–2.5	0.5	–2.55	0.70	EE1_WSD ease of use
–0.35	0.45	–2.89	0.48	–2.68	0.70	EE2_WSD quick adapt
–0.64	0.32	–4.03	0.33	–3.89	0.73	PE3_WSD improve QoL

concepts. According to Reckase (1979), the minimum variance to show the unidimensionality of the instrument is 20%; and the unexplained variance in the first contrast should not exceed 15% (Fisher 2007). The data in Table 6 demonstrates that the measures explain 36% of the raw variance, meeting the unidimensionality condition. No more than 15% of the unexplained variance exists in the first, second, third, fourth, or fifth contrasts. Therefore, this instrument reliably assesses the core concept of respondent's acceptance on lab-made devices. When a construct is unidimensional, it precisely captures and portrays the intended phenomenon without distortion from irrelevant factors influencing its evaluations.

Table 6 Unidimensionality by principal component analysis of residual (PCAR)

	Eigenvalue	Observed (%)		Expected (%)
Total raw variance in observations	32.8114	100.00		100.00
Raw variance explained by measures	11.8114	36.00		36.60
Raw variance explained by persons	5.365	16.40		16.60
Raw Variance explained by items	6.4464	19.60		20.00
Raw unexplained variance (total)	21	64.00	100%	63.40
Unexplained variance in 1st contrast	4.205	12.80	20%	
Unexplained variance in 2nd contrast	2.5296	7.70	12%	
Unexplained variance in 3rd contrast	2.1254	6.50	10.10%	
Unexplained variance in 4th contrast	1.7322	5.30	8.20%	
Unexplained variance in 5th contrast	1.6757	5.10	8%	

2.4 Discussion

Surveys that measure public acceptance of the lab-made device are important to understand how users perceived these devices. They provide valuable insights for manufacturers and researchers. Measuring acceptance helps with product improvement and confirmation of market demand. It also supports incorporating user-centered design principles.

The main goal of this study was to examine how receptive people are to a device developed in a lab. It monitors muscle health. The Unified Theory of Acceptance and Use of Technology (UTAUT) model provided the framework. Four new variables were added. Based on data analysis, the instrument is valid for use. The analysis of three key concepts in the context of Rasch Measurement Models (RMM) included summary statistics, item misfit, and unidimensionality. All-important indexes fell within accepted ranges outlined in Sect. 2.3.

2.5 Conclusion

In conclusion, applying the Unified Theory of Acceptance and Use of Technology (UTAUT) model provided valuable insight into how Bumiputera undergraduates in Malaysia accept in-lab-made technology to monitor muscle performance. Generation Z, a significant portion of Malaysia's population, plays a crucial role in the nation's social and economic development through their technology acceptance. This research employed the UTAUT model to understand the determinants of adoption within this demographic. It shed light on how performance expectancy, effort expectancy, social influence, facilitating conditions, perceived cost, perceived risk, perceived aesthetics, and perceived comfort shape acceptance. These findings can guide policy decisions, educational initiatives, and technological interventions tailored specifically for Generation Z. The aim is to foster a more inclusive and technologically proficient society in Malaysia. It is notable that the pilot study instrument was rigorously tested and validated. It demonstrated an ability to accurately measure the intended constructs. The instrument's reliability and validity established that the collected data provided a robust and dependable foundation. This further strengthened the UTAUT model as a valuable framework for assessing acceptance levels among undergraduates in Malaysia.

References

Adebesin F, Mwalugha R (2020) The mediating role of organizational reputation and trust in the intention to use wearable health devices: Cross-country study. JMIR mHealth uHealth 8. https://doi.org/10.2196/16721

Alagumalai S, Curtis DD, Hungi N (2005). Applied Rasch measurement: a book of exemplars. https://doi.org/10.1007/1-4020-3076-2

Bhandari U, Chang K, Neben T (2019) Understanding the impact of perceived visual aesthetics on user evaluations: An emotional perspective. Inf Manag 56:85–93. https://doi.org/10.1016/j.im.2018.07.003

Bond TG, Fox CM (2015) Applying the Rasch model: Fundamental measurement in the human sciences. https://doi.org/10.1111/j.1467-985x.2004.t01-2-.x

Boone WJ, JR S, MS Y (2014) Rasch analysis in the human sciences. https://doi.org/10.1111/j.1467-985x.2004.t01-2-.x

Chen J, Wang T, Fang Z, Wang H (2023) Research on elderly users' intentions to accept wearable devices based on the improved UTAUT model. Front Public Heal 10. https://doi.org/10.3389/fpubh.2022.1035398

Esquivel KM, Gillespie J, Kelly D, Condell J, Davies R, McHugh C, Duffy W, Nevala E, Alamäki A, Jalovaara J, Tedesco S, Barton J, Timmons S, Nordström A (2023) Factors influencing continued wearable device use in older adult populations: quantitative study. JMIR Aging 6. https://doi.org/10.2196/36807

Fisher WP (2007) Rating scale instrument quality criteria. Rasch Meas Trans 21:1095

Jeong J, Kim Y, Roh T (2021) Do consumers care about aesthetics and compatibility? The intention to use wearable devices in health care. SAGE Open 11. https://doi.org/10.1177/21582440211040070

Lee SC, Lee YC, Chiu EC (2023) Psychometric validation of the cognitive abilities screening instrument using Rasch analysis in people with dementia. Med (United States) 102:E34093. https://doi.org/10.1097/MD.0000000000034093

Lin C (2019) Applying the UTAUT model to understand factors affecting the use of e-books in Fujian, China. Univ Boras:1–87

Linacre JM (2016) A user's guide to Winstep mini step: Rasch-Model computer programs

Pandey S, Chawla D, Puri S, Jeong LS (2022) Acceptance of wearable fitness devices in developing countries: exploring the country and gender-specific differences. J Asia Bus Stud 16:676–692. https://doi.org/10.1108/JABS-11-2020-0456

Park E, Kim KJ, Kwon SJ (2016) Understanding the emergence of wearable devices as next-generation tools for health communication. Inf Technol People 29:717–732. https://doi.org/10.1108/ITP-04-2015-0096

Piculo dos Santos A, Moya Ferrari A, Orsi Mendola F, Eika Sandnes F (2020) Aesthetics and the perceived stigma of assistive technology for visual impairment. Disabil Rehabil Assist Technol:1–7

Raman P, Aashish K (2022) Gym users: an enabler in creating an acceptance of sports and fitness wearable devices in India. Int J Sport Mark Spons 23:707–726. https://doi.org/10.1108/IJSMS-08-2021-0168

Reckase MD (1979) Unifactor latent trait models applied to multifactor tests: results and implications. J Educ Stat 4:207. https://doi.org/10.2307/1164671

She M, Li Y, Tu D, Cai Y (2021) Computerized adaptive testing for sleep disorders: development of an item bank and validation in a simulated study. Eur J Heal Psychol 28:89–100. https://doi.org/10.1027/2512-8442/a000076

Shimura T, Sato S, Zalar P, Matsuhisa N (2022) Engineering the comfort-of-wear for next generation wearables. Adv Electron Mater. https://doi.org/10.1002/aelm.202200512

Suter W (2014) Introduction to educational research: a critical thinking approach. SAGE Publications, Inc; Fifth edition. https://doi.org/10.4135/9781483384443

Tjiptono F, Khan G, Yeong ES, Kunchamboo V (2020) Generation Z in Malaysia: The four 'E' generation. New Gener Z Asia Dyn Differ Digit:149–163. https://doi.org/10.1108/978-1-80043-220-820201015

Venkatesh V, Morris MG, Davis GB, Davis FD (2003) User acceptance of information technology: Toward a unified view. MIS q Manag Inf Syst 27:425–478. https://doi.org/10.2307/30036540

Whitehead AL, Julious SA, Cooper CL, Campbell MJ (2016) Estimating the sample size for a pilot randomised trial to minimise the overall trial sample size for the external pilot and main trial for a continuous outcome variable. Stat Methods Med Res 25:1057–1073. https://doi.org/10.1177/0962280215588241

Wilmskoetter J, Bonilha H, Hong I, Hazelwood RJ, Martin-Harris B, Velozo C (2019) Construct validity of the Eating Assessment Tool (EAT-10). Disabil Rehabil 41:549–559. https://doi.org/10.1080/09638288.2017.1398787

Wong MHY, Fenwick E, Aw AT, Lamoureux EL, Seah LL (2018) Development and validation of the Singapore thyroid eye disease quality of life questionnaire. Transl Vis Sci Technol, 7. https://doi.org/10.1167/tvst.7.5.14

Zhang D, Xu X, Yang X (2015) User satisfaction and its impacts on the use of a public bicycle system. Transp Res Rec J Transp Res Board 2512:56–65. https://doi.org/10.3141/2512-07

Automated Smart Dustbin (AuDust) Concept Design for Sport Arena

Nur Amira Zulkiflli, Phan Quoc Nguyen, Fatin Aliah Phang Abdullah, Leow Pei Ling, Azli Bin Yahya, Erni Yulianti, and Jaysuman Pusppanathan

Abstract In the fight against COVID-19, reducing unnecessary human contact and ensuring safety is crucial. This project introduces the Automated Smart Dustbin (AuDust), designed to minimize infection at crowded area such as stadium and other sports arena. AuDust, equipped with sensors and controlled by Arduino, opens without touch when detecting motion. It enhances safety for healthcare workers, patients, and frontliners by offering contactless waste disposal. Originally created for hospitals, AuDust also has potential applications in sports venues. The method involves programming, 3D designs using Solidworks, and simulation with TinkerCAD. Ultrasonic sensors detect people and waste levels, triggering appropriate actions. Experiments show AuDust's effectiveness, providing a safer waste disposal method to reduce infection spread.

N. A. Zulkiflli · A. B. Yahya · J. Pusppanathan (✉)
Department of Biomedical Engineering & Health Sciences, Faculty of Electrical Engineering, Universiti Teknologi Malaysia, Johor Bahru, Malaysia
e-mail: jaysuman@utm.my

P. Q. Nguyen
University of Law, Vietnam National University, Hanoi, Vietnam

F. A. P. Abdullah
Centre of Engineering Education, Universiti Teknologi Malaysia, Johor Bahru, Malaysia

L. P. Ling
Department of Control and Mechatronics Engineering (CMED), Faculty of Electrical Engineering, Universiti Teknologi Malaysia, Johor Bahru, Malaysia

A. B. Yahya
Medical Implant Technology Group (MediTEG), Institute of Human Centered Engineering (iHumEN), Universiti Teknologi Malaysia, Johor Bahru, Malaysia

E. Yulianti
Department of Science Education, Faculty of Mathematics and Natural Sciences, Universitas Negeri Malang, Jl. Semarang 5, Malang 65145, Indonesia

J. Pusppanathan
Sport Innovation & Technology Centre (SITC), Institute of Human Centered Engineering (iHumEN), Universiti Teknologi Malaysia, Johor Bahru, Malaysia

Z. Mohamed et al. (eds.), *Proceedings of the 2nd International Conference on Innovation and Technology in Sports, ICITS 2023, 27–28 November, Malaysia*, Lecture Notes in Bioengineering, https://doi.org/10.1007/978-981-97-3741-3_48

Keywords Automated Smart Dustbin (AuDust) · Health precautions · COVID-19 preventions · Sport venues

1 Introduction

In the midst of combating with covid-19 pandemic, people are finding ways to reduce unnecessary contact, keeping distance from each other, and other standard procedures (SOP) to avoid the spread of the virus. The World Health Organization states that; "The virus can also spread after infected people sneeze, cough on, or touch surfaces, or objects, such as tables, doorknobs and handrails. Other people may become infected by touching these contaminated surfaces, then touching their eyes, noses or mouths without having cleaned their hands first" (World Health Organization, https://www.who.int/emergencies/diseases/novel-coronavirus-019/question-and-answers-hub/qa-detail/coronavirus-disease-covid-19-how-is-it-transmitted).

As for post-covid era, people are now prone to gather for casual activities as usual. Certain crowded locations such as sports arena has higher risk for the spread and harder to control the high number of patients and frontliners working day and night. Besides, one of the high contributors to the current daily cases is from the working place and factory workers.

Problem statement: At such environment, what are the procedure for waste management? Medical personals or front-liners will have to collect and manage daily disposal waste from patient and visitors thus this will further expose them to the risk of getting infected.

This project is carried out as a part of the assignment for SMBE 3722 CDIO laboratory 3 and SEBB 2032 at the school of biomedical engineering & health Sciences, Universiti Teknologi Malaysia. The students are required to apply the physics concepts and C programming knowledge to produce a prototype (Zakaria et al. 2019). Due to the pandemic situation, the project has to be carried out on virtual platform. While some research can be done via finite element analysis simulation (Feng et al. 2019; Pusppanathan et al. 2015; Muji et al. 2013), this project was simulated on TinkerCAD online due to its powerful features for AUTODESK 3D drawing and Arduino simulation (Paul et al. 2018).

Through this teaching test by online virtual simulation experiments, the AuDust is able to reflect that the environment, contents, requirements, methods, procedures, and operation process of experiments truly reflected the process of hardware experiments and motivate users to participate (Yang et al. 2021).

2 Background

Automated Smart Dustbin (AuDust) is a smart dustbin equipped with a sensor which will open the lid without being touched by user. It will notify the user by the aid of an output which is a buzzer and LED. Providing a contactless waste collection disposal to reduce covid-19 infection. The design concept aims for a safer environment for medical workers, patients and front-liners as it is very useful and applicable to be installed at the main arena that deal with COVID-19 pandemic.

AuDust uses arduino, two units of ultrasonic sensors as input detector and two outputs: buzzer and LED. Ultrasonic sensor 2 is used to detect the level of waste inside the dustbin so that there will be no spilling of rubbish. Meanwhile, the UV sensor automatically opens the lid when it detects motion in a particular range so the doctors or medical staff can open the lid without being touched, thus will help to prevent further possible spread of any kind of disease and virus. Basically, our Smart Dustbin is targeted to main venue because this Dustbin is specially created for sports arena visitors and workers for safer environments.

In the realm of sports, innovation extends beyond the field and into the healthcare facilities that cater to athletes and sports enthusiasts. The Automated Smart Dustbin, or AuDust, initially designed to reduce COVID-19 transmission risks in crowded area, finds an unexpected yet valuable application in sports venues. This intelligent dustbin can serve as a contactless waste disposal solution for sports arenas, ensuring the safety of fans, athletes, and staff alike.

Equipped with sensors and automation technology, AuDust opens its lid without physical contact, preventing potential contamination. In a world where hygiene and health precautions have become paramount, the adaptability of such innovative solutions demonstrates their relevance not only in medical settings but also in sports arenas, where a safer environment is a top priority for all stakeholders.

3 Methodology

AuDust works like a normal dustbin except that it will open automatically with the help of a servo motor. The ultrasound sensors detect the present of people at a range of distance within 100 cm and activates the servo motor to open the dustbin lid. A buzzer & LED is mounted to indicate if the dustbin is full and will require personnel to dispose the waste. These operations are depicted (Fig. 1) as follows.

To establish this operation, a step-by-step program is constructed via a pseudo-code (see Fig. 2) before embarking into the actual programming for Arduino.

The prototyping design is realized via the use of Solidworks. This has been one of the major software which are used to design professional models for variety of things, and it has been always one of the best choices for designers, students and biggest companies to make the best assemblies, projects and products in which the user has uncountable tools and ways to reach to the desired result. Moreover, the

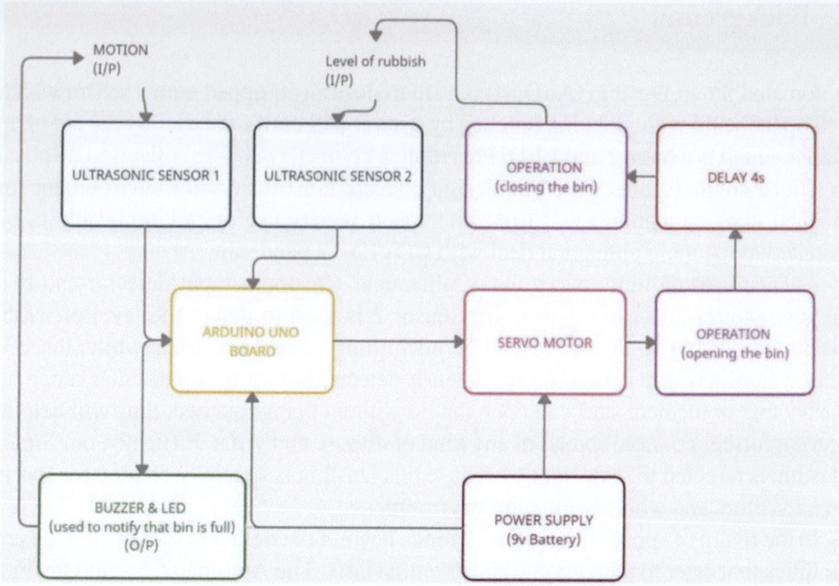

Fig. 1 Smart dustbin development workflow

```
Start
Switch ON
Set USsensor <= 0.10m
Set USsensor < 0.04 m
    Check condition: if USsensor <= 0.10m
    True : open lid, False : close lid
    Check condition: if USsensor < 0.04m
    True : buzzer ON & LED ON ,
    False: repeat 4
Switch OFF
```

Fig. 2 Pseudo-code design

accuracy and small considered details in this software play a significant role to have very accurate designs. The proposed concept design is as (see Fig. 3) follows.

A basic circuit diagram of the system is design using TinkerCAD which is available for free online for users to create & simulate 3D designs and Arduino program simulation. Figure 4 shows the overview of the program setup for Arduino simulation.

Fig. 3 Concept design in three-dimensional view

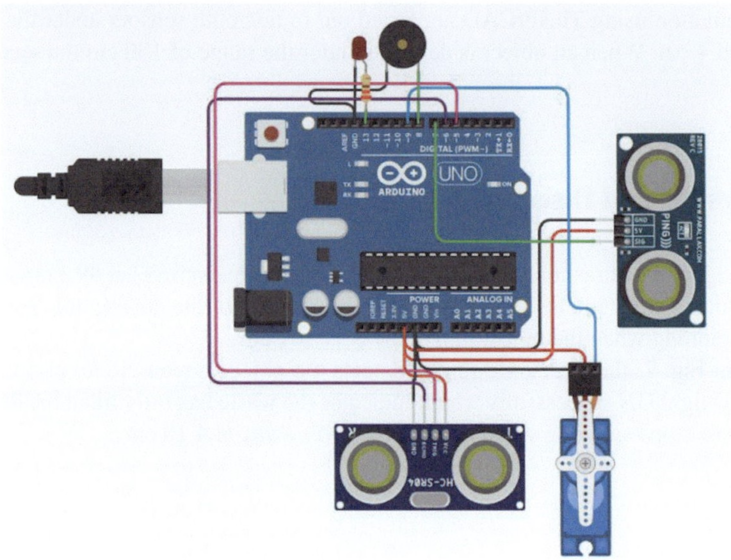

Fig. 4 Basic setup for smart dustbin functioning parts

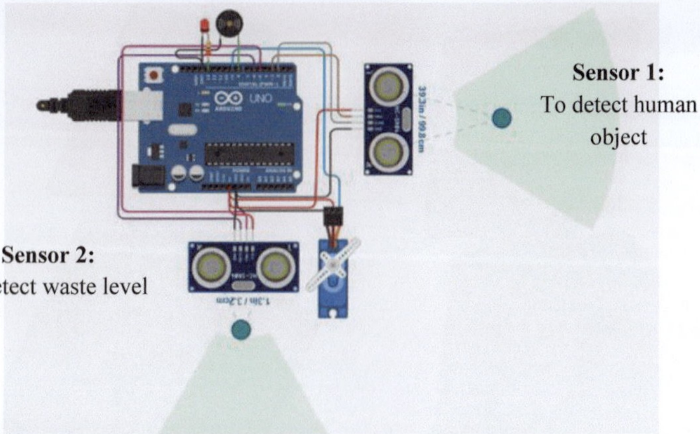

Fig. 5 Ultrasound sensor 1 & 2 sensing range 100 and 4 cm respectively

4 Experimental Setup

The ultrasound input sensors (see Fig. 5) are put into test for range within 100 cm. The system is design to position sensor 1 to detect human object within 100 cm and sensor 2 to detect the level of disposal inside the dustbin.

Simulation using TinkerCAD is carried out to test both sensors under the range 100 and 4 cm. When an object is detected under the range of 100 cm the servo will rotate.

5 Results and Discussion

From Fig. 6, the ultrasonic sensor detects the object (green circle) at 99.1 cm and the servo successfully rotates which indicates the opening of the dustbin lid. The servo stops spinning when the object moves away at 101 cm.

From Fig. 7, the ultrasonic sensor detects the object (green circle) at 4 cm and the LED light ON successfully which indicates the waste has fully filled the dustbin. The servo stops spinning when the object moves away at 4.10 cm.

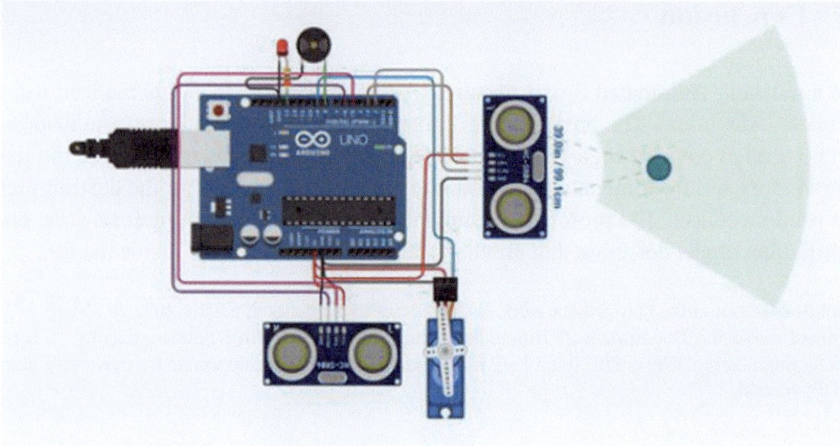

Fig. 6 Ultrasound sensing range at 100 cm

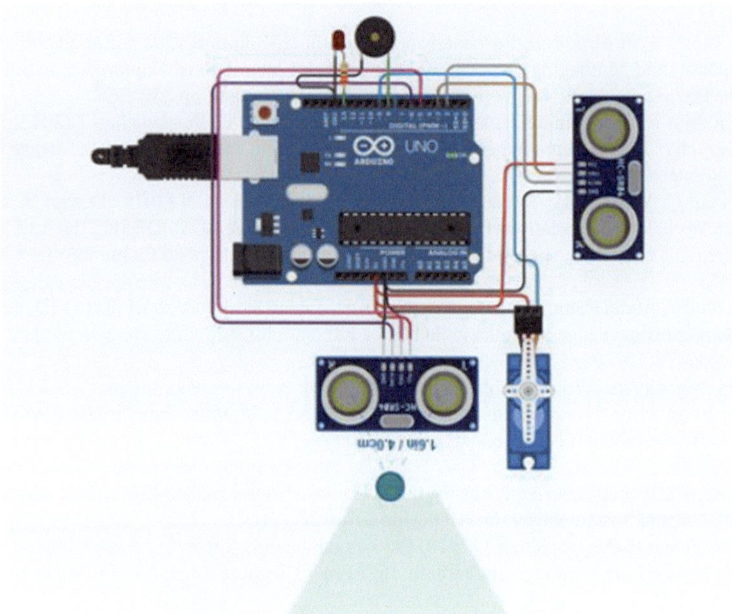

Fig. 7 Ultrasound sensing range at 4 cm

6 Conclusion

In a nutshell, Automated Smart Dustbin (AuDust) design concept is realized using TinkerCAD online. The prototype design aim to provide contactless waste disposal to be used at covid19 high-risk location such as hospital. This is to reduce the risk for visitors and frontliners to be exposed to the waste disposed into the dustbin such as used facemask. The prototype is simulated to test out the two input sensors; one for human object detection and another one to detect waste level inside the bin.

Acknowledgements This project work is carried out by students of SEBB 2032 & SMBE 3722 subject under the Department of Biomedical Engineering & Health Sciences, Faculty of Electrical Engineering, Universiti Teknologi Malaysia. The project is supported by university grant PDF2#4J648.

References

Feng H, Ling L, Ali M, Din S, Puspanathan J, Zulkiflli N, Rahim R (2019) Simulation study of two-phase fluid 3d imaging using lab-on-chip ECT. In: International Conference on Signal and Image Processing Applications (ICSIPA). IEEE, Kuala Lumpur, pp 259–263

Muji S, Rahim R, Rahiman M, Tukiran Z, Ayob N, Mohamad E, Puspanathan J (2013) Optical tomography: Image improvement using mixed projection of parallel and fan beam modes. Measurement 46(6):1970–1978

Paul B, Paul C, Varghese A, Sivasubramanian P, Shajoo S, Kurian N (2018) Design of a power feeder for elderly & simulation of motor circuit developed using AUTODESK TINKERCAD. In: International Conference on Circuits and Systems in Digital Enterprise Technology (ICCSDET), pp 1–13

Pusppanathan J, Abdul Rahim R, Phang FA, Abas KH, Sahlan S, Mohamad EJ (2015) 3D modelling of electrical capacitance and ultrasonic sensor integration using finite element method. Jurnal Teknologi 77(17)

World Health Organization. Corona virus disease. https://www.who.int/emergencies/diseases/novel-coronavirus-019/question-and-answers-hub/qa-detail/coronavirus-disease-covid-19-how-is-it-transmitted. Last accessed 9 July 2020

Yang H, Hsiao T, Zhu K, Yang Y, Guo L (2021) An exploration of online teaching based on arduino virtual simulation experiments. In: 3rd Eurasia Conference on Biomedical Engineering, Healthcare and Sustainability (ECBIOS), pp 126–128

Zakaria N, Phang F, Pusppanathan J (2019) Physics on the Go: A mobile computer-based physics laboratory for learning forces and motion. Int J Emerg Technol Learn (iJET) 14(24):167–183

Smart Electronic Health Monitoring System (SEHMoS)

Nur Amira Zulkiflli, Hadi Suwono, Fatin Aliah Phang Abdullah, Nina Diana Nawi, Yasmin Abdul Wahab, and Jaysuman Pusppanathan

Abstract The Smart Electronic Health Monitoring System (SEHMoS) addresses the challenge of infrequent health check-ups due to time constraints and financial limitations, which often result in the delayed detection of health issues. SEHMoS is a user-friendly, practical solution in the form of an adhesive gel pad sticker that can be easily attached to the body. It monitors body temperature and heart rate, with results displayed on an LCD screen. Amidst the ongoing pandemic, SEHMoS proves to be a valuable tool for self-examination, reducing unnecessary hospital visits and minimizing the risk of COVID-19 exposure. SEHMoS addresses these barriers by providing an affordable, non-invasive, and convenient monitoring solution for various demographics, including those with busy schedules, the elderly, individuals with disabilities, and frequent travelers. The system workflow involves temperature and heart rate sensors connected to the Arduino via jumper wires, with user interaction facilitated by a button. The results displayed on the LCD screen guide users through the process with the displayed temperature and heart rate readings.

N. A. Zulkiflli · J. Pusppanathan (✉)
Department of Biomedical Engineering & Health Sciences, Faculty of Electrical Engineering, Universiti Teknologi Malaysia, Johor Bahru, Malaysia
e-mail: jaysuman@utm.my

H. Suwono
Department of Biology, Faculty of Mathematics and Natural Sciences, Universitas Negeri Malang, Jl. Semarang 5, Malang 65145, Indonesia

F. A. P. Abdullah · N. D. Nawi
Centre of Engineering Education, Universiti Teknologi Malaysia, Johor Bahru, Malaysia

School of Education, Faculty of Social Sciences and Humanities, Universiti Teknologi Malaysia, Johor Bahru, Malaysia

Y. A. Wahab
Faculty of Electrical and Electronics Engineering Technology, Universiti Malaysia Pahang Al-Sultan Abdullah, Pekan, Malaysia

J. Pusppanathan
Sport Innovation & Technology Centre (SITC), Institute of Human Centered Engineering (iHumEN), Universiti Teknologi Malaysia, Johor Bahru, Malaysia

Z. Mohamed et al. (eds.), *Proceedings of the 2nd International Conference on Innovation and Technology in Sports, ICITS 2023, 27–28 November, Malaysia*, Lecture Notes in Bioengineering, https://doi.org/10.1007/978-981-97-3741-3_49

Keywords Health monitoring system · Routine check-up · Arduino · Tinkercad · Simulation-based prototype

1 Introduction

Health monitoring should be performed every month or at least once every 6 months however there are still so many people who do not do health routine check-up due to the lack of free time and financial problems which could not cover the cost of health checking every month. It is always safer for us to detect any problems earlier to prevent things getting worse. It has been proven that a lot of the elderly, the disabled, and those suffering from chronic diseases benefit from early detection of symptoms and have a greater opportunity to recover from diseases which they suffer. A device called Smart Electronic Health Monitoring System (SEHMoS) is developed as a solution for this problem, beside it is more practical, this monitoring device is also equipped with a service to check on body temperature and heart rate (Korhonen et al. 2003).

The purpose of using SEHMoS is to maintain a healthy life, reduce the number of people who die from silent killers, and it will be easier for doctors and hospitals to help us if they know our current condition. This device is very easy to use because this product is in the form of a sticker and can be easily attached on human body.

SEHMoS aims to promote a healthier lifestyle and enable early detection of health issues, which would hold great potential for sports and athletic performance. Maintaining peak physical condition is paramount for athletes, and SEHMoS can be a valuable tool for athletes and coaches alike. By continuously monitoring parameters like body temperature and heart rate, athletes can gain insights into their physiological state during training and competition.

In addition, the current pandemic significantly impacted sports events and activities, and monitoring health parameters became even more crucial for athletes and sports organizations to ensure safety and minimize the potential of spreading COVID-19 infection (Zakaria et al. 2019; Alsayaydeh et al. 2020).

2 Background Study

Silent killer is the nickname for a disease that comes suddenly without symptoms and can cause death such as heart disease, stroke, hypertension, diabetes, cancer and tumors. Although sometimes it comes without symptoms, the silent killer can be diagnosed early by doing medical checkup periodically. Based on research, there are five main reasons why many people still choose not to do the medical check-up frequently. That 52% of us do not do it because they feel healthy, 49% of people consider the cost is expensive, 26% were reluctant because of the fear of seeing the result, 23% of people rate the health test requires a lot of time and there is also a 10% that feel sure of himself does not have a serious illness (Yusuf et al. 2018).

The main objective of this project is to create a device that is very easy to use, cheap, safe, non-invasive and an analyzer system that can help a lot of people including the hospital itself, because it gives the opportunity to the people to avoid any risk by visiting the hospital to do routine medical check-up (Ogawa and Togawa 2003; Fezari et al. 2015). The parameters used by this system are body temperature and heart rate. Each of these parameters will be automatically censored by this tool and the results will be displayed onto the LCD screen. Thus, it will target:

(a) People who have limited time to do routine medical check-ups.
(b) The elderly and people with disabilities who find it difficult to go to the hospital alone.
(c) People who travel frequently and are careless in maintaining their health.
(d) This tool can be used by all people who need quick medical check-ups including children.

3 Methodology

Figure 1a shows the diagram of the project flow from idea generation to the documentation of the project while Fig. 1b shows the low-level workflow of the system:

The protype utilizes TMP36 temperature sensor and the Heart Rate sensor, showing on an LCD screen by using an Arduino UNO and a breadboard. Resistors and potentiometer were also used to finetune the prototype. The aim of this project is to build a temperature and heart rate health monitoring system. Due to the pandemic situation, this prototype is constructed on simulation-based using Tinkercad as in Fig. 2.

4 Three-Dimensional 3D Mockup Design

Figure 3 shows the final 3D model that was designed on Tinkercad. It has the LCD screen on the outside and on the inside of the brown box will be where the circuit of Arduino will be inserted. The white leather strap will attach itself to the sticker and can be applied onto our body.

4.1 Experimental Setup

Our product is created to mainly focus on patients with low-to-medium financial situations. Hence, the prototype of our project is simple enough to be applicable

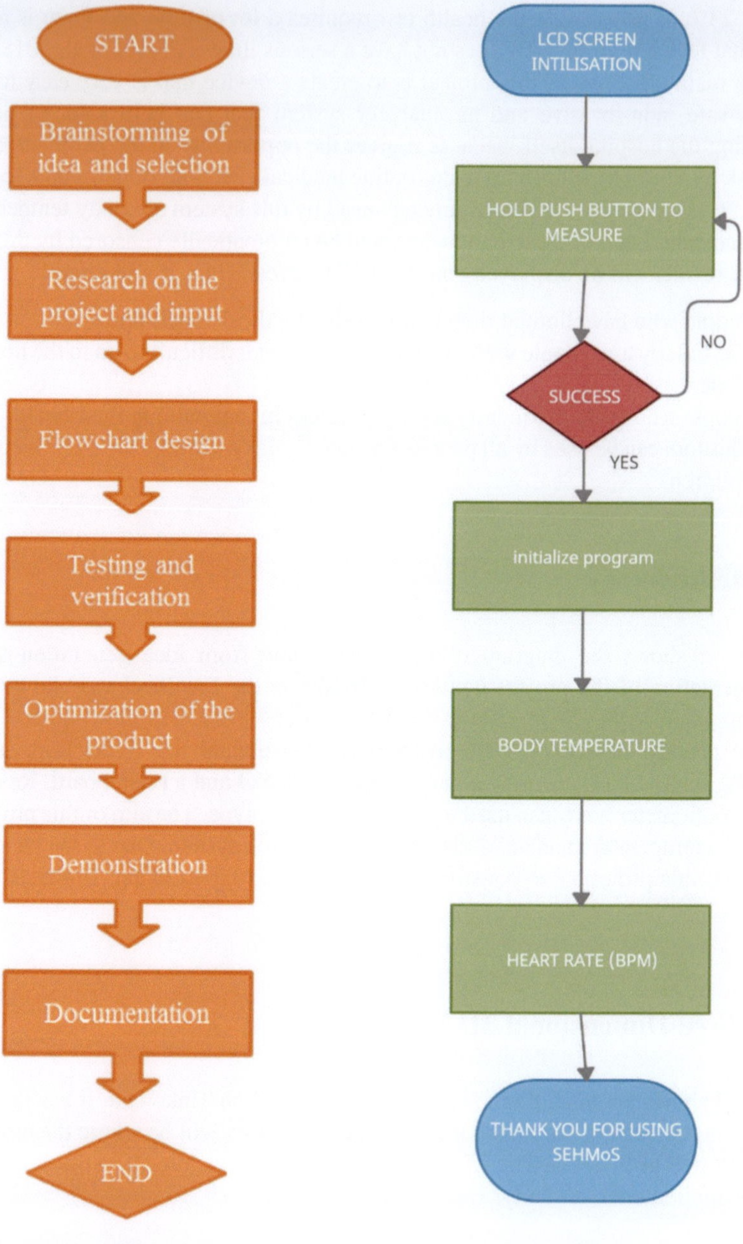

Fig. 1 Project overview **a** Idea generation to documentation, **b** low-level workflow

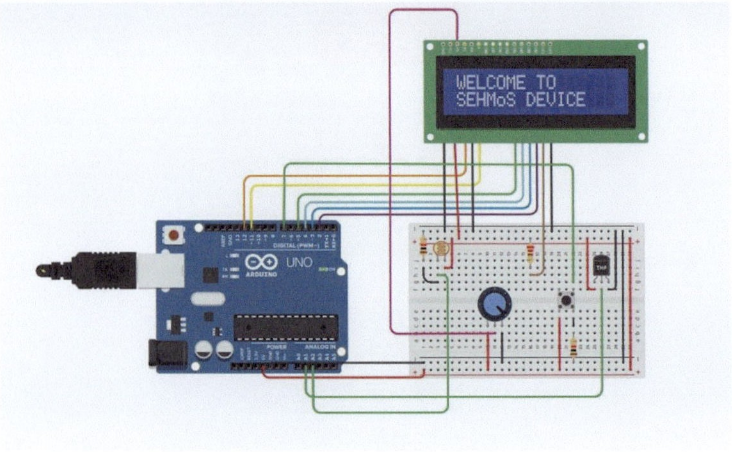

Fig. 2 Circuit setup using tinkercad

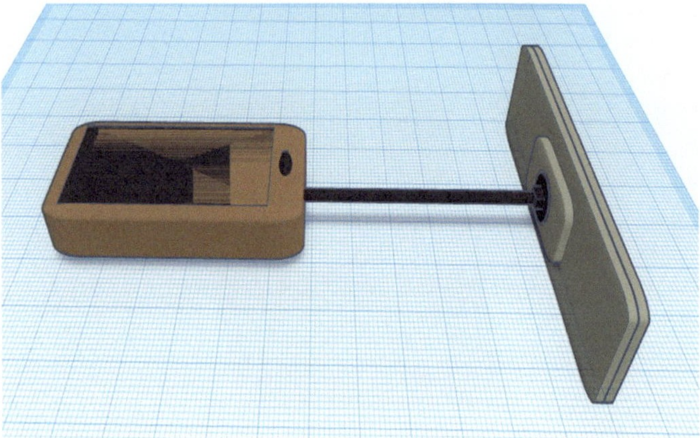

Fig. 3 3D model of SEHMoS

and easy for anyone to utilize it (Kadam et al. 2016). Figure 4 shows our system workflow.

As shown in the previous Tinkercad Circuit Design, a TMP36 Temperature sensor, and the Heart Rate sensor will be attached to a breadboard and connected with Arduino using Jumper Wires. Hence, they will be directly connected to the breadboard which will interface the patient and measure body temperature when the button is pressed. The workflow of the circuit design is shown in Fig. 5 follows:

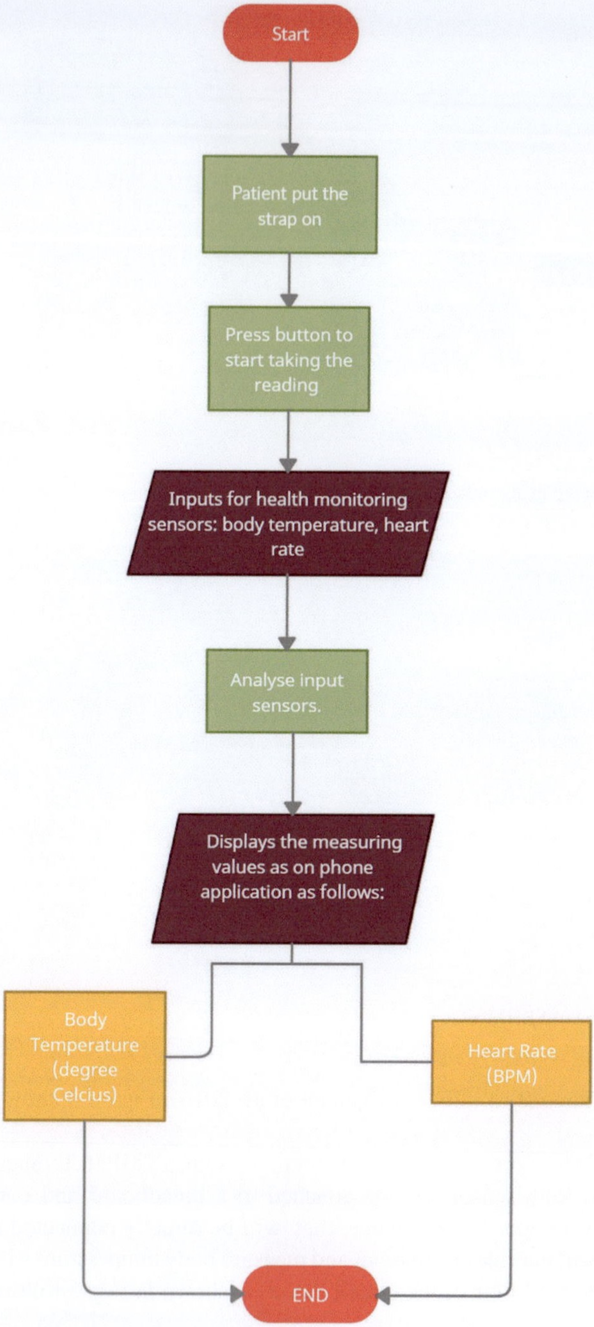

Fig. 4 Prototype system workflow

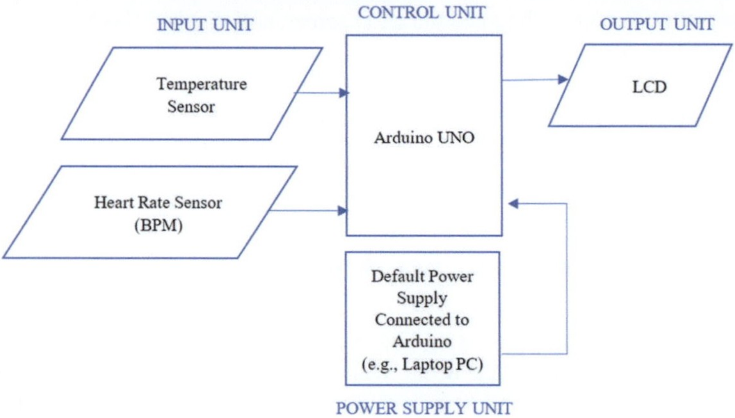

Fig. 5 Circuit design workflow

5 Results

In Fig. 6, it shows that the LCD will appear "Welcome to SEHMoS" to indicate the startup of the program. After that, the LCD will display "Loading" for preparation. Next, the LCD will display "Press the button to scan" and measurement can be commenced via the start button.

For the temperature measurement, it will change the value of the temperature if the user slides the temperature sensor. It is also the same during BPM calculations. It will change if the user turns the photoresistor. After both readings are taken, the

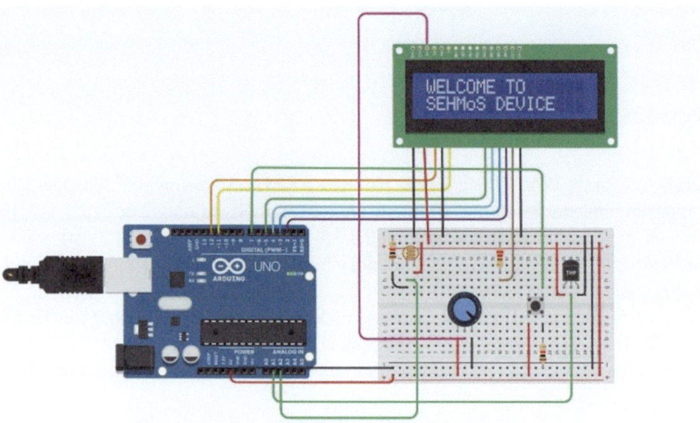

Fig. 6 SEHMoS system with LCD display screen

Arduino will begin uploading and the LCD screen will display "*THANK YOU FOR USING SEHMoS*".

When the ranges of temperature and heart rate is changed, the LCD screen will quickly display the result immediately. Moreover, a 1 s delay is set for each output to display after the one before thus, preventing both outputs to overlap with each other.

6 Conclusion and Future Recommendation

Smart Electronic Health Monitoring System (SEHMoS) is suitable for anyone who needs to check or monitor their health instantaneously. It is flexible and practical, equipped with a service to check on body temperature and heart rate. The SEHMoS application aims to lead users to a healthy lifestyle. This device offers easy use because it has an adhesive and washable gel pad. When affixed to the gel pad sticker, it can detect the patient's body temperature and heart rate, and it will display the results on the LCD screen. However, there are also some limitations in this prototype, such as:

(a) Users might feel uncomfortable sticking the sticker on their bodies.
(b) Over time, the sticker itself will not become sticky anymore.

For future improvement, the prototype of this project could be integrated with a phone app and by applying the IoT.

Acknowledgements This project work is an experiment work carried out by students of SMBE 2033 C Programming for Biomedical Engineering under the Department of Biomedical Engineering & Health Sciences, Faculty of Electrical Engineering, Universiti Teknologi Malaysia. The project is supported by university grant PDF2#4J648.

References

Alsayaydeh JAJ, Indra WA, Khang AWY, Hossain AKMZ, Shkarupylo V, Pusppanathan J (2020) The experimental studies of the automatic control methods of magnetic separators performance by magnetic product. ARPN J Eng Appl Sci 15

Fezari M, Rasras R, Emary I (2015) Ambulatory health monitoring system using wireless sensors node. Procedia Comput Sci 65:86–94

Kadam D, Patil S, Bormane D (2016) Wireless patient health monitoring system. Techno-Societal. Springer International Publishing, Noida, pp 415–418.

Korhonen I, Parkka J, Gils M (2003) Health monitoring in the home of the future. IEEE Eng Med Biol Mag 22(3):66–73

Ogawa M, Togawa T (2003) The concept of the home health monitoring. In 5th international workshop on enterprise networking and computing in healthcare industry (HealthCom). IEEE, Santa Monica, pp 71–73

Yusuf A, Zulkiflli F, Mustika I (2018) Development of monitoring and health service information system to support smart health on android platform. In: 4th International Conference on Nano Electronics Research and Education (ICNERE). IEEE, Hamamatsu, pp 1–6

Zakaria N, Phang F, Pusppanathan J (2019) Physics on the go: a mobile computer-based physics laboratory for learning forces and motion. Int J Emerg Technol Learn (iJET) 14(24):167–183

Wireless Timing System Using ESP-To-ESP Communication Protocol for Precision Timing in Sports

Muhammad Aqib Othman, Luqman Hakim Amri, Muhammad Nur Fahmy Azmi, Nur Aqilah Othman, and Mohd Hasnun Arif Hassan

Abstract In a sport where determining the winner depends on milliseconds, even the slightest variation is crucial. Traditionally, stopwatches have been used for timekeeping, but this method is prone to human error when starting and stopping the timer. Therefore, a system that operates autonomously, initiating the timer through sensors and stopping it once a second or endpoint sensor detects the movement necessary to halt the timer is required. The initial sensor's interaction to start the timer will be wireless, eliminating the need for Bluetooth, which has a limited range of only 10 m. An alternative to Bluetooth is using the internet as a means of communication, but this can be problematic in areas with no internet connectivity. To compensate for the short-range limitations of Bluetooth and the absence or poor quality of an internet connection, ESP-to-ESP communication protocol is one of the options. This system enables seamless communication between the sensors and the microcontroller for initiating and displaying the timer wirelessly. Additionally, the timer will automatically stop when triggered by the sensor at the end of the competition or race, eliminating the possibility of human error in timekeeping for any time-based competition. The prototype underwent field testing in tournaments and training sessions, with over 50 uses on a single occasion.

Keywords Timing gate · Wireless · Stopwatch

M. A. Othman · L. H. Amri · M. N. F. Azmi · N. A. Othman (✉) · M. H. A. Hassan
Universiti Malaysia Pahang Al-Sultan Abdullah, 26600 Pekan, Pahang, Malaysia
e-mail: nuraqilah@umpsa.edu.my

Z. Mohamed et al. (eds.), *Proceedings of the 2nd International Conference on Innovation and Technology in Sports, ICITS 2023, 27–28 November, Malaysia*, Lecture Notes in Bioengineering, https://doi.org/10.1007/978-981-97-3741-3_50

1 Introduction

1.1 Timing in Sports

Timing is a critical element in the world of sports, often serving as the fine line between success and failure. Whether it is the split-second precision required to execute a game-winning shot, the perfectly timed tackle to prevent a goal, or the strategic timing of a well-timed play, athletes and teams understand the immense impact that timing can have on their performance and outcomes. Different methods of measuring timing in sports include the use of timing gate systems, UHF RFID systems, real-time recording with video analysis, and electric timing systems. Timing gate systems are used to assess and enhance various aspects of athletic training, including speed, acceleration, responsiveness, power, and elevation (Mohd Jamil et al. 2023). UHF RFID systems utilize RFID reader or tags layouts to track and time athletes in different sports, with a focus on monitoring read probability and inaccuracy (Kolaja and Ehlerova 2019). Real-time recording with video analysis involves using image processing techniques, such as edge detection and Hough Transform, to calculate the time from a recorded video of a running race, providing a more precise alternative to estimating time with the human eye (Weng 2022). Electric timing systems, which combine synchronous electric clocks and high-speed motion picture cameras, have been developed for accurate timing in athletic events, with accuracy within milliseconds (Johansen and Tillaar 2022).

1.2 Timing Gates

Timing gates in sports are devices used to measure various aspects of athletic performance, such as speed, acceleration, and power. They work by using sensors, such as infrared sensors or photocells, placed at specific intervals along a track or course. When an athlete passes through these sensors, they trigger a timing system that records the time it takes for the athlete to travel between each sensor. This data can then be used to analyse the athlete's performance and track their progress over time. The accuracy and reliability of timing gates depend on factors such as the placement of the sensors and the number of timing checkpoints used (Mohd Jamil et al. 2023; Haugen et al. 2020). In some cases, magnet-based timing systems have been developed to provide more detailed timing analysis, allowing for the measurement of intermediate times at each gate (Fasel et al. 2019).

In addition to measuring time taken by the athletes, timing gates are used in sports to measure and analyse various mechanical parameters during sprinting activities such as maximal horizontal force (F_0), theoretical maximal velocity (v_0), maximal horizontal power (P_{max}), force–velocity slope (SFV), maximal ratio of force (RF_{max}), and index of force application technique (DRF) (Haugen et al. 2020). Timing gates are used to estimate maximum sprint speed by measuring split times at specific distances, such as 5 and 10 m (Zabaloy et al. 2021). They are also used to compare sprint mechanical parameters with other measurement devices, such as laser guns, to ensure accuracy and reliability (Tillaar et al. 2022). Timing gates have been validated against radar guns and GPS units for assessing maximum sprint velocity (V_{max}) in sports like rugby (Roe et al. 2017). Additionally, timing gates can be used in combination with global navigation satellite systems (GNSS) to accurately measure and analyse performance in alpine skiing (Supej and Holmberg 2011).

Several timing gate mechanisms used in sports include the dual-beamed timing gates covering the entire running course with 5 m intervals (Haugen et al. 2020), a photocell timing gate consisting of one start and a three-track timer (Apriantono et al. 2022), timing gates with a flying start distance to avoid premature triggering (Jovanović 2023), and a single-beam timing gate system used to measure sprint time at different starting distances (Altmann et al. 2015). The cost-effective timing gate system developed in one study utilized ESP32 Wi-Fi module as the microcontroller and infrared sensors as input. The dual-beamed timing gates in another study covered the entire running course with 5 m intervals and were used to compute various mechanical outputs in sprinting athletes. The new photocell timing (NPT) gate in a different study was compared to validated timing (VT) and manual hand-held timing (MHH) methods. Timing gates with a flying start distance were explored in a simulation study to assess their effects on bias and sensitivity in estimating short sprint parameters. Finally, a single-beam timing gate system was used to measure 5 m sprint time at different starting distances, with the accuracy of the initial timing gate being consistent across the distances.

Wireless timing gates in sports offer several advantages. They allow for the construction of a wireless communication network for sports timing training, improving the effective ability of sports (Mohd Jamil et al. 2023). They also provide a cost-effective and reliable solution for measuring speed and agility in athletes, enhancing athletic training in various aspects (Haugen et al. 2020). Additionally, wireless timing gates can ensure valid and reliable sprint mechanical outputs when properly set up, covering the entire acceleration phase (Zhang and Li 2020). However, there are also some disadvantages to using wireless timing gates. In the sport of canoe slalom, for example, the subjective decision of a human judge is still relied upon, as there is currently no automated system for gate detection (Ringwood et al. 2015). Furthermore, in short sprints, the use of flying start distances can introduce bias in estimated sprint parameters, requiring the use of correction models for more precise estimates (Jovanović 2023). In summary, wireless timing gates offer advantages such as improved training effectiveness and accurate measurement of athletic performance. However, challenges remain in certain sports where subjective judgments are still relied upon, and careful consideration is needed to minimize bias in certain timing scenarios.

Timing gates in sports offer several advantages and disadvantages. One advantage is that they provide a reliable and accurate way to measure speed, acceleration, and power in athletes, which is crucial for assessing and enhancing athletic training (Mohd Jamil et al. 2023). Additionally, timing gates can be used to standardize sprint starting techniques and monitor short sprint performance, providing precise parameter estimates (Jovanović 2023). However, the setup of timing gates can impact the mechanical outputs in sprinting athletes, with the need for timing gates to cover the entire acceleration phase for valid and reliable results (Haugen et al. 2020). Furthermore, the cost of developing a timing gate system can be high, which may limit its accessibility (Zabaloy et al. 2021). Despite these limitations, timing gates remain a valuable tool in sports for assessing and improving athletic performance.

Therefore, the objective of this study is to develop a cost-effective wireless timing gate system that can provide precise measurement of time taken between two points in a wide range of sports activities. For this purpose, the proposed timing gate system utilised an ESP-to-ESP communication protocol that provides a seamless wireless communication between the two checkpoints. In addition, the developed system needs to be user-friendly with automated timing system to minimise human-induced errors in initiating and stopping the timer, thereby enhancing the reliability of the timing process. The following sections elaborate the development of the wireless timing gate system, which includes the hardware used, and how the microcontroller was programmed to achieve its intended functionality.

2 System Development

2.1 *System Architecture and Hardware Configuration*

The wireless timing system is designed based on the ESP-to-ESP communication protocol, ensuring seamless and reliable communication between the sensors and the microcontroller. The system architecture consists of two main components: the initiating sensor (a push button at the starting point) and the endpoint sensor. The initiating sensor wirelessly triggers the timer, while the endpoint sensor, configured as a vibration sensor, marks the completion of the competition or race, automatically stopping the timer. The elapsed time is displayed on the LED dot matrix module at the endpoint. Using a single push button at the starting point, multiple endpoint sensors can represent the time taken for each lane in any competition game. A one-to-many ESP-to-ESP communication protocol is employed for this purpose, enabling the ESP-sender to communicate with multiple ESP-receivers.

Figure 1 depicts the schematic diagram for the starting point (sender), comprising an ESP32 microcontroller development board, push buttons to initiate and reset the timer, and a buzzer to indicate the timer starting. The schematic for the receiver unit is illustrated in Fig. 2, featuring another ESP32 development board to control and process the timer, a vibration sensor as an input to stop the timer, and an LED dot matrix display to showcase the time taken. ESPs in the sender and receiver communicate using a 2.4 GHz wireless signal, with the MAC addresses of the receivers indicated at the sender.

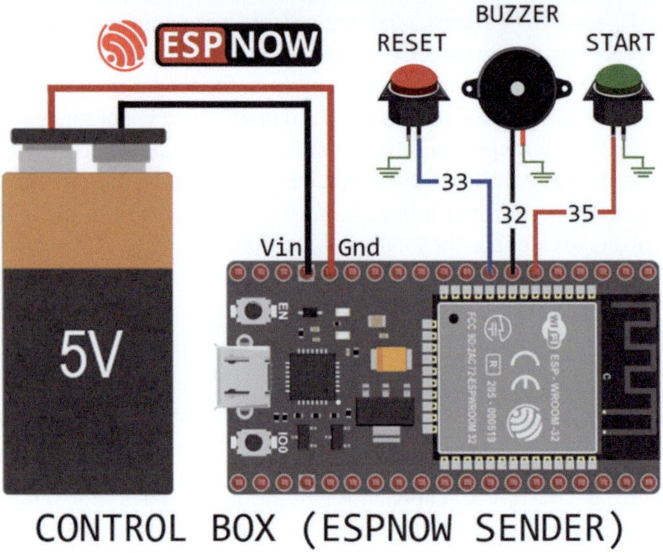

Fig. 1 Schematic diagram of the sender control unit at the starting point

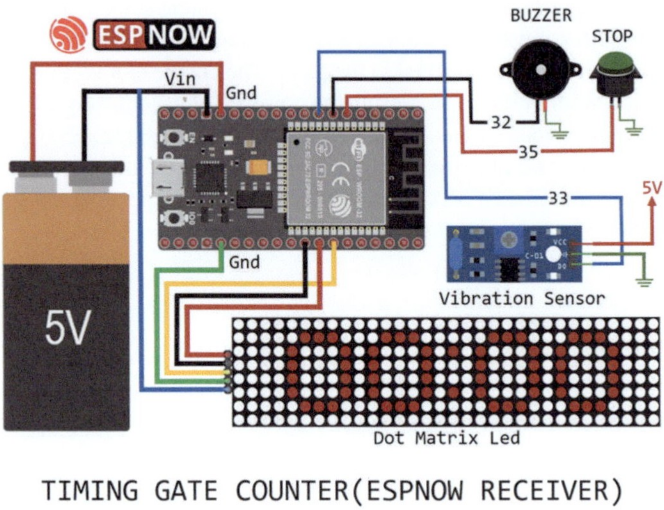

Fig. 2 Schematic diagram of receiver unit at the endpoint of the track

2.2 *ESP-To-ESP Communication Protocol*

The heart of the system lies in the ESP-to-ESP communication protocol, which allows the initiating sensor to wirelessly transmit a start signal to the endpoint sensor. The protocol is designed to be robust, enabling reliable communication even in environments with potential interference. Communication occurs via a 2.4 GHz Wi-Fi signal, with the associated MAC address of the receiver on the sender, making their communication more robust to signal interference and providing a wider range compared to Bluetooth. The signal range can be transmitted over 150 m, depending on the environment. The microcontroller is programmed to manage the timing logic, ensuring precise measurement of elapsed time. It utilizes timestamps from the initiation and endpoint sensors to calculate the total time duration accurately. A flowchart in Fig. 3 illustrates the step-by-step process of communication initiation, ensuring a systematic and error-free operation.

2.3 System Prototype

To visually explain the physical appearance of the development, Fig. 4 illustrates the starting box or sender of the system, while the endpoint sensor (receiver) and the LED display are depicted in Fig. 5.

Both designs in Figs. 4 and 5 are customised for a robotics tournament, namely the Amphibian Electric Vehicle (AEV) competition as shown in Fig. 6, where the winning robot is determined by its speed both on the ground and on the water surface. The miniature robot must make contact with the aluminium plate at the endpoint to trigger the vibration sensor attached to the plate. The choice of a vibration sensor for this application is deliberate, considering the presence of water at the end of the track, which may generate water waves, rendering infrared or ultrasonic sensors unsuitable. Other sensors that could potentially trigger the end-timer include proximity sensors, piezoelectric sensors, laser rays, or any other sensors, depending on the specific circumstances and the environment of the tournament.

3 Discussion

The ESP-to-ESP communication protocol employed in the prototype proves highly effective, overcoming the limitations of Bluetooth range and addressing internet connectivity issues. The system demonstrates reliable and low-latency communication, making it well-suited for time-sensitive sports applications. The prototype underwent field testing in tournaments and training sessions, with over 50 uses on a single occasion, and exhibited minimal errors, none of which significantly affected the timing accuracy.

The wireless timing system significantly mitigates the risk of human error associated with manual stopwatch timing. It ensures precise and consistent timekeeping for various sports competitions, providing a valuable tool for referees, coaches, and athletes, especially when determining the winner with a mere difference in milliseconds.

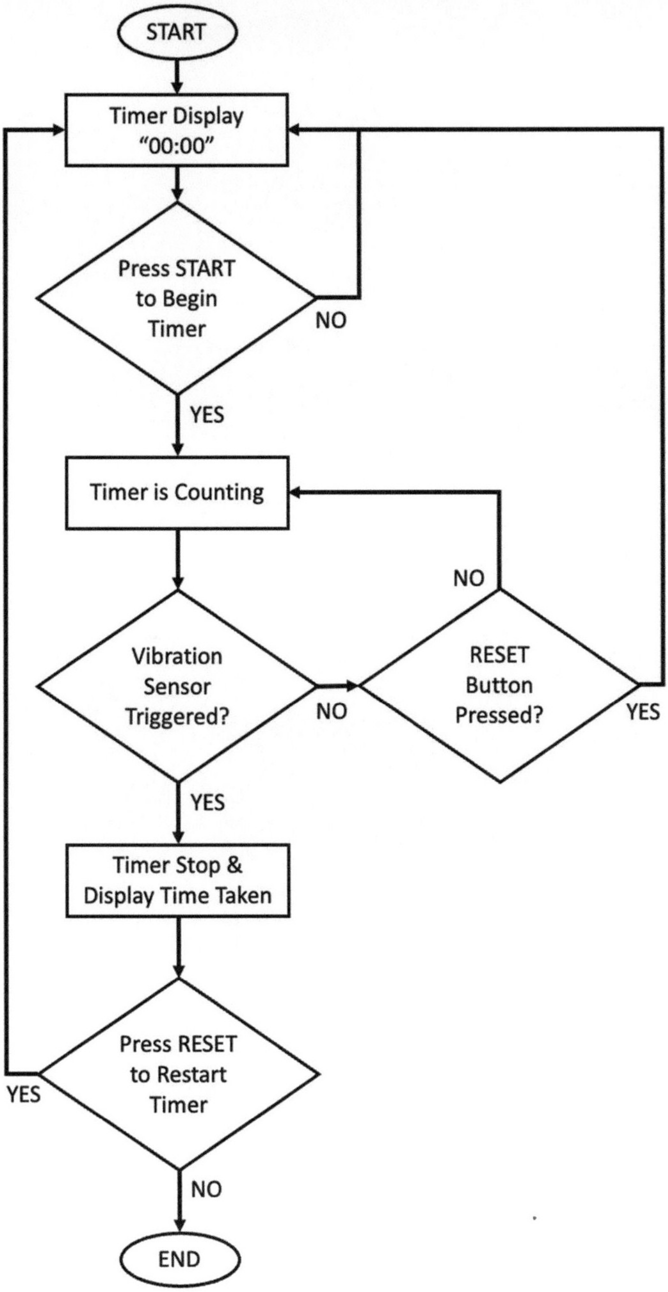

Fig. 3 Flowchart of the system for both sender (starting point) and receiver (endpoint)

Fig. 4 Control unit at the starting point, operated by the referee

Fig. 5 Sensor and display unit tailored for an Amphibian Electric Vehicle (AEV) tournament

Fig. 6 The prototype has been tested on field for an Amphibian Electric Vehicle (AEV) tournament

While the current implementation meets the immediate requirements of precision timing in sports, future enhancements could involve integration with a centralised display system or real-time data logging for post-event analysis. Additionally, exploring optimizations in power consumption and holder design could broaden the system's applicability.

4 Conclusion

In conclusion, this study aimed to develop a wireless, cost effective and auto-triggered timing gate, serving as an alternative to the traditional stopwatch method for recording time. The primary objective was to mitigate human errors associated with manual timekeeping. Through rigorous field testing, the prototype demonstrated excellent performance, exhibiting very minimal errors even after more than 50 uses on a single event.

While the current prototype has proven effective, there is still room for improvement. One potential enhancement is the integration of a wireless-triggered mechanism at the starting point. Options for achieving this include the use of infrared or Bluetooth remote control systems.

In future, an additional enhancement could involve incorporating an Internet of Things (IoT) framework for data collection and a centralised display system. This would enable the recording of tournament times, facilitating comprehensive performance analysis during or post-tournament, and providing insights for further refinement and optimization of the system.

Acknowledgements The authors would like to acknowledge Universiti Malaysia Pahang Al-Sultan

Abdullah (UMPSA) for providing financial assistance in this project under *Geran Inovasi Sosial dan Pemindahan Ilmu Komuniti* with grant number CDU230101.

References

Altmann S, Hoffmann M, Kurz G, Neumann R, Woll A, Haertel S (2015) Different starting distances affect 5-m sprint times. J Strength Cond Res, 29(8). https://doi.org/10.1519/JSC.0000000000000865

Apriantono T, Herman I, Adiprawita W, Winata B (2022) New photocell timing gates for measuring the maximum velocity of futsal players. Sport Sci Health 18(1). https://doi.org/10.1007/s11332-021-00783-2

Fasel B, Spörri J, Kröll J, Müller E, Aminian K (2019) A magnet-based timing tystem to detect gate crossings in Alpine ski racing. Sensors (Switzerland) 19(4). https://doi.org/10.3390/s19040940

Zhang HY, Li XX (2020) Analysis of the training method for the time-of-time of the movement based on the wireless. In: Lecture notes of the institute for computer sciences, social-informatics and telecommunications engineering, LNICST. https://doi.org/10.1007/978-3-030-51100-5_15

Haugen TA, Breitschädel F, Samozino P (2020) Power-force-velocity profiling of sprinting athletes: methodological and practical considerations when using timing gates. J Strength Cond Res, 34(6). https://doi.org/10.1519/JSC.0000000000002890

Johansen B, van den Tillaar R (2022) Athletics—The world's most measurable sport, but at what price? a comparison of fully automatic timing with times measured with standard mobile phones. Sensors 22(7). https://doi.org/10.3390/s22072774

Jovanović M (2023) Bias in estimated short sprint profiles using timing gates due to the flying start: simulation study and proposed solutions. Comput Methods Biomech Biomed Engin. https://doi.org/10.1080/10255842.2023.2170713

Kolaja J, Ehlerova JK (2019) Effectivity of sports timing RFID system, field study. In: 2019 IEEE international conference on RFID technology and applications, RFID-TA 2019. https://doi.org/10.1109/RFID-TA.2019.8892108

Mohd Jamil MF, Mohamed Z, Ibrahim MF, Mohd Ramli MH, Khusaini NS (2023) Prototype of IoT-based timing gate system for sports application. In: Lecture notes in bioengineering. https://doi.org/10.1007/978-981-99-0297-2_24

Ringwood J, Qian W, Fernandez JC (2015) A gate hit detection system for canoe slalom. In: 2015 26th Irish signals and systems conference, ISSC 2015. https://doi.org/10.1109/ISSC.2015.7163749

Roe G, Darrall-Jones J, Black C, Shaw W, Till K, Jones B (2017) Validity of 10-Hz GPS and timing gates for assessing maximum velocity in professional rugby union players. Int J Sports Physiol Perform, 12(6). https://doi.org/10.1123/ijspp.2016-0256

Supej M, Holmberg HC (2011) A new time measurement method using a high-end global navigation satellite system to analyze alpine skiing. Res Q Exerc Sport 82(3). https://doi.org/10.1080/02701367.2011.10599772

van den Tillaar R, Haugen ME, Falch HN (2022) A comparison of sprint mechanical parameters measured with timing gates and a laser gun. Front Sport Act Living 4. https://doi.org/10.3389/fspor.2022.877482

Weng Y (2022) Use edge detection method and matlab for time measurement in athletics field. In: ACM international conference proceeding series. https://doi.org/10.1145/3573428.3573484

Zabaloy S et al (2021) Estimation of maximum sprinting speed with timing gates: greater accuracy of 5-m split times compared to 10-m splits. Sport Biomech. https://doi.org/10.1080/14763141.2020.1838603

Integration of Sport Psychology and Performance Analysis in Behavioural Assessment During Sporting Competition

Philip Chun Foong Lew, Fawwaz Jamal, Arif Aziz, Jerry Gan, Hazwan Rahman, Musrizal bin Musanin, Sofwan Naim, Fatin Najihah Badarisam, and Hanif Hasan

Abstract The athletes' covert and overt behaviours, which include "action", "activity", "performance", "reaction", and "response", generates an amount of interest among researchers and practitioners in the field of sports. The purpose of this article is to explore the integrated insight of sport psychology and performance analysis in multiple sports, namely field hockey, badminton, football, rugby sevens, lawn bowling, and netball, during sporting events. The 5Is Behavioural Analysis process, which consisted of (1) Input, (2) Identify, (3) Insight, (4) Intervention, and (5) Impact, was established and applied by the authors through incorporating sport psychology's behavioural methodology of direct observation and measurement of actions with performance analysis. Interactive integration by each individual involved with the team was consistently conducted and optimally maintained through effective communication throughout the whole process. With the inclusion of behavioural measures, the process may eventually bridge the gap between practice and research. The potential of conducting a more comprehensive behavioural analysis on the respective opponents faced and establishing a behavioural database on the prospective reactions and responses on relevant situations may help to increase the possibility of converting previous unfavourable outcomes into potential future successes in the field of high-performance sport setting.

Keywords Sport psychology · Performance analysis · Behavioural analysis · Malaysia · High-performance

P. C. F. Lew (✉) · F. Jamal · A. Aziz · J. Gan · H. Rahman · M. Musanin · S. Naim · F. N. Badarisam · H. Hasan
High Performance Centre, National Sports Institute of Malaysia, Kuala Lumpur, Malaysia
e-mail: philiplew035@gmail.com

P. C. F. Lew
Centre for Health Research, University of Southern Queensland, Toowomba, Australia

Z. Mohamed et al. (eds.), *Proceedings of the 2nd International Conference on Innovation and Technology in Sports, ICITS 2023, 27–28 November, Malaysia*, Lecture Notes in Bioengineering, https://doi.org/10.1007/978-981-97-3741-3_51

1 Introduction

Behaviour can be technically defined as any glandular, muscular, or electrical activity of a living being (Martin and Pear 2011). It is related to anything that a person says or does. In the sport setting, behaviour of the participants, such as athletes and coaches, will include their "action", "activity", "performance", "reaction", and "response" (Luiselli and Reed 2011). It consists of overt and covert behaviours. The example of *overt behaviour*, i.e., visual, or physical responses that can be recognized by others, is shooting a ball whereas a performer feeling worried and thinking "I have not practiced enough" are examples of *covert behaviour*, i.e., internal within-brain cognition that may not readily identified by observers (Rachlin 2012). The influence of external and internal stimuli (which often known as a "cue" or "signal") may affect the individual's behaviour (Nideffer and Sagal 2006; Wilson et al. 2006). As performance is an important variable in sport (Bishop et al. 2018), integrating both subjective and objective measure (Saw et al. 2016) that can be identified, assess and track over time will be beneficial towards the goal setting process (Burton and Raedeke 2008) and overall performance setting (Cremades and Tashman 2014). Within the professional sports context, performance analysis has a number of key applications, including tactical and technical evaluation (Hughes and Bartlett 2002), analysis of movement and physical demands of both athletes and teams (Hughes 2004). Sport performance analysis can be accessed through the use of national analysis, implementing the identification and conducting analysis of relevant performance indicators and critical patterns that produced a successful outcome (Barlett 2001; Carling et al. 2009, 2014; Sarmento et al. 2013)

Nevertheless, despite the rise of behavioural sport psychology (Luiselli and Reed 2011; Martin 2015; Martin and Tkachuk 2000; Tkachuk et al. 2003) and acknowledging the importance of psychological aspects towards sporting performance, however there was insufficient resources on the impact and practical integration of performance analysis and sport psychology approaches in high-performance context.

2 Methodology

This article emphasized on the integrated insight of a sport psychology practitioner (first author) with the assigned performance analysts working collaboratively in various sports: field hockey (second, third and fourth author), badminton (fourth author), football, rugby sevens (fifth author), lawn bowling (sixth and seventh author) and netball (eighth author), during major sporting events that were held locally (e.g., South East Asian Games 2017 in Kuala Lumpur, Malaysia; Asian Champions Trophy 2016 in Kuantan, Pahang) and also internationally (e.g., 2019 Asia Pacific Lawn Bowling Championship in Gold Coast, Australia; Badminton World Federation World Championship 2023 in Copenhagen, Denmark). All the recorded footage taken during the tournaments were analyzed by video analysis software in the performance analysis field, SPORTSCODE. Notational analysis systems are used to gain insights into characteristics of athletes' performance during matches (Hughes et al. 2019).

The direct observation method was also administered concurrently by the sport psychology practitioner in a natural situation (Cohen et al. 2007; Gee and Sullivan 2006), which was conducted on match days at the competition venues. The recorded match videos were reviewed and allowed coders to stop, rewind, and pause the competitive action (Gee 2010; Kirker et al. 2000; Sheldon and Aimar 2001; Worrell 1980). It provided a more valid and reliable behavioural coding of the players to be documented.

3 Findings

Through incorporating the sport psychology's behavioural methodology of direct observation (Holder & Winter; Holder et al. 2017) and measurement of actions (Berkman et al. 2011; Berkman and Falk 2013) with the support of performance analysis (Tkachuk et al. 2003), the practitioner was capable of identifying and exploring the behavioural patterns of athletes in action during competitions.

The 5Is Behavioural Analysis process (refer Fig. 1), which comprised of (i) Input, (ii) Identify, (iii) Insight, (iv) Intervention, and (v) Impact, was being established and applied to explore *'why'* and *'how'* some behaviours emerge in performance contexts (McGarry 2009) and further understand the potential of individual and collective functional behaviour (Travassos et al. 2013) towards generating productive outcomes during sporting competitions.

To protect the integrity of the current teams and athletes' performance and administering available resources purely for an applied research study perspective, archival video recordings (Fonti et al. 2023) were selected by the first author to illustrate the examples throughout the behavioural analysis process.

3.1 Input

In this phase, relevant information and data were included through observational methodology (Sport psychology; Refer Fig. 2) and video recording (Performance

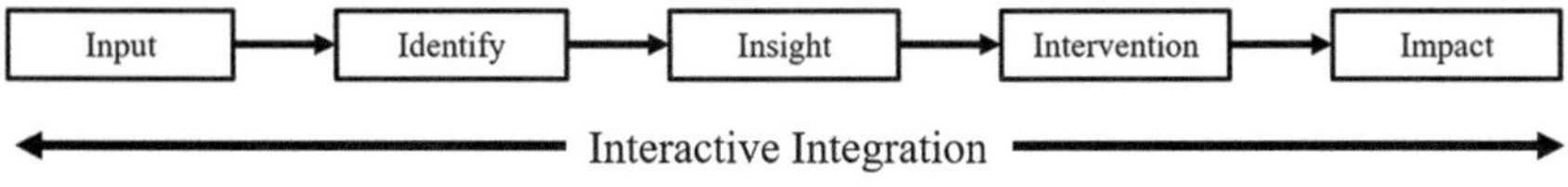

Fig. 1 The 5I behavioural analysis process

analysis; Refer Fig. 3). Match video replay can also be obtained through online platforms, such as YouTube (Refer Fig. 4). Identical to the input approach of performance analysis, the sport psychology practitioner can also capture additional audio-visual support (Refer Figs. 5 and 6) to be included in the feedback and discussion process, emphasizing on specific psychological elements, such as reaction after committing errors during competition. The inclusion archival dataset was the Malaysian men's field hockey match versus India in the 2016 Asian Champions Trophy (ACT) group stage.

Fig. 2 Sport psychology practitioner's direct observational method

Fig. 3 Match video recording of the performance analyst

Fig. 4 Match replay video on YouTube online platform

Fig. 5 Audio-visual recording support by sport psychology practitioner

3.2 *Identify*

Based on the data *'input'*, relevant behavioural patterns and individualized patterns of behaviour in response to certain competitive situations will be identified, coded and quantified by the sport psychology practitioner. Based on the archival data of the Malaysia v India match in the 2016 ACT, the Malaysian players showed great determination to challenge the higher ranked Indian team and had several good opportunities to score, forcing the Indian goalkeeper in action on more than one

Fig. 6 Example of audio-visual recording support on specific psychological aspect

occasions throughout the match. Each player played their role in supporting their team-mates (in attacking and defending) as the interaction among the players was consistently strong. In the first two quarters, the structure of the Malaysian defending area was well organized, as the Malaysian players were able to close the hot ball spot, made good tackles and stealing the ball in the process when involving in two- and three-men pressure. The Malaysian goalkeeper also made a couple of top-quality saves, especially from penalty corners. His assertiveness and dominant body language provided a presence of confidence to the team, specifically the defenders, throughout the match. Both sides traded blows through penalty corners, as India's opening goal in the first quarter was cancelled out by the Malaysian's effort in the second quarter. During the final two minutes towards the end of the match, a lack of concentration and a mixed-up in the positioning of the runners during the penalty corner routine provided India with the opportunity to score their winning goal.

3.3 *Insight*

Reference to the archival data of Malaysian's 1–2 lost to team India in the 2016 ACT group stage match, it was identified from previous phase (*'Identify'*) that several players were observed to have some heated discussion after the goal was conceded. Those behavioural patterns (i.e., showing frustration, arguments) were explored through the feedback of the players during post-match. It was understood that those undesirable reactions were due to the planned strategies not successfully being conducted, which led to the goal being conceded. Reducing tiny margins of errors during crucial moments of the match is pivotal in the high-performance setting on whether the athletes and team can successfully obtain the targeted outcome (e.g., winning the match). The players showed courage and eagerness to keep launching waves of attacks towards the Indian side and winning back balls in dangerous positions, in a bid to score the equalizer, however the opponent's ability and accuracy in making overhead passes to counter press enabled them to maintain their scoreline advantage until the end of the match.

3.4 Intervention

Upon obtaining the *'insight'* of the players on each relevant behavioural patterns, the sport psychology practitioner will work with the individuals on establishing appropriate psychological intervention and strategies, such as enhancement of focusing skills, strengthening resiliency, restructuring of disruptive self-talks, maximizing decision-making and effective application of psycho-regulation techniques during critical moments in matches. Those individualized psychological approaches will be integrated in their training programmes to enhance adaptability and acceptability in generating desirable behavioural responses.

3.5 Impact

Monitoring and evaluating the effectiveness (*'Impact'*) of the proposed intervention strategies in real-competitive situations are beneficial in identifying whether the overall behavioural patterns are moving towards the desirable direction and productive outcomes. The importance of managing the expectations and understanding that there is no one-size-fits-all approach will be integral in the possibility of prompting and shaping the interventions to become more effective towards achieving positive behavioural patterns.

As competitive outcome (i.e., game result) are among relevant indicators of the extent to which the intervention has been effective and successful, the Malaysian men's field hockey achieved the Gold medal in the 2017 Southeast Asian (SEA) Games, the runners-up position in the 2017 Men's Hockey Asia Cup and a 3–2 victory over team India in the Quarter-final match-up of the 2016–17 Hockey World League Semifinals London, which enabled the team to qualify successfully into the 2018 Men's Hockey World Cup. The team also managed to qualify to the Gold Medal match of the 2018 Asian Games by winning 3–2 over team India through a pulsating penalty shoot-out (End of regulation time score: 2–2). Despite the disappointment of not being able to obtain the Gold Medal and qualify to the 2020 Tokyo Olympic Games, the 2018 Asian Games outcome was still the best result in the nation's record at the Games.

Besides the implementation of (i) Input, (ii) Identify, (iii) Insight, (iv) Intervention, and (v) Impact, consistent *Interactive Integration* by all the individuals involved (e.g., athletes, coaches, support staffs) through effective communication is a critical component that required to be optimally maintained throughout the whole process of administering the 5I Behavioural Analysis process. It enhanced the accountability for everyone involved in the successful and effective implementation of the intervention that may lead to positive outcomes (Martin and Pear 2011).

4 Discussion

As more athletes and coaches have incorporated technology into their daily sport experiences (Lew et al. 2024), the inclusion of behavioural measures may eventually bridge the gap between practice and research in the field of elite sports context (Meredith et al. 2018). Monitoring the training process has evolved as a growing area of research in the last decades (Bourdon et al. 2017; Nässi et al. 2017), therefore the 5I Behavioural Analysis process could be implemented to convert an individual's potential into productive performance and been able to sustain quality performance into consistent outcomes. Nevertheless, more research studies and practical exercises should be conducted to enhance the generalizability of this approach in various types of sports across multicultural and multidisciplinary aspects. Including a 360° feedback of coaches, athletes, officials, support staffs, and a combination of nutritional and physiological indicators would be beneficial for the advancement of performance among elite athletes in preparation for competitions. Future application would be to conduct a more comprehensive longitudinal behavioural analysis on the respective opponents faced and establishing a behavioural database on the prospective reactions and responses on relevant situations. It would also be of interest to identify whether there is a clear relation between the players' work rate, the behaviours of the players in relation to the score-line and the perceived importance of the match.

Beyond sporting competition settings, the application of behavioural analysis could also be implemented in various conditions, such as press conferences or media interview sessions.

As reported by Adam Shergold on 26th January 2024 in Mail Online, psychologist Darren Stanton was invited to analyse the body language (Refer Fig. 7) of the outgoing Liverpool Football Club manager Jurgen Klopp's farewell interview on the Liverpool FC YouTube channel (2024).

The psychologist suggested that the manager's non-verbal communication during the interview indicated that he is tired, not just physically, but psychologically, due to lower tone, volume and pitch in his voice. He was observed to stroke his own legs a lot in the video, which is regarded as a gesture of self-assurance, like giving himself a hug to motivate himself to be courageous under combination of real sadness and disappointment, expressed through his micro-expressions of head bowed and mouth dropped.

Therefore, behavioural analysis can also be integrated into psychological approaches to identify and monitor psychological wellbeing and mental health status

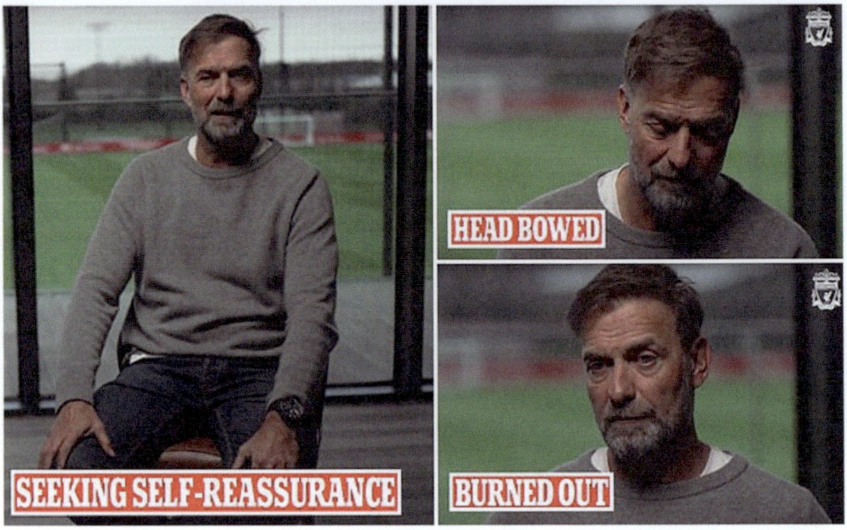

Fig. 7 Jurgen Klopp's farewell interview behavioural analysis on 26/1/2024

of individuals who are working in highly pressured environments, such as elite sports, military and competitive performance-based industry.

5 Conclusion

The integration of sport psychology's behavioural methodology and performance analysis through implementing the 5Is Behavioural Analysis process provides adequate support for practitioner to identify and explore the behavioural patterns of athletes in action during training and competitions. The applied process which consisted of (1) Input, (2) Identify, (3) Insight, (4) Intervention, and (5) Impact, may enhance the possibility of converting previous unfavourable outcomes into potential future success in high performance sporting context.

References

Arena A (2018) MAS versus INDIA|Hockey|Men's|Group Stage|Commonwealth Games 2018. YouTube. https://www.youtube.com/watch?v=1ud2N-Ch4gI

Bartlett R (2001) Performance analysis: can bringing together biomechanics and notational analysis benefit coaches? Int J Perform Anal Sport 1(1):122–126

Berkman ET, Falk EB (2013) Beyond brain mapping: using neural measures to predict real-world outcomes. Curr Dir Psychol Sci 22(1):45–50. https://doi.org/10.1177/0963721412469394/

Berkman ET, Falk EB, Lieberman MD (2011) In the trenches of real-world self-control: Neural correlates of breaking the link between craving and smoking. Psychol Sci 22:498–506

Bishop C, Turner A, Read P (2018) Effects of inter-limb asymmetries on physical and sports performance: a systematic review. J Sports Sci 36(10):1135–1144. https://doi.org/10.1080/02640414.2017.1361894

Bourdon PC, Cardinale M, Murray A, Gastin P, Kellmann M, Varley MC, Gabbett TJ, Coutts AJ, Burgess DJ, Gregson W, Bourd NT (2017) Monitoring athlete training loads: Consensus statement. Int J Sports Physiol Perform 12(Suppl. 2):161–170. https://doi.org/10.1123/ijspp.2017-0208

Burton D, Raedeke TD (2008) Sport psychology for coaches. Human Kinetics

Carling C, Reilly T, Williams AM (2009) Performance assessment for field sports. Routledge, Abingdon

Carling C, Wright C, Nelson LJ, Bradley PS (2014) Comment on 'Performance analysis in football: A critical review and implications for future research.' J Sports Sci 32(1):2–7. https://doi.org/10.1080/02640414.2013.807352

Cohen L, Manion L, Morrison KRB (2007) Research methods in education. Routledge, Oxon

Cremades JG, Tashman LS (eds) (2014) Becoming a sport, exercise, and performance psychology professional: A global perspective. Psychology Press

Fonti F, Ross JM, Aversa P (2023) Using sports data to advance management research: A review and a guide for future studies. J Manag 49(1):325–362

Gee CJ (2010) Using a direct observation methodology to study aggressive behavior in ice hockey: The good, the bad, and the ugly. J Behav Health Med 1:79–90

Gee C, Sullivan P (2006) Using a direct observation approach to study aggressive behaviour in hockey: some preliminary findings. Athl Insid: Online J Sport Psychology

Holder T, Winter S (2017) Experienced practitioners' use of observation in applied sport psychology. Sport Exerc Perform Psychol 6(1):6–19. https://doi.org/10.1037/spy0000072

Holder T, Winter S, Orr B (2017) Observation: Seeing athletes on the field. In: Taylor J (ed) Assessment in applied sport psychology. Human Kinetics, Champaign, IL, pp 115–126

Hughes MD (2004) Performance analysis—a 2004 perspective. Int J Perform Anal Sport 4:103–109

Hughes MD, Bartlett RM (2002) The use of performance indicators in performance analysis. J Sport Sci 20(10):739–754

Hughes M, Franks IM, Franks IM, Dancs H (2019) Essentials of performance analysis in sport. Routledge

Kirker B, Tennenbaum G, Mattson J (2000) An investigation of the dynamics of aggression: Direct observation in ice hockey and basketball. Res Q Exerc Sport 71:373–386

Lew PCF, Yusuf RB, How PN, Mat Nasir AB, Santhappan A, Lim JX, Vasuthevan S (2024) Integrating technology in sport psychology and application in Malaysian high-performance setting. In: Mohamed Z, Mohamad ZN, Sudin S, Ibrahim MF, Casson A (eds) Proceedings of the 2nd international conference on innovation and technology in sports, ICITS 2023, 27–28 Nov., Malaysia—innovation and technology in sports. Lecture Notes in Bioengineering. Springer, Singapore

Liverpool FC (2024) Why I've made the decision to leave liverpool|Jürgen Klopp|The Full Interview. YouTube. https://www.youtube.com/watch?v=mHYsAgAx5I4

Luiselli JK, Reed DD (eds) (2011) Behavioral sport psychology: Evidence based approaches to performance enhancement. Springer, New York. https://doi.org/10.1007/978-1-4614-0070-7

Martin J (2015) Behavior analysis in sport and exercise psychology. Behav Anal: Res Pract 15(2):148–151. https://doi.org/10.1037/bar0000018

Martin GL, Pear JJ (2011) Behavior modification: What it is and how to do it, 9th edn. Pearson-Prentice Hall, Upper Saddle River, NJ

Martin GL, Tkachuk GA (2000) Behavioral sport psychology. In: Austin J, Carr JE (eds) Behavioral sport psychology: Handbook of applied behavior analysis. Context Press, Reno, NV, pp 399–422

McGarry T (2009) Applied and theoretical perspectives of performance analysis in sport: Scientific issues and challenges. Int J Perform Anal Sport 9(1):128–140

Meredith SJ, Dicks M, Noel B, Wagstaff CRD (2018) A review of behavioural measures and research methodology in sport and exercise psychology. Int Rev Sport Exerc Psychol 11(1):25–46. https://doi.org/10.1080/1750984X.2017.1286513

Nässi A, Ferrauti A, Meyer T, Pfeiffer M, Kellmann M (2017) Psychological tools used for monitoring training responses of athletes. Perform Enhanc & Health 5(4):125–133. https://doi.org/10.1016/j.peh.2017.05.001

Nideffer RM, Sagal M (2006) Concentration and attention control training. In: Williams JM (ed) Applied sport psychology: Personal growth to peak performance, 5th edn. McGraw-Hill, New York, pp 382–403

Rachlin H (2012) Our overt behavior makes us human. The Behavior Analyst 35:49–57

Sarmento H, Pereira A, Campanico J, Anguera MT, Leitao J (2013) Soccer match analysis: A qualitative study with Portuguese first league coaches. In: Peters D, O'Donoghue P (eds) Performance analysis of sport IX. Routledge, London, pp 10–16. https://doi.org/10.4324/9780203080443

Saw AE, Main LC, Gastin PB (2016) Monitoring the athlete training response: Subjective self-reported measures trump commonly used objective measures: A systematic review. Br J Sports Med 50:281–291. https://doi.org/10.1136/bjsports-2015-094758

Sheldon JP, Aimar CM (2001) The role aggression plays in successful and unsuccessful ice hockey behaviours. Res Q Exerc Sport 72:304–309

Shergold A (2024) Head bowed, voice lowered, stroking his legs... Expert reveals what Jurgen Klopp's body language tells us during the Liverpool manager's bombshell interview. Mail Online

Tkachuk G, Leslie-Toogood A, Martin GL (2003) Behavioral assessment in sport psychology. Sport Psychol 17:104–117

Travassos B, Davids K, Araujo D, Esteves TP (2013) Performance analysis in team sports: advances from an ecological dynamics approach. Int J Perform Anal Sport 13(1):83–95. https://doi.org/10.1080/24748668.2013.11868633

Wilson VE, Peper E, Schmid A (2006) Strategies for training concentration. In Williams JM (Ed) Applied sport psychology: Personal growth to peak performance, 5th ed. McGraw-Hill, New York, pp 333–348

Worrell GL (1980) Aggression by high school basketball players: An observational study of the effects of opponents' aggression and frustration-inducing factors. Int J Sport Psychol 11:290–298

Integrating Technology in Sport Psychology and Application in Malaysian High-Performance Setting

Philip Chun Foong Lew, Rohana binti Yusuf, Peck Ngor How, Azura binti Mat Nasir, Aruna Santhappan, Jin Xuan Lim, and Sarvinna Vasuthevan

Abstract The integration of technology, such as biofeedback, neurofeedback, augmented reality, virtual reality, eye tracking, wearables and smartphones, provided an exciting new frontier in sport psychology domain. The main objective of this publication was to acknowledge and assess the conjunct technological approaches that were integrated in the applied psychology practices by the sport psychology practitioners of the National Sports Institutes of Malaysia in the high-performance setting. The focus group approach through a purposive sampling method was administered to obtain the relevant insights from all the participants. Eight components were presented based on the description of each participant with careful consideration on interpretive and representative in the difference of thoughts revealed individually and collectively. They are (i) Flip Flop Reflex Tester, (ii) concentration grid online, (iii) MUSE the brain sensing headband, (iv) In The Mood website, (v) mobile meditation application, (vi) Mood Board, (vii) behavioural analysis, and (viii) #PsychMapping. The continuous process of adapting, exploring, and integrating of emerging technologies, such as artificial intelligence, could significantly impact the integration with psychology practices in sport settings. More evidence-based studies, specifically in real-life competitive settings, could be generated on effectiveness of the respective technologies in service delivery process.

Keywords Application · Technology · Sport psychology · Malaysia · High performance

---

P. C. F. Lew (✉) · R. Yusuf · P. N. How · A. Mat Nasir · A. Santhappan · J. X. Lim · S. Vasuthevan
High Performance Centre, National Sports Institute of Malaysia, Kuala Lumpur, Malaysia
e-mail: philiplew035@gmail.com

P. C. F. Lew
Centre for Health Research, University of Southern Queensland, Toowomba, Australia

Z. Mohamed et al. (eds.), *Proceedings of the 2nd International Conference on Innovation and Technology in Sports, ICITS 2023, 27–28 November, Malaysia*, Lecture Notes in Bioengineering, https://doi.org/10.1007/978-981-97-3741-3_52

1 Introduction

According to Portenga et al. (2016), *"Performance psychology is the study and application of psychological principles of human performance to help people consistently perform in the upper range of their capabilities and more thoroughly enjoy the performance process."* (p. 6). Performance psychology is capable of helping individuals in their learning process towards performing better and more consistently in tasks where excellence is pivotal (Hays 2006, 2012). Considering sport to be a domain within performance psychology, this definition provides some insight into the unique field of sport psychology.

As stated by Aoyagi and Portenga (2010), "successful performance requires both the development and mastery of KSAs *(Knowledge, Skills and Ability)* and the capability to consistently and reliably deliver (i.e., perform) KSAs at the time of performance" (p. 254). Sport psychology comprises more than just the standard mental skills such as goal setting, relaxation, and imagery (Sebbens et al. 2012), therapeutic or clinical (Peterson et al. 2012) and competent practice can be achieved when psychological interventions are implemented within the professional philosophy and theoretical framework (Poczwardowski et al. 1998). Assessment is a key component of sport psychology support (McCann et al. 2002; Nideffer and Sagal 2001), which encompassed psychometric tests with objective measures (Chow and Gilson 2017), direct observation (Holder et al. 2017) and consultation (Taylor et al. 2017). Defined as a scientific and professional activity that consists of collecting, integrating and analyzing informational data on a particular individual (Anastasi and Urbina 1997), psychological assessment evaluates and understand the athletes to provide the practitioner with a mean in delivering quality, systematic and constructive applied sport psychology services (Taylor 2017). Through determining the fit of the assessment with the particular situation and the specific individual, it increases the possibility of obtaining a set of clearer and more interpretable results (Cid et al. 2022).

Innovative technologies are significantly influencing the sport psychology context (Giblin et al. 2016; Schack et al. 2014, 2020; Watson and Coker-Cranney 2018). Watson and Coker-Cranney (2018) highlighted that "*the incorporation of technology by sport psychology practitioners will continue to grow exponentially*" (p. 214). The development of new technologies to support sport performance (Hagan et al. 2018; Schack et al. 2014, 2020), such as biofeedback (Blumenstein and Hung 2016; Jiménez Morgan and Molina Mora 2017; Gross et al. 2018), neurofeedback (Gong et al. 2021; Rydzik et al. 2023), Vienna Test System (Ong 2015), motion-tracking, eye-tracking (Moran et al. 2018), augmented reality (Soltani and Maurice 2020) and virtual reality (Bideau et al. 2010; Bird 2020; Kittel et al. 2019; Page et al. 2019; Romeas et al. 2022), provide ways of monitoring athletes (Manley et al. 2012) and implementing more effective interventions (Siekanska et al. 2021) by integrating technology in psychological skills training (i.e., psycho-regulation, imagery). The innovative applied approach for complementary training may simulate real-life competitive situations in practice which enabled participants to experience the levels of anxiety in practice conditions during competitive action (Moran 2017). Through the implementation of technology, individuals may observe the correct execution of movements, identifying potential execution errors and generating coping strategies to rectify those issues (Strenge et al. 2020) and achieve performance optimization (Siekanska et al. 2021).

There is also a rapid rate of development and deployment of wearable technology (Watson and Coker-Cranney 2018; Windt et al. 2020), smartphones (Durand-Bush and DesClouds 2018) and low-tech wellness surveys on mobile devices (Nassi et al. 2017). Rist and Pearce (2017) highlighted the use of smartphones and emphasizing on utilizing the mindfulness smartphone application (Headspace), which verbally guide participants through the process of achieving mindfulness state, and the cognitive brain training smartphone applications (Cognifit), which consists of puzzles and games that targets on an individual's attention, working memory, and decision making. The benefits of smartphone-based mindfulness interventions' effectiveness at reducing stress and promoting well-being in athletes are also being identified in the study findings of Costalupes et al. (2021) and Shannon et al. (2019). Cotterill and Symes (2014) acknowledged that new technologies, such as smartphone applications, could potentially enhance the effectiveness of the sport psychology practitioners' service delivery through the productive sharing of information.

Among the advantages of online psychological support is promoting a sense of safety and anonymity (Price et al. 2020) and providing greater opportunities, flexibility and accessibility for individuals to engage with psychological support. The flexible nature of the devices and technology enhance capacity of integrating technology into a practitioner's consultancy approach (Cottrell et al. 2019). Nevertheless, it is highly important for a practitioner to be aware and manage the potential reliability, validity, legal and ethical considerations of integrating technology into practice (Price et al. 2020; Shei et al. 2022). Among the ethical risks associated with sport psychology support that incorporated with technology were confidentiality, the appropriateness of technology for specific services, competency, and the practitioners' own personal presence on social media sites (Bird and Harris 2019; Watson et al. 2014). By acknowledging the benefits and disadvantages of integrating technology in sport psychology practice, practitioners should consider all the possibilities and have a thoughtful approach when deciding whether it is necessary to implement technological advancement into their practice (Windt et al. 2020).

2 Methodology

The main objective of this publication was to identify and explore the commonly administered technological approaches that were integrated in their applied consultancy practice with the athletes and coaches by the sport psychology practitioners of the National Sports Institute of Malaysia in the high-performance setting.

Upon consideration, a purposive sampling method was conducted as the participants' qualification and experience in responding to the main objective was integral to the data collection process (Etikan et al. 2016). To ensure in-depth information is provided to obtain the desirable output, the focus group approach was selected as the most appropriate methodology to obtain the required information (McLafferty 2004; Sevilmis and Yıldız, 2022). The theme of this focus group was determined in accordance with the main objective of this study, which was to explore the integration and application of technology in sport psychology in Malaysian high-performance setting. In line with the five stages categories of focus group questioning suggested in Krueger (1997) on (i) opening questions, (ii) introductory questions, (iii) transition questions, (iv) key questions, and (v) ending questions, the moderator (i.e., first author) lead the groups of participants (i.e., second to sixth authors) to interact and respond on the specific theme as presented above. Each individual was provided ample opportunity to interact at an individual and group level on sharing their respective insight that was relevant to the theme (Michell 1999) during the discussion process. Upon the completion of the focus group process, appropriate categories and relevant coding on the information provided by the participants was assigned to achieve the objective of the study (Merriam and Tisdell 2015). All the data was identified, classified, synthesized, and harmonized with the correct category (Charmaz and Belgrave 2007).

3 Findings

With a careful consideration on interpretive and representative in the difference of thoughts revealed individually and as a group (Kitto et al. 2008), based on the insight provided by each of the participants on the integration of technology in sport psychology consultation and their respective application in Malaysian high-performance setting, eight relevant components were presented and described as follows:

3.1 Flip Flop Reflex Tester

This application was designed by Akkuzu Games, which is available for Android and iOS users, to access the reaction time of an individual, which is a measure of the quickness with which an organism responds to a particular stimulus, participants are required to select the lighted dots as swiftly as possible (refer Fig. 1). The total amount of time used will be presented upon the successful completion of the assessment. However, if the participant incorrectly selected the stimulus during the assessment, they are required to replay the whole session. Levels of difficulties are categorized into easy (3×3), normal (4×4) and hard (5×5). The speed and effectiveness of making decisions and initiating action may be influenced by an athlete's ability to react accordingly (Schmidt and Wrisberg 2008).

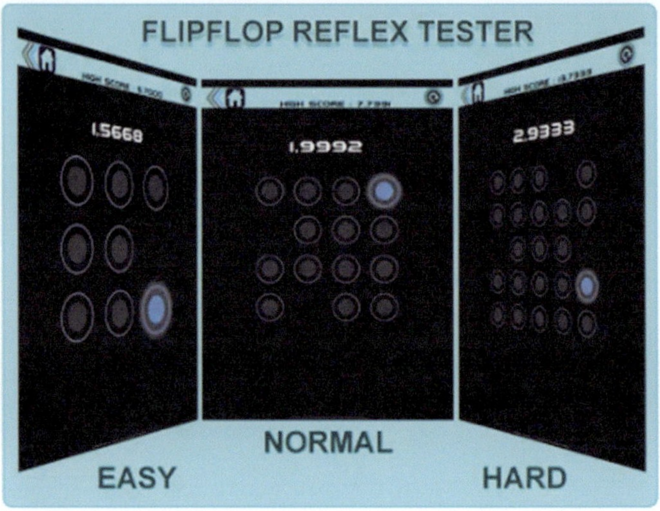

Fig. 1 Flip reflex tester application interface

3.2 *Concentration Grid Online*

The concentration grid exercise (Harris and Harris 1984) is traditionally used as an assessment of cognitive performance and effective in enhancing concentration (Greenlees et al. 2006; Schmid and Peper 1998; Weinberg 1989). The exercise consists of a 10 × 10 numerical grid, in which numbers from 00 to 99 are placed randomly into the 100 cells (Weinberg and Gould 1999) and respondents are required to check off numbers in numerical order with the aim of identifying as many numbers as possible in one minute. It was proposed that the exercise could develop the athlete's ability to ignore irrelevant distracting stimuli (Harris and Harris 1984) and enhance visual scanning for relevant information (Gill 2000). Moran (2017) reported that successful athletes were observed to be able of maintaining higher level of task-oriented focus in decision making and less distracted by irrelevant stimuli. The Concentration Grid Online, which could be accessed through the website link (http://zzzscore.com/1to50/en/#*)*, is a web/mobile device implementation of the exercise. Participants are requested to select in sequence each of the numbers in the randomly generated grid, from 1 to 50 (refer Fig. 2) and their performance time was presented during the completion of the exercise.

Based on findings of an applied study that was presented in the 5th International Seminar on Sports and Exercise Psychology 2018 (Lew 2018), the practitioner (first author) integrated the Flip Flop Reflex Tester and the Concentration Grid Online in the sport psychology training module for the national athletes towards their successful campaign in the 2017 Kuala Lumpur SEA Games. Through the implementation of technological application with sport psychology approaches, it was identified that

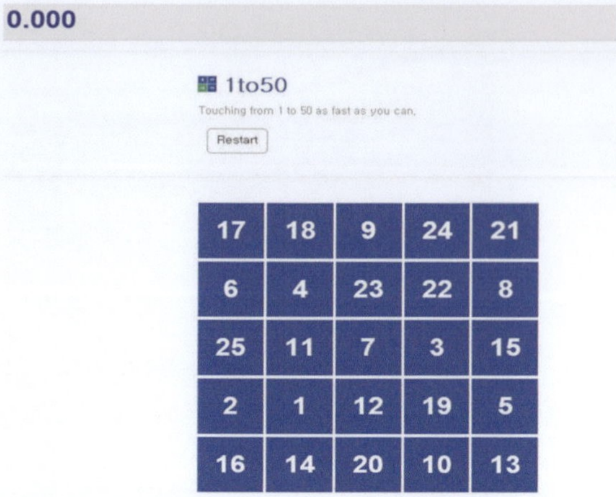

Fig. 2 Concentration grid online interface

athletes enjoyed the creative integrative practice which helped to enhance concentration, managing distraction, performing under pressure and better decision making in competitive elite sports situation.

3.3 *MUSE the Brain Sensing Headband*

The MUSE headband (refer Fig. 3) measures brain activity through electroencephalography (EEG) sensors and transmits the signal through Bluetooth to the participant's mobile device, which will be downloaded with an accompanying mobile app that automatically converts the EEG signal into brain state feedback (Baig 2014). The device distinguishes between how long you were "calm" (your mind was focused on the exercise), "active" (your mind wandered) or "neutral" (somewhere in the middle). The exercise (refer Fig. 4), which includes options such as Body, Breath, Heart, and Mind Meditation, varies in length (as selected by the user) and has been reported to reduce stress, anxiety and showed improvement in productivity and focus (Arsalan et al. 2019; Arsalan and Majid 2022; Krigolson et al. 2017). Upon the completion of the exercise, the MUSE will graph your data into scores (refer Fig. 5), indicating how the participant did in providing real-time performance feedback.

3.4 *In the Mood Website*

The Brunel Mood Scale (BRUMS; Terry and Lane 2010) comprises of six subscales (i.e., tension, depression, anger, vigour, fatigue, and confusion) of four items each. A Malay-language version of the BRUMS, known as the Malaysian Mood Scale (MASMS) was validated by the first author (Lew et al. 2022, 2023). Respondents

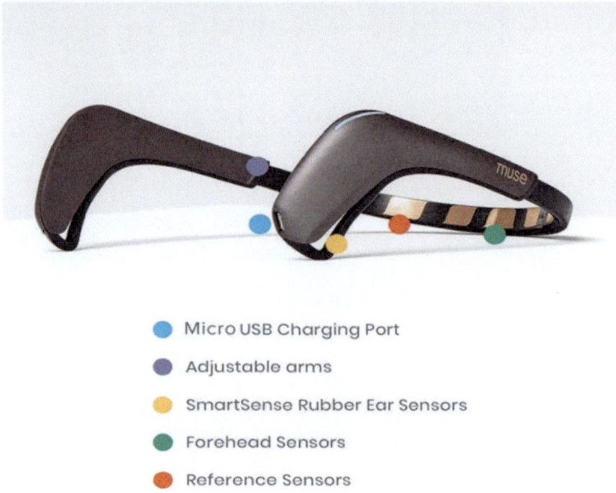

Fig. 3 MUSE headband device

can identify their mood states by completing the assessment form (refer Fig. 6) on the online mood profile website (www.moodprofiling.com), which was developed by Terry and Lim (2011). It generates an automated report (refer Fig. 7) on the interpretation of individual mood responses to the BRUMS (Terry and Lane 2010), which includes raw and standard scores, referring to the normative scores. A visual representation of the individualized mood profile and evidence-based mood regulation strategies (refer Fig. 8) is also being presented in the report (Terry et al. 2013, 2021).

3.5 Mobile Meditation Application

There is a significant rise in mobile-based mindfulness and meditation applications on the market. With an effectiveness at reducing stress and well-being promotion in athletes (Costalupes et al. 2021; Shannon et al. 2019), the Headspace app (refer Fig. 9) represents one of the most commonly used mindfulness apps with over 3.3 million subscribers and 70 million downloads at the time of writing (Curry 2023). It is also the most peer-reviewed mindfulness app in the efficacy of online-based mindfulness training (Jayewardene et al. 2017; Lim et al. 2015) and scored highest on engagement, functionality, and information quality (Mani et al. 2015). Therefore, *Headspace* was identified as a quality complementary psychological tool for the practitioners.

3.6 Mood Board

The Mood Board (Geh et al. 2023), which was co-developed by the first author to be administered in special need student's cohort, comprised of the psychological

Fig. 4 MUSE mind session

construct of the Malaysian Mood Scale (MASMS; Lew et al. 2022, 2023), Malaysian Emotion Regulation Questionnaire (M-ERQ; Lew et al. 2021), and Malaysian Perceived Stress Scale (M-PSS; Lew et al. 2021). The MASMS, M-ERQ and M-PSS are Malay-language translated and validated assessment tools of the Brunel Mood Scale (BRUMS; Terry and Lane 2010), Emotion Regulation Questionnaire (ERQ; Gross and John 2003) and Perceived Stress Scale (PSS; Cohen et al. 1983).

The core elements of the Mood Board which consists of mood, emotional regulation, and perceived stress were presented through:

Fig. 5 MUSE score

(i) The colours of red, yellow, and green using the traffic light analogy (refer Fig. 10)
(ii) Three emotions icons consist of smiling face with smiling eyes, anxious face, weary face (refer Fig. 11)
(iii) Numbering from one (1) to (5) with ascending order in sizes (refer Fig. 12).

The implementation of the Mood Board can be conducted physically and virtually, through online assessment template that participants can complete on their mobile devices.

3.7 Behavioural Analysis

Through incorporating the sport psychology's behavioural methodology of direct observation (Holder et al. 2017) and measurement of actions (Berkman and Lieberman 2011) with performance analysis (Tkachuk et al. 2003), the practitioner was capable of identifying and exploring the behavioural patterns of athletes in action during training and competitions through implementing the 5Is Behavioural Analysis process that was suggested in Lew et al. (2024). The applied process (refer Fig. 13) consisted of (1) Input, i.e., relevant information and data were included through observational (sport psychology) and video recording (performance analysis), (2) Identify, i.e., coding relevant behavioural patterns, (3) Insight, i.e., exploring the selected behavioural patterns, (4) Intervention, i.e., establishing appropriate intervention strategies such as enhancement of focusing skills and application of psychoregulation techniques during critical moments in matches, and (5) Impact, i.e., evaluating the effectiveness of the intervention strategies in real-competitive situations.

In The Mood

An online mood assessment based on the Brunel Mood Scale (BRUMS)

The Brunel Mood Scale (BRUMS)

Please take a moment to tell us a little bit about yourself.

Gender ○ Male ○ Female Age [Please select one]

Occupation [Please select one]

Ethnicity [Please select one]

Highest Education Achieved [Please select one]

Below is a list of words that describe feelings. Please read each one carefully. Then select the option that best describes *how you feel right now*. Make sure you answer every question.

1. Panicky	○ Not At All	○ A Little	○ Moderately	○ Quite A Lot	○ Extremely
2. Lively	○ Not At All	○ A Little	○ Moderately	○ Quite A Lot	○ Extremely
3. Confused	○ Not At All	○ A Little	○ Moderately	○ Quite A Lot	○ Extremely
4. Worn Out	○ Not At All	○ A Little	○ Moderately	○ Quite A Lot	○ Extremely
5. Depressed	○ Not At All	○ A Little	○ Moderately	○ Quite A Lot	○ Extremely
6. Downhearted	○ Not At All	○ A Little	○ Moderately	○ Quite A Lot	○ Extremely
7. Annoyed	○ Not At All	○ A Little	○ Moderately	○ Quite A Lot	○ Extremely
8. Exhausted	○ Not At All	○ A Little	○ Moderately	○ Quite A Lot	○ Extremely
9. Mixed-Up	○ Not At All	○ A Little	○ Moderately	○ Quite A Lot	○ Extremely
10. Sleepy	○ Not At All	○ A Little	○ Moderately	○ Quite A Lot	○ Extremely
11. Bitter	○ Not At All	○ A Little	○ Moderately	○ Quite A Lot	○ Extremely
12. Unhappy	○ Not At All	○ A Little	○ Moderately	○ Quite A Lot	○ Extremely
13. Anxious	○ Not At All	○ A Little	○ Moderately	○ Quite A Lot	○ Extremely
14. Worried	○ Not At All	○ A Little	○ Moderately	○ Quite A Lot	○ Extremely
15. Energetic	○ Not At All	○ A Little	○ Moderately	○ Quite A Lot	○ Extremely
16. Miserable	○ Not At All	○ A Little	○ Moderately	○ Quite A Lot	○ Extremely
17. Muddled	○ Not At All	○ A Little	○ Moderately	○ Quite A Lot	○ Extremely
18. Nervous	○ Not At All	○ A Little	○ Moderately	○ Quite A Lot	○ Extremely
19. Angry	○ Not At All	○ A Little	○ Moderately	○ Quite A Lot	○ Extremely
20. Active	○ Not At All	○ A Little	○ Moderately	○ Quite A Lot	○ Extremely
21. Tired	○ Not At All	○ A Little	○ Moderately	○ Quite A Lot	○ Extremely
22. Bad Tempered	○ Not At All	○ A Little	○ Moderately	○ Quite A Lot	○ Extremely
23. Alert	○ Not At All	○ A Little	○ Moderately	○ Quite A Lot	○ Extremely
24. Uncertain	○ Not At All	○ A Little	○ Moderately	○ Quite A Lot	○ Extremely

Please indicate why you are completing this mood assessment. [Please select one]

Submit

Share:

© 2021 P Terry, J Lim, & R Parsons-Smith. Design by Clint Mallet. All Rights Reserved.

Fig. 6 In the mood online website assessment form

Consistent Interactive Integration by all the individuals involved through effective communication is a critical component that is required to be optimally maintained throughout the whole process. With the inclusion of behavioural measures, it may eventually bridge the gap between practice and research (Meredith et al. 2017).

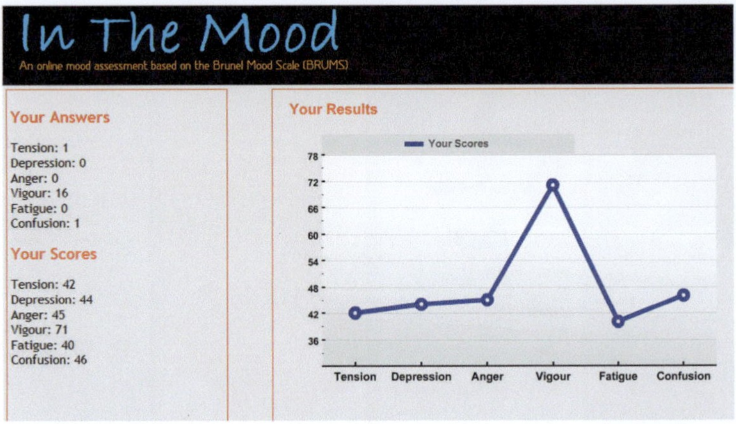

Fig. 7 In the mood automated report template

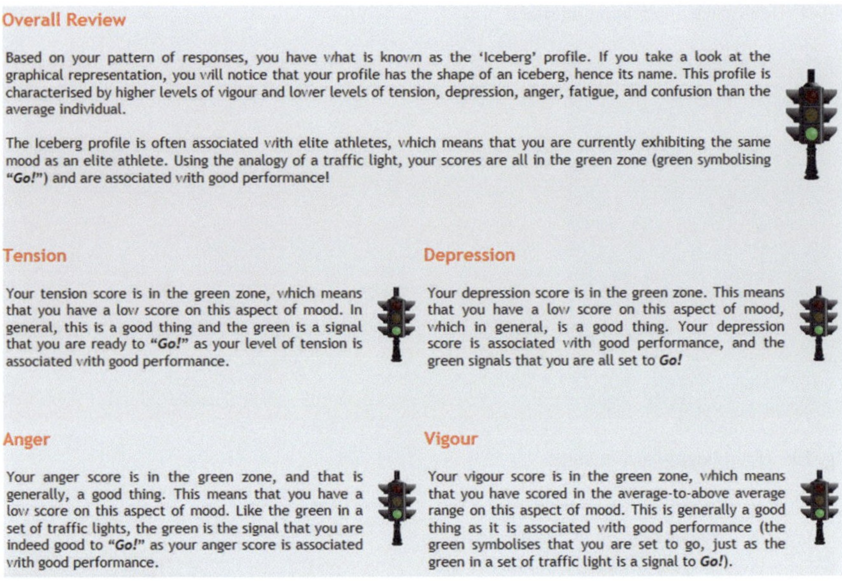

Fig. 8 In the mood suggested mood regulation strategies

3.8 #PsychMapping

The #PsychMapping is an exploratory interview framework by developed by Latinjak et al. (2021) that identifies the psychological reality of individuals and collectives through identification of external variables, personal descriptors, and psychological skills. Practitioners can note keywords and phrases based on the responses of

Fig. 9 Headspace application interface

Fig. 10 Mood board—traffic light

Fig. 11 Mood board—emotion icons

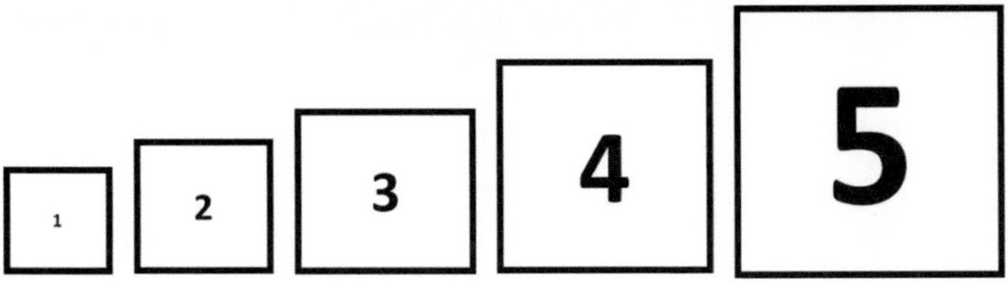

Fig. 12 Mood board—stress numbering

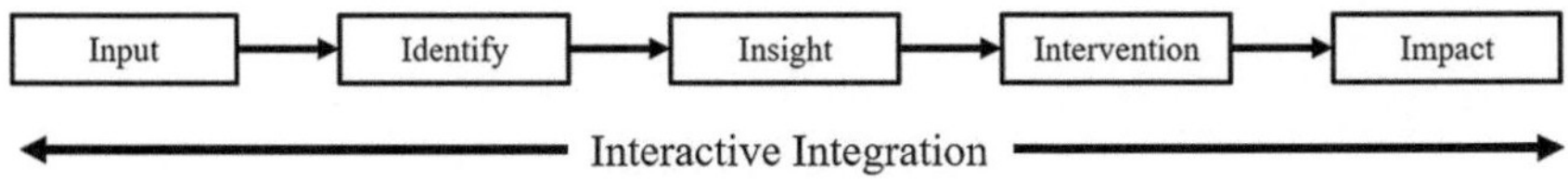

Fig. 13 The 5I behavioural analysis process

the clients (or also known as interviewee) on an empty layout of the map (refer Fig. 14). The process can be completed physically and virtually, through online Microsoft Powerpoint templates. Findings from an applied study (Lew and Latinjak 2022) highlighted that #PsychMapping can be a beneficial monitoring equipment in the sport psychology practice in elite sports setting and capable of generating individualized graphic representation with tangible information. The implementation of #PsychMapping was acknowledged as a beneficial addition in complementing applied performance psychology practices in the constructive process of self-monitoring and achieving better self-awareness towards managing influential performance factors.

4 Discussion

Each of the components presented through the findings from the focus group has its unique capacity to complement the integration of technology with sport psychology practice.

Nevertheless, from a psychological standpoint on better supporting the client, it is important for individuals to acknowledge that interpersonal skills are desirable characteristics to become a competent sport psychology practitioner (Thelwell et al. 2018; Woolway and Harwood 2019, 2020) in gaining an understanding into the client's emotions, feelings, thoughts, and behaviours (Harris and Birnbaum 2015).

Each consultant will have idiosyncratic preferences on which approaches and methods that works best for them (Cotterill and Symes 2014), therefore they should always have a clear understanding on what they are aiming to emphasize on, what are the requirement needed to achieve them and whether there are existing technology resources that can provide better support to their services.

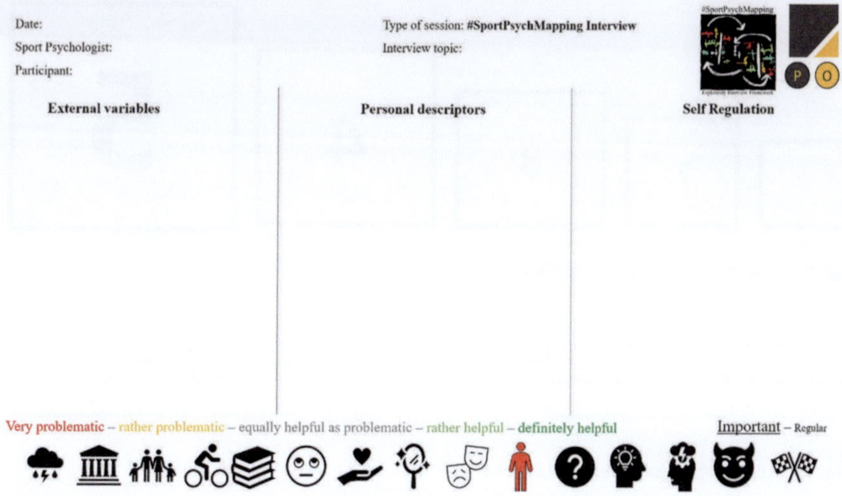

Fig. 14 Individualized #PsychMapping template

As highlighted in Grady (2023), the ongoing process of adapting, studying, and integrating emerging technologies, such as artificial intelligence, with sport psychology approaches could possibly enhance the impact of psychology in high-performance sport settings. The emerging of natural language model-based Artificial Intelligence, such as Chat Generative Pre-Trained Transformer (ChatGPT; Anderson et al. 2023; Dergaa et al. 2023; Szabo 2023) could have a major impact on the future potential technology application in the field of sport psychology that require extensive research and validation.

This publication provides multiple practical technological approaches that are capable of being integrated with respective sport psychology practices working with athletes and coaches. Nevertheless, the reliability, validity, legal and ethical considerations of their respective application (Bird and Harris 2019; Price et al. 2020; Shei et al. 2022; Watson et al. 2014) need to be recognized and considered in an individual and collective manner by the practitioner. Future research should further expand on generating evidence-based studies on effectiveness of the respective technologies in service delivery process, specifically in real-life competitive settings. The perception of the utility, differences in characteristics among sporting events, ease of use, methods of use, challenges in application, individualized status and specific demands would also be interesting topics to be explored and emphasized on to have a deeper understanding of their prospective impact.

5 Conclusion

As a conclusion, despite the overwhelming interest and growing use of technology, a competent practitioners should always be equipped with optimal interpersonal skills, a clear understanding of their objectives and being able to efficiently navigate the challenging equilibrium of associated benefits and limitations when integrating technology in the provision of sport psychology to maximize the productivity of the consulting process.

References

Anastasi A, Urbina S (1997) Psychological testing, 7th edn. Prentice Hall, Hoboken, NJ

Anderson N, Belavy DL, Perle SM, Hendricks S, Hespanhol L, Verhagen E, Remon AR (2023) AI did not write this manuscript, or did it? Can we trick the AI text detector into generated texts? The potential future of ChatGPT and AI in sports & exercise medicine manuscript generation. BMJ Open Sport Exerc Med 9:e001568. https://doi.org/10.1136/bmjsem-2023-001568

Aoyagi MW, Portenga ST (2010) The role of positive ethics and virtues in the context of sport and performance psychology service delivery. Prof Psychol Res Pract 41:253–259

Arsalan A, Majid M (2022) A study on multi-class anxiety detection using wearable EEG headband. J Ambient Intell Humaniz Comput 13(12):5739–5749

Arsalan A, Majid M, Butt AR, Anwar SM (2019) Classification of perceived mental stress using a commercially available EEG headband. IEEE J Biomed Health Inform 23(6):2257–2264

Baig EC (2014) Brainy muse headband: Wearable tech to calm you down. USA Today. https://www.usatoday.com/story/tech/columnist/baig/2014/08/12/muse-headband-wearable-tech-to-calm-you-down/13950367/

Berkman ET, Lieberman MD (2011) What's outside the black box?: The status of behavioural outcomes in neuroscience research. Psychol Inq 22:100–107

Bideau B, Kulpa R, Vignais N, Brault S, Multon F, Craig C (2010) Using virtual reality to analyse sports performance. IEEE Comput Graphics Appl 30:14–21

Bird JM (2020) The use of virtual reality head-mounted displays within applied sport psychology. J Sport Psychol Action 11(2):115–128. https://doi.org/10.1080/21520704.2018.1563573

Bird MD, Harris BS (2019) The ethical use of technology for clinical and performance enhancement services: Prevalence and perceptions among association for applied sport psychology certified consultants. J Clin Sport Psychol 13(1):56–71. https://doi.org/10.1123/jcsp.2017-0034

Blumenstein B, Hung ET-M (2016) Biofeedback in sport. In: Schinke RJ, McGannon KR, Smith B (eds) Routledge international handbook of sport psychology. Routledge/Taylor & Francis Group, pp 429–438

Charmaz K, Belgrave LL (2007) Grounded theory (2007). The blackwell encyclopedia of sociology. The Blackwell encyclopedia of sociology. Clinical Assessment of Malingering and Deception, Guilford, London. https://doi.org/10.1002/9781405165518.wbeosg070.pub2

Chow GM, Gilson TA (2017) Inventories: Using objective measures. In Taylor J (ed) Assessment in applied sport psychology. Human Kinetics, Champaign, IL, pp 83–99

Cid L, Monteiro D, Teixeira DS, Evmenenko A, Andrade A, Bento T, Vitorino A, Couto N, Rodrigues F (2022) Assessment in sport and exercise psychology: considerations and recommendations for translation and validation of questionnaires. Front Psychol 13:806176. https://doi.org/10.3389/fpsyg.2022.806176

Cohen S, Kamarck T, Mermelstein R (1983) A global measure of perceived stress. J Health Soc Behav 24(4):385–396. https://doi.org/10.2307/2136404

Costalupes B, Gilbert JN, Gilbert W, Coles MG (2021) A smartphone mindfulness-based intervention pilot study with competitive high school baseball players. J Kinesiol & Wellness, 9:63–72. https://doi.org/10.56980/jkw.v9i.76

Cotterill ST, Symes R (2014) Integrating social media and new technologies into your practice as a sport psychology consultant. Sport Exerc Psychol Rev 10(1):55–64

Cottrell C, McMillen N, Harris BS (2019) Sport psychology in a virtual world: Considerations for practitioners working in eSports. J Sport Psychol Action 10(2):73–81. https://doi.org/10.1080/21520704.2018.1518280

Curry (2023) Headspace revenue and usage statistics. Bus Apps. Available at https://www.businessofapps.com/data/headspace-statistics/ (Accessed 31 Oct 2023)

Dergaa I, Chamari K, Zmijewski P, BenSaad H (2023) From human writing to artificial intelligence generated text: examining the prospects and potential threats of ChatGPT in academic writing. Biol Sport 40(2):615–622. https://doi.org/10.5114/biolsport.2023.125623

Durand-Bush N, DesClouds P (2018) Smartphones: How can mental performance consultants help athletes and coaches leverage their use to generate more benefits than drawbacks? J Sport Psychol Action 9(4):227–238. https://doi.org/10.1080/21520704.2018.1496211

Etikan I, Musa SA, Alkassim RS (2016) Comparison of convenience sampling and purposive sampling. Am J Theor Appl Stat 5(1):1–4

Geh CL, Lew PCF, Mohamad SB (2023) Mood board: The psychological wellbeing assessment for special needs students in classroom setting. In: Proceedings of international conference on special education, vol 5. Retrieved from https://publication.seameosen.edu.my/index.php/icse/article/view/359

Giblin G, Tor E, Parrington L (2016) The impact of technology on elite sports performance. SensIa: J Mind Brain Cult,12(2):3–9. https://doi.org/10.7790/sa.v12i2.436

Gill DL (2000) Psychological dynamics of sport and exercise, 2nd edn. Human Kinetics, Champaign, IL

Gong A, Gu F, Nan W, Qu Y, Jiang C, Fu Y (2021) A review of neurofeedback training for improving sport performance from the perspective of user experience. Front Neurosci 15:638369. https://doi.org/10.3389/fnins.2021.638369

Grady B (2023) The role of information technology in enhancing sport psychology interventions for athlete development and performance optimization. Rev Psicol Del Deport (j Sport Psychol) 32(4):41–50

Greenlees I, Thelwell R, Holder T (2006) Examining the efficacy of the concentration grid exercise as a concentration enhancement exercise. Psychol Sport Exerc 7:29–39. https://doi.org/10.1016/j.psychsport.2005.02.001

Gross JJ, John OP (2003) Individual differences in two emotion regulation processes: Implications for affect, relationships, and well-being. J Pers Soc Psychol 85:348–362

Gross MJ, Bringer JD, Kilduff LP, Cook CJ, Hall R, Shearer DA (2018) A multimodal biofeedback protocol to demonstrate physiological manifestations of psychological stress and introduce heart rate variability biofeedback stress management. J Sport Psychol Action 9(4):216–226. https://doi.org/10.1080/21520704.2018.1496210

Hagan JE Jr, Schack T, Koester D (2018) Passion play. Embracing new scientific perspectives for improved sport psychology consulting. SOJ Psychology 4:1–5. https://doi.org/10.15226/2374-6874/5/1/00143

Harris B, Birnbaum R (2015) Ethical and legal implications on the use of technology in counselling. Clin Soc Work J 43(2):133–141. https://doi.org/10.1007/s10615-014-0515-0

Harris DV, Harris BL (1984) The athlete's guide to sport psychology: Mental skills for physical people. Leisure Press, New York

Hays KF (2006) Being fit: the ethics of practice diversification in performance psychology. Prof Psychol: Res Pract 37(3):223–232. https://doi.org/10.1037/0735-7028.37.3.223

Hays KF (2012) The psychology of performance in sport and other domains. In: Murphy SM (ed) The Oxford handbook of sport and performance psychology. Oxford University Press, New York, pp 24–45. https://doi.org/10.1093/oxfordhb/9780199731763.013.0002

Hays KF, Lesyk JJ (2014) Incorporating sport and exercise psychology into clinical practice. In: Van Raalte JL, Brewer BW (eds) Exploring sport and exercise psychology, 3rd edn. American Psychological Association, Washington, DC, pp 485–504

Holder T, Winter S, Orr B (2017) Observation: Seeing athletes on the field. In: Taylor J (ed) Assessment in applied sport psychology. Human Kinetics, Champaign, IL, pp 115–126

Jayewardene WP, Lohrmann DK, Erbe RG, Torabi MR (2017) Effects of preventive online mindfulness interventions on stress and mindfulness: A meta-analysis of randomized controlled trials. Prev Med Rep 5:150–159. https://doi.org/10.1016/j.pmedr.2016.11.013

Jiménez Morgan S, Molina Mora JA (2017) Effect of heart rate variability biofeedback on sport performance, a systematic review. Appl Psychophysiol Biofeedback 42:235–245. https://doi.org/10.1007/s10484-017-9364-2

Kittel A, Larkin P, Elsworthy N, Spittle M (2019) Using 360° virtual reality as a decision making assessment tool in sport. J Sci Med Sport 22:1049–1053

Kitto SC, Chesters J, Grbich C (2008) Quality in qualitative research. Med J Aust 188(4):243–246

Krigolson OE, Williams CC, Norton A, Hassall CD, Colino FL (2017) Choosing MUSE: validation of a low-cost, portable EEG system for ERP research. Front Neurosci 11:109. https://doi.org/10.3389/fnins.2017.00109

Krueger RA (1997) Developing questions for focus groups. Sage Press, London

Latinjak AT, Morelló-Tomás E, Figal-Gómez L (2021) #SportPsychMapping: An exploratory interview framework for sport and exercise psychology. Sport Psychol 35(3):240–249

Lew PCF, Latinjak AT (2022) #PsychMapping: an exploratory approach of applied performance psychology practices in Malaysia. In: 9th Asian South Pacific association of sport psychology international congress, Kuching, Sarawak, Malaysia

Lew PCF, How PN, Akmal S, Chong SK (2021) Validation of the Malaysian Emotion Regulation Questionnaire (M-ERQ) in sports & general population. In: 4th Malaysian international psychology conference, Virtual, Malaysia

Lew PCF, How PN, Akmal S, Chong SK (2021) Malaysian Perceived Stress Scale (M-PSS): psychometric characteristics among malaysian sport athletes & general population. In: International sport, health and emerging technologies summit conference, Virtual, Malaysia

Lew PCF, Parsons-Smith RL, Lamont-Mills A, Quartiroli A, Terry PC (2022) Development and validation of the Malaysian mood scale. FEPSAC 2022. In: 16th European congress of sport & exercise psychology abstract book, vol 541

Lew PCF, Parsons-Smith RL, Lamont-Mills, Terry PC (2023) Cross-cultural validation of the malaysian mood scale and tests of between-group mood differences. Int J Environ Res Public Health 20(4):3348. https://doi.org/10.3390/ijerph20043348

Lew PCF, Jamal F, Aziz A, Gan J, Rahman H, Musanin MB, Naim S, Badarisam FN, Hasan H (2024) The integration of sport psychology & performance analysis in behavioural assessment during sporting competition. In: Mohamed Z, Mohamad ZN, Sudin S, Ibrahim MF, Casson A (eds) Proceedings of the 2nd international conference on innovation and technology in sports, ICITS 2023, 27–28 Nov., Malaysia—innovation and technology in sports. Lecture notes in bioengineering. Springer, Singapore

Lew PCF (2018) Decision making in pressured competitive situation: a study from the field of elite sports in Malaysia. In: 5th international seminar on sports and exercise psychology (2018) Kuching. Sarawak, Malaysia

Lim D, Condon P, DeSteno D (2015) Mindfulness and compassion: An examination of mechanism and scalability. PLoS ONE 10(2):e0118221. https://doi.org/10.1371/journal.pone.0118221

Mani M, Kavanagh DJ, Hides L, Stoyanov SR (2015) Review and evaluation of mindfulness-based iPhone apps. J Med Internet Res Mhealth Uhealth 3(3):e82. https://doi.org/10.2196/mhealth.4328

Manley A, Palmer C, Roderick M (2012) Disciplinary power, the oligopticon and rhizomatic surveillance in elite sports academies. Surveill & Soc,10(3/4):303–319. https://doi.org/10.24908/ss.v10i3/4.4281

McCann SC, Jowdy DP, Van Raalte JL (2002) Assessment in sport and exercise psychology. In Van Raalte JL, Brewer BW (eds) Exploring sport and exercise psychology. American Psychological Association, pp 291–305. https://doi.org/10.1037/10465-014

McLafferty I (2004) Focus group interviews as a data collecting strategy. J Adv Nurs 48(2):187–194. https://doi.org/10.1111/j.1365-2648.2004.03186.x

Meredith SJ, Dicks M, Noel B, Wagstaff CRD (2017) A review of behavioural measures and research methodology in sport and exercise psychology. Int Rev Sport Exerc Psychol. https://doi.org/10.1080/1750984X.2017.1286513

Merriam SB, Tisdell EJ (2015) Qualitative research: A guide to design and implementation. John Wiley & Sons Press, Hoboken

Michell L (1999) Combining focus groups and interviews: telling how it is; telling how it feels. Sage Press, London

Moran A (2017) Attention and concentration training in sport. Ref Modul Neurosci Biobehav Psychol. https://doi.org/10.1016/B978-0-12-809324-5.05476-6

Moran A, Campbell M, Ranieri D (2018) Implications of eye tracking technology for applied sport psychology. J Sport Psychol Action 9(4):249–259. https://doi.org/10.1080/21520704.2018.1511660

Moran A, Toner J (2017) A critical introduction to sport psychology: A critical introduction. Taylor & Francis

Nassi A, Ferrauti A, Meyer T, Pfeiffer M, Kellmann M (2017) Psychological tools used for monitoring training responses of athletes. Perform Enhanc & Health 5(4):125–133. https://doi.org/10.1016/j.peh.2017.05.001

Nideffer R, Sagal M (2001) Assessment in sport psychology. Fitness Information Technology, Morgantown

Ong NCH (2015) The use of the Vienna test system in sport psychology research: a review. Int Rev Sport Exerc Psychol 8:1–20. https://doi.org/10.1080/1750984X.2015.1061581

Pagé C, Bernier P-M, Trempe M (2019) Using video simulations and virtual reality to improve decision-making skills in basketball. J Sports Sci 37(21):2403–2410. https://doi.org/10.1080/02640414.2019.1638193

Peterson K, Brown C, McCann S, Murphy SM (2012) Sport and performance psychology: A look ahead. In: Murphy SM (ed) The Oxford handbook of sport and performance psychology. Oxford University Press, New York, pp 741–753. https://doi.org/10.1093/oxfordhb/9780199731763.013.0040

Poczwardowski A, Sherman CP, Henschen KP (1998) A sport psychology service delivery heuristic: Building on theory and practice. Sport Psychol 12:191–207

Portenga ST, Aoyagi MW, Cohen AB (2016) Helping to build a profession: A working definition of sport and performance psychology. J Sport Psychol Action. https://doi.org/10.1080/21520704.2016.1227413

Price D, Wagstaff CRD, Thelwell RC (2020) Opportunities and considerations of new media and technology in sport psychology service delivery. J Sport Psychol Action. https://doi.org/10.1080/21520704.2020.1846648

Rist B, Pearce AJ (2017) Improving athlete mental training engagement using smartphone phone technology. Int J Soc Sci Humlty 7(3):138–142. http://www.ijssh.org/vol7/809-PS0008.pdf. https://doi.org/10.18178/ijssh.2017.V7.809

Romeas T, More-Chevalier B, Charbonneau M, Bieuzen F (2022) Virtual-reality training of elite boxers preparing for the Tokyo 2020 Olympics during the COVID-19 pandemic: A case study. Case Stud Sport Exerc Psychol 6:21–35

Rydzik Ł, Wąsacz W, Ambroży T, Javdaneh N, Brydak K, Kopańska M (2023) The use of neurofeedback in sports training: systematic review. Brain Sci 13(4):660. https://doi.org/10.3390/brainsci13040660

Schack T (2020 Mental representation in action. In: Tenenbaum G, Eklund RC (ed) Handbook of sport psychology. John Wiley & Sons, Ltd., Hoboken, NJ, pp 513–534. https://doi.org/10.1002/9781119568124.ch24

Schack T, Bertollo M, Koester D, Maycock J, Essig K (2014) Technological advancements in sport psychology. In: Papaioannou AG, Hackfort D (eds) Companion to sport and exercise psychology: global perspectives and fundamental concepts. Routledge, pp 953–965

Schack T, Hagan JE, Jr, Essig K (2020) New technologies in sport psychology practice. In: Hackfort D, Schinke RJ (eds), 2. The Routledge international encyclopedia of sport and exercise psychology. Applied and Practical Measures, p 14

Schmid A, Peper E (1998) Training strategies for concentration. In Williams JM (ed) Applied sport psychology: Personal growth to peak performance, 3rd ed. Mayfield, Mountain View, CA, pp 316–328

Schmidt RA, Wrisberg CA (2008) Motor learning and performance: A situation-based learning approach, 4th ed. Human Kinetics

Sebbens J, Andersen M, Hanrahan S (2012) Beyond performance: Training sport psychologists in Australia. InPsych. Retrieved from https://www.psychology.org.au/Content.aspx?ID=4990

Sevilmis A, Yıldız Ö (2022) Focus groups: A practical guide for sport education research. Int J Recreat Sport Sci 6(1):33–43. https://doi.org/10.46463/ijrss.1161584

Shannon S, Hanna D, Haughey T, Leavey G, McGeown C, Breslin G (2019) Effects of a mental health intervention in athletes: applying self-determination theory. Frontiers in Psychol 10:1875. https://doi.org/10.3389/fpsyg.2019.01875

Shei RJ, Holder IG, Oumsang AS, Paris BA, Paris HL (2022) Wearable activity trackers–advanced technology or advanced marketing? Eur J Appl Physiol 122:1975–1990. https://doi.org/10.1007/s00421-022-04951-1

Siekanska M, Bondar ZR, di Fronso S, Blecharz J, Bertollo M (2021) Integrating technology in psychological skills training for performance optimization in elite athletes: A systematic review. Psychol Sport Exerc 102008. https://doi.org/10.1016/j.psychsport.2021.102008

Soltani P, Morice AHP (2020) Augmented reality tools for sports education and training. Comput Educ 155(4):103923

Strenge B, Koester D, Schack T (2020) Cognitive interaction technology in sport—improving performance by individualized diagnostics and error prediction. Front Psychol 11:597913. https://doi.org/10.3389/fpsyg.2020.597913

Szabo A (2023) ChatGPT is a breakthrough in science and education but fails a test in sports and exercise psychology. Balt J Sport Health Sci, 1(128):25–40. https://doi.org/10.33607/bjshs.v127i4.1233

Taylor J (2017) Importance of assessment in sport psychology consulting. In: Taylor J (ed) Assessment in applied sport psychology. Human Kinetics, Champaign, IL, pp 3–14

Terry PC, Parsons-Smith RL, King R, Terry VR (2021) Influence of sex, age, and education on mood profile clusters. PLoS ONE 16(2):e0245341. https://doi.org/10.1371/journal.pone.0245341

Terry PC, Lane AM (2010) User guide for the brunel mood scale. Peter Terry Consultants, Toowoomba, QLD

Terry PC, Lim J (2011) In the mood: Development of an online mood profiling system. In: Turning a new page: A refreshing look at sports and exercise psychology from an Asian perspective. Proceedings of the 6th ASPASP Congress [CD-Rom]. Asian-South Pacific Association of Sport Psychology, Taipei, Taiwan

Terry PC, Lim J, Parsons-Smith RL (2013) In the mood: an online mood assessment based on the brunel mood scale. Available www.moodprofling.com

Thelwell RC, Wood J, Harwood C, Woolway T, Van Raalte JL (2018) The role, benefits and selection of sport psychology consultants: Perceptions of youth-sport coaches and parents. Psychol Sport Exerc 35:131–142. https://doi.org/10.1016/j.psychsport.2017.12.001

Tkachuk G, Leslie-Toogood A, Martin GL (2003) Behavioral assessment in sport psychology. The Sport Psychologist 17:104–117

Watson JC II, Coker-Cranney AM (2018) Introduction to the special issue: Using technology in applied sport psychology. J Sport Psychol Action 9(4):213–215. https://doi.org/10.1080/21520704.2018.1552446

Watson JC, Schinke R, Sampson JP (2014) Ethical issues affecting the use of teletherapy in sport and exercise psychology. In: Etzel EF, Watson JC, II (eds) Ethical issues in sport, exercise, and performance psychology. Fitness Information Technology, Morgantown, WV, pp 139–149

Weinberg RS (1989) Applied sport psychology: Issues and challenges. J Appl Sport Psychol 1:160–165

Weinberg RS, Gould D (1999) Personality and sport. Found Sport Exerc Psychol:25–46

Windt J, MacDonald K, Taylor D, Zumbo BD, Sporer BC, Martin DT (2020) "To tech or not to tech?" A critical decision-making framework for implementing technology in sport. J Athl Train 55(9):902–910

Woolway T, Harwood CG (2019) Gatekeepers' experiences of hiring a sport psychologist: A phenomenological study. J Appl Sport Psychol 31(4):474–493. https://doi.org/10.1080/10413200.2018.1484394

Woolway T, Harwood CG (2020) Consultant characteristics in sport psychology service provision: A critical review and future research directions. Int J Sport Exerc Psychol 18(1):46–63. https://doi.org/10.1080/1612197X.2018.1462230

MIX
Papier aus verantwortungsvollen Quellen
Paper from responsible sources
FSC® C105338

If you have any concerns about our products,
you can contact us on
ProductSafety@springernature.com

In case Publisher is established outside the EU,
the EU authorized representative is:
Springer Nature Customer Service Center GmbH
Europaplatz 3, 69115 Heidelberg, Germany

Printed by Libri Plureos GmbH
in Hamburg, Germany